THE ELEM[ENTS]

	Symbol	Atomic No.	Atomic Mass		Symbol	Atomic No.	Atomic Mass
Actinium	Ac	89	[227]†	Mercury	Hg	80	200.59
Aluminum	Al	13	26.98154	Molybdenum	Mo	42	95.94
Americium	Am	95	[243]†	Neodymium	Nd	60	144.24
Antimony	Sb	51	121.75	Neon	Ne	10	20.179
Argon	Ar	18	39.948	Neptunium	Np	93	237.0482‡
Arsenic	As	33	74.9216	Nickel	Ni	28	58.71
Astatine	At	85	[210]†	Niobium	Nb	41	92.9064
Barium	Ba	56	137.34	Nitrogen	N	7	14.0067
Berkelium	Bk	97	[249]†	Nobelium	No	102	[254]†
Beryllium	Be	4	9.0128	Osmium	Os	76	190.2
Bismuth	Bi	83	208.9804	Oxygen	O	8	15.9994
Boron	B	5	10.81	Palladium	Pd	46	106.4
Bromine	Br	35	79.904	Phosphorus	P	15	30.97376
Cadmium	Cd	48	112.40	Platinum	Pt	78	195.09
Calcium	Ca	20	40.08	Plutonium	Pu	94	[242]†
Californium	Cf	98	[251]†	Polonium	Po	84	[210]†
Carbon	C	6	12.011	Potassium	K	19	39.098
Cerium	Ce	58	140.12	Praseodymium	Pr	59	140.9077
Cesium	Cs	55	132.9054	Promethium	Pm	61	[147]†
Chlorine	Cl	17	35.453	Protactinium	Pa	91	231.0359‡
Chromium	Cr	24	51.996	Radium	Ra	88	226.0254‡
Cobalt	Co	27	58.9332	Radon	Rn	86	[222]†
Copper	Cu	29	63.546	Rhenium	Re	75	186.2
Curium	Cm	96	[247]†	Rhodium	Rh	45	102.9055
Dysprosium	Dy	66	162.50	Rubidium	Rb	37	85.4678
Einsteinium	Es	99	[254]†	Ruthenium	Ru	44	101.07
Erbium	Er	68	167.26	Samarium	Sm	62	150.4
Europium	Eu	63	151.96	Scandium	Sc	21	44.9559
Fermium	Fm	100	[253]†	Selenium	Se	34	78.96
Fluorine	F	9	18.99840	Silicon	Si	14	28.086
Francium	Fr	87	[223]	Silver	Ag	47	107.868
Gadolinium	Gd	64	157.25	Sodium	Na	11	22.98977
Gallium	Ga	31	69.72	Strontium	Sr	38	87.62
Germanium	Ge	32	72.59	Sulfur	S	16	32.06
Gold	Au	79	196.9665	Tantalum	Ta	73	180.9479
Hafnium	Hf	72	178.49	Technetium	Tc	43	98.906‡
Helium	He	2	4.00260	Tellurium	Te	52	127.60
Holmium	Ho	67	164.9304	Terbium	Tb	65	158.9254
Hydrogen	H	1	1.0079	Thallium	Tl	81	204.37
Indium	In	49	114.82	Thorium	Th	90	232.0381‡
Iodine	I	53	126.9045	Thulium	Tm	69	168.9342
Iridium	Ir	77	192.22	Tin	Sn	50	118.69
Iron	Fe	26	55.847	Titanium	Ti	22	47.90
Krypton	Kr	36	83.80	Tungsten	W	74	183.85
Lanthanum	La	57	138.9055	Uranium	U	92	238.029
Lawrencium	Lr	103	[257]†	Vanadium	V	23	50.9414
Lead	Pb	82	207.2	Xenon	Xe	54	131.30
Lithium	Li	3	6.941	Ytterbium	Yb	70	173.04
Lutetium	Lu	71	174.97	Yttrium	Y	39	88.9059
Magnesium	Mg	12	24.305	Zinc	Zn	30	65.38
Manganese	Mn	25	54.9380	Zirconium	Zr	40	91.22
Mendelevium	Md	101	[256]†				

* Only 103 elements are listed, as there is no international agreement for the names of elements 104–109.
† Mass number of most stable or best known isotope.
‡ Mass of most commonly available, long-lived isotope.

Chemistry
IMPACT ON SOCIETY

Chemistry

IMPACT ON SOCIETY

Melvin D. Joesten
Vanderbilt University

David O. Johnston
David Lipscomb College

John T. Netterville
Williamson County Schools

James L. Wood
Resource Consultants, Inc.

Saunders Golden Sunburst Series
Saunders College Publishing
Philadelphia   New York   Chicago
San Francisco   Montreal   Toronto
London   Sydney   Tokyo

Copyright © 1988 by W. B. Saunders Company

All rights reserved. No part of this publication may be reproduced or transmitted in any form or by any means, electronic or mechanical, including photocopy, recording, or any information storage and retrieval system, without permission in writing from the publisher.

Requests for permission to make copies of any part of the work should be mailed to: Permissions, Holt, Rinehart and Winston, 111 Fifth Avenue, New York, New York 10003

Text Typeface: ITC Cheltenham
Compositor: Progressive Typographers, Inc.
Acquisitions Editor: John Vondeling
Project Editor: Merry Post
Copy Editor: Martha Hicks-Courant
Art Director: Carol Bleistine
Art Assistant: Doris Roessner
Text Designer: Tracy Baldwin
Cover Designer: Lawrence R. Didona
Text Artwork: J & R Technical Services
Production Manager: Jo Ann Melody

Cover Credit: © Comstock, Inc.

Printed in the United States of America

Chemistry: Impact on Society

ISBN 0-03-008897-6

Library of Congress Catalog Card Number: 87-23548

789-032-987654321

PREFACE

Chemistry: Impact on Society is for students wishing a one-semester or one-quarter course in college chemistry. This text is unique in its understandable presentation of the origins and development of chemistry and in its overview of the present applications of chemistry and its future potential in human affairs. Enough factual information is presented to serve as a basis for understanding what the science is and how it works, but the emphasis throughout is that it does not take a chemist with vast amounts of memorized data to understand how chemistry affects our world.

No previous knowledge of chemistry is assumed or required in this presentation. However, the approach is sufficiently different to challenge and interest the student with a background in high school chemistry.

Throughout the text, emphasis is given to the impact chemical knowledge has had, and will have on the quality of human life. The basic assumptions of this presentation of chemistry are: (1) the chemical decisions necessary for the individual and society as a whole must be made by all affected if we are not to be controlled by special interest groups, and (2) a liberally educated person can understand the workings of the science without extensive study of its details and use this knowledge and understanding for decision making.

The book offers a blend of traditional chemistry and current topics. From the discovery of atomic weights to drugs used in the treatment of AIDS patients, from the discovery of the electron and the other subatomic particles to nuclear fission and nuclear fusion, and from the earliest beginnings of bonding theory to the structures of enzymes and DNA, chemistry is made relevant by current examples from today's world. Rather than just talking about the great discoveries, the science behind them is explained.

Learning aids include self-tests and matching sets with answers that provide students with a quick check on retention of information and comprehension of ideas. Boldface for new terms and concepts, along with marginal notes, adds emphasis to focus reader attention. Questions at the end of each chapter provide for additional study and opportunities for extended research.

Supplementary Material

Instructor's Manual

The instructor's manual includes (1) answers to questions at the end of chapters and (2) a list of references for each chapter that can be given to those students who want to delve more deeply into topics of interest to them.

Overhead Transparencies

A set of overhead transparencies of key figures and tables in the book will be provided to instructors who adopt this text.

Study Guide

Keith Harper, North Texas State University, has prepared a study guide to accompany our text. For each chapter in the text, this guide provides learning objectives, a review of the important concepts and terms, sample problems and additional questions with answers. The study guide also contains answers to half of the questions in the text and a list of pertinent references for each chapter.

Test Bank

A test bank with multiple choice questions for each chapter in the text is available to adopters of the text.

Acknowledgments

We are deeply grateful to the following persons for their prepublication reviews and suggestions: Frank Fazio, Indiana University of Pennsylvania; Norman Fogel, University of Oklahoma; Keith Harper, North Texas State University; Brian McGuire, Northeast Missouri State University; Thomas Ouellette, Washburn University; Salvatore Russo, Western Washington University; and V. P. Wystrach, Sacred Heart University.

Only through the expertise of the staff at Saunders College Publishing did these words get into print. Special thanks to Kate Pachuta and Merry Post for their cooperation and guidance as we moved from manuscript to galleys to proofs. To John Vondeling, Associate Publisher and supersalesman, we extend our deepest thanks for his encouragement and support.

Much help has come our way, but, of course, the responsibility for the contents of the text rests entirely on us.

As in all of our previous works, we dedicate this effort to our spouses and gratefully acknowledge their support and understanding during the preparation of this manuscript.

Melvin D. Joesten
David O. Johnston
John T. Netterville
James L. Wood

CONTENTS OVERVIEW

1 Impact of Science and Technology on Society   1
2 The Chemical View of Matter — The Search for Order in Apparent Disorder   18
3 Atoms — What Are They?   39
4 The Nuclear Atom — Uses and Dangers   65
5 Chemical Bonds — The Ultimate Glue   91
6 Reactivity Principles and Representative Reactions   109
7 Chemical Raw Materials and Products from the Earth, Sea, and Air   137
8 What Every Consumer Should Know About Energy   163
9 Introduction to Organic Chemistry: The Ubiquitous Carbon Atom   188
10 Organic Chemicals of Major Importance   205
11 Chemistry of Living Systems   238
12 Chemistry of Food Production   268
13 Nutrition: The Basis of Healthy Living   292
14 Toxic Chemicals   324
15 Water and Air Pollution   353
16 Medicines, Drugs, Drug Abuse   386
17 Chemical Formulations in Our Weekly Budget   419

Epilogue   445

Appendix A   The International System of Units (SI)   447

Appendix B   Topics for Themes   451

Answers to Self-Test Questions and Matching Sets   454

Index   463

CONTENTS

1
The Impact of Science and Technology on Society 1
What Is Science? 1
What Is Chemistry? 1
What Is Technology? 1
How Is Science Done? 3
How Do Scientists Communicate? 6
Where Are Science and Technology Taking Us? 8
Risks of Technology 11
What Is Your Attitude Toward Chemistry? 15
Summary 18

2
The Chemical View of Matter: The Search for Order in Apparent Disorder 18
Mixtures and Pure Substances 19
States of Matter 22
Definitions: Operational and Theoretical 22
Elements and Compounds 23
Chemical, Physical, and Nuclear Changes 25
The Language of Chemistry 27
The Structure of Matter 29
The Periodic Chart: Order in the Properties of the Elements 30
Measurement 35

3
Atoms: What Are They? 39
The Greek View of Atoms 39
Antoine-Laurent Lavoisier: The Law of Conservation of Matter in Chemical Change 40
Joseph Louis Proust: The Law of Constant Composition 41
John Dalton: The Law of Multiple Proportions 42
Dalton's Atomic Theory 42
Ideas About Atomic Weight 43
Atoms Are Divisible—Dalton and the Greeks Were Wrong 45
The Electron—The First Subatomic Particle Discovered 48
Protons—The Atom's Positive Charge 51
Neutrons—Neutral Particles Found in Most Atoms 51

The Nucleus—An Amazing Atomic Concept   52
Atomic Number—Each Element Has A Number   54
Isotopes—Dalton Never Guessed!   55
Where Are the Electrons?   57
The Bohr Model of the Atom   57

4
The Nuclear Atom: Uses and Dangers   65

Nuclear Reactions   66
Atomic Dating   68
Carbon-14 Dating   69
Artificial Nuclear Changes (Transmutations)   70
Nuclear Particle Accelerators   72
Transuranium Elements   72
Radiation Damage   73
Radon: A Deadly Gas   76
Nuclear Energy   78
Useful Applications of Radioactivity   82
Nuclear Waste: The Unsolved Problem   87

5
Chemical Bonds: The Ultimate Glue   91

Ionic Bonds   92
The Covalent Bond   98
Polar Bonds   100
Hydrogen Bonds   102
London Forces   105
Metallic Bonding   105

6
Reactivity Principles and Representative Reactions   109

Quantitative Changes in Weight and Energy During Chemical Change   110
Reaction Rates   110
Reversibility of Chemical Reactions   112
Chemical Equilibrium   113
Solutions   115
Acids and Bases   119
The pH Scale of Acidity and Alkalinity   120
Acid-Base Buffers   122
Oxidation and Reduction   124
Theory Broadens Redox Concepts   125
Electrolysis   126
Batteries   129
Corrosion: Unwanted Oxidation-Reduction   131
Fuel Cells   133

7
Chemical Raw Materials and Products from the Earth, Sea, and Air 137
Metals and Their Preparation 137
The Fractionation of Air 144
Silicon Materials Old and New 147
Sulfur and Sulfuric Acid 156
Sodium Hydroxide, Chlorine, and Hydrogen Chloride 157
Phosphoric Acid 158
Sodium Carbonate 159
Sources of New Materials 159

8
What Every Consumer Should Know About Energy 163
Fundamental Principles of Energy 166
Fossil Fuels 171
Novel, Nonmainstream Sources of Energy 174
Controlled Nuclear Reactions as a Source of Energy 175
Electricity: The Major Secondary Source of Energy 179
Controlled Nuclear Fusion 180
Solar Energy 181
Energy Prognosis 185

9
Introduction to Organic Chemistry: The Ubiquitous Carbon Atom 188
Why Are There So Many Organic Compounds? 189
Hydrocarbons 189
Hydrocarbon Nomenclature 194
Where Do Hydrocarbons Come From? 198
Refining Petroleum 199

10
Organic Chemicals of Major Importance 205
Functional Groups 205
More on Functional Groups in Nature 216
Optical Isomers 221
Organic Chemicals: Energy and Materials for Society 224
Coal 226
Coal Gasification 226
Industrial Catalysts 229
Key Chemicals 229

11
Chemistry of Living Systems 238
Carbohydrates 238
Proteins 244
Enzymes 248
Energy and Biochemical Systems 251
Nucleic Acids 256

12
Chemistry of Food Production 268
   Natural Soils Supply Nutrients to Plants 270
   Nutrients 272
   Fertilizers Supplement Natural Soils 275
   Protecting Plants in Order to Produce More Food 283

13
Nutrition: The Basis of Healthy Living 292
   The Setting of Modern Nutrition 292
   General Nutritional Needs 293
   Nutrient Requirements: RDA and USRDA 294
   Caloric Needs 296
   Individual Nutrients: Why We Need Them in a Balanced Amount 298
   Minerals 305
   Vitamins 308
   Food Additives 311

14
Toxic Chemicals 324
   Toxicity 324
   Other Toxins 336
   The Liver: The Body's Detoxification Factory 348
   Conclusions 349

15
Water and Air Pollution 353
   Water: A Special Compound 354
   Water: The Most Abundant Compound 355
   Public Use of Water 356
   The Scope of Water Pollutants 356
   Impact of Industrial Wastes on Water Quality 357
   What About the Future of Our Water Supplies? 363
   What Is the Atmosphere? 364
   What Is Air Pollution? 366
   Smog: Infamous Air Pollution 367
   Major Air Pollutants: Sulfur Dioxide and Nitrogen Oxides 371
   Acid Rain 374
   Organic Compounds As Air Pollutants 376
   Ozone: A Secondary Air Pollutant and Sunscreen 377
   Halogenated Hydrocarbons and the Ozone Layer 378
   Indoor Pollution 379
   Carbon Dioxide: An Air Pollutant . . . Or Is It? 382
   What Does the Future Hold for Clean Air? 382

16
Medicines, Drugs, and Drug Abuse 386
   Antacids 387
   Analgesics 388
   Aspirin Substitutes 391

Anti-Ulcer Drugs   392
Allergens and Antihistamines   392
Antimicrobial Drugs   394
Antiseptics and Disinfectants   398
Heart Disease Drugs   399
Anticancer Drugs   403
The Steroid Drugs   406
Drug Abuse   410
Drugs in Combinations   413
The Role of the FDA   414
Experimental Drugs with Promise   414

17
Chemical Formulations in Our Weekly Budget   419
Skin, Hair, and Nails: A Chemical View   419
Curling, Coloring, Growing, and Removing Hair   421
Skin Preparations for Health, Beauty, and Fragrance   427
Deodorants   430
Cleansing Agents   431
Toothpaste   439
Paints   440
Nail Polish and Polish Remover   442

Epilogue   445

Appendix A   The International System of Units (SI)   447

Appendix B   Topics for Themes   451

Answers to Self-Test Questions and Matching Sets   454

Index   463

Chemistry
IMPACT ON SOCIETY

CHAPTER 1

The Impact of Science and Technology on Society

What is Science?

Science can be defined in a number of ways. Perhaps the place to start is with the word *science*, which is derived from the Latin *scientia*, meaning "knowledge." Science is a human activity involved in the accumulation of knowledge about the universe around us. Pursuit of knowledge is common to all scholarly endeavors in the humanities, social sciences, and natural sciences. Historically, the natural sciences have been closely associated with our observations about nature, our physical and biological environment. Knowledge in this context is more than a collection of facts; it involves comprehension, correlation, and an ability to explain established facts, usually in terms of a physical cause for an observed effect.

Figure 1-1 is a classification for the natural sciences. There is no sharp distinction between the two groups of sciences or among members within a group. The scheme in Figure 1-1 should not be viewed as rigid, since new disciplines emerge that bridge areas at different levels; in the physical sciences, for example, there are biophysicists, geo-chemists, bioinorganic chemists, and chemical physicists. Some of these names define broad interdisciplinary fields; others refer to specialized subfields.

What is Chemistry?

There are many different kinds of science because there are many different ways to focus on the world around us. In this book we shall study chemistry, which is one of the physical sciences. Chemistry is concerned with the study of matter and changes in matter, particularly the changing of one form or kind of matter into another. Since matter is the material of the universe, every object we see or use is part of the chemical story. Your body is a sophisticated chemical factory in which hundreds of chemical reactions are occurring even as you read this page. Because chemistry is so intimately involved in every aspect of our contact with the material world, chemistry can be regarded as the central science, an integral part of our culture, having an unavoidable impact on our lives.

What is technology?

A deeper understanding comes when we distinguish among basic science, applied science, and technology. The difference between basic and applied science is determined by the motivation for doing the work. **Basic science** is the pursuit of knowledge about the universe with no short-term practical objectives for application; an example of **basic research** is research to answer the question "What is penicillin?" by determining

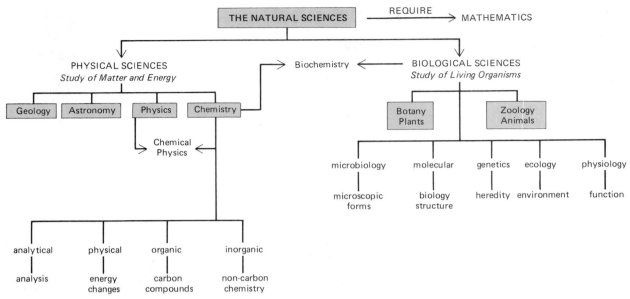

Figure 1-1
Organizational chart for the natural sciences, with emphasis on chemistry.

the molecular structure of penicillin. **Applied science** has well-defined, short-term goals related to solving a specific problem; for example, after the antibacterial action of penicillin had been discovered, scientists conducted **applied research** on the effectiveness of penicillin against different types of bacterial infections. Thus, both basic and applied scientific research produces new knowledge.

Technology is the use of scientific knowledge to manipulate nature for advantage. This may involve the production of new drugs, better plastics, safer automobiles, nuclear weapons, or chemical warfare agents. For example, engineers developed economical methods for the large-scale production of penicillin from a knowledge of the results of basic and applied research on the chemistry and biology of penicillin.

Technology, like science, is a human activity. Decisions about technological applications and priorities for technological developments are made by men and women; whether scientific knowledge is used to promote good or bad technological applications depends on those persons in industry and government who have the authority to make such decisions. **In a democratic society the voters may influence these decisions. Therefore, it is important to have an informed citizenry who can critically evaluate societal issues that are consequences of technology.**

The image most persons have of science is strongly influenced by their familiarity with technological advances and, in most instances, is their only view of scientific progress. It is important to get beyond the everyday image of technology in order to recognize the symbiotic relationship of science and technology. Modern science depends on technological advances, especially in the development of increasingly sophisticated instrumentation, to examine in greater depth unanswered questions about the universe. Examples of this interrelationship are given throughout this book.

Perhaps more easily seen are the advances in technology that occur whenever new scientific discoveries are made. Regardless of the type of scientific discovery, there is a delay between a discovery and its technological application. The incubation period

Symbiosis is the close association of two dissimilar things in a mutually beneficial relationship.

depends on (1) the rate of information transmittal, (2) recognition of the applicability of the discovery, (3) the invention of a technological application for the new science, and (4) the large-scale manufacture of the new invention. The incubation times for several ideas or scientific discoveries are given in Table 1–1. Although there are exceptions, innovations based on applied research tend to have a shorter time frame than those developed from basic research.

How is Science Done?

Scientific Method

The methodology of science is often summarized by the term *scientific method*. The scientific method is a logical approach to solving scientific problems. It includes (1) **observation,** or facts gathered by experiment; (2) **inductive reasoning** to interpret and classify facts by a general statement (law); (3) **hypothesis,** or speculation about how to explain facts or observations; (4) **deductive reasoning** to test a hypothesis with carefully designed additional experiments; and (5) **theory,** or a tested hypothesis or model to explain laws. Variations of this approach are often practiced in scientific research; imagination, creativity, and mental attitude are more important than actual procedure.

Inductive reasoning: from specific facts to generalization.

Deductive reasoning: from generalization to specific facts.

The strictest intellectual honesty is required in the collection of observable facts and in the effort to arrange these facts into a pattern that reveals the underlying cause of the *observed* behavior. The data normally must be collected under conditions that can be reproduced anywhere in the world. Then new data can be obtained to confirm or refute the correctness of the suggested pattern (geologists and astronomers may sometimes be excused from these rigid requirements). The results represent a unique type of

Another outline of the scientific method: observe, generalize, theorize, test, and retest.

Table 1–1
How Long It Has Taken Some Fruitful Ideas to Reach Technological Realization

Innovation	Conception	Realization	Incubation Interval (Years)
Antibiotics	1910	1940	30
Cellophane	1900	1912	12
Cisplatin, anticancer drug	1964	1972	8
Heart pacemaker	1928	1960	32
Hybrid corn	1908	1933	25
Instant camera	1945	1947	2
Instant coffee	1934	1956	22
Nuclear energy	1919	1945	26
Nylon	1927	1939	12
Photography	1782	1838	56
Radar	1907	1939	32
Recombinant DNA drug synthesis	1972	1982	10
Roll-on deodorant	1948	1955	7
Self-winding wristwatch	1923	1939	16
Videotape recorder	1950	1956	6
Xerox copying	1935	1950	15
X Rays in medicine	Dec. 1895	Jan. 1896	0.08
Zipper	1883	1913	30

objective truth that is independent of the differences in language, culture, religion, or economic status of the various observers. Such established truth is appropriately referred to as scientific fact.

More on Facts, Laws, Theories

A **scientific fact** is an observation about nature that usually can be reproduced at will. For example, wood readily burns in the presence of air at a sufficiently high temperature. If you have any doubt about this fact, it is easy enough to set up an experiment that will demonstrate the fact anew. You only need some wood, air, and a source of heat. The repeatability of a scientific fact distinguishes it from a historical fact, which cannot be reproduced. Of course, some scientific facts—such as the movement of heavenly bodies—are also historical facts and are not repeatable at will.

> A scientific fact can be verified independently of any particular observer.

Often a large number of related scientific facts can be summarized into broad, sweeping statements called natural **laws.** The law of gravity is a classic example of a natural law. This law—that all bodies in the universe have an attraction for all other bodies that is directly proportional to the product of their masses and inversely related to the square of their separation distance—summarizes in one sweeping statement an enormous number of facts. It implies that any object lifted a short distance from the surface of the Earth will fall back if released. Such a natural law can be established in our minds only by inductive reasoning; that is, you conclude that the law applies to all possible cases, since it applies in all of the cases studied or observed. A well-established law allows us to predict future events. When convinced of the generality of a scientific law, we may reason deductively on the basis of our belief that if the law holds for all related situations, it will hold for the events in question.

> A scientific law summarizes a large number of related facts. A scientific law predicts what *will* happen. A governmental law describes what people *should* or *should not* do.

The same procedure is used in the establishment of chemical laws, as can be seen from the following example. Suppose an experimenter carried out hundreds of different chemical changes in closed, leakproof containers, and suppose further that the containers and their contents were weighed before and after each of the chemical changes. Also, suppose that in every case the container and its contents weighed exactly the same before and after the chemical change had occurred. Finally, suppose that the same experiments were repeated over and over again, the same results being obtained each time, until the experimenter was absolutely sure that the facts were reproducible. It can be understood then that the experimenter would reasonably conclude, **"All chemical changes occur without any detectable loss or gain in weight."** This is indeed a basic chemical law and serves as one of the foundations of modern scientific theory.

After a natural law has been established, its explanation is sought by curious minds. Chemists are not satisfied until they have explained chemical laws logically in terms of the submicroscopic structure of matter. This is indeed a difficult process, and its progress has been painfully slow because of our lack of direct access into the submicroscopic structure of matter with our physical senses. All we can do is collect information in the macroscopic world in which we live and then try, by circumstantial reasoning, to visualize what the submicroscopic world must be like in order to explain our macroscopic world. Such a visualization of the submicroscopic world is called a **theoretical model.** If the theoretical model is successful in explaining a number of chemical laws, a major scientific theory is built around it. The atomic theory and the electron theory of chemical bonding are two such major theories, and both will be discussed in relation to chemical laws in later chapters.

> Theories are ideas or models used to explain facts and laws.
>
> Chemical theories use the concepts of atoms to explain chemical observations.

Consider again the chemical law concerning the conservation of weight in chemical changes. What is a possible theoretical model that could explain this law? If we assume that matter is made up of atoms, which are grouped in a particular way in a given pure substance, we can reason that a chemical change is simply the rearrangement of these atoms into new groupings without loss or destruction of atoms and, consequently, rearrangement into new substances. If the same atoms are still there, they should have the same individual characteristic weight; hence the law of conservation of weight is explained.

For a scientific theory to have much value, it must not only explain the pertinent facts and laws at hand, but also be able to explain or accommodate new facts and laws that are obviously related. If the theory cannot consistently perform in this manner, it is revised until it is consistent, or, if this is not possible, it must be discarded completely. You must not allow yourself to think that this process of trying to understand nature's secrets is nearing completion. The process is a continuing one.

In this book the word **hypothesis** is used to refer to a speculation about a particular event or set of data, and the word **theory** is reserved for the broad imaginative concepts that have gained wide acceptance by withstanding scrutiny and by their ability to explain facts and laws over a long period of time.

Experimental Methods

Discoveries come about through the observation of nature or by experimentation that can be categorized as trial and error, planned research, or accidental (serendipity).

Discovery by trial and error begins when one has a problem to solve and does various experiments in the hope that something desirable will emerge. The next set of experiments then depends on the results obtained. The discovery of the Edison battery by Thomas Edison's group is an example of discovery by trial and error. Edison's group performed more than 2000 experiments, each guided by the last, before settling on the composition of Edison's battery.

Discovery by planned research comes from carrying out specific experiments to test a well-defined hypothesis. The carcinogenic nature of some compounds is determined by progressing through a set pattern of experimental tests.

Discovery by accident may be a misnomer. The investigator is usually actively involved in investigating nature through experimentation, but "accidentally" finds some phenomenon not originally imagined or conceived. Thus the "accident" has an element of serendipity and is not seen unless the investigator is a trained observer. As Pasteur said, "Chance favors the prepared mind."

The discovery of one of the leading anticancer drugs, cisplatin, is an example of an "accidental" discovery. In 1964 Barnett Rosenberg and his co-workers at Michigan State University were studying the effects of an electric current on bacterial growth. They were using an electrical apparatus with platinum electrodes to pass a small alternating current through a live culture of *Escherichia coli* bacteria. After an hour, they examined the bacterial culture under a microscope and observed that cell division was no longer taking place. After thorough analysis of the culture medium and additional experimentation, they determined that traces of several different platinum compounds were produced during electrolysis from the reaction of the platinum electrodes with chemicals in the culture medium.

Careful observation was essential, since platinum electrodes are commonly regarded as inert or unreactive, and only a few parts of platinum compounds per million parts of culture medium were present. Additional testing indicated that a compound known as cisplatin was responsible for inhibiting cell division in *E. coli*. Approximately two years after its initial discovery, the Rosenberg group had the answer to the question "What caused the inhibition of cell division in *E. coli* bacteria?" At this point they had the idea that cisplatin might inhibit cell division in rapidly growing cancer cells. The com-

Planned research is based on fact and theory and is expected to yield predicted results.

Carcinogens are substances that cause cancer.

Figure 1–3 is an abstract of the first published report of the discovery of cisplatin.

Chemists represent the cisplatin molecule like this:

$$\begin{array}{c} Cl \\ \diagdown \\ Cl \end{array} Pt \begin{array}{c} NH_3 \\ \diagup \\ NH_3 \end{array}$$

Most scientific journals are publications in which the results of scientific research are reported or reviewed.

Abstracting journals publish summaries of scientific publications.

pound was tested as an anticancer drug, and in 1979 the Food and Drug Administration approved its use as such. The drug has now been proved to be effective alone or in combination with other drugs for the treatment of a variety of cancers.

Another interesting aspect of the story is that cisplatin was first prepared in 1845. Although its chemistry has been studied thoroughly since then, the biological effects of cisplatin and its inhibition of cell division were not discovered until "the accident" 120 years later.

How Do Scientists Communicate?

Scientific knowledge is cumulative, and progress in science and technology depends on access to this body of knowledge. Since the earliest beginnings of science, knowledge has been transmitted primarily by the written word. The invention of the printing press led to the development of scientific journals and other publications collectively known as the **scientific literature.** The explosive expansion of the scientific literature since the 1940s makes information management an essential part of modern science and technology.

The explosion of chemical literature can be illustrated by the growth rates of *Chemical Abstracts* (CA), which provides comprehensive coverage of chemical literature worldwide. Figure 1–2 shows the growth rate of papers abstracted per year. Note the sharp increase since 1947. The number of journals monitored by CA has grown from 400 in 1907 to over 12,000 in 1987.

An example of a CA abstract is shown in Figure 1–3. At present, CA publishes approximately 450,000 abstracts each year, 16% of which are patent abstracts. This is one measure of the ratio of scientific knowledge to technological application. CA indexes make this mountain of information accessible. Over 80 volumes make up the tenth Collective Index of Chemical Abstracts for the period 1977 to 1981, and this collective index covers over 2.5 million documents.

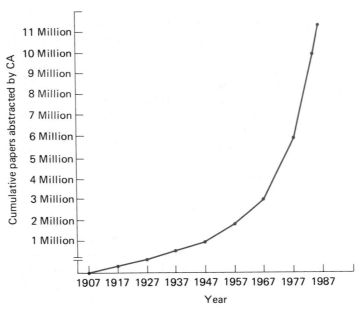

Figure 1–2
Growth rate of papers abstracted by *Chemical Abstracts*.

Figure 1-3
Example of an abstract found in *Chemical Abstracts*. This one was published in Volume 62, abstract 13543a, 1965. The "O atm" is actually O_2 atmosphere. It takes considerable understanding and experience to read abstracts quickly and efficiently. (This CA citation is copyrighted by The American Chemical Society and is reprinted by permission. No further copying is allowed.)

Inhibition of cell division in Escherichia coli by electrolysis products from a platinum electrode. Barnett Rosenberg, Loretta Van Camp, and Thomas Krigas (Michigan State Univ., Lansing). *Nature* 205(4972), 698-9(1965)(Eng). *E. coli* cells grown in a continuous culture app. with an esp. designed chamber contg. 2 Pt electrodes were subjected to an elec. field in an O atm. With an applied voltage of 500 to 6000 cycles/sec., cell division ceased and filamentous growth occurred. A similar effect was shown with a soln. of 10 ppm. $(NH_4)_2PtCl_6$. Pt(IV) was found in the electrolyzed medium in concns. of 8 ppm. Other transition metal ions, including Rh, in concns. of 1-10 ppm. inhibited cell division in *E. coli* while not interfering with growth.
Gloria H. Cartan

An example of how technology has aided science is in the retrieval of information from the massive volume of published works. The advent of the computer has resulted in the development of computer information retrieval as an efficient and accurate way to gain access to the literature. Two companies that provide databases for computer information retrieval are DIALOG Information Retrieval Service and Chemical Abstracts Service (CAS) ONLINE. DIALOG has more than 200 databases that cover natural sciences, social sciences, humanities, technology, engineering, medicine, business, government, law, economics, and current events.

Chemical Abstracts Service (CAS) ONLINE is the online equivalent of the printed *Chemical Abstracts*. The world's largest file of information on substances, it is updated weekly. Each registered substance is assigned a unique CAS Registry Number that permits retrieval of information about the substance's formula, structure, and name. A chemist can search by registry number, structure, or common name through more than 6.5 million substances cited in the chemical literature in the past 20 years and do so comprehensively and quickly; the search can take as little as 5 minutes, and most searches can be completed within 15 minutes.

The items you need to search database systems are shown in Figure 1-4. They are a telephone line, a modem, a computer terminal (microcomputers can be used), and an

The CAS Registry Number is the "Social Security Number" of the unique substance. Table salt (NaCl) is 14762-51-7. In 1986 there were 1076 abstracts that referred to table salt.

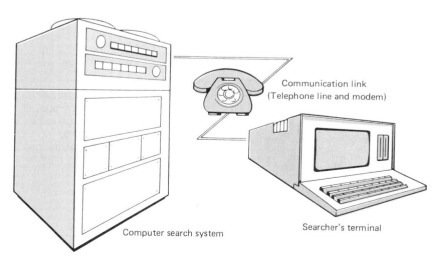

Figure 1-4
Equipment needed to perform a computer search of a database.

account with the information system. Regardless of your chosen profession, you should become familiar with computer information retrieval. It is essential for efficient information management. Leaders in any field will have to manage information with technical devices, since the human brain cannot process the information fast enough. Obviously, the growth of scientific information and the consequent technological application are facilitated by communication, and new knowledge and its application are growing exponentially.

Where are Science and Technology Taking Us?

Two major technological revolutions currently under way are the microelectronic revolution and the biotechnology revolution.

The Microelectronic Revolution

> The *small* chip stores and processes *vast* amounts of information at amazing speeds.

The chip, nickname for the integrated circuit, is a small slice of silicon that contains an intricate pattern of electronic switches (transistors) joined by "wires" etched from thin films of metal. Some are information storers called memory chips. Others combine memory with logic function to produce computer or microprocessor chips. These two applications would appear to make the chip, like the mind, capable of essentially infinite application. A microprocessor chip, for example, can provide a machine with decision-making ability, memory for instructions, and self-adjusting controls.

In everyday life we see many examples of the influence of the chip: digital watches; microwave oven controls; new cars with carefully metered fuel–air mixtures; hand calculators; cash registers that total bills, post sales, and update inventories; computers in a variety of sizes and capacity — all these make use of the chip. By looking at Figure 1–5, one can appreciate the technological advancement represented by the chip. A typical microprocessor chip holds 30,000 transistors but is small enough to be carried by a large ant.

> The transistor, smaller than the head of a match, replaced vacuum tubes ten or more cubic centimeters in volume.

The story of the chip starts with the invention of the transistor in 1947. The transistor is a semiconductor device that acts either as an amplifier or as a current switch. Although transistorized circuits were a tremendous improvement over vacuum tubes, large computer circuits using 50,000 or more transistors and similar numbers of diodes, capacitors, and resistors were difficult to build; computers had to be wired together in a continuous loop, and a circuit with 100,000 components could easily require 1 million soldered connections. The cost of labor for soldering and the chance for defects were

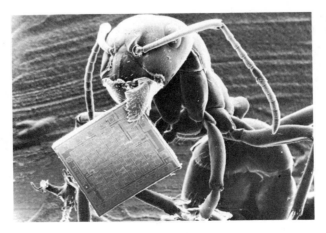

Figure 1–5
Photograph (17× magnification) of a typical microprocessor chip small enough to be carried by a large ant. (Courtesy of North American Philips Corporation.)

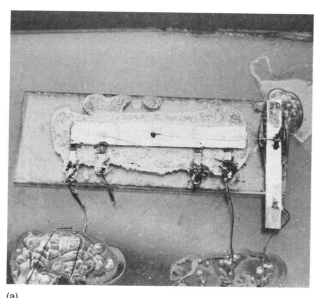 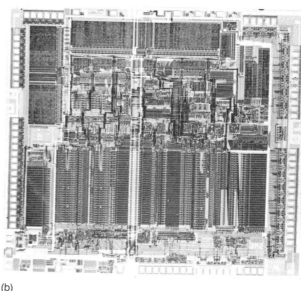

Figure 1–6
(a) Original integrated circuit built by Jack Kilby. (Courtesy of Texas Instruments, Inc.)
(b) Photomicrograph of a microprocessor chip with 200,000 transistors on one piece of silicon. (Courtesy of Motorola, Inc.)

both high. In the late 1950s, the Navy's newest destroyers required 350,000 electronic components and millions of hand-soldered connections. It was clear that the limit to the use of transistors in supercircuits was the number of individual connections that could be linked and maintained at one time.

In 1958 Jack Kilby at Texas Instruments and Robert Noyce at Fairchild Semiconductor, working independently, came up with the solution: make a semiconductor (silicon or germanium) in the transistor serve as its own circuit board. If all the transistors, capacitors, and resistors could be integrated on a single slice of silicon, connections could be made internally within the semiconductor and no wiring or soldering would be necessary. Figure 1–6 shows Kilby's handmade chip and a modern chip with 200,000 transistors on one piece of silicon.

The Biotechnology Revolution — Designer Genes

The biotechnology revolution began after the first successful gene splicing and gene cloning experiments produced recombinant DNA in the early 1970s. In Chapter 11 we shall discuss the biochemistry of DNA and the genetic code. The present discussion focuses on the potential of recombinant DNA technology to solve three of the world's greatest problems: hunger, sickness, and energy shortages.

The process for forming and cloning recombinant DNA molecules is outlined in Figure 1–7. The basic idea is to use the rapidly dividing property of common bacteria, such as *E. coli*, as a microbe factory for producing recombinant DNA molecules that contain the genetic information for the desired product. Rings of DNA called plasmids are isolated from the *E. coli* cell. The ring is cut open with a cutting enzyme, which also cuts the appropriate gene segment from the desired human, animal, or viral DNA. The new gene segment is spliced into the cut ring by a splicing enzyme. The altered DNA ring (recombinant DNA) is then reinserted into the host *E. coli* cell. Each plasmid is copied

Recombinant: capable of genetic recombination.

The control of life, not just the environment, is at hand!

"Cut" and "splice" are used figuratively, since we are really causing the breaking and making of chemical bonds (see Chapter 5).

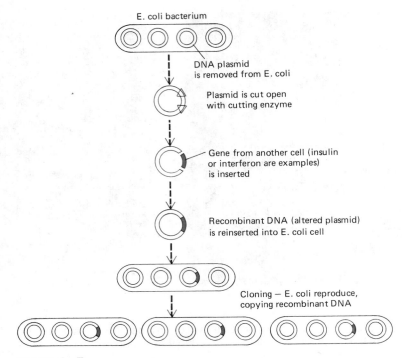

Figure 1-7
Synthesis and cloning of recombinant DNA molecules. DNA plasmids are small, circular, duplex DNA molecules that carry a few genes.

many times in a cell. When the *E. coli* cells divide, they pass on to their offspring the same genetic information contained in the parent cell.

The commercial potential of recombinant DNA technology was recognized very early, and several biotechnology companies were started by the scientists who had done key experiments in gene splicing and gene cloning. There are more than 200 biotechnology companies, but most of them are still doing development work and have no products to sell. The risk is high, but the payoff of successful ventures will also be high.

An early benefit of recombinant DNA was the biosynthesis of human insulin in 1978. Millions of diabetics depend on the availability of insulin, but many are allergic to animal insulin, which used to be the only source of commercial insulin. Biosynthesized human insulin is now being marketed by a firm called Genentech. Biotechnology firms are also producing human growth hormone, which is used in treating youth dwarfism, and interferon, which is a potential anticancer agent.

Genetic engineers are trying to modify crops so they will make more nutritious protein, resist disease and herbicides, and even provide their own fertilizers. Researchers are also trying to use recombinant DNA techniques to produce vaccines against diseases that attack livestock.

Strains of bacteria are being developed that will convert garbage, plant waste material such as cornstalks, and industrial wastes into useful chemicals and fuels. This not only helps solve the energy shortage but also provides a way to recycle wastes and lower the accumulation of solid wastes.

A decade ago, at the beginning of the recombinant DNA era, many people, including scientists working in the area, saw danger in biotechnology. Since *E. coli* is an

In 1985 Japanese scientists demonstrated that silkworm larvae can be used to biosynthesize human interferon, a potential anticancer drug.

intestinal bacterium, what if some of the genetically engineered *E. coli* escaped and found their way into people's intestines? These fears led to an 18-month moratorium on recombinant DNA research. Although the evidence to date shows that the *E. coli* used in recombinant DNA technology is too delicate to survive outside its environment, strict regulations are still being followed in experiments with genetically engineered bacteria.

There is also a deeper ethical concern. At present no genetic engineering of human genes is being done, but the capability is there. Should we raise a child's IQ from 80 to 100? If we do this, should we raise IQs from 120 to 160? Should we alter human genes to improve health, longevity, strength, and so forth? By altering life forms, whether plant, animal, or human, are we playing God? Debates on these questions have been going on since the beginning of biotechnology and will continue. The mechanism for resolving such important questions offers a new challenge to a free society.

Risks of Technology

We have described some of the benefits of technology, but we also need to examine related risks. *Bhopal, Challenger,* and *Chernobyl* are names associated with three catastrophic accidents that remind us of technological risks.

December 3, 1984: The worst chemical plant accident in history occurred in the early morning hours at a Union Carbide insecticide plant in Bhopal, India. Over 2,000 people were killed, and tens of thousands were injured when methyl isocyanate, a deadly gas used in the preparation of pesticides, escaped from a storage tank. Several violations of recommended safety procedures contributed to the disastrous leak, including an inoperative refrigeration system to keep the methyl isocyanate cool, an inoperative scrubbing tank to neutralize any escaping gas, and an inoperative flare tower designed to burn any gas that escaped from the storage tank.

> Union Carbide has announced that the Bhopal tragedy was not an accident but was an intentional act by an unhappy worker.

January 28, 1986: The space shuttle *Challenger* exploded 74 seconds after liftoff, killing the shuttle's seven crew members (Fig. 1-8). Officials at both the National Aeronautics and Space Administration and Morton Thiokol, the rocket manufacturer, recommended launch in spite of clear and repeated warnings from Thiokol engineers about problems with the plastic O-rings used to seal joints between casing sections in the booster rocket.

April 26, 1986: An explosion at the Chernobyl nuclear power plant near Kiev in the Soviet Union released large amounts of radioactive material into the atmosphere. About 25,000 persons were evacuated on April 27, and within a week a total of 135,000 had been evacuated from an 18-mile zone around the power plant. Two persons were killed in the explosion, and 29 died from radiation sickness during the next three months.

Andronik M. Petrosyants, chairman of the Soviet Committee for the Peaceful Uses of Atomic Energy, said, "The accident took place as a result of a whole series of gross violations of operating regulations by the workers." Ironically, the accident occurred in the course of a safety test. Workers had shut down automatic safety systems in order to test how long the reactor's turbine generators would continue to operate in the event of an unforeseen reactor shutdown. The six fatal errors, according to a Soviet report, were (1) shutting off the emergency cooling system, (2) lowering the reactor's power to a point at which the reactor was difficult to control, (3) exceeding flow rates by having all water circulation pumps on, (4) shutting off the automatic signal that shuts down the reactor when the turbines stop, (5) pulling out almost all the control rods from the core, and (6) shutting off safety devices that shut down the reactor if steam pressure or water levels become abnormal. When workers began the test by stopping power to the turbine, the reactor began immediately to overheat. Within seconds two explosions—

Figure 1–8
Photo of the Challenger space shuttle explosion. (UPI/Bettmann Newsphotos.)

one from steam pressure and one from the ignition of hydrogen gas produced from the reaction of steam with graphite, uranium fuel, and the zirconium alloy that encased the fuel—blew the roof off the reactor building and ignited more than 30 fires around the plant. The damaged reactor core and the graphite surrounding it began to burn at temperatures as high as 2500°C. The fire released into the atmosphere large amounts of radioactive material, which was borne by prevailing winds over the Ukraine and much of Europe. Other countries were not informed of the accident; the first indication that something was wrong was detected on April 28, when monitoring stations in Sweden recorded a sudden jump in levels of radioactive isotopes of krypton, xenon, iodine, cesium, and barium—the mixture that would be expected if a nuclear accident had occurred.

The long-term effects of the Chernobyl accident are difficult to predict. Estimates of an increase in cancer rates in the western Soviet Union are in the thousands. People evacuated from the Chernobyl area are subject to periodic examinations for signs of cancer and other illnesses caused by radiation. In addition, anyone who travels within 100 miles of the reactor site is tested for radioactivity. The full extent of contamination of food and water supplies is not known, but Swedish officials estimate that their country has so far lost at least $144 million in ruined food. Milk, vegetables, and water supplies will continue to be monitored for radioactive contamination.

All three of these accidents involved human error and lack of enforcement of recommended safety procedures. Years of excellent safety records in the chemical industry and the nuclear power industry do not eliminate the risks of chemical and nuclear technology. However, the risks can be reduced by the constant application of recommended safety procedures and by full utilization of technological developments.

The living must accept risk. The question is: How much?

Hazardous Wastes

Careless disposal of hazardous wastes has been the cause of many human health problems. Love Canal in Niagara Falls, Times Beach in Missouri, and Minamata Bay and Jinzu River in Japan are just a few locations where serious health problems have resulted from the improper disposal of hazardous wastes.

Love Canal, the neighborhood that in 1977 discovered it was built on a toxic chemical dump, was the first publicized example of the problems of chemical waste dumps. In the mid-1970s heavy rains and snows seeped into the dump and pushed an oily black liquid to the surface. The liquid contained at least 82 chemicals, 12 of which were suspected carcinogens.

Times Beach, Missouri, was bought by the U. S. Environmental Protection Agency (EPA) in 1983, and its 2200 residents were relocated because dioxins, a group of toxic chemicals produced in small amounts during the synthesis of a herbicide, were found in the soil at concentrations as high as 1100 times the acceptable level (Fig. 1–9).

In the 1950s tons of waste mercury were dumped into the bay at Minamata, Japan. In the next few years thousands of persons in the Minamata area suffered paralysis and mental disorders, and over 200 people died. Several years passed before it was determined that these people suffered from poisoning by methyl mercury compounds. Anaerobic bacteria in the sea bottom converted mercury to methyl mercury compounds, which were eaten by plankton. The methyl mercury compounds were carried up the food chain and eventually accumulated in the fatty tissue of fish, which are a major part of the Japanese diet. Intake of methyl mercury compounds reached levels that caused the sickness now known as Minamata disease.

Itai-Itai disease, which makes bones brittle and easily broken, is caused by cadmium poisoning and was also first observed on a major scale in Japan. *Itai* means "it hurts" in Japanese and graphically illustrates the pain associated with this disease. Many cases were observed downstream from a zinc-refining plant on the Jinzu River in Japan.

Figure 1–9
Type of protective gear worn by workers at hazardous waste sites. Here EPA workers remove a soil sample from the Times Beach, Missouri, site in November 1982. (Courtesy of Environmental Protection Agency.)

Cadmium is a byproduct of the zinc-refining industry and is used in various alloys and in nickel–cadmium rechargeable batteries.

What Is an Acceptable Risk?

Risk assessment for individuals involves a consideration of the likelihood or probability of harm and the severity of the hazard. Assessment of societal risks combines probability and severity with the number of persons affected. The science of risk assessment is still evolving, but it is clear that the importance of public perception of risks needs to be recognized before risk assessment can be quantified. Often there is little correlation between the actual statistics of risk and the perception of risk by the public or by individuals. For example, we are all aware that the risk of injury or death is much lower from traveling in a commercial airplane than from traveling in an automobile, yet all of us know persons who avoid airplane flights because of their fear of a crash.

What factors influence public perception of risk? Catastrophic accidents such as Bhopal and Chernobyl obviously affect public perception of risk. In addition, people tend to judge involuntary exposure to activities or technologies (such as living near a hazardous dump site) as riskier than voluntary exposure (such as smoking). In other words, persons rate risks they can control lower than those they cannot control.

No absolute answer can be provided to the question "how safe is safe enough?" Determining acceptable levels of risk requires value judgments that are difficult and complex, involving the consideration of scientific, social, and political factors. Over the years a number of laws designed to protect human health and the environment have been enacted to provide a basic framework for making decisions. The fact that three types of laws exist in this area adds to public confusion about risk assessment and its meaning.

Risk-based laws are zero-risk laws that allow no balancing of health risks against possible benefits. The Delaney Clause of the Federal Food, Drug, and Cosmetic Act is a risk-based law. It specifically bans the use of any intentional food additive that is shown to be a carcinogen in humans or animals, regardless of any potential benefits. The rationale for this law is the nonthreshold theory of carcinogenesis, which assumes that there is no safe level of exposure to a carcinogen.

The Safe Drinking Water Act, the Toxic Substances Control Act, and the Clean Air Act are **balancing laws;** they balance risks against benefits. The Environmental Protection Agency is required to balance regulatory costs and benefits in its decision-making activities. Risk assessments are used here. Chemicals are regulated or banned when they pose "unreasonable risks" to or have "adverse effects" on human health or the environment.

Technology-based laws impose technological controls to set standards. For example, parts of the Clean Air Act and the Clean Water Act impose pollution controls based on the best economically available technology or the best practical technology. Such laws assume that complete elimination of the discharge of human and industrial wastes into water or air is not feasible. Controls are imposed to reduce exposure, but true balancing is not attempted; the goal is to provide an "ample margin of safety" to protect public health and safety.

Risk Management

Those in responsible positions in business and government now have a greater awareness than they used to of the need to solve environmental problems associated with technological production and to assess the risks of technology. Most of our present environmental problems stem from decades of neglect. The Industrial Revolution brought prosperity, and little thought was given to the possible harmful effects of the technology that was providing so many visible benefits.

The chemical industry should take the lead in demonstrating its willingness to help solve the problems caused by its predecessors' lack of foresight. This should be done through cooperation, not confrontation. We need a science policy that is based on input from responsive leaders, both in industry and in government, who provide a forum from which to examine the facts and reach responsible decisions that lead to prompt action.

Is this possible? You may doubt it, but you have a responsibiliy to future generations to do your part in seeing that responsible action is taken. We cannot and should not "turn off" science and technology. Those who long for the "good old days" should remember what that means — diseases such as malaria, smallpox, and polio, which took many lives; no antibiotics for infections; and none of the modern fertilizers to increase crop yields needed to feed the world's population. You could add to this list many things that are of a humanitarian nature before you even start listing the technical advances that have raised the comfort level of our lives.

The control of chemical hazards is essential for everyone's well-being.

Risk management requires value judgments that integrate social, economic, and political issues with the scientific assessment of the risk. Determination of the acceptability of risk is a societal issue, not a scientific one. It is up to all of us to weigh the benefits against the risks in an intelligent and competent manner. The assumption of this text is that the wit to deal with environmental problems caused by uncontrolled technology is to be found in the educated public at large, not in the select group that stands to make a short-term financial or political profit. Always keep in mind that, except in the case of some radioactive wastes, the knowledge is available to "clean up" after any industrial operation; it is just a matter of cost, energy, and values.

Society must determine how much risk it will accept.

It is apparent that we need citizens to take responsibility for being informed about the technological issues that affect society. Albert Gore, Jr., U. S. senator from Tennessee, said in support of better science education,

> Science and technology are integral parts of today's world. Technology, which grows out of scientific discovery, has changed and will continue to change our society. Utilization of science in the solution of practical problems has resulted in complex social issues that must be intelligently addressed by all citizens. Students must be prepared to understand technological innovation, the productivity of technology, the impact of the products of technology on the quality of life, and the need for critical evaluation of societal matters involving the consequences of technology.

What Is Your Attitude Toward Chemistry?

Before beginning the study of chemistry and its relationship to our culture, each of us needs to examine our prejudices (if any) and attitudes about chemistry, science, and technology. Many nonscientists regard science and its various branches as a mystery and feel that they cannot possibly comprehend the basic concepts and consequent societal issues. Many also have a fear of unleashed chemicals **(chemophobia)** and a feeling of hopelessness about the environment. Many of these attitudes are a result of reading about harmful effects of technology, which are indeed tragic. However, what is needed is a full realization of both the benefits of and the risks from science and technology. In the analysis of the pluses and minuses, we need to determine why harmful effects occur and whether risks can be reduced for future generations as we seek advantages offered by the human understanding of nature. This book will give you the basics in chemistry, which we hope will give you a more satisfying life through understanding and a richer, healthier life through the ability to make wise decisions about personal problems and problems that concern our world.

Summary

How can we summarize this chapter as it concerns you, the student?

1. We need an informed citizenry to use and to evaluate scientific and technological advances.
2. To be informed about chemical problems requires a basic knowledge of what matter is really like and what matter does.
3. More sophisticated chemical problems require a deeper understanding of the workings (facts and theories) of chemistry.
4. You should be involved. As an educated person, you have a responsibility and a privilege.

SELF-TEST 1–A*

1. The ultimate test of a scientific theory is its agreement with _____.
2. Different workers in different countries who carry out a particular laboratory experiment in exactly the same way should get _____ result.
3. Chip is the nickname for _____.
4. The common bacterium used as the microbe factory in recombinant DNA technology is _____.
5. Arrange from most abstract to general to specific: laws, facts, theories.
    a. _____ b. _____ c. _____

SUBJECTIVE QUESTIONS

Rate the following as good, bad, or neither. Be prepared to state reasons for your rating.

1. Computer information retrieval _____
2. Nuclear energy _____
3. Coal as a fuel _____
4. Petroleum as a fuel _____
5. Birth control pills _____
6. Fertilizers _____
7. Plastic containers _____
8. Synthetic foods _____
9. Solar energy _____

* Use these self-tests as a measure of how well you understand the material. Take a test only after careful reading of the material preceding the test. Do not return to the text during the self-test, but reread entire sections carefully if you do poorly on the self-test on those sections. The answers to the self-tests are at the end of the text.

10. Recombinant DNA technology _____
11. Government control of scientific research _____
12. Government control of applied research _____

MATCHING SET

_____ 1. Scientific theory
_____ 2. Transmission of scientific knowledge
_____ 3. Chemistry
_____ 4. Unsatisfactory way to establish scientific truth
_____ 5. Computer information retrieval company
_____ 6. Scientific methods
_____ 7. Basis of scientific truth
_____ 8. Cisplatin

a. Appeal to authority
b. Study of matter and its changes
c. Used to study nature
d. Printed journals
e. Air
f. Observation
g. Anticancer agent
h. DIALOG
i. Used to explain facts and laws

QUESTIONS

1. Distinguish between theory and law in chemistry.
2. Give an example of a chemical fact.
3. Give an example of a chemical law.
4. How many times do you think a given experiment should yield the same result before a scientific fact is considered to have been established?
5. Suppose that, on returning from an afternoon ball game, a mother and her children discover the family car missing, even though the father rode to work with a neighbor that morning. The mother says to the children, "Don't worry, Dad must have had an unexpected need for the car and got it after lunch." Would you call the statement made by the mother a theory or a hypothesis? Why?
6. Persons often confuse science with scientism. Look up the definition of *scientism* in a dictionary and discuss why it is important to society that science not be confused with scientism.
7. What is the difference between pure science and technology?
8. A DPT shot is given to prevent diptheria, pertussis (whooping cough), and tetanus. Is a DPT shot for infants that occasionally causes death an acceptable risk?
9. Talk to a few professors and then answer the question, "how much trial-and-error research is involved in a university research project?"

CHAPTER 2

The Chemical View of Matter: The Search for Order in Apparent Disorder

Matter occupies space and has mass.

Although the astronomer's view shows space to be essentially empty, everywhere we look we see **matter.** More often than not the natural samples of matter we examine appear to be made up of different materials, and nature appears to be constantly changing material formations and life forms. Also, humans have demonstrated a remarkable ability to change the natural arrangements of matter.

Chemistry is the physical science primarily concerned with **the study of kinds of matter and the changes of one kind of matter into another with their associated energy changes.** The separation of natural mixtures of chemicals into pure and uniform substances is the start of an understanding of the different kinds of matter. Once the different kinds of matter are obtained and their properties are observed, the properties and changes are then understood in terms of structure too minute to experience with the five human senses.

Examples of energy are heat, light, sound, electricity, falling rock, burning fuel, and an electrical storm.

Figure 2-1
Lunar rock. The heterogeneous texture is similar to many earth crust mixtures. A metric scale is given to indicate size, and numbers and letters are given to identify sample.

Mixtures and Pure Substances

It is often easy to see the various ingredients in a **mixture,** such as in rocky or sandy soil or in a moon rock (Fig. 2–1). Such mixtures are **heterogeneous,** in that their texture is clearly uneven. Other mixtures appear at first glance to be homogeneous, but a closer look shows that they are heterogeneous. For example, air in a room may appear to be homogeneous, but a beam of bright light will reveal floating dust. Fat in milk, oil in mayonnaise, latex and pigments in paints, and water in Jello are other examples of heterogeneity that is revealed only with intense lighting or magnification.

Colloids are mixtures that appear to be homogeneous in normal lighting but, when an intense beam of light is directed through colloids, the light is scattered and reveals particles too small to be seen by the naked eye (Fig. 2–2). Truly homogeneous mixtures, called **solutions,** do exist. Examples of solutions are nitrogen and oxygen in air, sugar in water, and metal alloys. No amount of optical magnification will reveal a homogeneous solution to be heterogeneous.

When a mixture is separated into its components, the components are said to be *purified*. Most efforts at separation are unsuccessful in a single operation or step; repetition of the purification process results in better separation. Ultimately, purification methods result in **pure substances,** samples of matter that cannot be purified further. For example, if sulfur and iron powder are mixed, the iron can be separated from the sulfur by repeated stirrings of the mixture with a magnet. When the mixture is stirred the first time and the magnet is removed, much iron is removed with the magnet, leaving the sulfur in a higher state of purity. However, after just one stirring the sulfur may still have a dirty appearance due to a small amount of iron that remains. Repeated stirring with the magnet, or perhaps the use of a very strong magnet, will finally leave a bright yellow sample of sulfur that apparently cannot be purified further by this technique. In this purification process a property of the mixture, its color, is a measure of the extent of purification. The bright yellow color indicates that the sulfur has been purified.

Heterogeneous—of nonuniform texture; not the same at every observed point.
Homogeneous—of smooth texture, uniform throughout.

Mayonnaise is a colloidal dispersion of oil in water.

An alloy of brass is a mixture of copper and zinc. An alloy of sterling silver contains 7.5% dissolved copper.

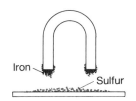

Purification separates the kinds of matter.

Figure 2–2
A colloid in a clear liquid or gas scatters light because of the relatively large size of the dispersed particles. In contrast, a solution, which is a dispersion at the molecular level, passes the light with no scatter.

However, to draw a conclusion on the basis of only one property of a mixture may be misleading, because other methods of purification might change some other properties of the sample. It is safe to call the sulfur a pure substance only when all possible methods of purification fail to change its properties. This assumes that all pure substances have a set of unchangeable properties by which they can be recognized. **A pure**

Over 6 million pure substances have been identified.

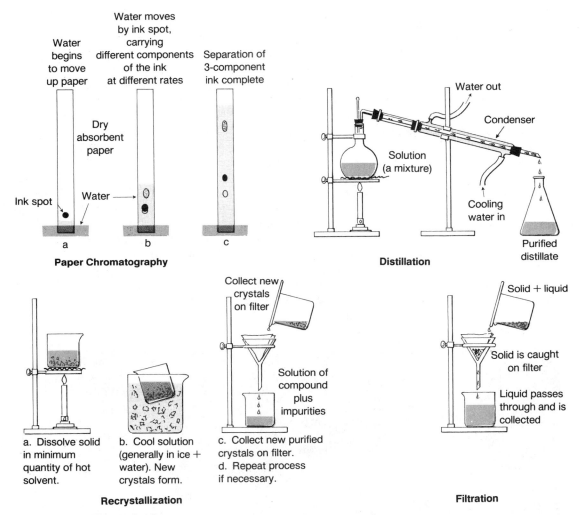

Figure 2–3
Four methods of purifying mixtures of elements and compounds. *Paper chromatography:* Because of the absorbent character of paper, water moves against gravity and carries ink dyes along its path. If the ink dyes move at different rates because of differing attractions to the paper, they are separated in the developed chromatogram. *Distillation:* Sodium chloride dissolves in water to form a clear solution. When heated above the boiling point (indicated by the thermometer), water vaporizes and passes into the condenser. Cool water injected into the glass jacket of the condenser circulates over the inner tube, causing the steam to liquefy and collect in the flask. In this simple example, pure water collects in the receiving flask, while the salt remains in the boiling flask. *Recrystallization:* This can be used to separate some solid mixtures. *Filtration:* The separation of an insoluble solid from a liquid by selective porosity of the filtering medium.

Figure 2–4
Distillation. Some of the most useful purification techniques copy processes in nature and date back to alchemical times. Distillation allows a more volatile substance to be separated from a less volatile one. In this case, alcohol is partially separated by evaporation from water and other ingredients. One distillation can produce a mixture that is 40% alcohol from one that is only 12%. Further distillations would produce an even better separation.

substance, then, is a kind of matter with properties that cannot be changed by further purification.

Most pure substances are not common naturally, but modern purification techniques have made possible the isolation of millions of them (Fig. 2–3 and 2–4). Common examples are refined sugar, table salt (sodium chloride), copper, sodium bicarbonate, nitrogen, dextrose, ammonia, uranium, and carbon dioxide. To date, more than 6 million pure substances have been isolated and catalogued.

Although they are relatively few in number, some very familiar materials are naturally occurring pure substances. Rain is very nearly pure water, except for small amounts of dust, dissolved air, and traces of pollutants. Gold, diamond, and sulfur are also found in nature in very pure form, though these substances are special cases. A human being, a complex assemblage of mixtures, lives in a world of mixtures—eating them, wearing them, living in houses made of them, and using tools made of them.

Most materials in nature are mixtures; a few are relatively pure substances.

Pure substances in nature result from natural processes closely akin to scientific separation techniques. Compare distillation with the natural water cycle.

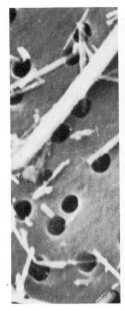

Scanning electron micrograph of particles of asbestos filtered from a sample of air by a small-pore filter.

Figure 2–5
The Orion Nebula. Light from such nebulae is produced in the plasma by highly energized atoms, molecules, and ions.

States of Matter

As we examine mixtures and pure substances, it is easy to recognize three of the four states of matter: **solids, liquids,** and **gases.** Solids have definite shapes and volumes; liquids have definite volumes but indefinite shapes; and gases take any shape or volume imposed by the containing vessel. **Plasmas** constitute a fourth state of matter. Although plasmas are not common in everyday experience, they may be the most prevalent state of matter in the universe (Fig. 2–5). A plasma is similar to a gas but is composed of freely moving charged particles, called **ions,** which dramatically respond to electric and magnetic forces. Natural materials in the plasma state include flames, the outer portion of the Earth's atmosphere, the atmosphere of the stars, much of the material in nebular space, and part of a comet's tail. The aurora borealis offers a dazzling display of matter in the plasma state streaming through a magnetic field.

An ion is a charged atom or group of atoms.

Definitions: Operational and Theoretical

The definition given earlier for a pure substance is an **operational** definition: "A pure substance is one whose properties cannot be changed by further purification efforts." Operational definitions result from the performance of operations or tests on matter and summarization of the results in a statement. For example, iron is a magnetic metal that melts at 1535°C, boils at 3000°C, and is 7.86 times more dense than water. These properties have been determined through the repeated application of heat to the pure substance and the measurement of temperatures, weights, and volumes. When all the properties of pure iron have been listed, we find that the pure substance has been characterized in a way that distinguishes it from all other pure substances.

Chemistry begins with observations and experiments.

A pure substance also can be defined theoretically in terms of a description of the molecules, atoms, and subatomic particles that compose the substance. For example, one important **theoretical** concept, discussed in Chapter 3, states that all atoms that contain exactly eight protons in the nucleus are oxygen atoms.

Both operational and theoretical definitions are important in chemistry, and both are used in this text. The theoretical definitions follow the development of the theories on which they are based.

SELF-TEST 2-A

1. Four common materials that cannot be pure substances are:
    a. _____ b. _____
    c. _____ d. _____
2. In the human experience, which usually comes first, the operational definition or the theoretical definition? _____
3. Four common materials that are very nearly pure substances are:
    a. _____ b. _____
    c. _____ d. _____
4. The properties of two different pure substances could all be identical.
    True ( ) or False ( )
5. A homogeneous mixture is a _____.
6. Match the following descriptions with the correct states of matter.
    _____ a. Of definite shape and volume   1. Gas
    _____ b. Of indefinite shape and volume   2. Liquid
    _____ c. Composed of charged particles   3. Solid
    _____ d. Of indefinite shape, definite volume 4. Plasma
7. Solutions may exist in solid, liquid, or gaseous states.   True ( ) or False ( )
8. A colloid is a true solution.   True ( ) or False ( )
9. The end of purification efforts results in _____.

Elements and Compounds

Experimentally, pure substances can be classified into two categories: those that can be broken down by chemical change into simpler pure substances and those that cannot. Table sugar (sucrose), a pure substance, decomposes when heated in an oven, leaving carbon (another pure substance) and evolving water. No chemical operation has ever been devised that will decompose carbon into simpler pure substances. Obviously, sucrose and carbon belong to two different categories of pure substances. Only 89 substances found in nature cannot be changed chemically to simpler substances; 20 others are available artificially. These 109 substances are called **elements.** Pure substances that can be decomposed into two or more different pure substances are referred to as **compounds.** Sucrose and water are two examples of compounds. Although there are only 109 known elements, there appears to be no practical limit to the number of compounds that can be made from the elements.

Elements are the basic building blocks of the universe and the world in which we live. Figure 2-6 shows the distribution of the elements in the crust of the Earth, the whole Earth, and the universe. Table 2-1 lists the properties of some common elements; a complete list of the elements is found inside the front cover of this text.

The cost of an elemental sample depends on the availability of the element in nature, the ease with which it can be reduced to its elemental form, and the desired level of purity for the sample (Table 2-2). Several elements are found in their elementary form in nature; examples include gold, silver, oxygen, nitrogen, carbon (graphite or diamond), copper, platinum, sulfur, and the noble gases (helium, neon, argon, krypton, xenon, and radon). However, many more elements are found chemically combined with other elements in the form of compounds.

Elements in compounds no longer show all their original, characteristic properties, such as color, hardness, and melting point. As an example, consider ordinary sugar, which is properly called sucrose. It is made up of three elements: carbon (which is a black powder), hydrogen (the lightest gas known), and oxygen (a gas necessary for respiration). Sucrose is completely unlike any of its three elements; the compound is a white crystalline powder that, unlike carbon, is readily soluble in water.

A distinction should be made between a compound of two or more elements and a mixture of the same elements. The two gases hydrogen and oxygen can be mixed in all proportions. However, these two elements can and do react chemically to form the compound water. Not only does water exhibit properties peculiar to itself and different from those of hydrogen and oxygen, it also has a definite percentage composition by weight (88.8% oxygen and 11.2% hydrogen). In addition to the distinctly different

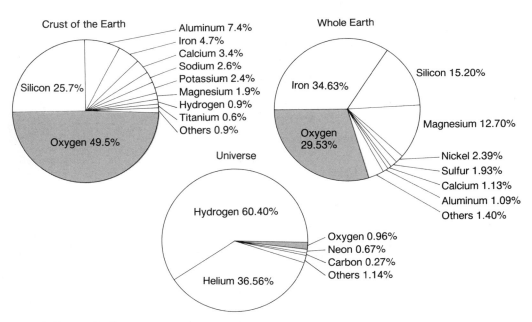

Figure 2-6
Relative abundance (by mass) of the most common elements in the Earth's crust, the whole Earth, and the universe. Note that the Earth's crust differs significantly from the cosmic array of elements.

Table 2–1
Some Common Elements*

Metals
A metal is a good conductor of electricity, can have a shiny or lustrous surface, and in the solid form usually can be deformed without breaking.

Name	Symbol	Properties of Pure Elements
Iron Latin, *ferrum*	Fe	Strong, malleable, corrodes in moist air
Copper Latin, *cuprum*	Cu	Soft, reddish-colored, ductile
Sodium Latin, *natrium*	Na	Soft, light metal, very reactive, low melting point
Silver Latin, *argentum*	Ag	Shiny, white metal, relatively unreactive, good conductor of electricity and heat
Gold Latin, *aurum*	Au	Heavy, yellow metal, very unreactive, ductile, good conductor
Chromium	Cr	Resistant to corrosion, hard, bluish-gray, brittle

Nonmetals
A nonmetal is often a poor conductor of electricity, normally lacks a shiny surface, and is brittle in crystal-solid form.

Hydrogen	H	Colorless, odorless, occurs as a very light gas (H_2), burns in air
Oxygen	O	Colorless, odorless gas (O_2), reactive, constituent of air
Sulfur	S	Odorless, yellow solid (S_8), low melting point, burns in air
Nitrogen	N	Colorless, odorless gas (N_2), rather unreactive
Chlorine	Cl	Greenish-yellow gas (Cl_2), very sharp choking odor, poisonous
Iodine	I	Dark purple solid (I_2), sublimes easily

* Chemists usually use the symbol rather than the name of the element. In addition to denoting the element, the chemical symbol has a very specialized meaning, which is described later in this chapter. A complete list of the elements with their symbols can be found inside the front cover of this book.

properties between compounds and their parent elements, there is this second difference between compounds and mixtures: **Compounds have a definite percentage composition by weight of the combining elements.**

Chemical, Physical, and Nuclear Changes

The three categories of material changes are physical, chemical and nuclear changes. In the study of chemistry it is useful to be familiar with both the operational and the theoretical definitions of each type of change.

In **physical changes,** pure substances are preserved, although the physical state may be altered, as in the case of melting ice. In theory, a physical change may break the gross arrangement of the overall structure of the theoretical particles (as in cutting a diamond) or even separate the particles (as in melting and vaporizing a metal), but the fundamental particles remain, and their tendency to interact with each other is retained.

A **chemical change** involves the disappearance of one or more pure substances and the appearance of one or more other pure substances. In theoretical terms, a chemical change produces a new arrangement of the atoms involved without a loss or gain in the number or kinds of atoms.

Table 2-2
Costs of Selected Elements*

Element	Form	Highest Purity (%)	Smallest Package (g)	Package Cost ($)	Gram Cost ($)
Aluminum	Wire	99.999	10.5	11.75	1.12
Arsenic	Mesh	99.9999	10	32.10	3.21
Barium	Stick	99	10	27.50	2.75
Beryllium	Flake	99.9	10	64.00	6.40
Boron	Powder	99.999	0.500	75.50	151.00
Calcium	Turnings	99	100	23.90	0.34
Cesium	Ingot	99.95	1	29.40	29.40
Copper	Rod	99.998	56	62.00	1.11
Gallium	Splatter	99.9999	1	11.25	11.25
Gold	Powder	99.999	0.250	29.40	117.60
Iodine	Crystal	99.999	20	22.90	1.14
Iridium	Powder	99.9	0.500	45.00	90.00
Iron	Powder	99.999	10	19.80	1.98
Lead	Foil	99.9995	28	17.40	0.62
Mercury	Liquid	99.999	500	41.40	0.08
Phosphorus	Lump	99.999	5	24.20	4.84
Platinum	Powder	99.99	1	56.50	56.50
Potassium	Ingot	99.95	1	27.50	27.50
Rhodium	Powder	99.99	1	117.50	117.50
Selenium	Pellets	99.9999	20	19.20	0.96
Silicon	Powder	99.999	5	12.00	2.40
Silver	Shot	99.9999	5	19.75	3.95
Sodium	Ingot	99.95	5	42.60	8.52
Sulfur	Powder	99.999	50	18.80	0.38
Zinc	Shot	99.999	25	27.50	1.10

The costs of pure samples of the elements are a function of the availability of the elemental source and the degree of difficulty in the isolation of the element in the purity desired.
* Prices published in 1986 by Aldrich Chemical Co., Milwaukee, Wisconsin. On the world's bullion markets, gold sold for about $12.30 per gram; silver sold for about $0.32 per gram.

Physically, a **nuclear change,** like a chemical change, produces new pure substances; however, nuclear changes are generally associated with the conversion of elements into other elements, radioactive emissions, and much larger energy transformations than those involved in chemical and physical changes. Furthermore, in many nuclear changes matter disappears into energy.

Consider the following examples:

Physical changes: Evaporation of a liquid, melting of iron, drawing of metal wire, freezing water, crystallization of salt from sea water, grinding or pulverization of a solid, cutting and shaping of wood, bending, breaking, and molding.

Chemical changes: Rusting iron, burning gasoline in an automobile engine, preparation of caramel by heating of sugar, preparation of iron from its ores, solution of copper in nitric acid, ripening and souring fruit, and decaying food.

Nuclear changes: Production of the elements in stars, splitting of uranium atoms in an atomic bomb, fusion of light atoms in a thermonuclear bomb, produc-

Nuclear changes are discussed in Chapter 4.

tion of radioisotopes in a nuclear reactor, and natural radiation from radioactive iodine or radon.

The Language of Chemistry

Symbols, formulas, and equations are used in chemistry to convey ideas quickly and concisely. These shorthand notations are merely a convenience and contain no mysterious concepts that cannot be expressed in words. Certain characters are used often, and a general familiarity with them is desirable.

A **chemical symbol** for an element is a one- or two-letter term, the first letter a capital and the second a lowercase letter. The symbol represents three concepts. First, the symbol stands for the element in general. H, O, N, Cl, Fe, and Pt are shorthand notations for the elements hydrogen, oxygen, nitrogen, chlorine, iron, and platinum, respectively, and it is useful and time-saving to substitute these symbols for the words themselves in describing chemical changes. Some symbols originate from Latin words (such as Fe, from *ferrum,* the Latin word for "iron"); others come from English, French, or German names. Second, the chemical symbol stands for a single atom of the element. The **atom** is the smallest particle of the element that can enter into chemical combinations. Third, the elemental symbol stands for a mole of the atoms of the element. The **mole** (the term is derived from the Latin for "a pile of" or "a quantity of") is the quantity of a substance that contains 602 sextillion identical particles. Just as a dozen apples would be 12 apples, a mole of atoms would be 602,200,000,000,000,000,000,000 atoms or 6.02×10^{23} atoms. How big is this number? A mole of textbooks like this one would cover the *entire surface of the continental states* to a height of 190 miles! To match the population density on Earth, a mole of people would require 150 trillion planets. It takes 134,000 years for a mole of water drops to flow over Niagara Falls at a flow rate of 112,500,000 gallons per minute. A mole is a very large number. A mole of very small atoms is usually a convenient amount for laboratory work. Thus, the symbol Ca can stand for the element calcium, or a single calcium atom, or a mole of calcium atoms. It will be evident from the context which of these is meant.

Atoms can unite (bond together) to form molecules. A **molecule** is the smallest particle of an element or a compound that can have a stable existence in the close presence of like molecules. One or more of the same kind of atom can make a molecule of an element. For example, two atoms of hydrogen bond together to form a molecule of ordinary hydrogen, and eight sulfur atoms form a single molecule. Subscripts in a chemical formula show the number of atoms involved: H_2 means that a hydrogen molecule is composed of two atoms, and S_8 means that a sulfur molecule is composed of eight atoms. The noble gases, such as helium, He, have monatomic molecules (monatomic = one atom).

When unlike atoms combine, as in the case of water (H_2O) or sulfuric acid (H_2SO_4), the formulas tell what the atoms are and how many of each are present in a molecule of the compound. For example, H_2SO_4 molecules are composed of two hydrogen atoms, one sulfur atom, and four oxygen atoms. A **formula** can stand not only for the substance itself, but also for one molecule of the substance or for a mole of such molecules, depending on the context.

When elements or compounds undergo a chemical change, the formulas, arranged in the form of a **chemical equation,** can present the information in a very concise fashion. For example, carbon can react with oxygen to form carbon monoxide. Like most solid elements, carbon is written as though it had one atom per molecule; oxygen exists as diatomic (two-atom) molecules, and carbon monoxide molecules contain two

Chemical symbols are abbreviations for the different elements.

A *mole* contains 6.022×10^{23} particles.

If 6.022×10^{23} pennies were distributed equally among the 5 billion inhabitants of the Earth, each man, woman, and child would have over $2 million to spend every hour of every day for the rest of his or her life!

A mole of water molecules in the liquid state occupies only about four teaspoonfuls. Are molecules small . . . or are they small?

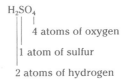

atoms, one each of carbon and oxygen. Furthermore, one oxygen molecule combines with two carbon atoms to form two carbon monoxide molecules. All this information is contained in the equation

$$2C + O_2 \longrightarrow 2CO$$

Chemical equations summarize information on chemical reactions in a concise fashion.

The arrow is often read "yields"; the equation then states the following information:

1. Carbon plus oxygen yields carbon monoxide.
2. Two atoms of carbon plus one diatomic molecule of oxygen yields two molecules of carbon monoxide.
3. Two moles of carbon atoms plus one mole of diatomic oxygen molecules yields two moles of carbon monoxide.

The number written before a formula, the **coefficient,** gives the amount of the substance involved, while the **subscript** is a part of the definition of the pure substance itself. For example, 2CO means either two molecules of carbon monoxide or two moles of these molecules, whereas CO_2 means a molecule or a mole of carbon dioxide, a very different substance.

SELF-TEST 2-B

1. The two large categories into which elements can be divided are (a) _____ and (b) _____. A half-dozen examples of each are:
    a. _____
    b. _____
2. How many elements are known presently? _____
3. A compound has properties that are a combination of the elemental properties.   True ( ) or False ( )
4. Four chemical changes not listed in this chapter are:
    a. _____
    b. _____
    c. _____
    d. _____
5. Four physical changes not listed in this chapter are:
    a. _____
    b. _____
    c. _____
    d. _____
6. What type of material change can result in the formation of new elements? _____

7. A chemical change always produces a new _____.
8. What element is most abundant
    a. In the crust of the Earth? _____
    b. In the bulk of the Earth? _____
    c. In the universe? _____

9. Consider the equation 2 Na + 2 HCl → H_2 + 2 NaCl. Explain what is meant by the symbols.

   a. Na _____

   b. 2 Na _____

   c. HCl _____

   d. → _____

   e. H_2 _____

   f. 2 NaCl _____

10. Name three concepts that are represented by a chemical symbol.

    a. _____

    b. _____

    c. _____

11. A chemical formula gives what two pieces of information?

    a. _____

    b. _____

12. Which is proper to change when you balance a chemical equation: a coefficient or a subscript? _____

13. How many atoms are in one molecule of each of the following substances?

    a. N_2O_5 _____

    b. $C_{12}H_{22}O_{11}$ _____

    c. $C_7H_5(NO_2)_3$ _____

14. A mixture of two elements can have many compositions, whereas a compound has a definite composition.   True ( ) or False ( )

The Structure of Matter

The structure of matter explains the properties of matter. Why does an element or compound have the properties it has? Why does one element or compound undergo a change that another element or compound will not undergo? Inanimate matter is the way it is because of the nature of its parts, just as a watch is what it is because of the nature of its individual parts. The most basic parts of matter must be very small. If we hope to understand the nature of matter, it is absolutely necessary that we have some understanding of these minute parts and how they are related to each other.

A very large portion of today's research in chemistry is aimed at elucidating the structure of matter. The theme of structure and related properties is of great interest because if we know exactly how and with what strength the minute parts of matter are put together, we may discover exact relationships between structure and properties. Armed with this understanding, we can make changes that result in new substances with predictable properties. Although we cannot yet predict chemical changes solely on the basis of structural characteristics, significant advances in that direction have been made. The success rate of such predictions greatly facilitates the efficiency of ongoing chemical discovery.

Samples of matter large enough to be seen, felt, and handled, and thus large enough for ordinary laboratory experiments, are called **macroscopic** samples, in contrast to

The causes of the properties of matter lie in the structure and composition of its parts.

Submicroscopic structures help explain chemistry.

microscopic samples, which are so small that they have to be viewed with the aid of a microscope. The structure of matter that controls the properties of matter must be at the **submicroscopic** level. Our senses have no direct access to this small world of structure, and any conclusions about it have to be based on circumstantial evidence gathered in macroscopic and microscopic experimentation (Fig. 2–7).

The Periodic Chart: Order in the Properties of the Elements

When the physical and chemical properties of the elements are considered, similar properties are found within groups of elements. For example, the elements in Table 2–1 are classified as metals or nonmetals. Most elements are metals, which generally are lustrous, malleable, ductile, and good conductors of heat and electricity. Less than one fifth of the elements are nonmetals, which are generally poor conductors of heat and electricity and, in the solid state, are brittle. With so many properties of the elements known, there are many ways to classify them into groups. However, there is only one

Figure 2–7
Direct observation stops at the microscopic level. Convinced of structure beyond the microscopic level, chemists employ circumstantial evidence to construct the world of molecules, atoms, and subatomic particles in the mind's eye.

classification system that correlates essentially all elemental classifications made to date: the Periodic Chart (Table) of the Elements (Fig. 2-8).

The periodic chart is so successful in interrelating the known properties of the elements that some have touted it as being the most important one-page document produced by the human race. Although others may consider this an overstatement, it is hard for the chemist to overemphasize the importance of the periodic chart for organizing knowledge, understanding relationships in chemical change, formulating theories to explain chemical phenomena, and predicting future chemical events. The periodic chart is so important and is so much a part of every facet of chemistry that it is only introduced in this chapter. It will be used as an instructional and learning tool in several other places in the book.

As early as 1817, Johann Wolfgang Döbereiner, a German chemist, saw trends and similarities within several groups of three elements each. First, he observed that lithium (Li), sodium (Na), and potassium (K) had similar properties. The oxides of these metals are readily obtained from ash and dissolve in water to form strong alkaline (basic) solutions; consequently, this group of elements became known as the **alkali metals.** Later information revealed that rubidium (Rb), cesium (Cs), and francium (Fr) belong to the same family, called Group IA on the modern periodic chart.

Calcium (Ca), strontium (Sr), and barium (Ba), known as the **alkaline earth metals,** constitute another group recognized by Döbereiner. They, along with beryllium (Be), magnesium (Mg), and radium (Ra), constitute Group IIA on our present chart. The oxides of the alkaline earth metals also yielded basic solutions in water but were not as soluble in water as the oxides of the alkali metals.

A third triad of elements recognized early was a group of nonmetals, chlorine (Cl), bromine (Br), and iodine (I). These elements were observed to be part of many salts and thus were given the name **halogen,** which means "salt former." Now we know that fluorine (F) and astatine (At) also belong to this family of elements, Group VIIA.

Observe that each family, or group, of elements constitutes a vertical column on the periodic chart. Other families of elements will be presented as we proceed through the study of chemistry.

In 1869 Dmitri Mendeleev, a Russian chemist and schoolteacher, took a major step toward understanding the relationship between the properties of the elements and the periodic chart as we know it today. In Germany, Lothar Meyer independently and concurrently presented essentially the same proposal. The idea of atomic weights will be presented in some detail in Chapter 3, but for the moment you can simply accept the idea that the atoms of the elements have mass (or weight) and that each element has a characteristic atomic weight in its natural occurrence. Although the idea of atomic weights was firmly established before 1860, there was general confusion about the individual atomic weight values for the elements. After 1860 our present set of atomic weights began to be established firmly. It occurred to Mendeleev and Meyer that if the elements were listed in order of their atomic weights, elements with similar properties would appear periodically (Fig. 2-9). In order to think along the same lines first thought by Mendeleev and Meyer, examine some of the information given in Table 2-3. First look at Element 3 and find two other elements down the line with similar properties. You will see that the Elements 3, 11, and 19 have similar properties. Follow by looking for other groups of elements in the list. The breakthrough in establishing the periodic chart came when order was found in the repetition of similar elements as the elements were listed according to their atomic weight.

We shall see in Chapter 3 that atomic number is a more fundamental property of the elements than atomic weight in grouping similar properties together. However, the

Atomic weights are developed and discussed in Chapter 3.

Periodic Table of the Elements

Period	Group IA	IIA	IIIB	IVB	VB	VIB	VIIB	VIII			IB	IIB	IIIA	IVA	VA	VIA	VIIA	Noble Gases
1	1 Hydrogen H 1.0079																	2 Helium He 4.0026
2	3 Lithium Li 6.939	4 Beryllium Be 9.0122											5 Boron B 10.811	6 Carbon C 12.0112	7 Nitrogen N 14.0067	8 Oxygen O 15.9994	9 Fluorine F 18.9984	10 Neon Ne 20.183
3	11 Sodium Na 22.9898	12 Magnesium Mg 24.312											13 Aluminum Al 26.9815	14 Silicon Si 28.086	15 Phosphorous P 30.9738	16 Sulfur S 32.064	17 Chlorine Cl 35.453	18 Argon Ar 39.948
4	19 Potassium K 39.098	20 Calcium Ca 40.08	21 Scandium Sc 44.956	22 Titanium Ti 47.90	23 Vanadium V 50.996	24 Chromium Cr 51.996	25 Manganese Mn 54.938	26 Iron Fe 55.847	27 Cobalt Co 58.933	28 Nickel Ni 58.71	29 Copper Cu 63.546	30 Zinc Zn 65.38	31 Gallium Ga 69.72	32 Germanium Ge 72.59	33 Arsenic As 74.922	34 Selenium Se 78.96	35 Bromine Br 79.904	36 Krypton Kr 83.80
5	37 Rubidium Rb 85.47	38 Strontium Sr 87.62	39 Yttrium Y 88.905	40 Zirconium Zr 91.22	41 Niobium Nb 92.906	42 Molybdenum Mo 95.94	43 Technetium Tc (99)	44 Ruthenium Ru 101.07	45 Rhodium Rh 102.905	46 Palladium Pd 106.4	47 Silver Ag 107.868	48 Cadmium Cd 112.40	49 Indium In 114.82	50 Tin Sn 118.69	51 Antimony Sb 121.75	52 Tellurium Te 127.60	53 Iodine I 126.904	54 Xenon Xe 131.30
6	55 Cesium Cs 132.905	56 Barium Ba 137.34	*57 Lanthanium La 138.91	72 Hafnium Hf 178.49	73 Tantalum Ta 180.948	74 Wolfram (Tungsten) W 183.85	75 Rhenium Re 186.2	76 Osmium Os 190.2	77 Iridium Ir 192.2	78 Platinum Pt 195.09	79 Gold Au 196.967	80 Mercury Hg 200.59	81 Thalium Tl 204.37	82 Lead Pb 207.19	83 Bismuth Bi 208.980	84 Polonium Po (210)	85 Astatine At (210)	86 Radon Rn (222)
7	87 Francium Fr (223)	88 Radium Ra (226)	**89 Actinium Ac (227)	104 — (261)	105 — (263)	106 — (263)	107 — (261)	108 — —	109 — (266)									

Atomic number → 11
Name → Sodium
Symbol → Na
Atomic weight → 22.9898

*Lanthanide series 6

58 Cerium Ce 140.12	59 Praseodymium Pr 140.907	60 Neodymium Nd 144.24	61 Promethium Pm (147)	62 Samarium Sm 150.35	63 Europium Eu 151.96	64 Gadolinium Gd 157.25	65 Terbium Tb 158.924	66 Dysprosium Dy 162.50	67 Holmium Ho 164.930	68 Erbium Er 167.26	69 Thulium Tm 168.934	70 Ytterbium Yb 173.04	71 Lutetium Lu 174.97

**Actinide series 7

90 Thorium Th 232.038	91 Protactinium Pa (231)	92 Uranium U 238.03	93 Neptunium Np (237)	94 Plutonium Pu (242)	95 Americium Am (243)	96 Curium Cm (247)	97 Berkelium Bk (247)	98 Californium Cf (251)	99 Einsteinium Es (254)	100 Fermium Fm (253)	101 Mendelevium Md (256)	102 Nobelium No (254)	103 Lawrencium Lr (257)

Figure 2–8
Modern Periodic Table of the Elements.

Table 2-3
Some Properties of the First 20 Elements

Element	Atomic Number	Description	Compound Formation* With Cl (or Na)	With O (or Mg)
Hydrogen (H)	1	Colorless gas; reactive	HCl	H_2O
Helium (He)	2	Colorless gas; unreactive	None	None
Lithium (Li)	3	Soft metal; low density; very reactive	LiCl	Li_2O
Beryllium (Be)	4	Harder metal than Li; low density; less reactive than Li	$BeCl_2$	BeO
Boron (B)	5	Both metallic and nonmetallic; very hard; not very reactive	BCl_3	B_2O_3
Carbon (C)	6	Brittle nonmetal; unreactive at room temperature	CCl_4	CO_2
Nitrogen (N)	7	Colorless gas; nonmetallic; not very reactive	NCl_3	N_2O_5
Oxygen (O)	8	Colorless gas; nonmetallic; moderately reactive	Na_2O, Cl_2O	MgO
Fluorine (F)	9	Greenish-yellow gas; nonmetallic; extremely reactive	NaF, ClF	MgF_2, OF_2
Neon (Ne)	10	Colorless gas; unreactive	None	None
Sodium (Na)	11	Soft metal; low density; very reactive	NaCl	Na_2O
Magnesium (Mg)	12	Harder metal than Na; low density; less reactive than Na	$MgCl_2$	MgO
Aluminum (Al)	13	Metal as hard as Mg; less reactive than Mg	$AlCl_3$	Al_2O_3
Silicon (Si)	14	Brittle nonmetal; not very reactive	$SiCl_4$	SiO_2
Phosphorus (P)	15	Nonmetal; low melting point; white solid; reactive	PCl_3	P_2O_5
Sulfur (S)	16	Yellow solid; nonmetallic; low melting point; moderately reactive	Na_2S, SCl_2	MgS
Chlorine (Cl)	17	Green gas; nonmetallic; extremely reactive	NaCl	$MgCl_2$, Cl_2O
Argon (Ar)	18	Colorless gas; unreactive	None	None
Potassium (K)	19	Soft metal; low density; very reactive	KCl	K_2O
Calcium (Ca)	20	Harder metal than K; low density; less reactive than K	$CaCl_2$	CaO

* The chemical formulas shown are lowest ratios. The molecular formula for $AlCl_3$ is Al_2Cl_6, and that for P_2O_5 is P_4O_{10}.

atomic number concept is more a refinement of Mendeleev's ideas than a rejection of them. Although the idea was at first vague, and the periods, or series, of elements were not all the same length, Mendeleev was able to predict properties of three undiscovered elements — scandium, gallium, and germanium — on the basis of the regularity of the properties of the known elements in each family. Table 2-4 lists some of the predicted properties of germanium and the subsequently measured values. The agreement is

			Ti = 50	Zr = 90	? = 180
			V = 51	Nb = 94	Ta = 182
			Cr = 52	Mo = 96	W = 186
			Mn = 55	Rh = 104,4	Pt = 197,4
			Fe = 56	Ru = 104,4	Ir = 198
			Ni = Co = 59	Pd = 106,6	Os = 199
H = 1			Cu = 63,4	Ag = 108	Hg = 200
	Be = 9.4	Mg = 24	Zn = 65,2	Cd = 112	
	B = 11	Al = 27.4	? = 68	Ur = 116	Au = 197?
	C = 12	Si = 28	? = 70	Sn = 118	
	N = 14	P = 31	As = 75	Sb = 122	Bi = 210?
	O = 16	S = 32	Se = 79,4	Te = 128?	
	F = 19	Cl = 35,5	Br = 80	J = 127	
Li = 7	Na = 23	K = 39	Rb = 85,4	Cs = 133	Tl = 204
		Ca = 40	Sr = 87,6	Ba = 137	Pb = 207
		? = 45	Ce = 92		
		?Er = 56	La = 94		
		?Yt = 60	Di = 95		
		?In = 75,6	Th = 118?		

Figure 2–9
Periodic Table of the Elements by Mendeleev. Although this is arranged perpendicular to the usual form of the modern chart, there are similarities between the two: periods of varying lengths (two elements in first period, seven in second, etc.) and families of elements that cut across the periods in a straight horizontal line. Also, notice the gaps left for undiscovered elements as well as the prediction of the atomic weights of those undiscovered elements.

remarkable. It was then easy for chemists to believe Mendeleev's first statement of the Periodic Law: "The properties of the elements are in periodic dependence on their atomic weights." Based on Mendeleev's theoretical predictions, the rush for discovery was on. Fifty years later, all gaps in the list of elements had been filled, and any hope for the discovery of new elements lay in the extension of the list, a process that continues today.

A few additional ideas should be noted as you look at a modern periodic chart shown in Figure 2–8. There are seven **periods** of elements that run across the chart in horizontal rows. Notice there are two elements in the first period, eight in the second and the third, eighteen in the fourth and the fifth, thirty-two in the sixth, and twenty-three in the seventh. The division between the metals and nonmetals is a stair-step line between Be and B, Al and Si, Ge and As, Sb and Te, and Po and At. The elements along

Table 2–4
Mendeleev's Prediction of the Properties of Germanium*

Property (Partial List)	Mendeleev's Prediction	Actual Value
Atomic weight	72	72.59
Density (g/cm^3)	5.5	5.32
Density of dioxide (g/cm^3)	4.7	4.23
Combining power	4	4
Atomic volume (cm^3)	13	13.6

this line are often referred to as **semimetals,** or **metalloids,** because their properties are intermediate between those of metals and nonmetals. Elements 57 through 70, the **lanthanide** series, and elements 89 through 102, the **actinide** series, are usually placed at the bottom of the chart as a footnote, since the chart would be too long for easy printing if they were placed in their correct positions after barium (No. 56) and radium (No. 88), respectively. Each element has a number, the **atomic number,** and the pattern of elements leaves no room for additional elements between 1 and 109.

Measurement

The establishment of natural facts and laws is dependent on accurate observations and measurements. Although one can be as accurate in one language as in another, or in one system of units as another, there has been an effort from the time of the French Revolution to have one simple system of measures embraced by all scientists and, indeed, the whole world. The hope was and is to facilitate communication.

The metric system, which was born of this effort, has two advantages. First, it is easy to convert from one unit to another, since subunits and multiple units differ only by factors of ten. Consequently, to change millimeters to meters, one has only to shift the decimal three places to the left. Compare the difficulty of the decimal shift with the problem of changing inches to miles. The second advantage, which has not been fully achieved, is that standards for measurements are defined by reproducible phenomena of nature rather than by the length of the king's foot or some other such changeable standard. For example, length is now defined in terms of the length of a particular wavelength of light, a number believed to be invariant.

Commonly used prefixes: milli — 0.001, centi — 0.01, kilo — 1000, micro — 10^{-6}, nano — 10^{-9}.

The International System of Units, abbreviated SI, was adopted in 1960 by the International Bureau of Weights and Measures. SI is an extension of the metric system, retaining its ease of unit conversion but doing a better job in the definition of units based on physical phenomena.

Appendix A gives further information about the SI system.

Seven fundamental units are required to describe what is now known about the universe. The concepts to be measured, the names of the units, and their symbols are:

1. Length   Meter   m
2. Mass   Kilogram   kg
3. Time   Seconds   s
4. Temperature   Degree Kelvin   K
5. Luminous intensity Candle   cd
6. Electric charge   Coulomb   C
7. Molecular quantity Mole   mol

Other units are derived from these seven units. For example, volume is defined in terms of cubic length.

In this book we employ six units that are commonly used in the chemical literature. Along with suitable conceptual definitions, the units needed are:

1. Meter   m   39.4 inches
2. Liter   L   1.06 quarts
3. Gram   g   0.0352 ounce
4. Celsius degree °C   Water boils at 100°C; water freezes at 0°C
5. Calorie   cal   Energy required to heat 1 g of water 1°C
6. Mole   mol   Number of atoms in 12 g of carbon-12 isotope

Temperature scales are defined in terms of the behavior of samples of matter. For example, the familiar Fahrenheit scale defines the temperature at which water freezes to be 32°F and that at which it boils to be 212°F; thus, there are 180 (212 − 32) Fahrenheit degrees between the freezing and boiling points of water. Most scientists prefer the Celsius scale, which defines the freezing point of water to be 0°C and the boiling point 100°C, resulting in a simple 100 degrees between the freezing and boiling points of water.

SELF-TEST 2-C

1. The SI system is more accurate than the English system of weights and measures.   True ( ) or False ( )
2. Put in order of decreasing size: microscopic, molecular, and macroscopic.
   _____, _____, _____
3. Are elements with similar chemical and physical properties arranged ( ) up and down a column or ( ) across a row of the periodic chart?
4. Look at the periodic chart and name an element that has properties similar to those of the element carbon. _____
5. Give the name of an element in the following families of elements:
   a. Alkali metals _____
   b. Alkaline earth metals _____
   c. Halogens _____
6. Who was the Russian and who was the German who first stated earlier versions of the periodic law relative to elemental properties? _____, _____
7. Is it likely that new elements will be discovered within the present list of elements? _____ Explain. _____
8. Is it likely that new elements will be discovered that will extend the present list of elements? _____ Explain. _____
9. How many periods of elements are given on the periodic chart? _____
10. What happens to the lengths of the periods as you move down the periodic chart? _____

MATCHING SET

_____   1. Produces a new type of matter   a. Filtration
_____   2. Air   b. Chemical change
_____   3. Unchanged by further purification   c. Element
_____   4. Used to separate a solid from a liquid   d. Pure substance

_____ 5. Cannot be reduced to simpler substances
_____ 6. SI system
_____ 7. Symbol for iron
_____ 8. Mole of atoms
_____ 9. Molecule containing three oxygen atoms
_____ 10. Carbon monoxide
_____ 11. 100 centimeters
_____ 12. Element symbol
_____ 13. Theoretical definition
_____ 14. Plasma
_____ 15. Solution

e. Mixture
f. Uses multiples of ten
g. O_3
h. CO
i. Fe
j. One meter
k. 6.02×10^{23} atoms
l. Based on the idea of atoms
m. Made of charged particles
n. The element, the atom, or a mole
o. Homogeneous mixture

QUESTIONS

1. Name as many materials as you can that you have used during the past day that were not changed chemically by artificial means.
2. In pottery making, an object is shaped and then baked. Which part of this process is chemical and which part is physical?
3. Identify the following as physical or chemical changes. Justify your answers in terms of operational definitions.
    a. Formation of snowflakes
    b. Rusting of a piece of iron
    c. Ripening of fruit
    d. Fashioning of a table leg from a piece of wood
    e. Fermenting of grapes
    f. Boiling of a potato in water
4. Name a physical change that food undergoes as you eat it. Name a chemical change that follows.
5. Would it be possible for two pure substances to have exactly the same set of properties? Give reasons for your answer.
6. Chemical changes can be both useful and destructive to human purposes. Cite a few examples of each kind with which you have had personal experience. Also give observed evidence that each is indeed a chemical change and not a physical change.
7. Name two pure substances that are used at the dinner table. Identify each as an element or a compound.
8. Classify each of the following as an element, compound, or mixture. Justify your answers.
    a. Mercury
    b. Milk
    c. Pure water
    d. A tree
    e. Ink
    f. Iced tea
    g. Pure ice
    h. Carbon
9. Which of the materials listed in Question 8 can be pure substances?
10. Why do theoretical definitions come after operational definitions in a particular concept?
11. Is it possible for the properties of iron to change? What about the properties of steel? Explain your answers.
12. Suggest a method for purifying water slightly contaminated with a dissolved solid.
13. Did most purification techniques arise from theory or practice? Illustrate with an example.
14. Define a solution as a special case in the general definition of a mixture.
15. What is the most abundant element in the universe?
16. Name an element that is a liquid at room temperature.
17. Given the following sentence, write a chemical reaction using chemical symbols that convey the same information. "One nitrogen molecule, containing two nitrogen atoms per molecule, reacts with three hydrogen molecules, each containing two hydrogen atoms, to produce two ammonia molecules, each containing one nitrogen and three hydrogen atoms." What advantage does the equation have over the words?
18. Aspirin is a pure substance, a compound of carbon, hydrogen, and oxygen. If two manufacturers produce equally pure aspirin samples, what can be said of the relative worth of the two products?
19. Is it possible to have a mixture of two elements and

also to have a compound of the same two elements? Cite an example.
20. How is the salt content of the sea related to the purity of rain water? What method of purification does nature employ in the purification of sea water to produce rain?
21. Name four forms of energy.
22. What is a difference between a chemical change and a nuclear change?
23. As your look up from this page, which are most abundant in your field of view: mixtures, compounds, or elements?
24. How is a plasma different from a gas?
25. Describe in words the chemical process that is summarized in the following equation:

$$2\ Na + Cl_2 \rightarrow 2\ NaCl$$

26. How many atoms are present in each of the following?
    a. One mole of He
    b. One mole of Cl_2
    c. One mole of CH_4
27. Name ten elements and give the symbol for each one.
28. Describe the meaning of each symbol and number in the chemical equation $2\ NO_2 \rightarrow N_2O_4$.
29. Would you think that tea in tea bags is a pure substance? Use the process of making tea to make an argument for your answer. How would your argument apply to instant tea?
30. Find and list as many pure substances as you can in a kitchen and in a laundry room.
31. What is the most fundamental assumption about structure and properties in chemical theory?
32. Consider the following equation:

$$2\ C_7H_{16} + 22\ O_2 \longrightarrow 14\ CO_2 + 16\ H_2O$$
Heptane   Oxygen   Carbon Dioxide   Water

Write out in words as much of the information represented in this equation as you can decipher. (Heptane is a component of gasoline.)
33. In the normal usage of the terms *atom* and *molecule*, which is composed of the other?
34. The number 12 is to a dozen and 144 is to a gross as 6.022×10^{23} is to a(n) ──────── .
35. If you had a mole of elephants, how many moles of elephant ears would you have? A mole of O_3 molecules contain how many moles of oxygen atoms? How many atoms?
36. Which English unit is closest to a liter? A meter?
37. How tall are you in meters?
38. What is your weight in kilograms?
39. Which is colder: 0°C or 0°F?
40. Convert the distance of your last auto trip to kilometers.
41. Library assignment: Look up other purification methods, such as sublimation, extraction, and zone refining, and tell in each case how the method is used in the separation of pure substances.

CHAPTER 3

Atoms: What Are They?

Imagine what the smallest thing in the universe is! How small can a sample of gold be? What are the building blocks of all matter? We have heard of atoms, but trying to fathom their minuteness is as challenging as trying to fathom the wholeness of the universe. One lures the mind to unseen smallness, the other to unseen largeness. If atoms are too small for us to observe, how do we know they exist? And if there were some way we could observe an individual atom, would there be some way we could probe inside it?

Answers to these questions are the focus of this chapter. Assembling the pieces of the puzzle has taken more than two thousand years, but much of the work has been done in the present era. What difference does it make what is inside atoms, or even whether they exist? The drive to discover is a strong force in certain individuals and cannot be discounted. Simply, it is thrilling to discover what makes nature tick. Knowledge of the atom leads to an understanding of bonding, chemical reactivity, light, and other intriguing phenomena. Perhaps the most powerful outcome of knowledge about atoms is our ability to predict accurately the properties of matter.

Democritus

The Greek View of Atoms

The ancient Greeks recorded the first theory of atoms. Leucippus and his student, Democritus (460–370 B.C.), argued for the concept of atoms. Democritus used the word *atom* (which literally means "uncuttable") to describe the ultimate particles of matter, particles that could not be divided further. He reasoned that in the division of a piece of matter, such as gold, into smaller and smaller pieces, one would ultimately arrive at a tiny particle of gold that could not be further divided and still retain the properties of gold. The atoms that Democritus envisioned representing different substances were all made of the same basic material. His atoms differed only in shape and size.

Democritus used his concept of atoms to explain the properties of substances. For example, the high density and softness of lead could be caused by lead atoms packed very closely together like marbles in a box and moving easily one over another. Iron was known to be less dense but quite hard. Democritus argued that the properties of iron resulted from atoms shaped like corkscrews, atoms that would entangle in a rigid but relatively lightweight structure. Although his concept of the atom was limited, Democritus did explain in a simple way some well-known phenomena, such as the drying of clothes, how moisture appears on the outside of a vessel of cold water, how an odor moves through a room, and how crystals grow from a solution. He imagined the scattering or collecting of atoms as needed to explain the events he saw. All atomic theory, even today's modern atomic theory, has been built on these original assumptions of Leucippus and Democritus: atoms, which we cannot see individually, are the cause of the phenomena that we can see.

Since the writings of Leucippus and Democritus have been destroyed, we know about their ideas only from recorded opposition to atoms and from a lengthy poem (55 B.C.) by the Roman poet Lucretius.

Plato (427–347 B.C.) and Aristotle (384–322 B.C.) led the arguments against the atom by asking to be shown atoms. They also argued that the idea of atoms was a challenge to God. If atoms could be used to explain nature, they said, there would be no need for God. For centuries most of those in the mainstream of enlightened thought rejected or ignored the atoms of Democritus.

Ideas about atoms drifted in and out of philosophical discussions for about 2200 years without playing a major role. Galileo (1564–1642) reasoned that the appearance of a new substance through chemical change involved a rearrangement of parts too small to be seen. Francis Bacon (1561–1626) speculated that heat might be a form of motion by very small particles.

Although the idea that heat is an expression of molecular motion was proposed three and a half centuries earlier by Roger Bacon, it was not until 1620 that Francis Bacon wrote his book *New Organon,* which put experimental science in the most refined and scholarly terms and made it possible for other scholars to accept it.

It was John Dalton (1766–1844), an English schoolteacher, who forcefully revived the idea of the atom. More clearly than any before him, Dalton was able to explain general observations, experimental results, and laws relative to the composition of matter. Dalton was particularly influenced by the experiments of two Frenchmen, Antoine Lavoisier (1743–1794) and Joseph Louis Proust (1754–1826). Let's look at the major contributions of these two experimentalists before we examine Dalton's theory.

Antoine-Laurent Lavoisier: The Law of Conservation of Matter in Chemical Change

There are many reasons why Antoine-Laurent Lavoisier has been acclaimed the Father of Chemistry. For example, Lavoisier was the first to use systematic names for the elements and a few of their compounds. Although he made other contributions, his most notable achievement was to show the importance of very accurate weight measurements of chemical changes. His work began the process of establishing chemistry as a quantitative science.

Lavoisier weighed the chemicals in such changes as the decomposition of mercury oxide by heat into mercury and oxygen.

$$2\ HgO \xrightarrow{heat} 2\ Hg + O_2$$

Mercury oxide   Mercury   Oxygen

Antoine-Laurent Lavoisier

Very accurate measurements showed that the total weight of all the chemicals involved remained constant during the course of the chemical change. Similar measurements on many other chemical reactions led Lavoisier to the summarizing statement now known as the **Law of Conservation of Matter:** *Matter is neither lost nor gained during a chemical reaction*. In other words, if one were to weigh all the products of a chemical reaction — solids, liquids, and gases — the total would be the same as the weight of the reactants. In an atomic view, a chemical reaction was just a recombination of atoms. As a further example of the law of conservation of matter, consider Figure 3–1.

The Personal Side

In spite of his success, Lavoisier had problems and disappointments. His highest goal, to discover a new element, was never achieved. He lost some of the esteem of his colleagues when he was accused of saying the work of someone else was his own. In 1768 he invested half a million francs in a private firm retained by the French government to collect taxes. He used the earnings (about 100,000 francs a year) to support his research. Although Lavoisier was not actively engaged in tax collecting, he was brought to trial as a "tax-farmer" during the French Revolution. Along with his father-in-law and other tax-farmers, Lavoisier was guillotined on May 8, 1794, just two months before the end of the revolution.

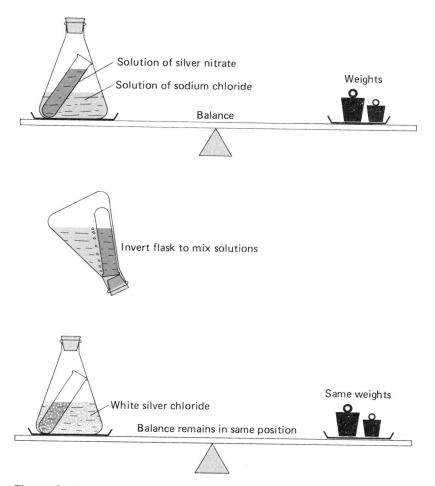

Figure 3-1
Mixing a solution of sodium chloride with a solution of silver nitrate produces a new substance, solid silver chloride, but the total weight of matter remains the same.

Joseph Louis Proust: The Law of Constant Composition

Following the lead of Lavoisier, several chemists investigated the quantitative aspects of compound formation. One such study, made by Proust in 1799, involved copper carbonate. Proust discovered that, regardless of how copper carbonate was prepared in the laboratory or how it was isolated from nature, it always contained five parts of copper, four parts of oxygen, and one part of carbon by weight. His careful analyses of this and other compounds led to the belief that any given compound has an unvarying composition. These and similar discoveries are summarized by the **Law of Constant Composition:** *In a compound, the constituent elements are always present in a definite proportion by weight.*

The Personal Side

Proust's generalization has been verified many times for many compounds since its formulation, but its acceptance was delayed by controversy. Comte Claude Louis Berthollet (1748-

1822), an eminent French chemist and physician, believed and strongly argued that the nature of the final product was determined by the amount of reacting materials one had at the beginning of the reaction. The running controversy between Proust and Berthollet reached major proportions, but more careful measurements supported Proust. Proust showed that Berthollet had made inaccurate analyses and had purified his compounds insufficiently—two great errors in chemistry.

Unlike Lavoisier, Proust saved his head during the French Revolution. Proust fled to Spain, where he lived in Madrid and worked as a chemist under the sponsorship of Charles IV, King of Spain. When Napoleon's army ousted Charles IV, Proust's laboratory was looted and his work came to an end. Later, Proust returned to his homeland, where he lived out his life in retirement.

Compounds have constant composition. Mixtures may have variable composition.

Pure water, a compound, is always made up of 11.2% hydrogen and 88.8% oxygen by weight. Pure table sugar, another compound, always contains 42.11% carbon by weight. Contrast these with 14-carat gold, a mixture that should be at least 58% gold, from 14% to 28% copper, and from 4% to 28% silver by weight. This mixture can vary in composition and still be properly called 14-carat gold, but a compound that is not 11.2% hydrogen and 88.8% oxygen is not water.

John Dalton: The Law of Multiple Proportions

John Dalton (1766–1844) made a quantitative study of different compounds made from the same elements. Such compounds differed in composition from each other, but each obeyed the law of constant composition. Examples of this concept are the compounds carbon monoxide, a poisonous gas, and carbon dioxide, a product of respiration. Both compounds contain only carbon and oxygen. Carbon monoxide has carbon and oxygen in proportions by weight of 3 to 4. Carbon dioxide has carbon and oxygen in proportions by weight of 3 to 8. Note that for equal amounts of carbon (three parts), the ratio of oxygen in the two compounds is 8 to 4, or 2 to 1.

Methane is the main component of natural gas. Ethylene is the only component of polyethylene.

In 1803, after analyzing compounds of carbon and hydrogen such as methane (in which the ratio of carbon to hydrogen is 3 to 1 by weight) and ethylene (in which the ratio of carbon to hydrogen is 6 to 1 by weight) and compounds of nitrogen and oxygen, Dalton first clearly enunciated the **Law of Multiple Proportions:** *In the formation of two or more compounds from the same elements, the weights of one element that combine with a fixed weight of a second element are in a ratio of small whole numbers (integers) such as 2 to 1, 3 to 1, 3 to 2, or 4 to 3.*

Dalton's Atomic Theory

Why do the laws of conservation of matter, constant composition, and multiple proportions exist? How can they be explained? John Dalton employed the idea of atoms and endowed them with properties that enabled him to explain these chemical laws (Fig. 3–2).

The Personal Side

Whereas Lavoisier is considered the father of chemical measurement, Dalton is considered the father of chemical theory. Dalton, a gentle man and a devout Quaker, gained acclaim because of his work. He made careful measurements, kept detailed records of his research, and expressed them convincingly in his writings. However, he was a very poor speaker and was not well received as a lecturer. He began teaching in a Quaker school when only 12 years

John Dalton

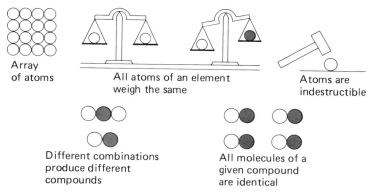

Figure 3-2
Features of Dalton's atomic theory. Isotopes have been ignored.

old, discovered a basic law of physics, the law of partial pressure of gases, and helped found the British Association for the Advancement of Science. He kept over 200,000 notes on meteorology. When Dalton was 66 years old, some of his admirers sought to present him to King William IV. Dalton resisted because he would not wear the court dress. Since he had a doctor's degree from Oxford University, the scarlet robes of Oxford were deemed suitable, but a Quaker could not wear scarlet. Dalton, being colorblind, saw scarlet as gray, so he was presented in scarlet to the court but in gray to himself. This remarkable man was, in fact, the first to describe color blindness. Despite his accomplishments, he shunned glory and maintained he could never find time for marriage.

Here are the major points of Dalton's atomic theory, presented in modernized statements:

1. Matter is composed of indestructible* particles called atoms.
2. All atoms of a given element have the same properties, such as size, shape, and weight,† which differ from the properties of atoms of other elements.
3. Elements and compounds are composed of definite arrangements of atoms, and chemical change occurs when the atomic arrays are rearranged.

Dalton's theory was successful in explaining the three laws of chemical composition and reaction (Fig. 3-3).

Ideas About Atomic Weight

John Dalton's idea about unique atomic weights for the atoms of the different elements naturally generated interest in the search for the atomic weight characteristic of each element. It was not until after 1860 that chemists developed a consistent set of atomic weights, although several notable attempts were made before that, including an early attempt by Dalton himself.

There were differences of opinion on whether the formula of water was HO or H_2O, whether hydrogen gas was H_2 or H, and whether oxygen gas was O_2 or O. Water is 88.8%

---

* Radioactive atoms are self-destructive. Dalton had no knowledge of this phenomenon.
† We now know that all atoms of the same element do not necessarily have the same weight. The idea of isotopes is introduced later in this chapter.

Law	Statement of Law	Explanation of Law
Law of Conservation of Matter	Matter is neither lost nor gained in a chemical change.	A chemical change is the result of a new arrangement of the same atoms present initially; hence, the weight is the same before and after the change.
Law of Constant Composition	When two or more elements combine to form a given compound, the ratio of the weights of the elements involved is always the same.	The smallest unit of a compound is a molecule. It has a fixed ratio of atoms, hence a fixed ratio of weights. Any larger sample of this compound would merely represent a multiple of the weights in the same ratio.
Law of Multiple Proportions	In the formation of two or more compounds from the same elements, the weights of one element that combine with a fixed weight of a second element are in a ratio of integers such as 2:1, 3:1, 3:2, or 4:3.	If, for example, the first compound has a ratio of one atom of C to one atom of O (above), and a second compound has a ratio of one atom of C to two atoms of O (below), then for a fixed number of atoms of C, the ratio of atoms of O (and the weights of O) is a ratio of integers: 1:2.

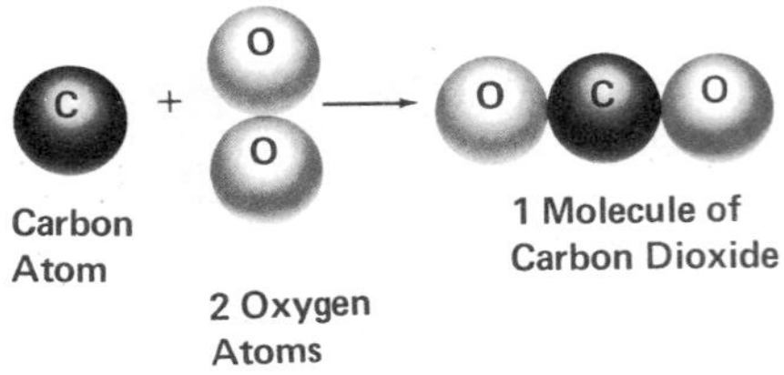

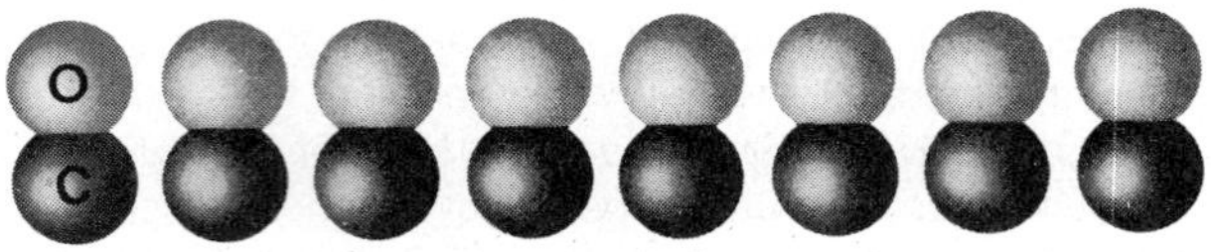

Figure 3–3
John Dalton's explanation of three laws of chemistry in terms of atoms.

oxygen and 11.2% hydrogen by weight — a firmly established experimental fact by that time. If water is HO, as Dalton argued (because of his belief that the simplest formula is likely to be the correct one), then the weight of an oxygen atom should be about eight times that of a hydrogen atom:

$$\frac{\text{weight of an oxygen atom}}{\text{weight of a hydrogen atom}} = \frac{88.8}{11.2} = \frac{7.9}{1}$$

If the formula for water is H_2O, as the scientist Amedeo Avogadro (1776–1856) had proposed in 1811, then one oxygen atom is 88.8% of the molecule but two hydrogen atoms are 11.2%. Each hydrogen atom would be $\frac{1}{2}(11.2\%)$ or 5.6%. With the formula H_2O, then, an oxygen atom would be about 16 times heavier than a hydrogen atom:

$$\frac{\text{weight of an oxygen atom}}{\text{weight of a hydrogen atom}} = \frac{88.8}{5.6} = \frac{15.9}{1}$$

The fact that an oxygen atom is about 16 times heavier than a hydrogen atom does not tell us the weight of either atom. These are **relative weights** in the same way that a grapefruit may weigh twice as much as an orange. This information gives the weight of neither the grapefruit nor the orange. However, if a specific number is *assigned* as the weight of any particular atom, this fixes the numbers assigned to the weights of all other atoms. The standard for comparison of relative atomic weights was for many years the weight of the oxygen atom, which was taken as 16.0000 atomic weight units. This allowed the lightest atom, hydrogen, to have an atomic weight of 1.008, or approximately 1.

By 1862, 18 years after Dalton's death, a consistent set of atomic weights was generally agreed on and used. The modern set of atomic weights (inside the back cover) is based on the assignment of the weight of a particular kind of carbon atom, the carbon-12 atom, as exactly 12 atomic weight units. On this scale, an atom of magnesium (Mg), with an atomic weight of about 24, has twice the weight of a carbon-12 atom. An atom of titanium (Ti), with an atomic weight of 48, has four times the weight of a carbon-12 atom.*

The modern set of atomic weights was adopted by scientists worldwide in 1961.

Atoms are Divisible—Dalton and the Greeks Were Wrong

Dalton's concept of the indivisibility of atoms was severely challenged by the subsequent discoveries of radioactivity and cathode rays and was even in conflict with some previously known electrical phenomena such as static electric charge.

Static Electric Charge

Electric charge was first observed and recorded by the ancient Egyptians, who noted that amber, when rubbed, attracted light objects. A bolt of lightning, a spark between a comb and hair in dry weather, and a shock on touching a doorknob are all results of the discharge of a buildup of electric charge.

The two types of electric charge had been discovered by the time of Benjamin Franklin (1706–1790). He named them positive (+) and negative (−) because they appear as opposites, in that they can neutralize each other. The existence and nature of the two kinds of charge, and their effects on each other, can be shown with a simple electroscope (Fig. 3–4). When a hard rubber rod is rubbed vigorously with silk and allowed to touch the lightweight balls, the balls spring apart immediately. The touching allows the rod and the balls to share the same type of charge (positive). If the rod is brought near one of the balls, the ball moves away from the rod. This movement indicates that *like charges repel*.

---

* What we are talking about here is really atomic mass. Atomic weight, an uncorrected misnomer from the past, persists today in, for example, the "Tables of Atomic Weights" published in most chemistry textbooks. If they were really atomic weights, they would change value wherever the force of gravity changes on earth (less at the equator, more at the poles). Instead, we have only one table for the whole world, which means atomic weights are really atomic masses. In this text, we shall use both terms: *atomic weight* because it is a practice of chemistry and a term possibly more familiar to students, and *atomic mass* where it seems necessary to clarify the thought.

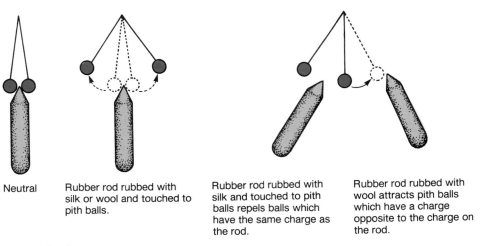

Figure 3-4
Effects of charged matter on other charged matter. Like charges repel; unlike charges attract.

If the same rod is now rubbed vigorously with wool and brought near the charged balls, they move toward the rod. The opposite type of charge is now on the rod. The generalization is: *Unlike charges attract, and like charges repel.*

Natural Radioactivity

See Chapter 4 for a discussion of radioactivity, its discovery, and how it is applied in today's world.

The discovery of natural radioactivity, a spontaneous process in which some natural materials give off very penetrating radiations, indicated that atoms must have some kind of internal structure. Henri Becquerel (1852–1908) discovered this property in natural uranium and radium ores in 1896. His student, Marie Curie (1867–1934), isolated the radioactive element radium and some of its pure compounds. It turns out that radioactivity is characteristic of the elements, not the compounds, and that about 25 elements are naturally radioactive.

Three types of radiation were discovered. These were named alpha (α), beta (β), and gamma (γ) by Ernest Rutherford, a New Zealander working at McGill University in Montreal, Canada. Alpha and beta rays were shown to be deflected by electric and magnetic fields, but gamma rays were not affected.

Gamma rays have no detectable mass and are more like light. Alpha particles have a mass of 4 on the carbon-12 atomic weight scale, positive charge, and low penetrating power (they will not penetrate skin, for example). In the arrangement shown in Figure 3–5, they are attracted toward the negatively charged plate. Beta particles have a mass of 0.0005 on the carbon-12 atomic weight scale, negative charge (they are attracted toward the positive plate), and enough penetrating power to go through kitchen-

Table 3-1
Summary of Properties of Alpha Particles, Beta Particles, and Gamma Rays

	Charge	Relative Mass	Symbols
Alpha particle	Positive (+2)	4	α, $^4_2\alpha$, ^4_2He
Beta particle	Negative (−1)	0.0005	β, $^{\ 0}_{-1}\beta$
Gamma ray	Neutral (0)	0	γ, $^0_0\gamma$

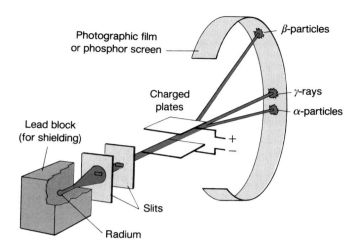

Figure 3-5
Separation of alpha, beta, and gamma rays by an electric field.

strength aluminum foil. Gamma rays are a type of electromagnetic energy like light and X rays, but more penetrating. Gamma rays can penetrate a considerable thickness of aluminum and even thin sheets of lead. They are not deflected at all by charged plates.

Cathode rays, as we shall see subsequently, are similar to beta rays in that both are composed of negatively charged particles, with identical charges and masses.

The discoveries of natural radiation, cathode rays, and electric charge are evidence that atoms can be divided and may even divide spontaneously. The smallest atom is 1836 times more massive than the beta or cathode-ray particle. Therefore, beta (and cathode-ray) particles appear to be subatomic in origin.

The properties of the three fundamental subatomic particles are summarized in Table 3-1.

SELF-TEST 3-A

1. Two Greek philosophers who were influential in advocating the concept of atoms were _____ and _____ .
2. The Greek approach to the "discovery" of atoms can best be described as:
   a. Experimentation
   b. Philosophy (use of logic)
   c. Direct observation of atoms
   d. Consistent explanation of well-known, established laws of nature
   e. Deductive reasoning
3. The law of conservation of matter states that matter is neither lost nor _____ in a _____ reaction.
4. The law of multiple proportions explains the existence of compounds like _____ and _____ .
5. a. Assume that you are a chemist of many years ago. Your field of study is compounds composed of nitrogen and oxygen only. You know about several. One contains 16 g of oxygen for every 14 g of nitrogen, while another contains

32 g of oxygen for every 14 g of nitrogen. Your assistant discovers what he claims is a new compound of nitrogen and oxygen. On analysis, the compound is found to contain 8 g of oxygen for every 14 g of nitrogen. Has your assistant discovered a new compound, or is it one of the others? _____

b. What is the ratio by weight of oxygen in these compounds for a given weight of nitrogen? _____

c. What fundamental law of chemistry is illustrated by a comparison of the compounds? _____

6. According to Dalton's atomic theory, what happens to atoms during a chemical change? Select one:

    a. Atoms are made into new and different kinds of atoms.

    b. Atoms are lost.

    c. Atoms are gained.

    d. Atoms are recombined into different arrangements.

7. According to Dalton's atomic theory, a compound has a definite percentage by weight of each element because:

    a. All atoms of a given element weigh _____.

    b. All molecules of a given compound contain a definite number and kind of _____.

8. Like charges _____; unlike charges _____.

9. The three types of radiation from a radioactive element such as radium are _____, _____, and _____, of which _____ pass through an electric field without being deflected.

The Electron—The First Subatomic Particle Discovered

The first ideas about electrons came from experiments with cathode-ray tubes. A forerunner of neon signs, fluorescent lights, and TV picture tubes, a typical cathode-ray tube is a partially evacuated glass tube with a piece of metal sealed in each end (Fig. 3–6). The pieces of metal are called electrodes. The electrode given a negative charge is called the **cathode,** and the one given a positive charge is called the **anode.**

If a sufficiently high electrical voltage is applied to the electrodes, an electrical discharge can be created between them. This discharge, called a **cathode ray,** causes gases and fluorescent materials to glow and heats metal objects in its path to red heat. Cathode rays travel in straight lines and cast sharp shadows. Unlike light, however, cathode rays are attracted to a positively charged plate in a discharge tube—an almost certain indication that cathode rays are negatively charged. In addition, cathode rays are deflected by magnetic fields.

Careful microscopic study of a screen that emits light when struck by cathode rays shows that the light is emitted in tiny, random flashes. Thus, not only are cathode rays negatively charged, but they also appear to be composed of particles, each one of which produces a flash of light upon collision with the material of the screen. The cathode-ray particles became known as **electrons.**

Cathode rays are streams of the negatively charged particles called electrons.

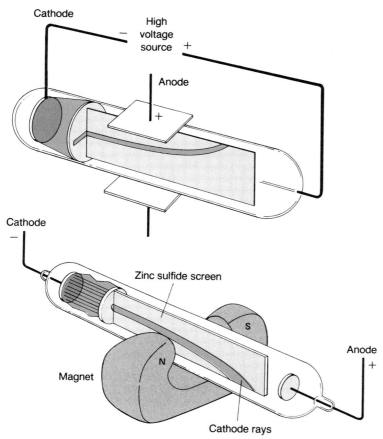

Figure 3-6
Deflection of a cathode ray by an electric field and by a magnetic field. When an external electric field is applied, the cathode ray is deflected toward the positive pole. When a magnetic field is applied, the cathode ray is deflected from its normal straight path into a curved path.

By using a specially designed cathode-ray tube, Sir Joseph John Thomson applied electric and magnetic fields to the rays (Fig. 3-7). Using the basic laws of electricity and magnetism, he determined the **charge-to-mass ratio** of the electrons. Thomson was able to measure neither the absolute charge nor the absolute mass of the electron, but he established the ratio between the two numbers and made it possible to calculate either one if the other could ever be measured. What Thomson did was like showing that a peach weighs 40 times more than its seed. What is the weight of the peach? What is the weight of the seed? Neither is known, but if it can be determined by other means that the peach weighs 120 g, then the weight of the seed, by ratio, must be 3 g.

An important part of Thomson's experimentation was his use of 20 different metals for cathodes and of several gases to conduct the discharge. Every combination of metals and gases yielded the same charge-to-mass ratio for the cathode rays. This led to the belief that electrons are common to all of the metals and gases used in the experiments, and probably to all atoms in general. Thus it appeared that the electron was a fundamental atomic building block.

Thomson discovered the charge-to-mass ratio of the electron.

Electrons are present in all the elements.

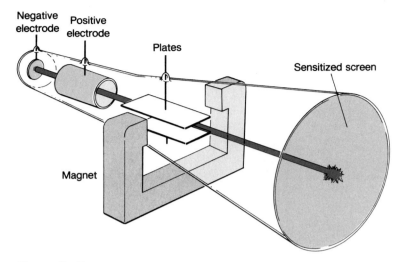

Figure 3-7
J. J. Thomson experiment. An electric field, applied by plates, and a magnetic field, applied by a magnet, cancel each other's effects to allow a cathode ray (electron beam) to travel in a straight line.

The Personal Side

Sir Joseph John Thomson (1856–1940) was a scientist chiefly working in mathematics until he was elected Cavendish Professor at Cambridge University in 1884. He was not skilled at experimental techniques, but his ability to suggest experiments and interpret their results led to the discovery of the electron. In 1897 Thomson wrote in the *Philosophical Magazine,* "We have in the cathode rays a new state, a state in which the subdivision of matter is carried very much further than in the ordinary gaseous state—this matter being the substance from which the chemical elements are built up." Thomson won the Nobel prize in 1906 and was knighted in 1908. Seven of his research assistants later won Nobel prizes for their own research work. Thomson was buried in Westminster Abbey near the grave of Sir Isaac Newton.

Robert Andrews Millikan measured the fundamental charge of matter—the charge on an electron—in 1909. A simplified drawing of his apparatus is shown in Figure 3-8.

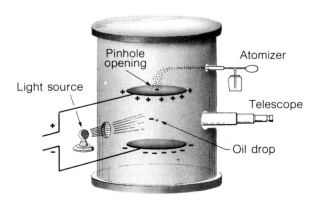

Figure 3-8
Millikan's oil drop experiment for determining the charge on an electron. The pull of gravity on the drop is balanced by the upward electrical force. This drawing does not include an X-ray source for charging the droplets.

The experiment consisted of measuring the electrical charge carried by tiny drops of oil that were suspended in an electrical field. By means of an atomizer, oil droplets were sprayed into the test chamber. As the droplets settled slowly through the air, high-energy X rays were passed through the chamber to charge the droplets negatively (the X rays caused air molecules to give up electrons to the oil). By using a beam of light and a small telescope, Millikan could study the motion of a single droplet. When the electric charge on the plates was increased enough to balance the effect of gravity, a droplet could be suspended motionless. At this point, the gravitational force would equal the electrical force. Measurements made on the droplet while it was in its motionless state were inserted into equations for the forces acting on the droplet, and Millikan was thus able to calculate the total charge carried by the oil droplet.

Millikan measured the charge on an electron.

Millikan found different amounts of negative charge on different drops, but the charge measured each time was always a whole-number multiple of a very small basic unit of charge. The *largest* common divisor of all charges measured by this experiment was 1.60×10^{-19} coulomb (the coulomb is a charge unit). Millikan assumed this to be the fundamental charge, which is the charge on the electron.

Only a whole number of electrons may be present in a sample of matter.

With a good estimate of the charge on an electron and the ratio of charge-to-mass as determined by Thomson, the very small mass of the electron could be calculated. The mass of an electron is 9.11×10^{-28} g. On the carbon-12 relative scale, the electron would have a weight of 0.000549 atomic weight units. The negative charge on an electron of -1.60×10^{-19} coulomb is set as the standard charge of -1.

Protons — The Atom's Positive Charge

The first experimental evidence of a fundamental positive particle came from the study of **canal rays,** which are produced by a special type of cathode-ray tube (Fig. 3–9). In the canal-ray tube, the cathode is perforated, and the tube contains a gas at very low pressure. When high voltage is applied to the tube, cathode rays can be observed between the electrodes as in any cathode-ray tube. On the other side of the perforated cathode, a different kind of a ray is observed. These rays are attracted to a negative plate brought alongside the rays; the rays must therefore be composed of positively charged particles. When Thomson's method for measuring the charge-to-mass ratio of these positive particles was used, different gases were found to give different charge-to-mass ratios. When hydrogen gas was used, the largest charge-to-mass ratio was obtained, indicating that hydrogen provides the positive particles with the smallest mass. This particle was considered to be the fundamental positively charged particle of atomic structure and was called a **proton** (from Greek for "the primary one").

Canal rays are streams of positive ions derived from the gases present in the discharge tube.

The lightest atom is the hydrogen atom.

The production of canal rays is caused by high-energy electrons moving from the negative cathode to the positive anode, hitting the molecules of gases occupying the tube. Electrons are knocked from some atoms by the high-energy electrons, leaving each molecule with a positive charge. The positively charged molecules are then attracted to the negative electrode. Since the electrode is perforated, some of the positive particles go through the holes or channels (hence the name *canal rays*).

The mass of the proton is 1.67261×10^{-24} g, which is 1.00727 relative weight on the carbon-12 scale. The charge of $+1$ on the proton is equal in size but opposite in effect to the charge on the electron.

Neutrons — Neutral Particles Found in Most Atoms

Masses of atoms indicated that neutral particles with about the mass of the proton must be present in the atom in addition to protons and electrons. However, this third type of

52 Atoms: What Are They?

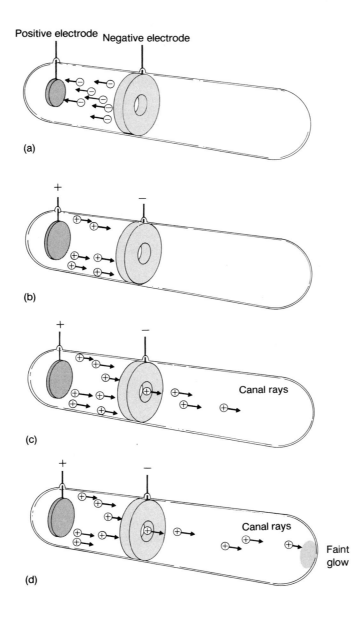

Figure 3–9
(a) Electrons rush from the negative electrode to the positive electrode as a result of high voltage. (b) Electrons collide with gas molecules to produce positive ions, which are accelerated toward the negative electrode. (c) Some of the positive ions escape capture by the electrode and rush through the opening as a result of their kinetic energy. (d) Some of the positive ions in the positive ray collide with gas molecules to produce a characteristic glow and strike the end of the glass tube to produce a luminous spot.

particle in the atom proved difficult to find. Since the particle has no charge, the usual methods of detecting small particles could not be used.

In 1932 James Chadwick devised a clever experiment that produced neutrons by a nuclear reaction and then detected them by having the neutrons knock hydrogen ions, a detectable species, out of paraffin.

A neutron has no electrical charge and has a mass of 1.67492×10^{-24} g, which is a relative weight of 1.00867 on the carbon-12 scale.

Paraffin is the hydrocarbon that seals home-canned preserves.

The Nucleus — An Amazing Atomic Concept

In 1909 two of Ernest Rutherford's students, Hans Geiger and Ernest Marsden, carried out some experiments in which a very thin gold foil was bombarded by alpha particles

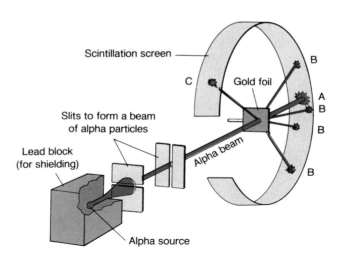

Figure 3–10
Rutherford's gold foil experiment. A cylindrical scintillation screen is shown for simplicity; actually, a movable screen was employed. Most of the alpha particles pass straight through the foil to strike the screen at point A. Some alpha particles are deflected to points B, and some are even "bounced" backward to points such as C.

from a radioactive source (Fig. 3–10). As they had expected, the paths of most of the alpha particles were only slightly changed as they passed through the gold foil. To their amazement, however, a few of the alpha particles were deflected at extreme angles, and some even "bounced" back toward the source. Rutherford expressed his astonishment by stating that he would have been no more surprised if someone had fired a 15-inch artillery shell into tissue paper and then found the shell in flight back toward the cannon.

What allowed most of the alpha particles to pass through the gold foil in a rather straight path? According to Rutherford's interpretation, the atom is mostly empty space and therefore offers little resistance to alpha particles (Fig. 3–11).

Alpha particles are scattered by the nuclei of the gold atoms.

Alpha-particle scattering can be explained if the nucleus occupies a very small volume of the atom.

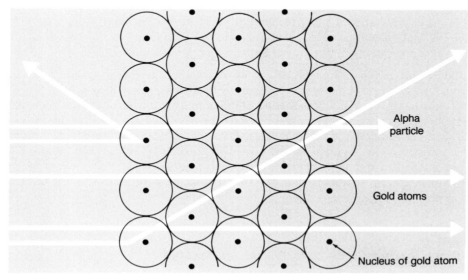

Figure 3–11
Rutherford's interpretation of how alpha particles interact with atoms in a thin gold foil line-up. Actually, the gold foil was about 1000 atoms thick. For illustration purposes, points are used to represent the gold nuclei, and the path widths of the alpha particles are drawn much larger than scale.

What caused a few alpha particles to be deflected? According to Rutherford's interpretation, concentrated at the center of the atom is a **nucleus** containing most of the mass of the atom and all of the positive charge. Rutherford theorized that, when an alpha particle passes near the nucleus, the positive charge of the nucleus repels the positive charge of the alpha particle, and the path of the smaller alpha particle is deflected. The closer an alpha particle comes to a target nucleus, the more it is deflected. Those alpha particles that meet a nucleus head-on bounce back toward the source as a result of the strong positive–positive repulsion, since the alpha particles do not have enough energy to penetrate the nucleus.

Rutherford's calculations, based on the observed deflections, indicated that the nucleus is a very small part of an atom. His calculations showed that an atom occupies about a trillion times more space than does a nucleus; the radius of an atom is about 10,000 times greater than the radius of its nucleus. Thus, if a nucleus were the size of a baseball, then the edges of the atom would be about one third of a mile away, and most of the space in between would be absolutely empty.

Since the nucleus contains most of the mass and all of the positive charge of an atom, the nucleus must be composed of the most massive atomic particles, the protons and neutrons. The electrons are distributed in the near-emptiness outside the nucleus.

Truly, Rutherford's model of the atom was one of the most dramatic interpretations of experimental evidence to come out of this period of significant discoveries.

Lord Rutherford (Ernest Rutherford, 1871–1937) was Professor of Physics at Manchester when he and his students discovered the scattering of alpha particles by matter. Such scattering led to the postulation of the nuclear atom. For this work he received the Nobel prize in 1908. In 1919, Rutherford discovered and characterized nuclear transformations.

Atomic Number — Each Element Has a Number

The **atomic number** of an element indicates the number of protons in the nucleus of the atom, which is the same as the number of electrons outside the nucleus. The two types of particles must be present in equal numbers for the atom to be neutral in charge. Note that the periodic table of the elements, inside the back cover, is an arrangement of the elements consecutively according to atomic number. Beginning with the atomic number 1 for hydrogen, there is a different atomic number for each element.

$$\begin{pmatrix} \text{Number of} \\ \text{electrons} \\ \text{per atom} \end{pmatrix} = \begin{pmatrix} \text{Number of} \\ \text{protons} \\ \text{per atom} \end{pmatrix} = \begin{pmatrix} \text{Atomic number} \\ \text{of the} \\ \text{element} \end{pmatrix}$$

The **atomic mass** of a particular atom is the sum of the masses of the protons, neutrons, and electrons in that atom. Since an electron has such a small mass, the atomic mass is very nearly the sum of the masses of the protons and neutrons in the nucleus. Both protons and neutrons have masses of approximately 1.0 on the atomic weight scale (Table 3–2). Hydrogen, with an atomic weight of 1, must be composed of one proton (and no neutrons) in the nucleus and one electron outside. Helium has an atomic number of 2 and an atomic weight of 4. The atomic number of 2 indicates two protons and two electrons per atom of helium. The atomic weight of 4 means that, in addition to the two protons in the nucleus, there are two neutrons.

Table 3–2
Summary of Properties of Electrons, Protons, and Neutrons

	Relative Charge	Relative Mass	Location
Electron	−1	0.00055	Outside the nucleus
Proton	+1	1.00727	Nucleus
Neutron	0	1.00867	Nucleus

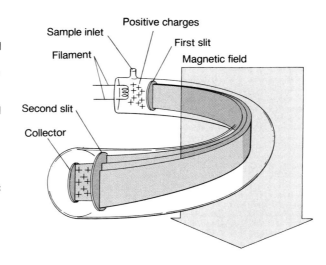

Figure 3-12
Mass spectrometer. The sample to be studied is injected near the filament. Electrodes (not shown) subject the sample to an electron beam that ionizes a part of the sample by knocking electrons from neutral atoms or molecules. The electrodes are arranged to accelerate positive ions toward the first slit. The positive ions that pass the first slit are immediately put into a magnetic field perpendicular to their path and follow a curved path determined by the charge-to-mass ratio of the ion. A collector plate behind the second slit detects charged particles passing through the second slit. The relative magnitudes of the electrical signals are a measure of the numbers of different kinds of positive ions.

$$\begin{pmatrix} \text{Approximate} \\ \text{number of} \\ \text{neutrons} \\ \text{per atom} \end{pmatrix} = \begin{pmatrix} \text{Atomic weight} \\ \text{of the} \\ \text{element} \end{pmatrix} - \begin{pmatrix} \text{Atomic number} \\ \text{of the} \\ \text{element} \end{pmatrix}$$

A notation frequently used to show the atomic mass (also called **mass number**) and atomic number of an atom uses subscripts and superscripts to the left of the symbol:

$$^{19}_{9}\text{F}$$

Atomic mass ↗ ← Symbol of the element
Atomic number ↗

Symbolism refers to individual isotopes.

For an atom of fluorine, $^{19}_{9}\text{F}$, the number of protons is 9, the number of electrons is also 9, and the number of neutrons is $19 - 9 = 10$.

Isotopes — Dalton Never Guessed!

Many of the elements, when analyzed by a special type of canal-ray tube called a *mass spectrometer* (Fig. 3-12), are found to be composed of atoms of different masses (Fig. 3-13). Atoms of the same element having different atomic masses are called **isotopes** of that element.

The element neon is a good example to consider. A natural sample of neon gas is found to be a mixture of three isotopes of neon:

$$^{20}_{10}\text{Ne} \quad ^{21}_{10}\text{Ne} \quad ^{22}_{10}\text{Ne}$$

The fundamental difference between isotopes is the different number of neutrons per atom. All atoms of neon have 10 electrons and 10 protons; about 90% of the atoms have 10 neutrons, some have 11 neutrons, and others have 12 neutrons. Because they

Isotopes are atoms of the same element with different numbers of neutrons.

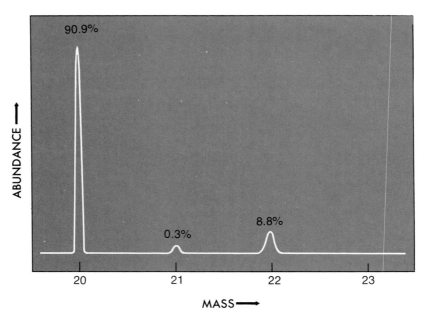

Figure 3–13
Mass spectrum of neon (+1 ions only). The principal peak corresponds to the most abundant isotope, neon-20. Percent relative abundance is shown. (From W. L. Masterton and E. J. Slowinski: *Chemical Principles.* Philadelphia, Saunders College Publishing, 1973.)

have different numbers of neutrons, they must have different masses. Note that all the isotopes have the same atomic number. They are all neon.

There are only 109 known elements, yet more than 1000 isotopes have been identified, many of them produced artificially. Some elements have many isotopes; tin, for example, has ten natural isotopes. Hydrogen has three isotopes, and they are the only three that are generally referred to by different names: ^1_1H is called protium, ^2_1H is called deuterium, and ^3_1H is called tritium. Tritium is radioactive. The natural assortment of isotopes, each having its own distinctive atomic mass, results in fractional atomic weights for the elements.

> The weighted average of the atomic weights of the isotopes in a natural mixture is the noninteger atomic weight of the element.

SELF-TEST 3–B

1. Isotopes of an element are atoms that have nuclei with the same number of _____ but different numbers of _____.
2. The nucleus of an atom occupies a relatively (large or small) fraction of the volume of the atom.
3. The positive charges in an atom are concentrated in its _____.
4. The negatively charged particles in an atom are _____, the positively charged particles are _____, and the neutral particles are _____.
5. In a neutral atom there are equal numbers of _____ and _____.

6. The number of protons per atom is called the _____ number of the element.
7. The mass of the proton is _____ times the mass of the electron.
8. An atom of arsenic, $^{75}_{33}$As, has _____ electrons, _____ protons, and _____ neutrons.
9. Positive (canal) rays obtained with different gases are (different/identical), while the cathode rays obtained using different cathodes are (different/identical).
10. Cathode rays are composed of a universal constituent of matter named _____.
11. The two fundamental particles revealed by studies using gas discharge (cathode-ray) tubes are _____ and _____.
12. All atoms of a given element are exactly alike.   True ( ) or False ( )

Where Are The Electrons?

Two major theories have been presented concerning the position, movement, and energy of electrons in an atom. The Bohr theory of the hydrogen atom was put forth in 1913 by Niels Bohr. This theory, extended and modified by Erwin Schrödinger and others in 1926, is referred to as the wave mechanical theory.

The Bohr Model of the Atom

In Bohr's concept, electrons revolve around a nucleus in definite orbits, much as planets revolve around the sun. He equated classical mathematical expressions for the force tending to keep the electron traveling in a straight line and the force tending to pull the electron inward (the positive-to-negative attraction between a proton and an electron).

In a revolutionary sort of way, Bohr suggested that electrons stay in rather stable orbits and can have only certain energies within a given atom. According to Bohr, an electron can travel in one orbit for a long period or in another orbit some distance away for a long period, but it cannot stay for any measurable time between the two orbits. A rough analogy is provided by books in a bookcase. Books may rest on one shelf or on another shelf for very long periods but cannot rest between shelves. When a book is moved from one shelf to another shelf, the potential energy of the book changes by a definite amount. According to Bohr's theory, when an electron moves from one orbit to another, its energy changes by a definite amount, called a **quantum** of energy.

Energy is required to separate objects attracted to each other. For example, energy is required to lift a rock from the Earth, to separate two magnets, or to pull a positive charge away from a negative charge. Bohr suggested that a very definite and characteristic amount of energy is required to move an electron from one energy level to another. The energy added to move an electron farther from its nucleus is stored in the system as potential energy (energy of position). Thus, the electron has more energy when it is in an orbit farther from the nucleus and less energy when it is in an orbit close to the nucleus. When the electron passes from an outer orbit to an inner orbit, energy is emitted from the atom, generally in the form of light energy (a quantum of light).

Bohr used the idea of electrons moving up and down a "bookcase" of energy levels corresponding to orbits to explain the observable bright-line emission spectrum of hydrogen. A **spectrum** is the display produced when light is separated or dispersed into its component colors. The spectrum of white light is the rainbow display of separated

Energy of matter in motion is kinetic *energy.*

Energy stored in matter is potential *energy.*

Bohr assumed that an atom can exist only in certain energy states.

Dispersed light produces a spectrum.

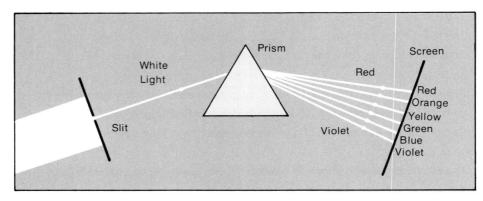

Figure 3–14
Spectrum from white light produced by refraction in a glass prism. The different colors blend into one another smoothly.

In 1900 Max Planck proposed that light and other forms of electromagnetic radiation (X rays, microwaves, radio waves) are discrete packets of energy, which are called *quanta* or *photons*. Bohr used Planck's quantum theory to explain the emission spectrum of hydrogen.

Bohr's theoretical calculations for hydrogen spectral lines agreed amazingly well with experimental data.

colors shown in Figure 3–14. An **emission spectrum** is observed when the light emitted by atoms energized by a flame or an electric arc is allowed to pass through a narrow vertical slit and then through a prism of glass or quartz. If sunlight or light from a white-hot solid is dispersed, all of the colors of the rainbow are seen; this is a **continuous emission spectrum.** However, if the light from an energized gaseous element is dispersed, only colored lines are produced; this is a **bright-line emission spectrum** (Fig. 3–15).

According to Bohr, the light forming the lines in the bright-line emission spectrum of hydrogen comes from electrons moving toward the nucleus after having first been energized into orbits farther from the nucleus (Fig. 3–16). A movement between two particular orbits involves a definite quantum of energy. Each time an electron moves from one orbit to another orbit closer to the nucleus, energy loss occurs and a quantum of light having a characteristic energy (and color) is emitted. Transitions toward the nucleus between two outer orbits emit quanta having smaller characteristic energies. Transitions from an orbit to orbits near the nucleus emit quanta having larger characteristic energies.

Not only could Bohr explain the cause of the lines in the bright-line emission spectrum of hydrogen, he also calculated the expected wavelengths of the lines. He expressed the results of his calculations in the alternate view of the nature of light, its

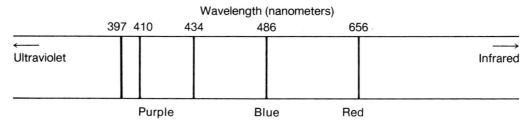

Figure 3–15
The visible portion of the hydrogen bright-line emission spectrum. Frequency and energy increase to the left; wavelength increases to the right. There are 7 more lines to the left in the ultraviolet range and 13 more lines to the right in the infrared range.

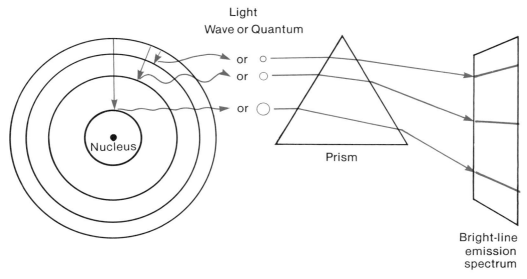

Figure 3–16
The formation of light according to the atomic theory. Electrons previously energized to higher energy levels make transitions back toward the nucleus (three transitions are shown). The decrease in potential energy during a transition is transformed into light. Transitions between different energy levels produce different-sized quanta of light. When the quanta are dispersed by being passed through a prism, the bright lines can be seen on film or on a spectroscope. Billions of transitions per second make each line bright enough (sufficient same-sized quanta emitted) for each line to be detected.

wave nature. The wave properties of light—wavelength, frequency, and speed—are considered in Figure 3–17 as they apply to any phenomenon possessing wave properties (water waves, sound waves from violin strings, radio waves, and light).

Niels Bohr had tied the unseen (the interior of the atom) with the seen (the observable lines in the hydrogen spectrum)—a fantastic achievement. The Bohr theory was accepted almost immediately after its presentation, and Bohr was awarded the Nobel prize in physics in 1922 for his contribution to the understanding of the hydrogen atom.

Atom Building Using the Bohr Model

Recall that the atomic number is the number of protons in the nucleus of an atom. Since atoms are neutral, the number of electrons in orbits around the nucleus must equal the number of protons in the nucleus. Orbits are numbered with integers, beginning with 1 for the orbit closest to the nucleus. The maximum number of electrons per orbit is $2n^2$, where n is the number of the orbit:

Orbit	Maximum Number of Electrons
1	2
2	8
3	18
4	32
5	50

A general, overriding rule to the preceding numbers is that the outside orbit can have no

Niels Bohr

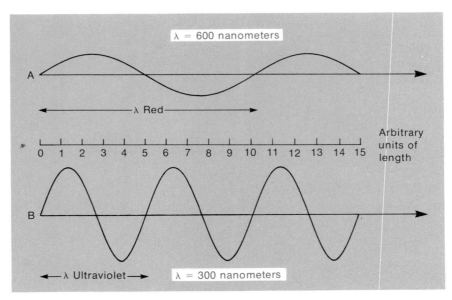

Figure 3–17
The wave theory of light considers light to be waves of wavelength λ (lambda) vibrating at right angles to their path of motion. Red light, A, completes one vibration or wave in the same distance and time that involve two complete waves of ultraviolet light, B. Ultraviolet light has a shorter wavelength than violet light. A nanometer is 10^{-9} meter.

more than eight electrons. When electrons are placed in orbits as close to the nucleus as possible, the electrons are said to be in their **ground state.**

You might like to follow along in Figure 3–18 as the building-up process is described. Hydrogen, with atomic number 1, has one electron. In its ground state, this electron is in the first orbit. The two electrons of helium are in its first orbit, since the first orbit can have a maximum of two electrons.

The Personal Side

Niels Bohr (1885–1962) received his doctor's degree the same year that Ernest Rutherford announced his discovery of the atomic nucleus, 1911. After studying with Thomson and Rutherford in England, Bohr formulated his model of the atom. Bohr returned to the University of Copenhagen and as Professor of Theoretical Physics, directed a program that produced a number of brilliant theoretical physicists. He received the Nobel prize for physics in 1922.

During World War II, having done all he could to get Jewish physicists to safety, Bohr was still in Denmark when Hitler's army suddenly occupied the country in 1940. In 1943, to avoid imprisonment, he fled to Sweden. There he helped to arrange the escape of nearly every Danish Jew from Hitler's gas chambers. He was later flown to England in a tiny plane, in which he passed into a coma and nearly died from lack of oxygen.

Bohr went on to the United States, where until 1945 he worked with other physicists on the atomic bomb development at Los Alamos. His insistence upon sharing the secret of the atomic bomb with other allies, in order to have international control over nuclear energy, so angered Winston Churchill that he had to be restrained from ordering Bohr's arrest. Bohr worked hard and long on behalf of the development and use of atomic energy for peaceful purposes. For his efforts, he was awarded the first Atoms for Peace prize in 1957. He died in Copenhagen on November 18, 1962.

Placement of Electrons in Ground State

Element	Atomic Number	Bohr Model			
Hydrogen (H)	1	1p	1)e		
Helium (He)	2	2p 2n	2)e		
Lithium (Li)	3	3p 4n	2)e	1)e	
Beryllium (Be)	4	4p 5n	2)e	2)e	
Boron (B)	5	5p 6n	2)e	3)e	
Carbon (C)	6	6p 6n	2)e	4)e	
Nitrogen (N)	7	7p 7n	2)e	5)e	
Oxygen (O)	8	8p 8n	2)e	6)e	
Fluorine (F)	9	9p 10n	2)e	7)e	
Neon (Ne)	10	10p 10n	2)e	8)e	
Sodium (Na)	11	11p 12n	2)e	8)e	1)e
Magnesium (Mg)	12	12p 12n	2)e	8)e	2)e
Aluminum (Al)	13	13p 14n	2)e	8)e	3)e
Silicon (Si)	14	14p 14n	2)e	8)e	4)e
Phosphorus (P)	15	15p 16n	2)e	8)e	5)e
Sulfur (S)	16	16p 16n	2)e	8)e	6)e
Chlorine (Cl)	17	17p 18n	2)e	8)e	7)e
Argon (Ar)	18	18p 22n	2)e	8)e	8)e
Potassium (K)	19	19p 20n	2)e	8)e	8)e 1)e
Calcium (Ca)	20	20p 20n	2)e	8)e	8)e 2)e

Figure 3–18
Electron arrangements of the first 20 elements. The nuclear contents of a typical isotope are shown.

For all atoms of other elements, two electrons are in the first orbit, and the other electrons of the atoms are assorted into higher numbered energy levels. In atomic-number order, lithium through neon, two electrons are placed in the first orbit (which fill it), and one, two, three, and so on to eight electrons (for Ne) are placed into the second orbit. Eight electrons fill the second orbit.

Sodium (Na), with 11 electrons, has the first two orbits filled with 2 and 8 electrons, respectively, and has 1 electron in the third orbit. Each succeeding element in atomic-number order, magnesium through argon, adds one more electron to the third orbit of its atoms.

At argon (Ar), the maximum of eight electrons in the outside orbit comes into play. When 19 electrons are present, as in an atom of potassium (K), the first orbit has 2 electrons, and the second orbit has 8 electrons. The third orbit could have the other 9 electrons if it were not the outside orbit; instead, the electronic arrangement is 2-8-8-1. Calcium (Ca), with 20 electrons per atom, has an electronic arrangement of 2-8-8-2.

Beginning with scandium (Sc), atomic number 21, and continuing through zinc (Zn), atomic number 30, 10 electrons are added to the third orbit to complete its maximum of 18. Zinc has the electronic arrangement 2-8-18-2.

You might pause in your reading here and predict the ground-state electronic arrangement of gallium (Ga), atomic number 31, and rubidium (Rb), atomic number 37, using this system.

What, then, is an atom really like? We have Dalton's concept of an atom as a hard sphere similar to a small billiard ball. We have Bohr's concept of the atom as a small three-dimensional solar system with a nucleus and electrons in paths called *orbits*. More modern theories, not discussed here, have more detail; according to them, orbits have suborbits called *orbitals,* and we are given approximate spaces where electrons exert their greatest influence in an atom.

Why present all these theories? They help us understand the phenomena we observe. We simply use whatever detail is necessary to explain what we see. For example, the simpler Dalton concept adequately explains many properties of the gaseous, liquid, and solid states. Most bonding between atoms of the light elements can be explained by application of the orbits of Bohr. The shapes of molecules and the arrangement of atoms with respect to each other can be explained by electron pairing and repulsions. In the explanations given in this text, we shall follow the principle that simplest is best.

SELF-TEST 3-C

1. Light under some conditions has properties of _____ and under other conditions exhibits the properties of _____.
2. When light is dispersed into the different colors composing the light, a _____ is produced.
3. According to Bohr's theory, light of characteristic wavelength is produced as an electron passes from an orbit ( ) closer to or ( ) farther from the nucleus to an orbit ( ) closer to or ( ) farther from the nucleus.
4. If one compares the atom to the solar system, as in the Bohr concept of the atom, the nucleus is like the _____ and the electrons are like the _____.
5. Electrons in orbits as close to the nucleus as possible are said to be in their _____.
6. The maximum number of electrons in the $n = 3$ orbit is _____.

MATCHING SET

1. Atomic mass
2. Unlike electrical charges
3. $2n^2$
4. Nucleus
5. Electron
6. ^{22}Ne and ^{20}Ne
7. Atomic number
8. Quantum
9. Gamma ray
10. Particles in an H atom
11. Neutron

a. Attract
b. Equal to number of protons in nucleus
c. Have the same mass number
d. Cathode-ray particle
e. Neutrons plus protons
f. A small, definite amount of energy
g. Proton and electron
h. Uncharged elementary particle
i. Contains most of the mass in an atom
j. Maximum number of electrons in an orbit
k. A form of radiant energy
l. Isotopes
m. Charge on the nucleus

QUESTIONS

1. What kinds of evidence did Dalton have for atoms that the early Greeks (Democritus, Leucippus) did not have?
2. How does Dalton's atomic theory explain:
    a. The law of conservation of matter?
    b. The law of constant composition?
    c. The law of multiple proportions?
3. Although there may be no very reliable way to check the conservation of matter in a large explosion of dynamite, what leads us to believe that the law of conservation of matter is obeyed?
4. Describe the potential energy relationship as an electron moves from a high-energy orbit to a low-energy orbit.
5. What experimental evidence indicates that
    a. Cathode rays have considerable energy?
    b. Cathode rays have mass?
    c. Cathode rays have charge?
    d. Cathode rays are a fundamental part of all matter?
    e. Three isotopes of neon exist?
    f. Atoms are not indestructible?
6. What was the importance of the alpha-particle–scattering experiment using gold foil?
7. Why was Thomson's charge-to-mass ratio determination for electrons significant, although he did not determine either the charge or the mass of the electron?
8. What part do electrons play in producing positive rays?
9. How do the following discoveries indicate the inadequacy of the Daltonian model of atoms?
    a. Cathode rays
    b. Positive rays
    c. Nucleus
    d. Natural radioactivity
    e. Isotopes
10. Characterize the three types of emissions from naturally radioactive substances with respect to charge, relative mass, and relative penetrating power.
11. Explain what the following terms mean:
    a. Isotopes of an element
    b. Atomic number
    c. Alpha emitter
12. If electrons are a part of all matter, why are we not electrically shocked continually by the abundance of electrons about and in us?
13. There are more than 1000 kinds of atoms, each with a different weight, yet there are only 109 elements. How does one explain this in terms of subatomic particles?
14. A common isotope of lithium (Li) has a mass of 7. The atomic number of lithium is 3. What are the constituent particles in its nucleus?
15. An element has 12 protons in its nucleus. How many electrons do the atoms of this element possess?

16. An isotope of atomic mass 60 has 33 neutrons in its nucleus. What is its atomic number, and what are the name and chemical symbol of the element?
17. The element iodine (I) occurs naturally as a single isotope of atomic mass 127; its atomic number is 53. How many protons and how many neutrons does it have in its nucleus?
18. An element with an atomic number of 8 is found to have three isotopes with atomic masses of 16, 17, and 18. How many protons and neutrons are present in each nucleus? What is the element?
19. Suppose Millikan had determined the following charges on his oil drops:
    1.33×10^{-19} coulomb
    2.66×10^{-19} coulomb
    3.33×10^{-19} coulomb
    4.66×10^{-19} coulomb
    7.92×10^{-19} coulomb
    What do you think his value for the electron's charge would have been?
20. What is a quantum? What is a photon?
21. Discuss, in quantum terms, how a ladder works.
22. Distinguish between atomic number and atomic weight.
23. How does the Bohr theory explain the many lines in the spectrum of hydrogen although the hydrogen atom contains only one electron?
24. True or False
    a. If compounds conform to the law of multiple proportions, they must necessarily conform to the law of constant composition.
    b. If compounds conform to the law of constant composition, they must necessarily conform to the law of multiple proportions.
25. According to John Dalton's concept, were atoms more like billiard balls, cotton-puff balls, tennis balls, or small solar systems?
26. How are fluorescent lights and TV picture tubes related to the study of the atom?
27. If you found the number of wheels received by an assembly plant to be twice the number of motors, what type of vehicle would you assume to be assembled there? Of what chemical law does this remind you?
28. One hundred grams of pure water on analysis yield 88.8 grams of oxygen.
    a. How many grams of hydrogen would be produced from this sample of water?
    b. What is the percentage of oxygen in the water sample?
    c. How many grams of oxygen will be found in 1 gram of water?
29. Which of the following is an empirical (observable) fact?
    a. Water is 88.8% oxygen by weight.
    b. Water molecules contain one atom of oxygen each.
30. Krypton is the name of Superman's home planet and also that of an element. Look up the element krypton, and list its symbol, atomic number, atomic weight, and electronic arrangement.
31. Explain in your own words why alpha particles are deflected in one direction in an electrical field while beta particles are deflected in the opposite direction.

The Nuclear Atom: Uses and Dangers

CHAPTER 4

The nuclear age is often considered to have started either in the late 1800's and early 1900s with the discovery of radioactive elements by Becquerel and Curie or in 1945 with the first explosion of the atomic bomb. Actually, radioactivity has been a part of the universe since the beginning. The stars are gigantic thermonuclear reactors, and all the planets, moons, and other solid objects in the universe, like the earth and its moon, are thought to contain various radioactive elements. It is true that early atomic theory said nothing about radioactivity, but that was because radiation cannot be directly detected by the five senses. It took the maturing of the sciences — with such diverse discoveries as how to produce a vacuum, photographic film, fluorescent materials, electricity, and magnetic fields — to lead to the knowledge that some atoms spontaneously disintegrate and, in the process, produce radiation.

In February 1896, Henri Becquerel, was experimenting in France with materials exposed to the recently discovered X rays. He exposed uranium potassium sulfate to X rays and then positioned the compound on a photographic plate, which was exposed as if by light. Becquerel found by accident that the X-ray exposure was not needed for the uranium compound to expose the photographic plate. In further experiments all uranium compounds, and even the metal itself, were found to expose photographic plates spontaneously. Becquerel showed that uranium and its salts emitted radiation, which was capable of causing ionization in the air. He showed this by charging gold leaves in a vacuum jar (Fig. 4–1) and then discharging them by bringing the uranium samples near the jar. Both positive and negative charges were neutralized this way.

Wilhelm Roentgen discovered X rays in 1895 and was awarded the Nobel prize for this work in 1901.

The Personal Side

Soon after Becquerel's discovery of uranium's radioactivity, Marie Curie, also working in France, studied the radioactivity of thorium (an alpha emitter) and began to search systematically for new radioactive elements. She showed that the radioactivity of uranium was an atomic property — that is, its radioactivity was proportional to the amount of the element present and was not related to the particular compound present. Her experiments indicated that other radioactive elements were present with certain uranium samples. By painstaking technique, she and her husband, Pierre Curie, separated the element radium from uranium ore. Radium was found to have an activity over 1 million times greater than that of uranium. Atomic spectra were used to help characterize the new element. In 1903 Marie and Pierre Curie shared the Nobel prize in physics with Henri Becquerel for their discoveries.

By 1899 Ernest Rutherford had shown that at least two types of radiation were emitted from uranium. One type was alpha (α) radiation. Rutherford found that alpha rays could be stopped by thin pieces of paper and had a range of only about 2.5 cm to

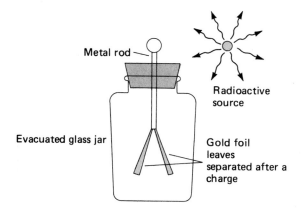

Figure 4–1
Electroscope used to detect and measure electrical charge.

8.5 cm in air before being absorbed. The other form of radiation was beta (β) radiation. Beta rays were capable of penetrating far greater distances in air. Rutherford later found alpha-ray particles to be identical to helium nuclei (He^{2+}) and beta particles to be the same as electrons (e^-).

Gamma rays, like X rays, are part of the electromagnetic spectrum.

In 1900 Paul Villard characterized a third form of natural radiation, gamma (γ) rays. These, he found, were not streams of particles, but rather had the general characteristics of light or X rays. Gamma rays are extremely penetrating; they are capable of passing through over 9 inches of steel and about 1 inch of lead. Figure 4–2 compares the penetrating ability of the three forms of natural radiation.

Isotopes were discussed in Chapter 3.

After the discovery of natural radioactivity in uranium, thorium, and radium, many other elements were found to have radioactive isotopes. All the elements above bismuth (atomic number 83) and a few below bismuth have naturally occurring radioactive isotopes. Let's now turn our attention to some of the characteristics of nuclear reactions.

Nuclear Reactions

The isotope of uranium with atomic mass 238 is an alpha emitter. When the $^{238}_{92}U$ nucleus gives off an alpha particle, made up of two protons and two neutrons, four units of atomic mass and two units of atomic charge are lost. The resulting nucleus has a mass of 234 and a nuclear charge of 90. Atoms containing 90 protons in the nucleus are atoms of thorium, not uranium. This spontaneous nuclear reaction then has changed an atom of

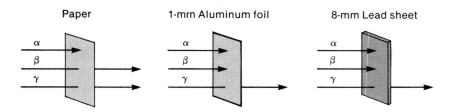

Figure 4–2
Penetrating ability of alpha (α), beta (β), and gamma (γ) radiation. Gamma rays even penetrate an 8-mm lead sheet. Skin will stop alpha rays but not beta rays.

one element into an atom of another element and is an example of the **transmutation** of elements.

The decomposition of the $^{238}_{92}U$ nucleus is stated briefly by the following nuclear equation:

$$^{238}_{92}U \longrightarrow {}^{4}_{2}He + {}^{234}_{90}Th$$

If the characterized alpha emission were not proof enough that this reaction occurs, additional evidence is supplied by the fact that $^{234}_{90}Th$ is always found with $^{238}_{92}U$ in natural ore deposits and almost always in the concentration predicted by the rates of the uranium and thorium decay.

atomic mass→ $^{238}_{92}U$
atomic number→

Thorium-234 is also radioactive. However, this nucleus is a beta emitter, which leads to an interesting question: How can a nucleus containing protons and neutrons emit a beta particle, which is an electron? It has been established that an electron and a proton can combine outside the nucleus to form a neutron. Therefore, the reverse process is proposed to occur in the nucleus. A neutron decomposes, giving up an electron and changing itself into a proton:

$$^{1}_{0}n \longrightarrow {}^{1}_{1}H + {}^{0}_{-1}e + \text{energy}$$

Since the mass of the electron is essentially zero compared with that of the proton and neutron, the nucleus would maintain essentially the same mass but would now carry one more positive charge (a proton instead of one of the neutrons). This nucleus is no longer thorium, since thorium has only 90 protons in the nucleus; it is now a nucleus of element 91, protactinium (Pa). The reaction is:

$$^{234}_{90}Th \longrightarrow {}^{234}_{91}Pa + {}^{0}_{-1}e + \text{energy}$$

Gamma radiation may or may not be given off simultaneously with alpha or beta rays, depending on the particular nuclear reaction involved. Since gamma rays involve no charge and essentially no mass, the emission of a gamma photon cannot alone account for a transmutation event.

A photon is a quantum of energy.

The decay of uranium-238 is extremely slow compared with the decay of thorium-234. The rate of decay can be represented by a characteristic **half-life.** A half-life represents the period required for half of the radioactive material originally present to undergo transmutation.

No matter how much of a radioactive substance is present at the beginning, only half of it remains at the end of one half-life.

The half-life is independent of the amount and chemical form of radioactive material present and is determined only by the type of radioactive nucleus present in the sample. For example, in the preceding reaction, the half-life of $^{234}_{90}Th$ is 24 days. This means that half of the thorium will remain unreacted after 24 days. In another 24 days, half of the half ($\frac{1}{4}$) will remain. This process continues indefinitely, with half of the $^{234}_{90}Th$ remainder decaying each 24 days.

Figure 4–3 illustrates graphically how the concept of half-life works for a radioactive isotope. Some half-lives are extremely long, and others are extremely short. The half-life for the $^{238}_{92}U$ alpha decay is 4.5 billion years. As one would expect, relatively large amounts of $^{238}_{92}U$ can be found in nature, whereas only trace amounts of $^{234}_{90}Th$ are present.

The radioactive decay of $^{234}_{90}Th$ into $^{234}_{91}Pa$ is the second step in a series of nuclear decays that starts with $^{238}_{92}U$. After 14 decays the series ends with a stable, nonradioactive isotope of lead, $^{206}_{82}Pb$. This decay series is called the **uranium series.** Table 4–1 gives the half-lives of the isotopes in the uranium series. Two other natural decay series exist that are similar to the uranium series, but they start out with a different isotope and proceed through a different set of radioactive decay products. The **thorium series** begins with $^{232}_{90}Th$ (a different isotope from the two thorium isotopes that occur in the

Figure 4–3
Half-life. The rate of decay of a radioactive substance is such that in a given period—the half-life for the species—half of the original number of atoms will be gone regardless of the number present at the start. In this graph the number of atoms remaining is plotted against time. At the end of one half-life period the original number, N, is reduced to $\frac{1}{2}$N. After two of these periods, the number is reduced to half of $\frac{1}{2}$N, or $\frac{1}{4}$N.

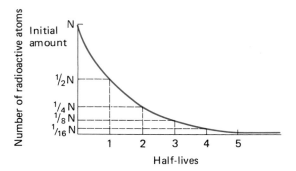

Alpha particles are $^4_2\text{He}^{2+}$ ions.

uranium series) and ends with stable $^{206}_{82}\text{Pb}$. A third series, called the **actinium series,** begins with $^{235}_{92}\text{U}$ and ends with $^{207}_{82}\text{Pb}$. Most of the naturally occurring radioactive isotopes are members of one of these three decay series.

Atomic Dating

The concept of radioactive half-life was almost immediately recognized as a useful tool for measuring the age of radioactive materials. As early as 1905, Rutherford suggested during a lecture at Yale University that one indication of age would be the measurement of helium formed from alpha particles produced by radioactive decay.

$$^4_2\text{He}^{2+} \text{ (alpha)} + 2e^- \longrightarrow {}^4_2\text{He} \text{ (monatomic gas)}$$

At about the same time, Bertram Boltwood suggested that the $^{238}_{92}\text{U}/^{206}_{82}\text{Pb}$ ratio could be measured to date rock samples. To understand how this method works, consider the

Table 4–1
Half-Lives of the Naturally Occurring Radioactive Elements in the Uranium-238 ($^{238}_{92}\text{U}$) Series

Isotope	Type of Disintegration	Half-Life
^{238}U	α	4.5 billion years
^{234}Th	β	24.1 days
^{234}Pa	β	1.18 minutes
^{234}U	α	250,000 years
^{230}Th	α	80,000 years
^{226}Ra	α	1620 years
^{222}Rn	α	3.82 days
^{218}Po	α,β	3.05 minutes
^{214}Pb	β	26.8 minutes
^{214}Bi	α,β	19.7 minutes
^{210}Tl	β	1.32 minutes
^{210}Pb	β	22 years
^{210}Bi	β	5 days
^{210}Po	α	138 days
^{206}Pb	Stable	

fact that 1.00 g of $^{238}_{92}$U in its half-life of 4.5 billion years would leave 0.50 g of $^{238}_{92}$U and in the process produce 0.43 g of $^{206}_{82}$Pb. The time required for all the other decay steps after the breakdown of $^{238}_{92}$U is relatively short. Hence, the rate of lead formation is controlled by this *rate-determining* step. The amount of $^{206}_{82}$Pb can be calculated on the basis of the fact that one $^{238}_{92}$U atom is converted into one $^{206}_{82}$Pb atom and by use of the conversion

$$(207.21 \text{ g Pb}/238.07 \text{ g U}) \times 0.50 \text{ g U} = 0.43 \text{ g Pb}$$

Now, if the ratio of 0.50 g of $^{238}_{92}$U to 0.43 g of $^{206}_{82}$Pb is found in a uranium ore, it would follow that the rock is 4.5 billion years old, the half-life of $^{238}_{92}$U.

> The rate-determining step is the slowest step in a multistep process.
>
> Atomic weights:
> Pb — 207.21
> U — 238.07

Carbon-14 Dating

Cosmic rays (interstellar radiation) are composed of many forms of very high-energy particles such as H$^+$ and He^{2+}. Many of these particles enter the Earth's atmosphere every second. Cosmic rays undergo nuclear reactions with stable nuclei in the upper atmosphere to produce slow-moving neutrons. These neutrons can react with $^{14}_{7}$N nuclei present in nitrogen molecules in the upper atmosphere to produce a radioactive isotope of carbon, $^{14}_{6}$C, which has a half-life of 5730 years.

$$^{1}_{0}n + {}^{14}_{7}N \longrightarrow {}^{14}_{6}C + {}^{1}_{1}H$$

The $^{14}_{6}$C decays by a beta emission.

$$^{14}_{6}C \longrightarrow {}^{0}_{-1}e + {}^{14}_{7}N$$

Radioactive $^{14}_{6}$C in the compound carbon dioxide mixes with the ordinary carbon dioxide in the atmosphere and in turn is incorporated into the structure of all living matter through natural food chains. Upon the death of the organism, the intake of food ceases and the natural level of radioactive carbon present within the structure begins to decrease at the rate of 50% every 5730 years. The realization of this fact led Professor Willard F. Libby, of the University of Chicago, to postulate that radioactive $^{14}_{6}$C could be used to date ancient artifacts derived from living matter such as parchment, cloth, and wood carvings. The $^{14}_{6}$C remaining in the artifact would have to be compared with the normal isotopic ratio of $^{12}_{6}$C, $^{13}_{6}$C, and $^{14}_{6}$C.

> Libby received a Nobel prize in 1960 for $^{14}_{6}$C dating.

An important assumption is that the flow of $^{14}_{6}$C into the biosphere is constant over time. According to several studies, radioactive carbon is indeed slowly mixed with its nonradioactive isotopes. The assumption appears true that the rate of production of $^{14}_{6}$C has been essentially constant over the past several thousand years. Recently, growth rings on sequoia and bristlecone pine trees have been measured accurately and compared with $^{14}_{6}$C dates. According to these experiments, $^{14}_{6}$C production has fluctuated, particularly during the first millenium B.C. In other words, perfect agreement does not exist between the tree-ring age and the $^{14}_{6}$C found in each ring. Nevertheless, radioactive $^{14}_{6}$C dates do compare reasonably well with dates obtained by other methods (Table 4–2).

> Assumptions are made in radiodating.

SELF-TEST 4–A

1. Two names associated with the discovery of natural radioactivity are
   _____ and _____ .

Table 4-2
Comparison of Ages* of Various Artifacts over a Span of 3500 Years by Radiocarbon Dating and Other Methods†

Material	Radiocarbon Age	Age By Another Method‡
Mammalian remains from middle of an Inca temple	450 ± 150 years	444 ± 25 years
Sequoia tree ring	930 ± 100	880 ± 15
Wood from Roman ship	2030 ± 200	1990 ± 3
Charcoal from Etruscan tomb	2730 ± 240	2600 ± 100
Wood from Egyptian tomb of Zoser	3979 ± 350	4650 ± 75

* These ages are given as the age in 1950.
† This table is taken from the *McGraw-Hill Encyclopedia of Science and Technology*, Vol. 11. New York, McGraw-Hill, 1971, p. 291.
‡ Other methods include tree-ring dating and chronological methods.

2. The three forms of natural radioactivity are _____, _____, and _____.
3. Of these three forms, _____ is the most penetrating, while _____ is the least penetrating.
4. When an element is transformed into another element by a nuclear reaction, this process is called a nuclear _____.
5. If we begin with a sample of some radioactive element and 12 days later only half of that element is still present (the rest has disintegrated), the 12-day period represents the _____ of that element.
6. What radioisotope could be used to date artifacts about 2000 years old? _____
7. What radioisotope could be used to measure the age of the earth? _____

Artificial Nuclear Changes (Transmutations)

In 1919 Rutherford produced the first artificial nuclear change by bombarding nitrogen gas with alpha particles. He determined that high-energy protons were given off, and he assumed the nuclear reaction to be

$$^{14}_{7}N + ^{4}_{2}He \longrightarrow [^{18}_{9}F] \longrightarrow ^{17}_{8}O + ^{1}_{1}H$$

where $^{18}_{9}F$ is an unstable nucleus. Natural fluorine consists exclusively of the isotope $^{19}_{9}F$. Since both the $^{17}_{8}O$ and the hydrogen nuclei are stable, the products show no further tendency to undergo nuclear change.

Following Rutherford's original transmutation experiment, there was considerable interest in discovering new nuclear reactions. Many isotopes were subjected to beams of high-energy particles. As you might guess, numerous reactions were found; for example, bombardment of beryllium with alpha particles produced carbon:

One of the dreams of the alchemists (1200–1700 A.D.) was to transmute base metals such as lead and iron into gold. The dream was discarded only after the acceptance of Dalton's indestructible atom.

$$^{9}_{4}\text{Be} + ^{4}_{2}\text{He} \longrightarrow ^{13}_{6}\text{C} \longrightarrow ^{12}_{6}\text{C} + ^{1}_{0}\text{n}$$
(Neutron)

Although the $^{12}_{6}\text{C}$ produced in this reaction is stable, the neutron is given off with sufficient energy to provoke additional nuclear reactions in nuclei with which the neutron collides. It was just this nuclear reaction that was used by James Chadwick in 1932 to prove the existence of the previously postulated neutron.

An interesting question arises as to why the alpha particles were scattered by the gold foil in Rutherford's gold-foil experiment (Chapter 3) while the same alpha source can produce a nuclear change with a smaller atom such as $^{9}_{4}\text{Be}$ (Fig. 4–4). The answer lies in the fact that the charge on the gold nucleus is +79, whereas the charge on the beryllium nucleus is +4. Most of the alpha particles emitted from natural radioactive decay do not have enough energy to penetrate a heavy, positively charged nucleus such as that of gold. Therefore, if artificial nuclear reactions are to be studied for the heavier elements, the kinetic energy of the subatomic projectile particles must be increased.

Recall that like charges repel and unlike charges attract.

Not all nuclear reactions produce stable isotopes. If $^{25}_{12}\text{Mg}$ is bombarded with an alpha source, a radioactive isotope of aluminum, $^{28}_{13}\text{Al}$, is produced that does not exist in nature:

$$^{25}_{12}\text{Mg} + ^{4}_{2}\text{He} \longrightarrow ^{29}_{14}\text{Si} \longrightarrow ^{28}_{13}\text{Al}^{*} + ^{1}_{1}\text{H}$$

An asterisk is often used to denote a radioactive isotope in nuclear equations. Like naturally occurring radioactive isotopes, radioactive isotopes such as $^{28}_{13}\text{Al}$ have charac-

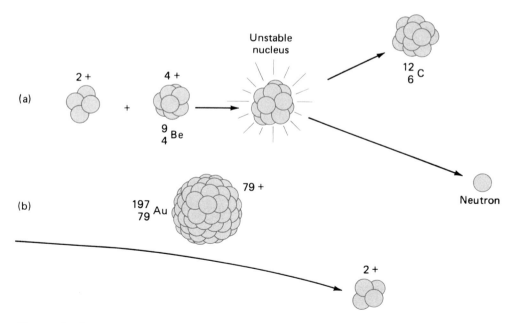

Figure 4–4
(a) A beryllium nucleus (Be) is struck by an alpha particle, which has sufficient energy to overcome the repulsions of like charges. A nuclear reaction then occurs, producing a carbon atom and a neutron. *(b)* In Rutherford's gold-foil experiment (the experiment that suggested the nuclear atom), the alpha particles were not energetic enough to penetrate the gold nucleus (Au) and were deflected.

teristic half-lives. The half-life of $^{28}_{13}$Al is relatively short, only 2.3 minutes. The $^{28}_{13}$Al nucleus emits a beta particle and becomes a stable isotope of silicon:

$$^{28}_{13}\text{Al}^* \longrightarrow \,^{28}_{14}\text{Si} + \,^{0}_{-1}\text{e}$$

Nuclear Particle Accelerators

Many designs were tried in the quest for faster and faster (more energetic) particles to bombard atomic nuclei. All these devices accelerated charged particles through electric potentials. For example, if an electron "sees" a positive charge on an electrode, it will begin to move toward that electrode. As the electron moves from point A to point B, its energy increases by an amount that can be expressed as the voltage difference between the two points. If electrons are attracted to an electrode that is 1000 volts more positive, the kinetic energy will increase by 1000 electron volts. If positively charged particles such as protons are used, the charges on the electrodes are reversed.

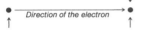

At point B, electron has energy of 1000 eV

Point A potential = 0 volts

Point B potential = 1000 volts.

Some of the largest particle accelerators can accelerate charged particles such as protons and lower-mass nuclei to near the speed of light. At these speeds the charged particles have energies of 10 billion electron volts, a sufficient energy to penetrate the nuclei of the heaviest atoms.

The early particle accelerators were used to make the first artificial elements (next section). The latest particle accelerators, such as the one shown in Figure 4–5, are being used to study the composition of subatomic particles such as the proton and the neutron.

Transuranium Elements

The heaviest known element before 1940 was uranium. The invention of the nuclear particle accelerator to obtain high-energy particles made it possible for these particles to react with heavy nuclei and to form even more massive nuclei. Thus, **transuranium elements** with atomic numbers greater than 92 were prepared.

(a)

(b)

Figure 4–5
Particle accelerator at the Fermi National Laboratory, near Batavia, Illinois. (a) Aerial view of the main accelerator. (b) Magnets in the interior of the main accelerator, which is 4 miles in circumference and 1.27 miles in diameter. (Fermi Lab Photo.)

In 1940 at the University of California, E. M. McMillan and P. H. Abelson prepared element 93, the synthetic element neptunium (Np). The experiment involved directing a stream of high-energy deuterons (^2_1H) onto a target of $^{238}_{92}U$. A deuteron is the nucleus of an isotope of hydrogen with one neutron and one proton. The initial reaction was the conversion of $^{238}_{92}U$ to $^{239}_{92}U$.

$$^{238}_{92}U + ^2_1H \longrightarrow ^{239}_{92}U + ^1_1H$$

Uranium-239 has a half-life of 23.5 minutes and decays spontaneously to the element neptunium by the emission of beta particles.

$$^{239}_{92}U \longrightarrow ^{239}_{93}Np + ^{\;\;0}_{-1}e$$

Neptunium is also unstable, with a half-life of 2.33 days; it converts into a second new element, plutonium.

$$^{239}_{93}Np \longrightarrow ^{239}_{94}Pu + ^{\;\;0}_{-1}e$$

Plutonium-239, like neptunium, is radioactive, with a half-life of 24,100 years. Because of the relative values of the half-lives, very little neptunium could be accumulated, but the plutonium could be obtained in larger quantities. $^{239}_{94}Pu$ is important as fissionable material, since atomic bombs can be made with it as well as with naturally occurring $^{235}_{92}U$. The names of neptunium and plutonium were taken from the mythological names Neptune and Pluto, in the same sequence as the planets Uranus, Neptune, and Pluto.

Although Neptune and Pluto are the last of the known planets in the solar system, their namesakes are not the last in the list of elements. The rush of transuranium experiments that followed produced additional elements: americium (Am), curium (Cm), berkelium (Bk), californium (Cf), einsteinium (Es), fermium (Fm), mendelevium (Md), nobelium (No), lawrencium (Lr), and elements 104, 105, 106, 107, and 109 (see Table 4–3 for recommended names). Obviously, many of these new elements were named after countries, states, cities, and people. Reactions employed in the production of the transuranium elements are given in Table 4–3. As accelerators with greater and greater energy capabilities are produced, even more nuclear reactions should be available for study.

The first synthetic element was prepared in 1940.

Plutonium is used to make atomic bombs and is also one of the most toxic elements known.

Radiation Damage

We are bombarded constantly by radiation from a number of sources. This radiation includes cosmic rays, medical X rays, radioactive fallout from countries that do nuclear testing, and naturally occurring, widespread radioisotopes. Most radiation damage is too slight to be noticed immediately, although its very presence should be regarded as one of the hazards of everyday life.

As we have seen, a radioisotope either disintegrates into a stable species or becomes part of a decay series. In a sample of radioactive matter large enough to measure, there will be many disintegrations over a given time if the half-life is short or few disintegrations over the same interval if the half-life is long.

Three principal factors render a radioactive substance dangerous: (1) the number of disintegrations per second, (2) the half-life of the isotope, and (3) the type or energy of the radiation produced. In addition, radiation can be very damaging if the radioactive substance is of a chemical nature such that it can be incorporated into a food chain or otherwise enter a living organism.

Radioactive disintegrations are measured in **curies** (Ci); one Ci is 37 billion disintegrations per second. A more suitable unit is the microcurie (μCi), which is 37,000

Table 4-3
Nuclear Reactions Used to Produce Some Transuranium Elements*

Element	Atomic Number	Reaction
Neptunium, Np	93	$^{238}_{92}U + ^{1}_{0}n \longrightarrow ^{239}_{93}Np + ^{0}_{-1}e$
Plutonium, Pu	94	$^{238}_{92}U + ^{2}_{1}H \longrightarrow ^{238}_{93}Np + 2\,^{1}_{0}n$
		$^{238}_{93}Np \longrightarrow ^{238}_{94}Pu + ^{0}_{-1}e$
Americium, Am	95	$^{239}_{94}Pu + ^{1}_{0}n \longrightarrow ^{240}_{95}Am + ^{0}_{-1}e$
Curium, Cm	96	$^{239}_{94}Pu + ^{4}_{2}He \longrightarrow ^{242}_{96}Cm + ^{1}_{0}n$
Berkelium, Bk	97	$^{241}_{95}Am + ^{4}_{2}He \longrightarrow ^{243}_{97}Bk + 2\,^{1}_{0}n$
Californium, Cf	98	$^{242}_{96}Cm + ^{4}_{2}He \longrightarrow ^{245}_{98}Cf + ^{1}_{0}n$
Einsteinium, Es	99	$^{238}_{92}U + 15\,^{1}_{0}n \longrightarrow ^{253}_{99}Es + 7\,^{0}_{-1}e$
Fermium, Fm	100	$^{238}_{92}U + 17\,^{1}_{0}n \longrightarrow ^{255}_{100}Fm + 8\,^{0}_{-1}e$
Mendelevium, Md	101	$^{253}_{99}Es + ^{4}_{2}He \longrightarrow ^{256}_{101}Mv + ^{1}_{0}n$
Nobelium, No	102	$^{246}_{96}Cm + ^{12}_{6}C \longrightarrow ^{254}_{102}No + 4\,^{1}_{0}n$
Lawrencium, Lr	103	$^{252}_{98}Cf + ^{10}_{5}B \longrightarrow ^{257}_{103}Lr + 5\,^{1}_{0}n$
Unnilquadium, Unq	104[a]	$^{242}_{94}Pu + ^{22}_{10}Ne \longrightarrow ^{260}_{104}? + 4\,^{1}_{0}n$
Unnilpentium, Unp	105[a]	$^{249}_{98}Cf + ^{15}_{7}N \longrightarrow ^{260}_{105}? + 4\,^{1}_{0}n$
Unnilhexium, Unh	106[a]	$^{249}_{98}Cf + ^{18}_{8}O \longrightarrow ^{263}_{106}? + 4\,^{1}_{0}n$
Unnilseptium, Uns	107[a]	$^{209}_{83}Bi + ^{54}_{24}Cr \longrightarrow ^{262}_{107}? + ^{1}_{0}n$
Unnilennium, Une	109[a,b]	$^{209}_{83}Bi + ^{58}_{26}Fe \longrightarrow ^{266}_{109}? + ^{1}_{0}n$

[a] The International Union of Pure and Applied Chemistry (IUPAC) has recommended three-letter symbols for the elements of atomic number 104 and higher. The symbols are based on the numerical roots *nil* (0), *un* (1), *bi* (2), *tri* (3), *quad* (4), *pent* (5), *hex* (6), *sept* (7), *oct* (8), and *enn* (9). For example, the symbol for element 104 is Unq, a combination of the first letter of each numerical root. The name is obtained by adding "ium" to the numerical roots. Hence, element 104 is Unnilquadium.

[b] Element 109 was discovered in Germany in August 1982. It was prepared by a technique known as cold fusion. Only a single atom of this element was detected.

There is a normal background radiation from natural causes.

disintegrations per second. One curie of a radioisotope is a potent sample if the energy per disintegration is large enough to cause a biochemical change. Normal background radiation to the human body is 2 to 3 disintegrations per second.

The unit **roentgen** is used to measure the intensity of X rays or gamma rays. One roentgen is the quantity of X-ray or gamma-ray radiation delivered to 0.001293 g of air, such that the ions produced in the air carry 3.34×10^{-10} coulomb of charge. A single dental X ray represents about 1 roentgen.

Damage by radiation is due to ionization caused by the fast-moving particles colliding with matter and by the excitation of matter by gamma and X rays, which in turn produce ionization. Neutrons are produced in nuclear explosions, in nuclear reactions, and by background cosmic radiation. A neutron does not produce ionization per se but instead imparts its kinetic energy to atoms, which in turn may ionize or break away from the atom to which they are bonded. Neutrons render many engineering materials, such as plastics and metals, structurally weak over long periods as a result of the decay caused by breaking chemical bonds.

Neutrons can damage metals, causing structural failure. This is a severe problem in nuclear reactors.

Biological tissue is harmed easily by radiation. A flow of high-energy particles may cause destruction of a vital enzyme, hormone, or chromosome needed for life of a cell. In general, those cells that divide most rapidly are most easily harmed by radiation (Fig. 4-6).

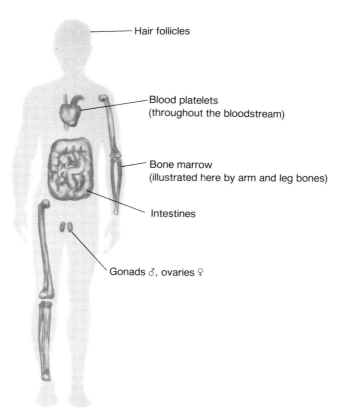

Figure 4-6
The fast-dividing cells within the body are the ones most harmed by radiation. These include cells in bone marrow, white cells, platelets of the blood, and cells lining the gastrointestinal tract, hair follicles, and gonads. In addition, the lymphocytes (cells producing the immune responses) are easily killed by radiation.

Whole-body radiation effects are divided into **somatic effects,** which are confined to the population exposed, and **genetic effects,** which are passed on to subsequent generations. A unit of measurement of radiation density is helpful in measuring the effect of radiation on tissue. The **rad** is defined as 100 ergs of energy imposed on a gram of tissue. Whole-body doses of radiation of up to 150 rads produce scarcely any symptoms, whereas doses of 700 rads produce death. Intermediate doses produce vomiting, diarrhea, fatigue, and loss of hair. Often the somatic effects are delayed. Perhaps the best studied of the delayed effects are the incidences of cancer related to exposure to radiation. It has been estimated that 11% of all leukemia cases and about 10% of all forms of cancer are attributable to background radiations. Certainly an individual who is exposed to a higher than normal level of radiation over a considerable length of time increases the chances of cancer. The alteration of normal cells to cancerous cells caused by radiation is undoubtedly a series of changes, since in almost all cases the onset of cancer lags behind the exposure to radiation by an induction period of 5 to 20 years.

The genetic effects of radiation are the result of radiation damage to the germ cells of the testes (sperm) or the ovaries (egg cells). Ionization caused by radiation passing

One *rad* is roughly the energy absorbed by tissue exposed to one roentgen of gamma rays.

1 joule = 10^7 ergs
1 calorie = 4.184×10^7 ergs

through a germ cell may break a DNA strand or cause it to be altered in some other way. When this damaged DNA is replicated (i.e., when the DNA structure is copied during cell division; see Chapter 11), the result may be the transmission of a new message to successive generations, a **mutation.** Every type of laboratory animal on which radiation damage experiments have been performed has responded with an increased incidence of mutation. Therefore, the necessity of protecting the population of childbearing age from radiation should be apparent. Theoretically at least, one photon or one high-energy particle can ionize a chromosomal DNA structure and produce a genetic effect that will be carried for generations. Table 4-4 shows average doses of radiation to the soft tissues and gonads from a variety of sources.

Radon: A Deadly Gas

Radon is the heaviest member of the noble gas family of elements (He, Ne, Ar, Kr, Xe, Rn). Radon-222, the most common isotope of radon, is radioactive, with a half-life of 3.82 days. It is a product of the uranium decay series and is a direct result of the naturally occurring radium-226 isotope:

$$^{226}_{90}Ra \longrightarrow ^{222}_{88}Rn + ^{4}_{2}He$$

Therefore, radon tends to occur in the ground, where there are naturally occurring deposits of uranium and its radioactive decay products.

When radon decays, it produces alpha particles and another short-lived isotope, polonium-218.

Table 4-4
Average Dose of Radiation to Soft Tissues and Gonads from Surroundings*

Source	Dose to Gonads Per Year (Rad)
Natural Background	
Cosmic rays	0.028
Local gamma rays	0.047
Radon in air	0.001
Potassium-40	0.019
Carbon-14	0.001
Other sources	0.002
Subtotal	0.098
Man-Made	
Medical X rays	0.100
Luminous watch dials	0.001
Occupational exposure	0.002
Television sets	0.001
Fallout from weapons test	0.001
Subtotal	0.105
Total	0.203

*This table is taken from the *McGraw-Hill Encyclopedia of Science and Technology*, Vol. 11. New York, McGraw-Hill, 1971, p. 250.

Medical X rays account for about 50% of the average radiation dosage shown here.

$$^{222}_{88}Rn \longrightarrow {}^{218}_{86}Po + {}^{4}_{2}He$$

Polonium-218 in turn decays to the next isotope in the uranium decay series (Table 4–1).

Being a member of the noble gas family, radon is chemically very nonreactive. Although the half-life of radon-222 is short, more continues to form as the longer lived radium-226 (the half-life of which is 1620 years) continues to decay.

When buildings are built over soil or rock containing heavy radioactive elements that decay to radium-226, radon gas is produced, some of which seeps through minute fissures in the rock or soil and migrates into the air in these buildings. It has been estimated that perhaps 8 million homes in the United States are affected by radon contamination. So far, radon has been detected in homes in 30 states. Areas of New Jersey and Pennsylvania have the highest radon levels recorded so far (Figure 4–7).

Breathing air containing small amounts of radon appears to increase one's chances of developing lung cancer, presumably caused by the alpha particles from the decay of the radon isotope. Alpha particles can cause the disruption of chemical bonds in DNA strands in lung cells, which may cause the cancerous growth of these cells. Although the amounts of radon in the air of affected homes is not high by normal standards of radiation measurement (generally less than 4 picocuries, or 14,800 disintegrations per second), many homes today are tightly sealed by weather stripping and insulation. This means the radon that enters the house does not leave quickly.

Pico means 10^{-12}.

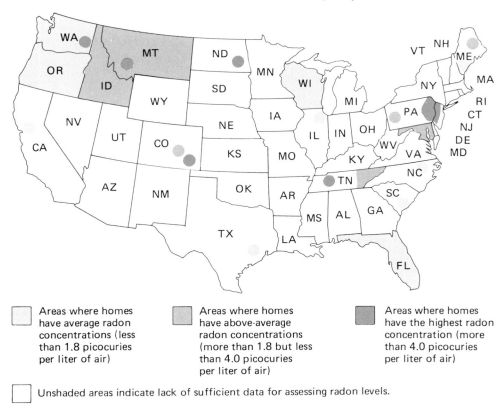

Figure 4–7
States in which radon concentrations are the highest. Radon has been detected in homes and other buildings in over 30 states.

No one is certain about how to deal effectively with this radon problem. Making homes radon-free is costly, since the very ground the homes are built on is the source of the problem. Perhaps land could be tested for the presence of radon before a new house is built on it. Maybe sales contracts on existing homes will require radon tests.

SELF-TEST 4-B

1. The first scientist to produce an artificial transmutation was _____.
2. To make particles like the alpha particle react with other nuclei of heavier atoms, the _____ of the alpha particles must be increased.
3. Can a neutron be accelerated by a charged particle accelerator?   Yes ( ) No ( )
4. An element that has an atomic number greater than that of uranium is called a _____ element.
5. How many elements are known? _____
6. The radiation measurement unit _____ is used to measure the intensity of gamma rays.
7. The radiation measurement unit _____ is used to describe the actual number of disintegrations per second from a sample of radioactive material.
8. The radiation measurement unit _____ is used to describe the amount of energy absorbed by a gram of human tissue.
9. What element in soil and rock causes the production of radon gas? _____

Nuclear Energy

Few issues in the past four decades have captured the awe, imagination, and scrutiny of mankind to quite the extent that nuclear energy has. Nuclear energy has been acclaimed on the one hand as the source of all our energy needs and accused on the other hand of being our eventual destroyer.

The splitting of heavy atomic nuclei is called **fission;** in this process, vast amounts of energy are released. The combination of small atomic nuclei to make heavier nuclei is called **fusion.** Consider the energy contrast between combustion of a fossil fuel and a nuclear fusion reaction. When one mole (6.02×10^{23} molecules, or 16 g) of methane is burned, over 200 kcal of heat are liberated:

$$CH_4 + 2\ O_2 \longrightarrow CO_2 + 2\ H_2O + 211 \text{ kilocalories (kcal/mole } CH_4)$$

In contrast, a lithium nucleus can be made to react with a hydrogen nucleus to form two helium nuclei in a nuclear reaction. The energy released per mole of lithium in this reaction is 23,000,000 kcal. This means that 7 g of lithium and 1 g of hydrogen produce 100,000 times more energy through fusion of nuclei than 16 g of methane and 64 g of oxygen produce by electron exchange.

$${}^{7}_{3}Li + {}^{1}_{1}H \longrightarrow 2\ {}^{4}_{2}He + 23{,}000{,}000 \text{ kcal/mole of } {}^{7}_{3}Li$$

Energy changes associated with nuclear events may be many thousands of times larger than those associated with chemical events.

Atomic mass ⟶ ${}^{7}_{3}Li$
Atomic number ⟶

Realizing that nuclear changes could involve giant amounts of energy relative to chemical changes for a given amount of matter, Otto Hahn, Fritz Strassman, Lise Meitner, and Otto Frisch discovered in 1938 that $^{235}_{92}U$ is fissionable. The dream of controlled nuclear energy subsequently became a reality, followed by the atomic bomb and nuclear power plants. In the 1950s it was hoped that nuclear energy would soon relieve the shortage of fossil fuels. To date this has not been accomplished, although the production of nuclear energy has grown very rapidly in recent years. The use of nuclear energy to generate electricity is much more utilized in Europe than in the United States; nuclear reactors supply 65% of electricity in France, versus 11% in the United States.

Fission Reactions

Fission can occur when a thermal neutron (with a kinetic energy about the same as that of a gaseous molecule at ordinary temperatures) enters certain heavy nuclei with an odd number of neutrons ($^{235}_{92}U$, $^{233}_{92}U$, $^{239}_{94}Pu$). The splitting of the heavy nucleus produces two smaller nuclei, two or more neutrons (an average of 2.5 neutrons for $^{235}_{92}U$), and much energy. Typical nuclear fission reactions may be written as follows:

$$^{235}_{92}U + ^{1}_{0}n \longrightarrow ^{141}_{56}Ba + ^{92}_{36}Kr + 3\, ^{1}_{0}n + \text{energy}$$

$$^{235}_{92}U + ^{1}_{0}n \longrightarrow ^{103}_{42}Mo + ^{131}_{50}Sn + 2\, ^{1}_{0}n + \text{energy}$$

Note that the same nucleus may split in more than one way. The fission products, such as $^{141}_{56}Ba$ and $^{92}_{36}Kr$, emit beta particles ($_{-1}^{0}e$) and gamma rays ($_{0}^{0}\gamma$) until stable isotopes are reached.

$$^{141}_{56}Ba \longrightarrow \,^{0}_{-1}e + ^{0}_{0}\gamma + ^{141}_{57}La$$

$$^{92}_{36}Kr \longrightarrow \,^{0}_{-1}e + ^{0}_{0}\gamma + ^{92}_{37}Rb$$

The products of these reactions emit beta particles, as do their products. After several such steps, stable isotopes are reached: $^{141}_{59}Pr$ and $^{90}_{40}Zr$, respectively.

The neutrons emitted can cause the fission of other heavy atoms if they are slowed down by a moderator, such as graphite. For example, the three neutrons emitted in the first preceding reaction could produce fission in three more uranium atoms, the nine neutrons emitted by those nuclei could produce nine more fissions, the 27 neutrons from these fissions could produce 81 neutrons, the 81 neutrons could produce 243, the 243 neutrons could produce 729, and so on. This process is called a **chain reaction** (Fig. 4–8), and it occurs at a maximum rate when the uranium sample is large enough for most of the neutrons emitted to be captured by other nuclei before passing out of the sample. A sufficiently large sample to sustain a chain reaction is termed the **critical mass.**

In the atomic bomb the critical mass is kept separated into several smaller subcritical masses until detonation, at which time the masses are driven together by an implosive device. It is then that the tremendous energy is liberated and everything in the immediate vicinity is heated to a temperature of 5 to 10 million degrees. The sudden expansion of hot gases explodes everything nearby and scatters the radioactive fission fragments over a wide area. In addition to the movement of gases, the tremendous vaporizing heat makes the atomic bomb devastating.

There is no danger of an atomic explosion in the uranium mineral deposits in the Earth for two reasons. First, uranium is not found pure in nature — it is found only in compounds, which in turn are mixed with other compounds. Second, less than 1% of the uranium found in nature is fissionable $^{235}_{92}U$. The other 99% is $^{238}_{92}U$, which is not fissionable

Fission is the breakup of heavy nuclei.

Note in nuclear reactions that the sum of the atomic numbers on the left side of the equation equals the sum of the atomic numbers on the right side of the equation. This is also true for the atomic masses.

$^{1}_{0}n$ *represents a neutron.*

$^{0}_{-1}e$ *or* $^{0}_{-1}\beta$ *represents a beta particle.*

A low-energy neutron will disrupt some large nuclei.

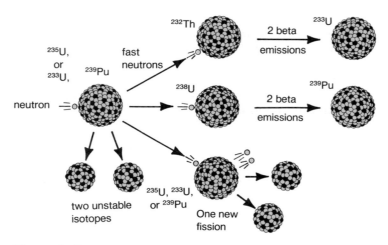

Figure 4-8
A chain reaction. A thermal neutron collides with a fissionable nucleus, and the resulting reaction produces three additional neutrons. These neutrons can either convert nonfissionable nuclei such as $^{232}_{90}$Th to fissionable ones or cause additional fission reactions. If enough fissionable nuclei are present, a chain reaction will be sustained.

The separation of uranium isotopes had to precede the control of atomic energy.

by thermal neutrons. In order to make nuclear bombs or nuclear fuel for electrical generation, a purification enrichment process must be performed on the uranium isotopes, increasing the relative proportion of $^{235}_{92}$U atoms in a sample. Ordinary uranium such as that found in ores is only 0.711% $^{235}_{92}$U.

It is interesting to note that fission products can be found in the Gabon Republic of West Africa; this indicates that a uranium ore deposit "went critical" about 150,000 years ago. At that time the natural uranium-235 content would have been higher than it is now.

Mass Defect: The Ultimate Nuclear Energy Source

What is the source of the tremendous energy of the fission process? It ultimately comes from the conversion of mass into energy, according to Einstein's famous equation $E = mc^2$, where E is energy that results from the loss of an amount of mass, m, and c^2 is the speed of light squared. If separate neutrons, electrons, and protons are combined to form any particular atom, there is a loss of mass called the **mass defect.** For example, the calculated mass of one $^4_2$He atom from the masses of the constituent particles is 4.032982 amu:

Mass that is lost leaves in the form of energy: $E = mc^2$.

amu = atomic mass unit.

$6.02 \times 10^{23} \dfrac{\text{amu}}{\text{gram}}$
or
$\dfrac{1 \text{ mol amu}}{\text{gram}}$

$2 \times 1.007826 = 2.015652$ amu, mass of two protons and two electrons

$2 \times 1.008665 = \underline{2.017330}$ amu, mass of two neutrons

total = 4.032982 amu, calculated mass of one $^4_2$He atom

Since the measured mass of a $^4_2$He atom is 4.002604 amu, the mass defect is 0.030378 amu:

4.032982 amu
−4.002604 amu
0.030378 amu, mass defect

Because the atom is more stable than the separated neutrons, protons, and electrons, the atom is in a lower energy state. Hence, the 0.030378 amu lost per atom would be released in the form of energy if the $^4_2$He atom were made from separate protons, electrons, and neutrons. The energy equivalent of the mass defect is called the **binding energy,** which is a measure of the energy necessary to separate the nucleus into its parts.

Atoms with atomic numbers between 30 and 63 have a greater mass defect per nuclear particle than very light elements or very heavy ones (Fig. 4-9). This means the most stable nuclei are the middle-weight ones found in the atomic number range from 30 to 63.

Separated nuclear particles have more mass than when combined in a nucleus.

Because of the relative stabilities, it is in the intermediate range of atomic numbers that most of the products of nuclear fission are found. Therefore, when fission occurs, the resulting smaller, more stable nuclei contain less mass per nuclear particle. In the process, mass must be changed into energy. This energy gives the fission process its tremendous energy. It takes only about 1 kg of $^{235}_{92}$U or $^{239}_{94}$Pu undergoing fission to be equivalent to the energy released by 20,000 tons (20 kilotons) of ordinary explosives like TNT; furthermore, the atomic fragments from the 1 kg of nuclear fuel weigh 999 g, so only one tenth of 1% of the mass is actually converted to energy. The fission bombs dropped on Japan during World War II contained approximately this much fissionable material.

Intermediate-sized nuclei tend to have the greatest nuclear stability.

Fusion Reactions

When very light nuclei such as H, He, and Li are combined, or **fused,** to form an element of higher atomic number, energy must be given off consistent with the greater stability of the elements in this intermediate atomic number range (Fig. 4-9). This energy, which

Fusion is the combination of very light nuclei.

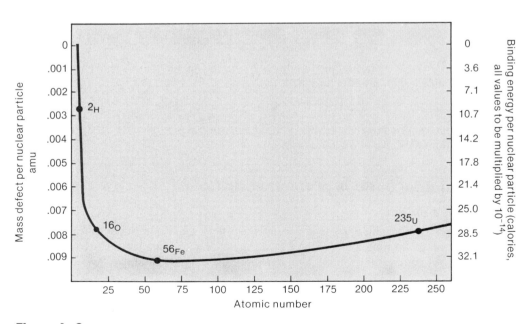

Figure 4-9
Mass defect for different nuclear masses. The most stable nuclei center around $^{56}_{26}$Fe, which has the largest mass defect per nuclear particle.

^2_1H = Deuterium
^3_1H = Tritium
$_{+1}^0e$ = Positron

comes from a decrease in mass, is the source of the energy released by the Sun and by hydrogen bombs. Typical examples of fusion reactions are:

$$4\,^1_1H \longrightarrow\,^4_2He + 2\,_{+1}^0e + 26.7\text{ Mev for four }^1_1H\text{ fused}$$

$$^2_1H + ^2_1H \longrightarrow\,^3_2He + ^1_0n + 3.2\text{ Mev}$$

$$^2_1H + ^2_1H \longrightarrow\,^3_1H + ^1_1H + 4.0\text{ Mev}$$

$$^3_1H + ^2_1H \longrightarrow\,^4_2He + ^1_0n + 17.6\text{ Mev}$$

The net reaction for the last three reactions given here is:

$$5\,^2_1H \longrightarrow\,^4_2He + ^3_2He + ^1_1H + 2\,^1_0n + 24.8\text{ Mev for five }^2_1H\text{ fused}$$

Materials for fusion reactions are available in enormous amounts.

Deuterium is a relatively abundant isotope — out of 6500 atoms of hydrogen in sea water, for example, 1 is a deuterium atom. What this means is that the oceans are a potential source of fantastic amounts of deuterium. There are 1.03×10^{22} atoms of deuterium in a single liter of sea water. In a single cubic kilometer of sea water, therefore, there would be enough deuterium atoms with enough potential energy to equal the energy produced by the burning of 1360 billion barrels of crude oil, approximately the total amount of oil originally present in this planet.

The plasma state is described in Chapter 2.

Fusion reactions occur rapidly only when the temperature is of the order of 100 million degrees or more. At these high temperatures atoms do not exist as such; instead there is a **plasma,** consisting of unbound nuclei and electrons, in which nuclei merge or combine. In order to achieve the high temperatures required for the fusion reaction of the hydrogen bomb, a fission bomb (atomic bomb) is first set off.

One type of hydrogen bomb depends on the production of tritium (^3_1H) in the bomb. In this type, lithium deuteride ($^6_3Li\,^2_1H$, a solid salt) is placed around an ordinary $^{235}_{92}U$ or $^{239}_{94}Pu$ fission bomb. The fission is set off in the usual way. A ^6_3Li nucleus absorbs one of the neutrons produced and splits into tritium, ^3_1H, and helium, ^4_2He.

$$^6_3Li + ^1_0n \longrightarrow\,^3_1H + ^4_2He$$

The temperature reached by the fission of $^{235}_{92}U$ or $^{239}_{94}Pu$ is sufficiently high to bring about the fusion of tritium and deuterium:

$$^3_1H + ^2_1H \longrightarrow\,^4_2He + ^1_0n + 17.6\text{ Mev}$$

A 20-megaton bomb usually contains about 300 pounds of lithium deuteride along with the plutonium or uranium fission bomb.

Useful Applications of Radioactivity

Food Irradiation

Consider the importance of killing harmful pests that would otherwise destroy our food during storage. In some parts of the world stored-food spoilage claims up to 50% of the food crop. In our society refrigeration, canning, and chemical additives lower this figure considerably.

Still, there are problems with food spoilage. Food protection costs amount to a sizable fraction of the final cost of food. Food irradiation using gamma-ray doses from sources such as ^{60}Co and ^{137}Cs is commonly used in European countries, Canada, and Mexico. The U.S. Food and Drug Administration (FDA) has been reluctant to allow this form of food preservation, but changes seem to be coming soon. Foods may be pasteurized by irradiation to retard the growth of organisms such as bacteria, molds, and yeasts.

This irradiation prolongs shelf life under refrigeration in much the same way that heat pasteurization protects milk. Normally chicken has a three-day refrigerated shelf life. After irradiation, chicken may have a three-week refrigerated shelf-life. The FDA may soon permit irradiation up to 100 kilorads for the pasteurization of foods.

Radiation levels in the 1- to 5-megarad range sterilizes; that is, it kills every living organism. Foods irradiated at these levels keep indefinitely when sealed in plastic or aluminum-foil packages. The FDA is unlikely to approve irradiation sterilization of foods in the near future because of potential problems caused by as yet undiscovered, but possible, "unique radiolytic products." It is feared that irradiated foods might contain chemical substances capable of causing genetic damage. To prove or disprove the presence of these substances, animal feeding studies using irradiated foods are presently being conducted.

> There are 1000 kilorads in 1 megarad.

Even with these uncertainties, over 40 classes of foods are presently being irradiated in 24 countries. In the United States only a small number of foods may be irradiated (Table 4–5).

Recent findings regarding the potentially harmful health effects of several common agricultural fumigants have indicated that irradiation of fruits and vegetables could be an effective alternative to some of these fumigants. The agricultural products involved could be picked, packed, and readied for shipment. After that, the entire shipping container could be passed through a building containing a strong source of radiation (Fig. 4–10). This type of sterilization is safe for workers because it does not expose them to harmful chemicals (see Chapter 13), and it protects the environment because it does not involve the risk of contamination of water supplies with toxic chemicals (Chapter 14).

> Ethylene dibromide (EDB) has been used widely in fumigating fruits. Now EDB is suspected to cause cancer and damage to human reproductive organs. Because of this toxicity, EDB has been banned by the U.S Environmental Protection Agency.

Materials Testing

The radioisotope ^{60}Co, a gamma-ray emitter, has proved quite useful in the testing of metal castings in industry. Contained within an aluminum thimble, the cobalt radioisotope can be placed inside a casting after a piece of photographic film has been positioned on the outside of the object (Fig. 4–11). The gamma rays penetrate the metal part and make observable any structural flaws in the metal by exposing the photographic film. The intensity of the gamma rays passing through the flawed portion of the casting is different from the intensity passing through the rest of the metal. After development, the photographic film can be examined for the presence of any flaws. Aviation safety has

Table 4–5
Examples of Irradiated Foodstuffs

Food	Purpose	Status
Potatoes	Retardation of sprouts	FDA approved
Wheat	Insect disinfection	FDA approved
Wheat flour	Insect disinfection	FDA approved
Spices	Retardation of microbe growth	FDA approved
Grapefruit	Mold control	For export
Strawberries	Mold control	For export
Fish	Microbe control	For export
Shrimp	Microbe control	For export

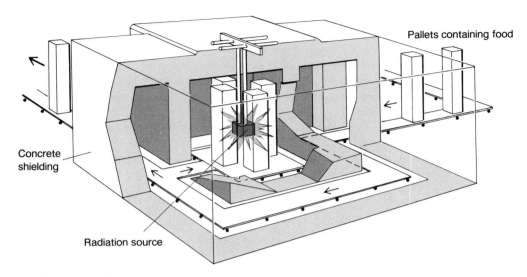

Figure 4–10
A typical commercial food irradiator. Boxes of food are conveyed into the shielded chamber and around the radiation source (center). When not in use, the source can be lowered into a pool of water below.

been increased by the use of radiation detection of flaws and structural weaknesses in structural members of aircraft.

Radioactive Tracers

Because radioisotopes act chemically in a manner almost identical to that of the nonradioactive isotopes of that element, chemists have been using radioactive isotopes as **tracers** in various chemical reactions since their use was discovered in 1945. Several common radioisotopes used as tracers are listed in Table 4–6. For example, plants are known to take up the element phosphorus from the soil through their roots. The use of the radioactive phosphorus isotope ^{32}P, a beta emitter, presents a way not only of detecting the uptake of phosphorus by a plant but also of measuring the speed of uptake

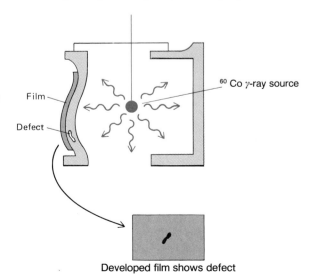

Figure 4–11
A gamma-ray source to detect defects in cast metal parts. The developed photographic film is more strongly exposed where gamma rays passed through a defect.

Table 4-6
Radioisotopes Used as Tracers

Isotope	Half-Life	Use
^{14}C	5730 years	CO_2 for photosynthesis research
^{3}H	12.26 years	Tag hydrocarbons
^{35}S	86.7 days	Tag pesticides, air flow
^{32}P	14.3 days	Phosphorus uptake by plants
^{131}I	8.05 days	Medical purposes

under various conditions. Plant biologists can grow hybrid strains of plants that can absorb phosphorus quickly and then test this ability with the radiophosphorus tracer. This type of research leads to faster maturing crops, better yields per acre, and more food or fiber at less expense.

One can measure important characteristics of a pesticide by tagging the pesticide with short–half-life radioisotopes and applying it to a test field. Following the tagged pesticide can provide information on its tendency to accumulate in the soil, be taken up by the plant, and accumulate in runoff surface water. This is done with a high degree of accuracy by counting of the radioactive disintegrations of the tracer radioactive isotope. After these tests are completed, the radioisotopes in the tagged pesticides decay to a harmless level in a few days or a few weeks because of the short half-lives of the species used. This type of research leads to safer, more effective pesticides.

Medical Imaging

Radioisotopes are also used in **nuclear medicine** in two distinctly different ways, diagnosis and therapy. In the diagnosis of internal disorders and other maladies, physicians need information on the locations of disorders. This is done by **imaging**, a technique in which the radioisotope, either alone or combined with some other chemical, accumulates at the site of the disorder. There, acting like a homing device, the radioisotope disintegrates and emits its characteristic radiation, which is detected. The detectors in modern medical diagnostic instruments are controlled by computers, which not only determine where the radioisotope is located in the patient's body but also construct an image of the area within the body where radioisotopes are concentrated (Fig. 4-12).

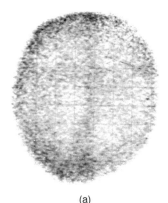
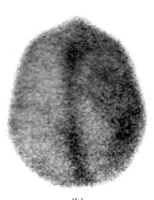

(a)   (b)

Figure 4-12
Brain scans using radioactive technetium-99m. (a) Scan of a normal brain. (b) Scan of an abnormal brain showing an accumulation of radioisotope in a region of suspected tumor growth.

Table 4–7
Diagnostic Radioisotopes

Radioisotope	Name	Half-Life (Hours)	Uses
^{99m}Tc*	Technetium-99m	6	As TcO_4^- to the thyroid, brain, kidneys
^{201}Tl	Thallium-201	21.5	To the heart
^{123}I	Iodine-123	13.2	To the thyroid
^{67}Ga	Gallium-67	78.3	To various tumors and abscesses

* The technetium-99m isotope is the one most commonly used for diagnostic purposes. The *m* stands for "metastable," a term explained in the text.

Four of the most common diagnostic radioisotopes are listed in Table 4–7. All of these are made by using a particle accelerator in which heavy charged nuclear particles are made to react with other radioisotopes or stable atoms. Each of these radioisotopes produces gamma radiation, which in low doses is less harmful to the tissue than ionizing radiations such as beta or alpha rays.

By the use of special carriers, these radioisotopes can be made to accumulate in specific areas of the body. For example, the pyrophosphate ion, $P_4O_7^{4-}$, a polyatomic ion, can bond to the technetium-99m radioisotope; together they accumulate in the skeletal structure where abnormal bone metabolism is taking place. Such investigations often pinpoint bone tumors.

The technetium-99m radioisotope is metastable (denoted by the letter *m*). Metastable isotopes lose energy by disintegrating to a more stable version of the same isotope.

$$^{99m}Tc \longrightarrow {}^{99}Tc + \gamma$$

Technetium-99m, like the other common diagnostic radioisotopes, has a short half-life, which means that the radioactivity does not linger for an unacceptably long period in the patient's body. Examples of the uses of technetium-99m in the diagnosis of brain tumors and thyroid tumors are shown in Figures 4–12 and 4–13, respectively. When a diagnosis involves technetium-99m, the radioisotope is washed with a salt solution from a

Figure 4–13
Thyroid scan using technetium-99m. The darkened area on the left side is a tumor. (Courtesy of the Department of Radiology, Vanderbilt University.)

cartridge containing molybdenum-99 (half-life, 66 hours). The molybdenum radioisotope is constantly producing technetium isotopes.

$$^{99}\text{Mo} \longrightarrow {}^{99m}\text{Tc} + _{-1}^{0}\text{e} + \gamma$$

The longer half-life of the molybdenum isotope makes it possible to ship the technetium in a form that will ensure the arrival of technetium-99m with sufficient strength to still be useful.

Therapeutic radioisotopes are generally beta emitters, which are produced in nuclear reactors by the bombardment of stable isotopes with neutrons. Two common therapeutic radioisotopes are ^{131}I (iodine-131) and ^{32}P (phosphorus-32). The most common use of iodine-131 is in the treatment of thyroid cancers. A patient drinks a solution of ^{131}I ions as potassium iodide. The iodine, as iodide ion, then makes its way to the thyroid, where the beta rays produced by the ^{131}I radioisotopes destroy the cancerous thyroid cells. Of course, healthy cells are also destroyed, but not in sufficient numbers to destroy all of the healthy tissue.

Nuclear medicine, the use of radioisotopes for therapeutic and diagnostic purposes, has established itself in medical practice throughout the world. Over 2000 hospitals in the United States are licensed to use radioisotopes. Discoveries of new substances to carry radioisotopes to specific sites in the body offer one of the most promising areas of research. Another is the use of computers for imaging those sites in the body where the radioisotopes are concentrated.

Still another useful application of nuclear radiation in medicine is the heating effect caused by radiation. This heat can be used to generate electricity, which in turn can power a cardiac pacemaker. Pacemakers powered by plutonium-238 (half-life, 89 years) have been used in humans in France since 1970 and in the United States since 1972. The electricity generated by the plutonium thermoelectric source is sent as a pulse directly to the ventricles of the heart at a preset rate. Because of the relatively long half-life of the plutonium source, these pacemakers can remain in the patient for longer periods without the need for additional surgery than can pacemakers powered by batteries.

It is hoped that these and other beneficial uses of radioisotopes will allow nuclear science to save far more lives than nuclear bombs have or ever will have destroyed.

Nuclear Waste: The Unsolved Problem

Radioactive isotopes are produced in nuclear fission reactors. The radiation levels produced by many of these isotopes are dangerous to life. Compounding the danger is the fact that some of the half-lives extend the active lifetime of the radioactive wastes into thousands of years. If all nuclear power and weapons research had come to an abrupt halt after the Three Mile Island incident in 1978, most of the nation's high-level nuclear waste stockpiled up to that point would be as deadly today, and nearly so a thousand years from now. Radioactive decay must run its course even if it takes thousands of years. There is no known way to speed up the process or shut it off. This is a problem with no apparent solution.

The persistent half-lives are not the only problem. In the process of producing nuclear energy, a large quantity of radioactive waste is made. According to the Department of Energy, 71 million pounds of radioactive waste were discharged into the air, water, and ground from 1946 to 1983 at seven facilities, three at Oak Ridge, Tennessee, one at Paducah, Kentucky, and three in Ohio (Fernald, Piketon, and Ashtabula). Additional wastes were discharged from other facilities.

The waste gases were mostly radioactive isotopes of the noble gases, such as krypton-85 (half-life, 10.76 years). Tritium, an isotope of hydrogen (^3_1H; half-life, 12.26 years), is also discharged in gaseous form, principally as water vapor. As the hydrogen part of a water molecule, tritium may be taken up into biological food chains.

Radioactive isotopes in liquid wastes are generally converted to solid form through precipitation, thereby decreasing the volume to one tenth or less of the original volume. Then the solid is stored. The 51 million pounds of radioactive wastes buried in the ground at the Y-12 plant at Oak Ridge are buried in unlined trenches and covered with dirt—no casks, no containers. The extent to which groundwater will spread the radioactive materials over the next several centuries is not known.

If we cannot control radioactive decay and we have these huge buildups of radioactive wastes, what can be done? Proposed solutions to these problems center on the theme, "One person's solution is another person's pollution (problem)." The premise is to get the wastes as far away from us as possible. Fanciful schemes, such as rocketing the nuclear waste into the Sun or deep space or burying it in the deep oceans, have been largely dismissed as too expensive and risky. The Department of Energy is committed to putting on-line in 1998 underground storage chambers carved in salt beds, clay, or rock. According to the plan, the waste will first go into specially built containers designed to withstand shock and corrosion for thousands of years. Scientists believe this is the best of the alternatives available.

Even if the storage method planned by the Department of Energy is adequate, there is still another problem: that of transporting the radioactive wastes from the end-user site to the storage site. In some cases this requires transportation of the wastes across most of the United States. If a truck, train, or ship is used, how can the transportation be done safely? Some of the transportation is through heavily populated centers.

The danger in transporting nuclear wastes is the spread of dangerous radioactivity, not a nuclear explosion.

Of the myriad problems associated with nuclear waste, there is a unique challenge to mankind. It is not the size of the problem in terms of the amount of wastes to be stored or their transportation, but rather the complete lack of control of the rate of radioactive decay. Chemical theory and technology exist for solving the problems of air pollution, water pollution, and chemical wastes, despite the size and complexity of these problems. But nothing can be done to change the rate at which radioactive decay occurs. Outside of health problems, this may be the most perplexing and frustrating problem to be faced by mankind today.

SELF-TEST 4-C

1. Name three uses of radioactive isotopes. _____, _____, and _____
2. Which radioisotope is used for examining metal castings? (a) ^{60}Co, (b) ^{32}P, (c) ^{67}Ga, (d) ^{99m}Tc _____
3. Name a radioisotope that might be useful as a tracer in agricultural research. (a) ^{32}P, (b) ^{14}C, (c) ^{3}H, (d) all of these _____
4. The process of concentrating a radioisotope at a particular site of the body in order to locate and measure the extent of a disorder is called: (a) radiotherapy, (b) imaging, (c) sterilization. _____

5. In the symbol for the radioisotope technetium-99m, the *m* stands for (a) middle, (b) mathematical, (c) metastable. _____
6. With its half-life of approximately 6 hours, how much technetium-99m would remain 18 hours after injection into a patient? (a) one eighth of the original dose, (b) one half of the original dose, (c) one sixth of the original dose, (d) one fourth of the original dose. _____
7. If two radioisotopes were available for diagnosis, worked equally well, and decayed by giving off gamma rays, but one had a half-life of 13 hours while the other had a half-life of 6 hours, which one would you recommend? (a) 13-hour half-life isotope or (b) 6-hour half-life isotope _____

MATCHING SET

_____ 1. Somatic effect
_____ 2. 1 microcurie
_____ 3. $^{14}_{6}C$
_____ 4. M. Curie
_____ 5. Genetic effect
_____ 6. $^{238}_{92}U$ dating
_____ 7. Element 109
_____ 8. Neptunium
_____ 9. $^{218}_{84}Po$
_____ 10. James Chadwick
_____ 11. Half-life
_____ 12. E. Rutherford
_____ 13. Bone marrow
_____ 14. P. Villard
_____ 15. ^{99m}Tc
_____ 16. ^{60}Co

a. Intake stops when organism dies
b. Effect of radiation on the general population
c. First suggested by Rutherford
d. First transuranium element
e. Radiation damage to DNA
f. Tissue easily damaged by radiation
g. Radioisotope used in medical diagnosis
h. Time required for half of the nuclei to disintegrate
i. 37,000 disintegrations per second
j. Used to detect defects in metal castings
k. Discovered radium
l. Characterized gamma rays
m. Latest synthetic element
n. Studied alpha and beta rays
o. Discovered neutron
p. Alpha decay product of $^{222}_{86}Rn$
q. Can penetrate 8 cm of lead
r. Half-life of 0.5 sec

QUESTIONS

1. Describe Becquerel's discovery of natural radioactivity.
2. What element did Marie Curie discover? What experimental techniques did she use?
3. What does the symbol $^{11}_{5}B$ mean?
4. In general, how have the synthetic transuranium elements been produced?
5. Complete or supply the following nuclear equations:
   a. $^{1}_{1}H + ^{35}_{17}Cl \rightarrow ^{4}_{2}He + ?$
   b. Beta emission of $^{60}_{27}Co$
   c. Alpha emission of $^{238}_{90}Th$
   d. $^{1}_{0}n + ^{60}_{28}Ni \rightarrow ? + ^{1}_{1}H$
   e. $^{2}_{1}H + ^{1}_{1}H \rightarrow ? + ^{0}_{0}\gamma$
   f. $^{238}_{92}U + ^{12}_{6}C \rightarrow ? + 4\,^{1}_{0}n$

6. If the two nuclei $^{209}_{83}Bi$ and $^{58}_{26}Fe$ were fused together, what radioisotope would be produced?
7. If 1 curie of radioactive isotope represents 37 billion disintegrations per second, how many disintegrations per second occur in a sample of 1 microcurie?
8. If 1 curie of a radioisotope that decays to stable products is held in a container, how many curies of radiation would be present after five half-lives?
9. If a radium atom ($^{226}_{88}Ra$) loses one alpha particle per atom, what element is formed? What is its atomic weight? What is its atomic number?
10. What are the important assumptions made in radiocarbon dating?
11. Name a method by which the age of rocks can be determined. What assumptions are made in this method?
12. What error would be introduced into the age determination of a tree ring if the amount of cosmic rays had been double their present value at the time the tree grew that ring?
13. What is meant by "delayed somatic effect"? Give an example.
14. Look up the origin of the word *mutation* and explain why the word *transmutation* was an apt choice to describe the changing of one element into another.
15. What is the difference between a thermal neutron and a high-energy neutron resulting from cosmic radiation?
16. The ^{99}Mo (half-life, 67 hours) canisters used to prepare solutions of ^{99m}Tc have an effective life of about one week. Can you suggest a reason for this based on half-lives of radioisotopes?
17. Suggest a therapeutic use for the gamma-emitting radioisotope ^{60}Co.
18. Suggest an experiment using a radioisotope tracer in agriculture.
19. Iodine-123 (half-life, 13.2 hours) is used to measure iodine uptake by the thyroid gland. If 1 mg is injected into a patient's bloodstream, how long will it take for the radioisotope to be reduced to less than 1 μg?
20. Ask for an interview with a radiologist at your local hospital. Ask him or her to tell you about the greatest benefits and risks of the use of radioisotopes.
21. Read a recent magazine or newspaper account of the radon gas problem in your state or a nearby state. Write a short summary.
22. Describe how radon gas is produced and how it gets inside homes.
23. Describe in your own words the differences between nuclear energy and fossil fuel energy.
24. Complete the following equations:

$^{235}_{92}U + ^{1}_{0}n \longrightarrow ^{141}_{56}Ba + ^{92}_{36}Kr + $ _____

$^{235}_{92}U + ^{1}_{0}n \longrightarrow ^{103}_{42}Mo + ^{131}_{50}Sn + $ _____

25. Describe how a nuclear chain reaction takes place. Be sure to include the term *critical mass*.
26. Add the masses of two protons (1.007826 amu) and two neutrons (1.008665 amu) to arrive at the mass of the $^{4}_{2}He$ isotope. The measured mass of $^{4}_{2}He$ is 4.002604 amu. Explain this difference.
27. What is meant by the term *binding energy*?
28. The isotope $^{56}_{26}Fe$ has one of the highest binding energies. Would you expect to see this isotope used as a fuel in a nuclear reactor?
29. The isotope $^{235}_{92}U$ has one of the lowest binding energies for the heavier elements. Would you expect to see this isotope used as a fuel in a nuclear reactor? If so, what kind: fission or fusion?
30. Complete the following reactions:

$^{2}_{1}H + ^{2}_{1}H \longrightarrow ^{3}_{2}He + $ _____

$^{2}_{1}H + ^{3}_{1}H \longrightarrow ^{4}_{2}He + $ _____

$4\, ^{1}_{1}H \longrightarrow ^{4}_{2}He + 2$ _____

CHAPTER 5

Chemical Bonds: The Ultimate Glue

What holds matter together? Why does glue stick? What causes pieces of hard candy to stick together? Why is a diamond so hard; why is wax soft? Or, in reverse, why do things break or fall apart? Why is table salt so brittle? Why does paint peel? Why do some substances melt at a rather low temperature, while others melt at higher temperatures?

Chemical bonds hold atoms, molecules, and ions together.

These and similar questions can be answered logically and the answers will be consistent with experimental evidence if we think of matter as one atom bound to another (Fig. 5–1). Granted, it is a little hard to consider the Empire State Building or the Washington Monument or a living organism as a conglomeration of atoms bonded one to the other. But large pieces of matter, even the Rocky Mountains, conform to the same fundamental principles of nature as a small crystal of sugar or salt.

Most of the reasons for matter bonding to matter (or atom bonding to atom) can be summarized by two concise notions:

1. Unlike charges attract.
2. Electrons tend to exist in pairs.

Couple these two ideas (one empirical, one theoretical) with the proximity requirement that only the outer electrons of the atoms (the *valence electrons*) interact, and you have the basic concepts that explain how atoms in over 6 million compounds bond to each other. Just how the different atoms use these principles to bond atom to atom is the subject of this chapter. We shall see that the action of an atom in the formation of a bond is dictated by its atomic structure and generalized by its position in the periodic table.

Various interactions of the atoms cause the formation of five major types of chemical bonds. The type of bonds, along with some common materials in which they occur, are:

1. Ionic bonding: Salts, such as table salt (sodium chloride); and metal oxides, such as lime, iron rust, ruby, and sapphire
2. Covalent bonding: Molecular compounds, such as water, methane, and sugar; and polymers, such as polyethylene
3. Intermolecular bonding
    a. Hydrogen bonding: Water, ammonia, DNA, and proteins
    b. London forces: Liquid helium and solid CO_2 (dry ice)
4. Metallic bonding: Metals and alloys

As always, it is the properties of the substances that dictate and verify the related theories. Properties such as chemical reactivity, volatility (ability to pass into the gaseous state), melting point, electrical conductivity, and color often give some indication of how atoms are bonded to each other. For example, since melting involves atoms or

Bonding theories must explain the observed behavior of chemicals.

Figure 5-1
A water molecule is often represented by a ball-and-stick model. This model tells which atoms are bonded together and the angle involved but gives no information as to why the atoms are bonded in a particular pattern, the relative sizes of the atoms, or the actual distances between them.

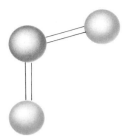

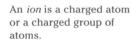

CO_2 changes readily from a solid to a gas *(sublimes)*.

molecules becoming less firmly bound to their neighbors, a high melting point implies that a solid is held together by very stable chemical bonds. As we shall see shortly, compounds composed of a network of tightly bound ions or atoms tend to have relatively high melting points. The volatility of a substance also indicates how strongly molecules are attracted to each other. For example, in the case of carbon dioxide, CO_2, we must assume that the bonding between molecules (intermolecular bonding) is slight, since it takes relatively little energy to break up solid CO_2 (dry ice).

In the ensuing discussion of chemical bonds, major emphasis is placed on accounting for the properties of a given substance by the bonds that hold that substance together.

Ionic Bonds

A large category of compounds forms hard, brittle crystalline solids with relatively high melting points. When melted or in solution, these compounds conduct electricity well, but when solid they do not conduct. If in solution or melted, they often react quickly with each other. Compounds with these properties are known as **ionic compounds.** Examples of ionic compounds are sodium chloride (NaCl), magnesium fluoride (MgF_2), and calcium oxide (CaO).

An *ion* is a charged atom or a charged group of atoms.

All the properties of these compounds can be explained if the compounds are assumed to be composed of charged atoms (called **ions**) rather than neutral atoms. X-ray and mass spectrographic studies of these kinds of compounds strongly indicate that ions exist.

How do atoms become ions, and which atoms are most likely to form ions? A guiding principle that will help answer these questions becomes apparent when we examine a group of relatively inert gaseous elements: the noble gases. You are probably familiar with helium and neon in this group. Helium is used to fill weather balloons and blimps, and neon is used in "neon" lights. Argon, krypton, xenon, and radon are less well known members of the noble gas family of elements. Until 1962, compounds of noble gases were unknown. Now, a number of compounds have been made, including XeF_4, XeF_6, $XeOF_4$, XeO_3, and KrF_2, but attempts to prepare stable compounds with He, Ne, and Ar have been unsuccessful so far. By any criterion, this family of elements is relatively inert compared with the other 82 natural elements.

The inertness of the noble gases must be related to their electronic structure. When the electronic arrangements of the noble gases are written out, two features are apparent:

He	2			
Ne	2	8		
Ar	2	8	8	
Kr	2	8	18	8
Xe	2	8	18	18	8

First, all lower energy levels containing any electrons at all are filled to capacity. Second, except for He, the highest numbered energy level contains 8 electrons. These electronic arrangements appear to be particularly stable. Perhaps if other elements achieved these electronic structures, they would also be stable chemically. This seems to be the case for a large group of elements.

The Group IA (or alkali metal) elements, Li, Na, K, Rb, and Cs, have one more electron than do the noble gases He, Ne, Ar, Kr, and Xe, respectively. If each of these metals loses an electron from each atom, the species would have the noble gas electronic arrangement. For example, Li has one valence electron in the ground-state arrangement. The loss of one electron gives Li the electronic structure of He. A Li atom with only two electrons and three protons has a charge of 1+. A "charged atom," such as Li$^+$, or a charged group of atoms, such as the sulfate group (SO_4^{2-}), is an **ion**.

Elements in Group IIA of the periodic table (the alkaline earth metals) have two valence electrons. Thus, for Mg, Ca, Sr, and Ba to take on the noble gas structure, the atoms of each element must lose two electrons each. The loss of two electrons would leave two protons in the nucleus unneutralized, so each ion would have a charge of 2+. To remove the third electron requires the breakup of pairs of electrons in a lower, main energy level. This takes considerably more energy.

Valence electrons are in the highest stable energy level.

The removal of electrons from metals and the consequent formation of positive ions can be depicted in varying degrees of detail. It is shown here for Group IA, with sodium (Na, atomic number 11) as the example:

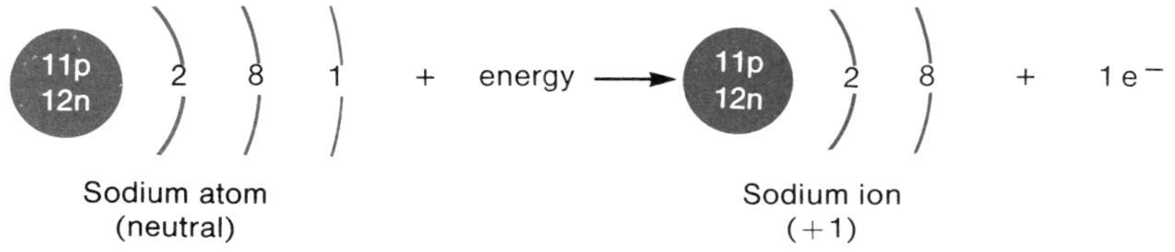

[Diagram: Sodium atom (neutral) 11p/12n with shells 2, 8, 1 + energy → Sodium ion (+1) 11p/12n with shells 2, 8 + 1 e⁻]

or

$$Na + energy \longrightarrow Na^+ + e^-$$
$$2\text{-}8\text{-}1 \qquad\qquad 2\text{-}8$$

or, simply

$$Na \longrightarrow Na^+ + e^-$$

Here it is shown for Group IIA metals, with Mg as the example:

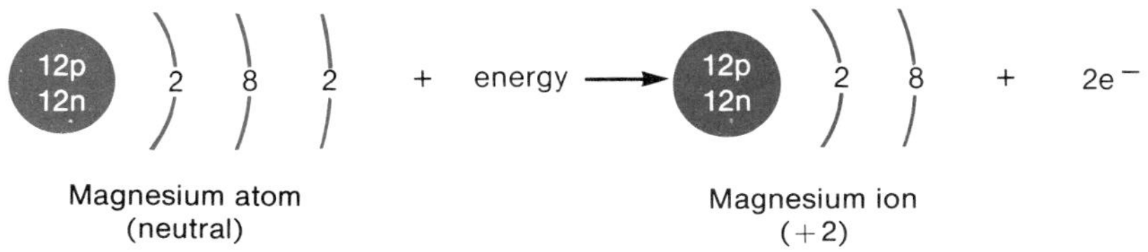

[Diagram: Magnesium atom (neutral) 12p/12n with shells 2, 8, 2 + energy → Magnesium ion (+2) 12p/12n with shells 2, 8 + 2 e⁻]

or

$$Mg + energy \longrightarrow Mg^{2+} + 2e^-$$
$$2\text{-}8\text{-}2 \qquad\qquad 2\text{-}8$$

or, simply

$$Mg \longrightarrow Mg^{2+} + 2e^-$$

It occurs in a similar fashion for Group IIIA metals. Al is used as the example:

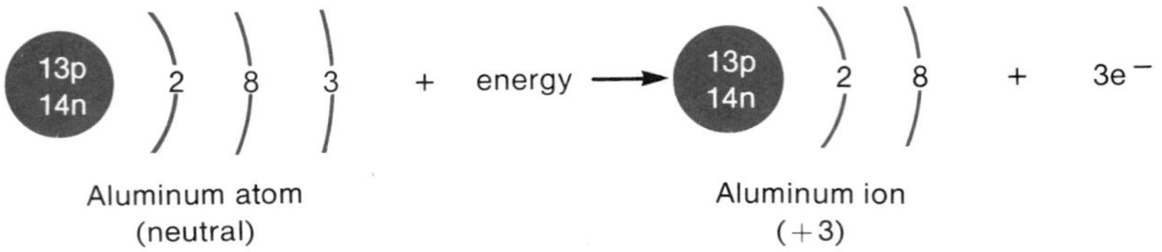

Aluminum atom (neutral) → Aluminum ion (+3)

or

$$Al + \text{energy} \longrightarrow Al^{3+} + 3e^-$$
2-8-3   2-8

or, simply

$$Al \longrightarrow Al^{3+} + 3e^-$$

It becomes more difficult to remove each succeeding electron from a given atom. With the loss of each electron, the net positive charge builds up, and this charge helps hold the remaining electrons more securely.

Some simple stable ions with a noble gas electronic configuration are listed in Table 5–1.

In summary, positive ions are formed when metal atoms lose one electron (Group IA), two electrons (Group IIA), or three electrons (Group IIIA) to nonmetal atoms. The resulting ions have the same electronic arrangement as a noble gas.

Positive ions are stabilized by the presence of negative ions. The neutralization of charge stabilizes the charge on both types of ions. Stable negative ions can be produced by atoms that have six or seven valence electrons. These atoms may gain enough electrons to achieve the noble gas structure. For example, atoms of Group VIIA elements (the halogens) have seven valence electrons and need one each to have the electronic arrangement of a noble gas. If atoms of F, Cl, Br, and I gain one electron each, the resulting ions, F^-, Cl^-, Br^-, I^-, have the same electronic arrangement as Ne, Ar, Kr, and Xe, respectively.

The Group VIA elements (O, S, and Se) need to gain two electrons for each atom to achieve the electronic structure of a noble gas. The excess of two electrons per atom produces a 2^- ion.

The gain of electrons by the nonmetals, like the loss of electrons by the metals, can be depicted in various degrees of detail. It is shown below for Group VIIA, with fluorine used as an example:

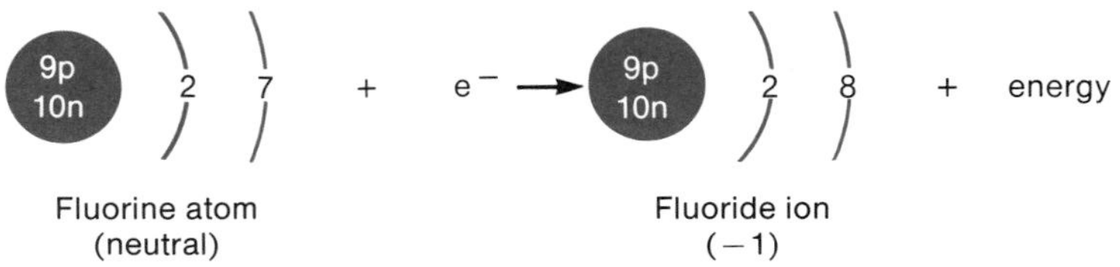

Fluorine atom (neutral) → Fluoride ion (−1)

$$F + e^- \longrightarrow F^- + \text{energy}$$
2-7   2-8

$$:\!\ddot{F}\!\cdot + e^- \longrightarrow :\!\ddot{F}\!:^-$$

$$F + e^- \longrightarrow F^-$$

Table 5-1
Electronic Configurations of the Noble Gases and Ions with Identical Configurations

Species	Configuration
He, Li^+, Be^{2+}, H^-	2
Ne, Na^+, Mg^{2+}, F^-, O^{2-}	2-8
Ar, K^+, Ca^{2+}, Cl^-, S^{2-}	2-8-8
Kr, Rb^+, Sr^{2+}, Br^-, Se^{2-}	2-8-18-8
Xe, Cs^+, Ba^{2+}, I^-, Te^{2-}	2-8-18-18-8

Notice that an electron is on the left side of each equation, and the electron is therefore gained. The third depiction is very informative, and involves **electron dot formulas.** Only the valence electrons are represented around the symbol for the element. The electrons are placed at north, east, south, and west positions adjacent to the symbol. The method clearly shows the pairing of electrons as well as the attainment of eight valence electrons when the negative ion is formed.

Electron dot formulas were devised in 1916 by Gilbert Newton Lewis (1875-1946).

How an oxygen atom gains two electrons to become an oxide ion (O^{2-}) can be depicted in several ways, shown earlier:

$$O + 2e^- \longrightarrow O^{2-} + energy$$
$$2\text{-}6 \qquad\qquad 2\text{-}8$$

$$\cdot \ddot{O}: + 2e^- \longrightarrow :\ddot{O}:^{2-}$$

$$O + 2e^- \longrightarrow O^{2-}$$

As the number of electrons added to an atom is increased, it becomes more difficult for each succeeding electron to enter. Each electron that comes into an atom after the first must enter against an existing net negative charge.

In summary, nonmetals in the presence of metals tend to gain one, two, or three electrons from metal atoms to form negative ions, which have all valence electrons paired and have the stable eight-electron arrangement of noble gases.

A memory aid tied to the periodic table is given in Figure 5-2 for ions formed by the elements.

A chemical formula of a compound is electrically neutral. Electrical neutrality requires an equal number of positive and negative charges in the crystal of the compound: two F^- for each Ca^{2+}; three O^{2-} for two Al^{3+}. In a crystal of table salt the Na^+ ions and Cl^- ions are held in place by electrical attraction between unlike charges. Furthermore, the ratio of sodium ions to chloride ions must be 1 to 1 if the compound is to be neutral. The simplest formula is found to be NaCl, so the theory is consistent with observations.

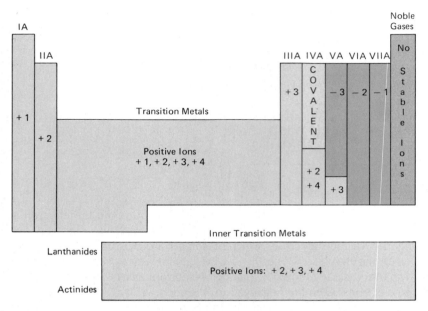

Figure 5–2
The periodic table and the formation of ions.

The crystalline structure of NaCl is shown in Figure 5–3. Note that each sodium ion is attracted by all six chloride ions around it. Similarly, each chloride ion is attracted by all six sodium ions around it. There is *no unique molecule* in ionic structures; no particular ion is attached exclusively to another ion.

When atoms become ions, their properties are drastically altered. For example, a collection of bromine molecules is red, but Br^- ions contribute no color to a crystal of a compound. A chunk of sodium atoms is soft, metallic, and violently reactive with water, but Na^+ ions are stable in water. A large collection of chlorine molecules constitutes a greenish-yellow, poisonous gas, but chloride ions (Cl^-) produce no color in compounds and are not poisonous. In fact, sodium and chloride ions in the form of table salt can be put on tomatoes without fear of a violent reaction. When atoms become ions, atoms obviously change their nature.

The *electrical conductivity* of melted ionic compounds is based on the movement of free ions to oppositely charged poles when an electrical field is imposed. The movements of the ions transport charge, or electric current, from one place to another. In a rigid solid, the immobile ions are not free to move, and the solid does not conduct electricity.

The *hardness* of ionic compounds is caused by the strong bonding between ions of unlike charge. The strong bonds require much energy to separate the ions and allow the freer movement of the melted state. Much energy means *higher melting points*, which are characteristic of ionic compounds.

Ionic compounds are *brittle* because the structure of the solid is a regular array of ions. Take, for example, the structure of sodium chloride (NaCl) (Fig. 5–3). If a plane of ions is shifted just one ion's distance in any direction, identically charged ions are now next to each other. This causes repulsion — there is no attraction — and the crystalline solid breaks. Sodium chloride cannot be hammered into a thin sheet; it shatters instead.

Ionic Bonds 97

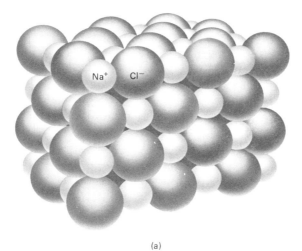

Figure 5-3
Structure of sodium chloride crystal. *(a)* Model showing the relative sizes of the ions. *(b)* Ball-and-stick model showing cubic geometry.

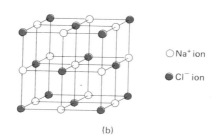

SELF-TEST 5-A

1. Charged atoms are called _____.
2. The attraction between positive and negative ions produces a(n) _____ bond.
3. A sodium atom loses _____ electron(s) in achieving a noble gas configuration.
4. What is the correct formula for calcium iodide (Ca and I)? _____
5. Which ion gained an electron in its formation: Na^+ or Cl^-? _____
6. Electrons in the outer orbit may be called _____ electrons.
7. Positive ions are formed from neutral atoms by ( ) losing or ( ) gaining electrons.
8. Negative ions are formed from neutral atoms by ( ) losing or ( ) gaining electrons.
9. Predict the number of electrons lost or gained by the following atoms in forming ions. Indicate whether the electrons are gained or lost.

    Rb _____   S _____
    Ca _____   Mg _____
    K _____   Br _____

The Covalent Bond

The covalent bond is formed by the sharing of electrons, rather than by the transferring of electrons. The strength of the bond comes from the attraction of the same electrons by both nuclei.

Millions of known compounds are composed of nonmetals only. These compounds contain only elements with 4, 5, 6, or 7 valence electrons per atom. Nonmetal atoms have too many valence electrons to lose in order to achieve a noble gas electron structure by electron loss. Nonmetals bond by pairing electrons and in the process often achieve a noble gas electron structure. An electron pair shared between two nuclei constitutes a **covalent bond.** Examples of substances with covalent bonds include carbon dioxide (CO_2), table sugar ($C_{12}H_{22}O_{11}$), DNA (deoxyribonucleic acid, our heredity material), and diamond (a form of the element carbon).

The simplest example of electron sharing is in diatomic molecules of gases such as H_2, F_2, and Cl_2. A hydrogen atom has only one electron. An electron pair can be achieved if two hydrogen atoms share their single electrons. If dots and x's are used to represent valence electrons and the element symbol is used to represent the rest of the atom, the sharing of electrons can be represented by placing the two electrons between the symbols:

$$H \text{\scriptsize{x}} H$$

Since both atoms have an equal attraction for the pair of electrons, the two nuclei are held together and a chemical bond is formed.

Sharing electrons often, but not always, leads to species in which each element has a noble gas electronic structure. Except for H, this means that four pairs of electrons will be in the valence shell. If we limit the electron dot representations to include only the valence electrons, and place the shared pairs of electrons between the symbols, the F_2 molecule would be shown as having one pair of shared electrons:

$$\overset{xx}{\underset{xx}{\text{x}F\text{x}}} + \cdot \overset{..}{\underset{..}{F:}} \longrightarrow \overset{xx}{\underset{xx}{\text{x}F\text{x}}}\overset{..}{\underset{..}{F:}}$$

Fluorine Molecule

A double covalent bond is formed by the sharing of two pairs of electrons by two nuclei. A molecule of carbon dioxide has two double bonds:

$$\cdot \overset{\cdot}{\underset{\cdot}{C}} \cdot + 2 \, \overset{xx}{\underset{x}{\times O \text{x}}} \longrightarrow \overset{xx}{\underset{xx}{O \text{x}}} : C : \overset{xx}{\underset{xx}{\text{x} O}}$$

Similarly, the nitrogen molecule N_2 has three pairs of shared electrons:

$$\overset{xx}{\underset{x}{\times N \text{x}}} + \cdot \overset{..}{N} \cdot \longrightarrow \text{x}N\overset{x}{\underset{x}{:}}N:$$

Nitrogen Molecule

The pairs of electrons that are not included between the symbols are called **nonbonding electrons.** The electron dot structures of several other molecules are shown in Figure 5-4.

If a line is used to represent a pair of bonding (shared) electrons, F_2 and N_2 are represented as:

$$F-F \quad N\equiv N$$

The F_2 molecule is bonded by a single bond and the N_2 molecule by a triple bond.

The rule that only eight electrons are in the valence shell of a bonded atom (often called the **octet rule**) is not a hard and fast one; indeed, many compounds do not have

---

Electrons are all the same. Dots and x's are used to distinguish the sources of the electrons.

A line between two atoms, as in H—H, represents a bonding pair of electrons.

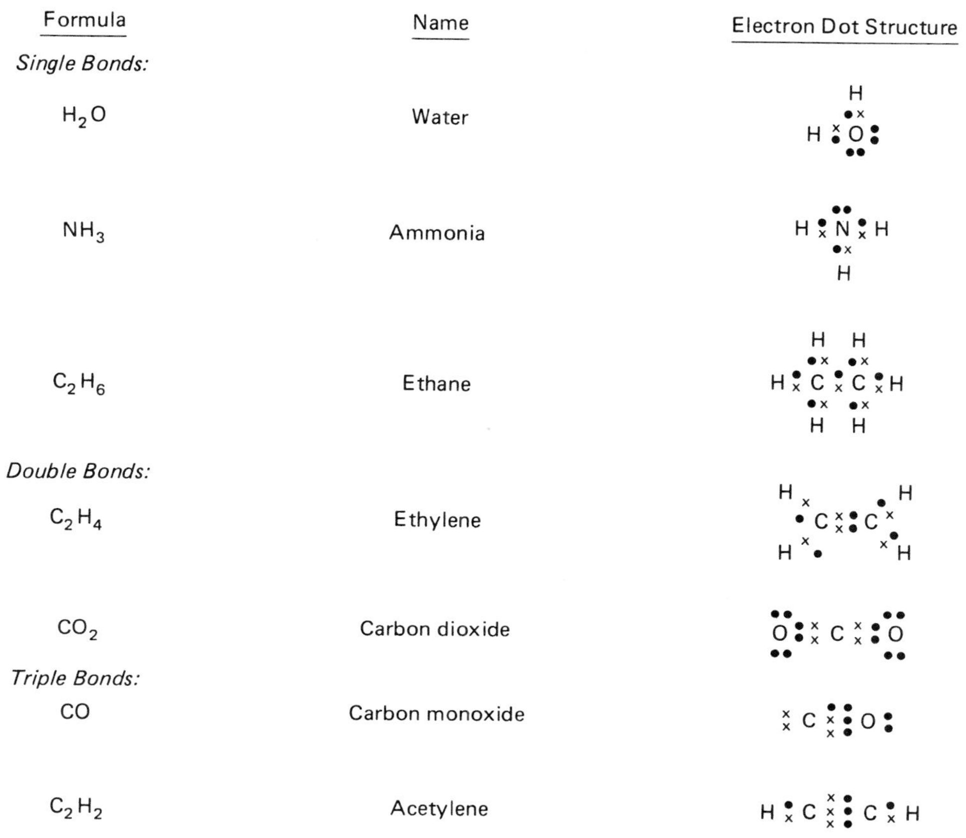

Formula	Name	Electron Dot Structure

Single Bonds:

H_2O	Water	H:Ö: with H attached
NH_3	Ammonia	H:N:H with H below
C_2H_6	Ethane	H:C:C:H with H's above and below

Double Bonds:

C_2H_4	Ethylene	H₂C::CH₂
CO_2	Carbon dioxide	:Ö::C::Ö:

Triple Bonds:

CO	Carbon monoxide	:C:::O:
C_2H_2	Acetylene	H:C:::C:H

Figure 5–4
Electron dot structures of some molecules containing single, double, and triple bonds.

an octet of electrons around one of the atoms in the molecule. For example, boron trichloride, BCl_3, is a stable compound, although the metalloid boron has only three pairs of electrons in its valence shell:

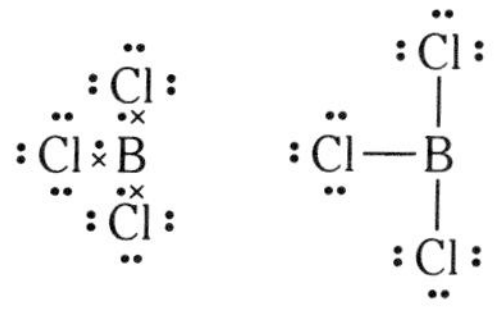

$$\begin{array}{c} :\ddot{C}l: \\ :\ddot{C}l:B \\ :\ddot{C}l: \end{array} \quad \begin{array}{c} :\ddot{C}l: \\ | \\ :\ddot{C}l-B \\ | \\ :\ddot{C}l: \end{array}$$

The fundamental premise of chemical bonding is the tendency for electrons to pair and not the invariable grouping of eight electrons. In fact, Gilbert Newton Lewis, who is generally credited with developing the octet rule, realized the limitations of this theory. He wrote in 1923, "The electron pair, especially when it is held conjointly by two atoms, and thus constitutes the chemical bond, is the essential element in chemical structure."

Have you ever used trisodium phosphate (TSP, Na_3PO_4) or blue vitriol [copper (II) sulfate, $CuSO_4$]? These are common substances sold in hardware stores and elsewhere for cleaning floors (Na_3PO_4) and killing algae in ponds ($CuSO_4$). Both substances have the properties of ionic compounds. When Na_3PO_4 is dissolved in water, sodium ions (Na^+) and phosphate ions (PO_4^{3-}) are formed. Copper (II) sulfate forms copper ions

(Cu^{2+}) and sulfate ions (SO_4^{2-}) in water. Since the PO_4^{3-} and SO_4^{2-} ions are composed of nonmetal atoms only, the P—O and S—O bonds are covalent bonds. If the P and S atoms are surrounded by the oxygen atoms in electron dot structures, the experimentally correct arrangement of atoms is represented. Circles are used to represent the electrons transferred from the metal (Na, Cu) atoms to the PO_4^{3-} (addition of three electrons) and SO_4^{2-} (addition of two electrons). The bonds marked "coordinate covalent" are formed by one atom supplying both electrons for the shared bond.

The bonds between Na^+ and PO_4^{3-} and between Cu^{2+} and SO_4^{2-} are ionic bonds.

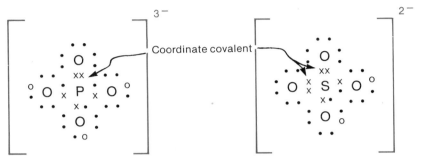

The phosphate ion and the sulfate ion are examples of **polyatomic** (many-atom) **ions,** which are held intact by covalent bonds. Polyatomic ions are stable in solution and in their melted and crystalline compounds. A few common examples are listed in Table 5-2.

Polar Bonds

In a molecule like H_2 or F_2, where both atoms are alike, there is equal sharing of the electron pair in the bond, making the bond a **nonpolar** covalent bond. Where two different atoms are bonded covalently, there is unequal sharing of the electron pair, since the shared pair of electrons is shifted toward the more electronegative (electron-attracting) element. In any one period of the periodic table, the element with the smallest atom needing the fewest electrons to pair its electrons and complete the bonding shell is the most electronegative element. Electronegativity generally increases as one moves up and to the right in the periodic table (Fig. 5-5). Unequal sharing of electron pairs leads to **polar** bonds.

In a covalent bond between two different atoms, the more electronegative element pulls the electron blanket toward itself. The negative portion of the bond is in the region of the more electronegative atom, and the positive portion is in the region of the less electronegative atom. For example, in the molecule HF, the bonding pair of electrons is

The electronegativity of an atom is a measure of its ability to attract electrons to itself in a compound. The most electronegative atom is fluorine.

In a *polar bond,* there is unequal sharing of the bonding electrons.

Table 5-2
A Few Polyatomic Ions

Ammonium	NH_4^+	Hypochlorite	ClO^-	Chromate	CrO_4^{2-}
Acetate	$CH_3CO_2^-$	Chlorate	ClO_3^-	Silicate	SiO_3^{2-}
Nitrate	NO_3^-	Perchlorate	ClO_4^-	Phosphate	PO_4^{3-}
Nitrite	NO_2^-	Carbonate	CO_3^{2-}	Arsenate	AsO_4^{3-}
Hydroxide	OH^-	Sulfate	SO_4^{2-}		

Polar Bonds   101

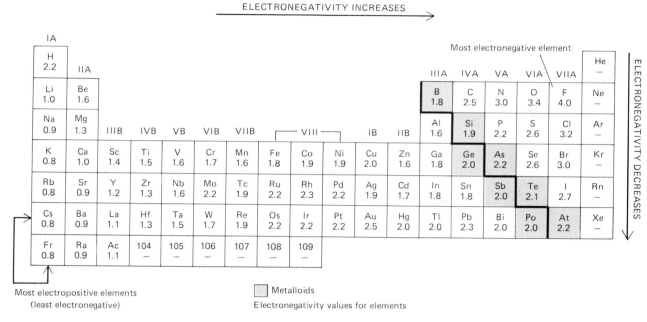

Figure 5–5
Some electronegativity values in a periodic table arrangement.

controlled more by the fluorine atom than by the hydrogen atom (Figs. 5–6 and 5–7). Part of the negative "blanket" of charge has been pulled off the hydrogen atom, partially uncovering the positive hydrogen nucleus.

Polar bonds fall between the extremes of pure covalent and ionic bonds as far as separation of charge is concerned. In a pure covalent bond, there is no charge separation; in ionic bonds there is complete separation; and in polar bonds the separation falls somewhere in between.

Figure 5–6
Fluorine pulls the negative electron blanket toward itself and exposes the nucleus of hydrogen. The shifting of electrons toward one atom forms a polar bond.

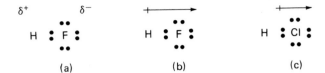

Figure 5–7
Polar bonds in HF and HCl molecules. *(a)* (Delta plus-fractional plus charge) and (delta negative-fractional negative charge) are used to indicate poles of charge. In *(b)* and *(c)*, an arrow is used to indicate electron shift, the arrow has a plus tail to indicate partial positive charge on the hydrogen atom. Note the longer arrow for the HF structure, which indicates a greater degree of polarity than in HCl.

Hydrogen Bonds

When hydrogen is attached to a highly electronegative atom like fluorine, oxygen, or nitrogen, the conditions are right for a very important type of intermolecular, positive-to-negative attraction called "hydrogen bonding." The bond is produced by the attraction arising between a slightly positive hydrogen atom on one molecule and a very electronegative atom (N, O, or F) on another molecule (or at another location on the same molecule if the molecule is big enough to bend back on itself). The shifting of an electron pair toward very electronegative nitrogen, oxygen, or fluorine causes these atoms to take on a partial negative charge. The hydrogen bond, then, is a "bridge" between two highly electronegative atoms with a hydrogen atom bonded covalently to one of the electronegative atoms and electrostatically (positive-to-negative attraction) to the other electronegative atom. Hydrogen bonds have a strength of about one tenth to one fifteenth that of an average single covalent bond.

Hydrogen bonds are found in many substances. They are responsible for such phenomena as hard candy getting sticky, cotton fabrics taking longer to dry than nylons, lanolin softening skin, the ultimate shape of proteins and enzymes, and a host of apparent anomalies in the nature of water. Each phenomenon is dealt with in this text, but here we examine some of the strange things about water that result from H-bonding.

One of the peculiar properties of water is that its solid form, ice, floats (Fig. 5–8). In contrast, most solids sink in their own liquid form. It is fortunate that ice floats. If ice sank to the bottom of ponds and rivers, water would continue to freeze in subfreezing weather until the pond or river became solid. Besides having a devastating effect on aquatic life, this would cause the water cycle to be diminished, and the lack of water would destroy plant life.

Ice floats because water expands when it freezes. The expansion is caused by hydrogen bonding between angular (bent) water molecules. The angular structure of the water molecule comes from the tetrahedral arrangement of four pairs of electrons in the valence shell of an atom (Fig. 5–9). Like four like-charged table tennis balls tied to strings, four pairs of electrons tied to a nucleus position as far away from each other as physically possible. The bond angle for the maximum separation of electron pairs is 109.5°, the tetrahedral angle. If an atom is bonded at each electron pair, as in CH_4, the molecular structure is **tetrahedral** (a four-sided shape, each side formed by a plane with three electron pairs [or three H's in the case of CH_4]). Diamond is a very hard

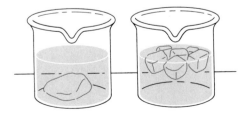

Figure 5–8
Solid paraffin (on the left) sinks in its liquid while solid water (ice, on the right) floats in its liquid. Why the difference?

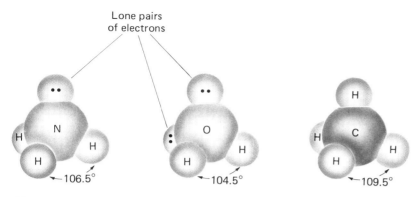

Figure 5-9
Those little molecules in water, natural gas (methane), and ammonia have something in common. Methane has only bonding pairs of valence electrons: water and ammonia have bonding and nonbonding pairs of valence electrons. The nonbonding pairs distort the tetrahedral structure slightly.

substance, due in part to its network of interconnected tetrahedral carbon atoms (Fig. 5-10).

If only three atoms bond to the four pairs of valence electrons (as in NH_3), the molecular structure is **triangular pyramidal** (a three-sided pyramid with a base). Electrons not involved in the bonding (**nonbonding,** or **lone pairs,** of electrons) are not detected by structure-determining probes (such as X rays) and are not included as part of the molecular structure.

The oxygen atom in a water molecule has four pairs of valence electrons. The

H⟋Ö⟍H structure is **angular,** with two pair of nonbonding electrons and two pair bonding with two hydrogen atoms. The **angular** structure of the water molecule and hydrogen bonding between water molecules means that each water molecule can be hydrogen-bonded to a maximum of four other water molecules. Imagine that liquid water is made up of hydrogen-bonded clusters of water molecules that are in constant motion. The number of water molecules per cluster and the speed of cluster motion are temperature dependent.

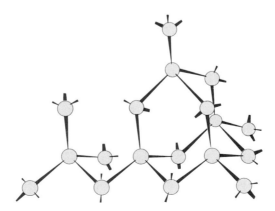

Figure 5-10
The tetrahedral arrangement of the carbon atoms in a diamond crystal. Each atom has four nearest neighbors, which are arranged about it at the corners of a regular tetrahedron.

The molecular action used to form the ice structure is an "everything-in-its-place" action. As water is cooled, the rapidly moving clusters of water molecules slow down and at the freezing point, the clusters hook together to give the three-dimensional, expanded structure of ice shown in Fig. 5–11. The more open structure makes ice less dense than water. The melting of ice breaks about 15% of the hydrogen bonds, and this collapses the structure shown in Fig. 5–11 to a more dense liquid.

Another strange property of water is its relatively high boiling point. Almost all the hydrogen compounds of oxygen's neighbors and family members are gases at room temperature: CH_4, NH_3, H_2S, H_2Se, H_2Te, PH_3, HCl. But H_2O is a liquid. To transform a molecule into the vapor state, the molecule must absorb energy to free itself from other molecules. Since liquid water is made up of hydrogen-bonded clusters of water molecules, energy must be supplied to break the hydrogen bonds. (A halfback with three or four tacklers hanging onto him has a similar problem freeing himself.) Not all the hydrogen bonds are broken, however, and aggregates of water molecules exist in liquid water even near 100°C. As water is heated, thermal agitation disrupts the hydrogen bonding until, in water vapor, there is only a small fraction of the number of hydrogen bonds that are found in liquid or solid water. If strong intermolecular bonding, such as hydrogen bonding, is absent, substances generally boil according to their molecular masses. Larger molecular masses require a higher temperature to boil, mostly because the larger electron clouds are more easily distorted and lead to stronger intermolecular London forces.

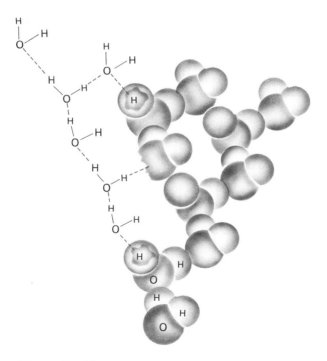

Figure 5–11
Hydrogen bonding in the structure of ice. The hydrogen bonds are indicated by the dashed lines. Hydrogen bonding is not as extensive in liquid water as it is in ice.

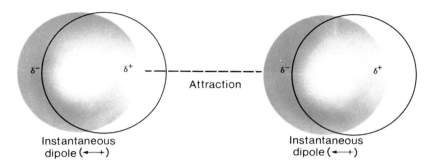

Figure 5–12
An illustration of London forces. One instantaneous dipole interacts with another in a neighboring atom.

Hydrogen bonding plays a key role in the chemistry of life. In later chapters we shall discuss hydrogen bonding in connection with the properties of a number of substances, such as carbohydrates (cotton), proteins, and DNA.

London Forces

Forces between molecules that cannot be explained by polar attractions or hydrogen bonds are often explained by a weak attraction known as **London forces** (named after Fritz London, who proposed the temporary dipole concept in 1937). The two electrons in helium, for example, are nonbonding, and yet there is a slight attraction between two helium atoms. This slight attraction allows helium to become a solid at $-272\,°C$ under high pressure.

London forces can arise from a variety of causes. For instance, as two atoms or molecules approach each other, intermolecular interactions cause a temporary shifting of the electron clouds. An uneven electron distribution in an atom makes the atom itself temporarily polar, producing a **dipole** (meaning "two poles"). The temporary poles on two adjacent atoms can interact with each other, resulting in a momentary attractive force (Fig. 5–12). The existence of the solid state of many nonpolar molecular substances (such as oxygen, nitrogen, and helium) can be explained by London forces.

Metallic Bonding

Metals have some properties totally unlike those of other substances. For example, most metals are good electrical conductors; they are shiny solids and have relatively high melting points (with a few notable exceptions, such as mercury). Any theory of the bonding of metal atoms must be consistent with these properties. Structural investigations of metals have led to the conclusion that metals are composed of regular arrays or lattices of metal ions in which the bonding electrons are loosely held. The loosely held electrons can be made to move rather easily through the lattice upon application of an electric field. In this way the metal acts as a conductor of electricity. As a consequence of this movement of bonding electrons, we cannot use an electron dot structure to describe satisfactorily the bonding in a metal.

Loosely held electrons are found in metals. They can move freely and are not confined to the area between any particular pair of atoms.

SELF-TEST 5-B

1. a. An example of a molecule containing covalent bonding where the electrons are equally shared between the atoms is _____.
   b. One where they are unequally shared is _____.
2. The number of electrons shared in a triple covalent bond is _____.
3. There are _____ covalent bonds in an ammonia (NH_3) molecule.
4. Which atom cannot form a double bond, fluorine or sulfur? _____
5. a. How many valence (bonding) electrons are thought to be involved in molecules containing covalently bound atoms of period 2 and 3 elements? _____
   b. This is known as the _____ rule.
   c. Is this rule true most of the time or all of the time? _____
6. Which is the most electronegative of all of the elements? _____
7. Hydrogen bonding occurs when hydrogen is bound to atoms of _____, _____, or _____.
8. Which molecule is a polar molecule: H_2, Cl_2, or HCl? _____
9. Which substance has ionic bonds: PCl_5, CCl_4, $MgCl_2$, or Cl_2O? _____
10. Which substance has intermolecular hydrogen bonds: CH_4, H_2O, or H_2? _____
11. What is the molecular structure of the water molecule? _____
12. The water molecule has _____ bonding pairs and _____ nonbonding pairs of electrons.
13. The ammonia molecule (NH_3) has _____ bonding pairs and _____ nonbonding pairs of electrons.

MATCHING SET

_____ 1. Double covalent bond
_____ 2. Ionic bonds
_____ 3. Single covalent bond
_____ 4. Metallic bonding
_____ 5. London forces
_____ 6. Noble gas
_____ 7. Covalent bonds
_____ 8. NaCl
_____ 9. Metal ion
_____ 10. Hydrogen bonds
_____ 11. NH_3

a. An electrically neutral arrangement of covalently bonded atoms
b. Shared electrons
c. Require O, F, or N
d. Positive ions attracted to negative ions
e. Ionic compound
f. Electrons free to move
g. Element with two or eight valence shell electrons
h. Covalent compound
i. Attraction between nonpolar neutral particles
j. K^+
k. Four electrons shared
l. Two electrons shared

QUESTIONS

1. Diamond (a form of carbon) has a melting point of 3500°C, whereas carbon monoxide (CO) has a melting point of −207°C. What does this suggest about the kinds of bonding found in these two substances?
2. Is Ca^{3+} a possible ion under normal chemical conditions? Why or why not?
3. Match the electronic configurations that would be expected to lead to similar chemical behavior. The numbers denote the numbers of electrons in the orbits.
   a. 2-2
   b. 2-5
   c. 2-8-8-1
   d. 2-8-2
   e. 2-1
   f. 2-8-5
4. Fluorine (atomic number 9) has an electronic configuration of 2-7. How many electrons will be involved in the formation of a single covalent bond?
5. What kind of bond (ionic, pure covalent, or polar covalent) is likely to be formed by the following pairs of atoms?
   a. A Group IA element with a Group VIIA element
   b. A Group VIA element with a Group VIIA element
   c. Two chlorine atoms
   d. An element with low electronegativity and an element with high electronegativity
   e. Two elements with about the same electronegativity
   f. Two elements with the same electronegativity
6. How many electrons would there be in an iodide ion, I^-?
7. Write the electron dot structures for the fluoride ion, F^-, the chloride ion, Cl^-, and the bromide ion, Br^-.
8. Draw the electron dot structure for water. Based on bonding theory, why is water's formula not H_3O?
9. Draw electron dot structures for the following molecules or ions:
   a. NF_3
   b. CCl_4
   c. C_2Cl_2
   d. OF_2
   e. H_2S
   f. CO
   g. N_2H_4
   h. CH_4
   i. Br_2
   j. HCl
   k. BCl_3
   l. PH_3
   m. SiH_4
   n. IBr
   o. ClO_3^-
   p. SO_3^{2-}
   q. NH_4^+
   r. OH^-
   s. AsO_4^{3-}
10. The members of the nitrogen family, N, P, As, and Sb, form compounds with hydrogen: NH_3, PH_3, AsH_3, and SbH_3. The boiling points of these compounds are

SbH_3	−17°C	PH_3	−87.4°C
AsH_3	−55°C	NH_3	−33.4°C

    Comment on why NH_3 doesn't follow the downward trend of boiling points.
11. Match the substances listed below with the type of bonding responsible for holding units in the solid together.

Solid krypton (Kr)	Ionic
Ice	Covalent
Diamond	Metallic
CaF_2	Hydrogen
Iron	London forces

12. Predict the general kind of chemical behavior (loss, gain, or sharing of electrons) you would expect from atoms with the following electron arrangements:
    a. 2-8-1
    b. 2-7
    c. 2-4
13. Boron trichloride has the electron dot formula

    :Cl:
    :Cl:B:Cl:. What does this tell you about the octet rule even for Period 2 elements?
14. How is an ionic bond formed? How is a covalent bond formed?
15. A compound will not conduct electricity when melted, and it melts at 46°C, a low melting point. What type of bond holds atom to atom in this compound?
16. What ions would probably be formed by Br, Al, Ba, Na, Ca, Ga, I, S, O, Mg, K, At, Fr, all Group IA metals, and all Group VIIA nonmetals?
17. How are ionic solids held together?
18. A compound will conduct electricity when melted, but it is rather hard to melt. What types of bonds are in this compound?
19. A substance is composed of carbon only, or of two nonmetals. The substance has a high melting point and is very hard. What kinds of bonds hold the atoms of the substance together?
20. In which case would hydrogen bonding be most extensive: liquid water, water vapor, or ice?
21. Predict the formulas of compounds consisting of the following elements.

Nonmetal	Metal	Ba	Al	K
Cl				
O				
S				
N				
I				

22. Which is harder to break, an ordinary covalent bond or a hydrogen bond?
23. What is the direction of energy transfer in a bond-making process? In a bond-breaking process?
24. Liquid water consists of water molecules held together by covalent O—H bonds, and the water molecules are held together loosely by hydrogen bonds. When water boils, which type of bond breaks first?
25. Select the polar molecules from the following list, and explain why they are polar: N_2, HCl, CO, NO.
26. One use of calcium chloride is to keep down dust on a road. It is an ionic compound with the formula $CaCl_2$. What does this formula mean?
27. Octane is a component of gasoline that, if not burned, is emitted into the atmosphere. It is a covalent compound and has the formula C_8H_{18}. Tell what this formula means.
28. If covalent bonding is to white as ionic bonding is to black, how would polar covalent bonding be represented? On this scale, how could a more polar molecule be distinguished from a less polar one?
29. Aluminum does not "rust" and crumble like iron because it forms a thin coating of aluminum oxide on its surface, which prevents further combination with oxygen. Use atomic theory to predict the formula for the ionic compound formed between aluminum and oxygen.
30. Explain why atoms in Groups I and II of the periodic table readily form positive ions, whereas atoms in Groups IV and V do not.
31. A hypothetical element has the symbol J and six valence electrons:

    $:\ddot{J}\cdot$

    a. Write the formula for the expected compound when this element combines with hydrogen.
    b. If element J were to form an ion, what would be the charge on this ion?
    c. Element J would be in what group of the periodic table?
    d. Write the electron dot formula for the compound formed between J and chlorine.

CHAPTER 6

Reactivity Principles and Representative Reactions

Matter in the universe is going from a state of high orderliness to one of complete disorder, according to the big bang theory. Chemical reactions are driven by this unwinding of the material clock. Spontaneous reactions, such as the burning of fuels, occur at the expense of natural order, while nonspontaneous reactions, such as the charging of an automobile battery, increase order but occur only when driven by spontaneous processes, the result being more disorder on the driving end than order on the driven one. Our sun is winding down and, in the process, driving many physical and chemical processes on the surface of the Earth. For example, photosynthesis, a driven process, produces fuels which, when coupled with the oxygen in the atmosphere, provide the chemical energy for many of the natural chemical events that are so important to our lives.

The unwinding of the universe is phenomenological, and after observing the phenomena, one may wonder why. We turn to the world of atoms and molecules to understand these unfolding events. Chemical reactions occur because of interactions among atoms, ions, and molecules. These interactions result from the electrical forces within and between these fundamental particles. In the Earth environment, essentially all the atoms, except for the noble gases, are already tied into bonded groups of atoms. The grouping may be elemental, as with H_2, N_2, and O_2, but it is more likely to be in compound form, as in H_2O, CO_2, and $C_{12}H_{22}O_{11}$ (sucrose).

Isolated atoms can simply combine to form interatomic bonds, but it is more common for existing bonds in relatively unstable arrangements of atoms to break, yielding new bonding arrangements with higher stabilities under the reaction conditions. For example, both hydrogen (H_2) and oxygen (O_2) are stable gases under room conditions. They can be mixed without any reaction occurring, unless energy is supplied. But if a mixture of hydrogen and oxygen is sparked, an explosion occurs and water is formed:

$$2\,H_2 + O_2 \longrightarrow 2\,H_2O + \text{energy}$$

Although much energy is released, energy is required to break existing bonds before new ones can be formed (see Chapter 5).

In this chapter we look at some of the factors that control the regrouping of atoms (chemical reactions) and illustrate with two major classifications of reactions: acid-base and oxidation-reduction.

Quantitative Changes in Weight and Energy During Chemical Change

Atomic theory provided the explanation for the conservation of mass: atoms are not lost or created in chemical change (see Fig. 3-1). The same number of atoms, even in rearranged form, has the same weight. A benefit of the law of conservation of matter is that it gives us the ability to predict exactly how much of one chemical will react with another and how much product will be formed in a chemical reaction. The balanced chemical equation is a quantitative statement about a chemical reaction. For example, in the balanced chemical equation

$$2H_2 + O_2 \longrightarrow 2 H_2O$$

$$\text{2 moles} \quad \text{1 mole} \quad \text{2 moles}$$
$$\text{2 (2 g)} \quad \text{1 (32 g)} \quad \text{2 (18 g)}$$

Review the discussion of chemical equations given in Chapter 2.

A mole of a substance is its molecular weight in grams.

multiplying the coefficient times the molecular weight in grams (1 mole) for hydrogen, oxygen, and water shows that 4 g of hydrogen and 32 g of oxygen will produce 36 g of water. Fractions or multiples of moles can be used for predictions as long as the balanced chemical equation is followed. For example, 0.5 moles of hydrogen (1.0 g) will combine with 0.25 moles of oxygen (8.0 g) to give 0.5 moles of water (9.0 g)

The heat evolved in a particular chemical reaction is exactly the same as the heat required for the reverse reaction. For example, if 1 mole of water, 18 g, is produced from hydrogen and oxygen under specified conditions, 68.300 kcal of heat are always produced at the same time. Furthermore, the same amount of energy is required to return the mole of water to the hydrogen and oxygen from which the water was made. The first law of thermochemistry, as stated by Antoine Laurent Lavoisier, is: *The heat evolved in a given chemical reaction is equal to the heat absorbed in the reverse reaction.* It never varies! Heat changes in chemical reactions are just as quantitative as weight relationships.

Lavoisier is often acclaimed as the "Father of Chemistry."

In 1840 G. H. Hess proposed a second law of thermochemistry: *The heat involved in a chemical reaction under specified conditions is independent of the intermediate reactions involved.* Carbon can be burned to form carbon dioxide, and 94 kcal of heat are emitted per mole of carbon dioxide formed:

$$C + O_2 \longrightarrow CO_2 + 94 \text{ kcal}$$

Carbon monoxide can be burned to carbon dioxide with the liberation of 68 kcal of heat per mole of carbon dioxide formed:

$$2 CO + O_2 \longrightarrow 2 CO_2 + 136 \text{ kcal (or 68 kcal per mole of } CO_2)$$

How many kilocalories would be released in the formation of 1 mole of carbon monoxide from carbon and oxygen? It is the difference between 94 kcal and 68 kcal, or 26 kcal per mole of CO. Through the observation of many chemical reactions, it becomes evident that a definite energy is associated with each chemical reaction, and quantities of energy are quite predictable. Energy relationships between reactants and products are important in understanding energy from fuels (Chapter 8) and caloric needs for body metabolism (Chapter 11).

Reaction Rates

How fast do you digest the food you eat? How fast can iron be made from iron ore? How fast does iron rust? How fast does gasoline burn? The whole notion of how fast or how

Figure 6 – 1
The biochemical processes of decomposition occur more rapidly at higher than at lower temperatures. Half of the peach shown in the photograph was refrigerated, while the other half was kept warm. The refrigerated half on the right shows little discoloration, whereas the other one shows typical signs of decay.

slow chemical reactions proceed can be expressed quantitatively by **reaction rates:** *The rate of a reaction is the change in the amount of chemical substances appearing (products) or disappearing (reactants) per unit of time.* For example, if we consider the burning of sulfur to produce sulfur dioxide,

$$S + O_2 \longrightarrow SO_2$$

we can measure the rate of the reaction by the amount of SO_2 formed per minute or by the amount of S or O_2 consumed per minute.

Three major factors affect the rate of a particular chemical reaction: temperature, concentration, and catalysis. Increasing the *temperature* will cause a reaction to proceed faster, and lowering the temperature will decrease the rate of the reaction. This principle is used in cooking foods (a roast cooks at a faster rate at a higher temperature) and in preserving foods (foods spoil less quickly if refrigerated). Figure 6 – 1 illustrates the effects of temperature on the reactions that take place in a slice of fruit exposed to air. From theory we reason that fast-moving molecules at high temperatures come together often and with great collision energies to cause atomic rearrangements.

Increasing the *concentration* of a reactant will cause a reaction to proceed faster, whereas increasing the concentration of a product will slow the reaction speed. In the reaction of sulfur with oxygen, if air replaces oxygen, the reaction will proceed at a slower rate, since air is a mixture of about one part oxygen and four parts nitrogen. Figure 6 – 2 contrasts the oxidation of steel wool in air and in pure oxygen. A theoretical explanation of the concentration effect on reaction rates is presented in Figure 6 – 3. More reactant particles result in more collisions; hence, more particles react. The concentration effect is related to particle size and the mixing of small particles. A sack of flour, for example, will not burn in an ordinary fire, but the same flour distributed as dust in the air of a building has the potential to blow the building apart.

A *catalyst* causes a reaction to proceed faster without being permanently consumed by the reaction. For example, in the manufacture of sulfuric acid, it is necessary to convert sulfur dioxide to sulfur trioxide:

$$2\ SO_2 + O_2 \longrightarrow 2\ SO_3$$

If the pure reactants are mixed, the reaction is far too slow to be of commercial use. However, if some oxides of nitrogen are introduced, the reaction proceeds at a rapid rate to produce the desired sulfur trioxide, and the nitrogen oxides are still there in the end to be used over and over again. The catalyst offers an easier reaction pathway, or molecular rearrangement sequence, to produce the new atomic combinations in the

Raising the temperature speeds up chemical reactions. For many reactions, a temperature rise of 10°C doubles the reaction rate.

Increasing the concentration of reactants speeds up a reaction.

Figure 6-2
Effect of concentration on reaction rate. Steel wool held in the flame of a gas burner is oxidized rapidly. It is in contact with air, which is 20% oxygen. When the red hot metal is placed in pure oxygen in the flask, it oxidizes much more rapidly.

Enzymes are biochemical catalysts (see Chapter 11).

product. Amazingly, biological systems rapidly produce catalysts on demand and then destroy them after a particular chemical need is met.

Reversibility of Chemical Reactions

Chemical reactions are capable of going forward or backward. Arranged atoms can be rearranged.

Most chemical reactions are capable of being reversed under suitable conditions. An example is the heating of calcium hydroxide to produce lime (CaO):

$$Ca(OH)_2 \xrightarrow{heat} CaO + H_2O$$

The CaO can easily be returned to $Ca(OH)_2$ with the addition of water.

Many reversible reactions are important to human life. One of them is involved in the transport of atmospheric oxygen from the lungs to various parts of the body. This task is carried out by hemoglobin, a complex compound found in the blood. While in the lungs, hemoglobin takes up oxygen to form oxyhemoglobin:

$$\text{Hemoglobin} + O_2 \rightleftharpoons \text{Oxyhemoglobin}$$

The double arrows indicate a reversible reaction.

The oxyhemoglobin is then carried by the bloodstream to various parts of the body, where it releases oxygen for use in metabolic processes (Fig. 6-4). If carbon monoxide, a deadly poison, is present, it will react with hemoglobin. This reaction between carbon monoxide and hemoglobin is not easily reversed and thus will block normal oxygen transport.

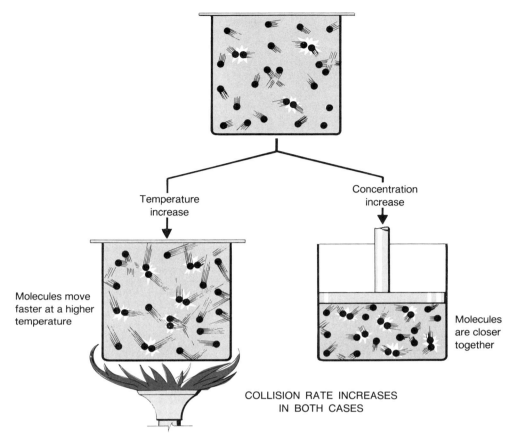

Figure 6-3
Effects of temperature and concentration on rates of chemical reactions. At higher temperatures, more collisions occur between molecules, and a greater percentage of the collisions produce chemical reactions. (The molecules move faster at higher temperatures and, if kept in the same volume, will strike more often. Moving faster, they will also have more energy to cause structural changes to occur.) At higher concentrations (no temperature change), more collisions occur, but the percentage of effective collisions remains the same.

Chemical Equilibrium

Chemicals do not always react to form products with the complete extinction of the reactants. We may get the idea that all chemical reactions go to completion when we watch a piece of wood "burn up." However, in many reactions both reactants and products are present at constant, but not necessarily the same, concentration levels. When reversible reactions reach the point at which the forward reaction is proceeding at the same rate as the reverse reaction, the amount of chemicals present remains constant because a particular chemical is produced as fast as it is consumed. At that point we have **chemical equilibrium.**

A chemical change is at chemical equilibrium when products are produced at the same rate at which the products are consumed in reproducing

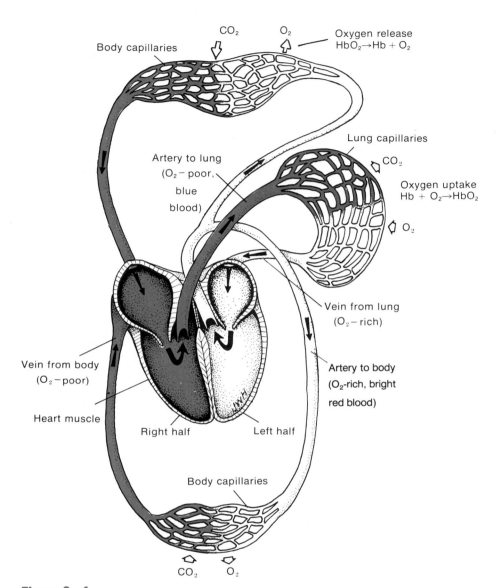

Figure 6-4
Simplified diagram of human circulation. The heart (shown in front view) is divided into two parallel halves. The right half pumps oxygen-poor blood to the lungs; the left half pumps oxygen-rich blood to the body. Hb = hemoglobin; HbO₂ = oxyhemoglobin.

reactants. Consider the equilibrium among limestone ($CaCO_3$), lime (CaO), and carbon dioxide in a closed container. If dry limestone is placed in a vacuum, carbon dioxide gas will soon appear in the container with the limestone. The reaction is

$$CaCO_3 \rightleftharpoons CaO + CO_2$$

After the carbon dioxide builds up to a constant concentration level, the system is at equilibrium and the carbon dioxide level does not rise further. We know that the reaction is proceeding in both directions when the system is at equilibrium because we can

The theory and uses of radioactive elements as tracers are discussed in Chapter 4.

introduce radioactive carbon in either the reactant or the product and, using radiation-detecting devices, trace it through the reaction in either direction.

SELF-TEST 6-A

1. As natural events occur spontaneously, _____ is increased in the universe.
2. Isolated atoms, when colliding with other atoms, are likely to form chemical _____ in the formation of molecules and ions.
3. If 1 kg of water is decomposed by electricity, what would be the combined weight of the hydrogen and oxygen produced? _____
4. Energy is released in the formation of bonds between atoms. If 85 kcal are released as a given set of isolated atoms form bonded species, how many kilocalories would be the minimal requirement to return to the isolated atoms? _____
5. The heat involved in a chemical reaction under specified conditions is _____ of the intermediate reactions involved.
6. Name three factors that affect the rate of a chemical reaction. _____, _____, and _____
7. Give an example of a chemical catalyst. _____
8. a. All chemical reactions can be reversed in theory.   True ( ) or False ( )?
    b. All known chemical reactions have been reversed in the laboratory.   True ( ) or False ( )?
9. What is the same in a chemical reaction at chemical equilibrium?
    a. The weight of the reactants equals the weight of the products.
    b. The number of atoms on the reactant side equals the number of atoms on the product side.
    c. The rate of the forward reaction is equal to the rate of the reverse reaction.
    d. The number of molecules on the reactant side is equal to the number of molecules on the product side.
10. There is physical evidence that the reaction stops at chemical equilibrium.   True ( ) or False ( )?

Solutions

Much of the chemistry around us occurs in **aqueous,** or water, solutions. In general, a **solution** is a homogeneous mixture. The special case of a liquid solution involves a liquid medium in which there is an even dispersion of other particles from dissolved gases, liquids, or solids. Salt or sugar in water, oil paint in turpentine, grease in gasoline, and oxygen in fish tank water are common examples of solutions.

In a solution, the **solvent** is the substance present in the greater amount and the **solute** is the dispersed substance. In a glass of tea, water is the solvent, and sugar, most

Aqueous solutions are water solutions.

A solution is a homogeneous mixture of atoms, small ions, or small molecules.

Solute particles may be ions or molecules.

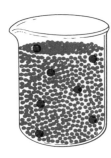

Figure 6-5
The molecular level of a sugar solution in water. The large circles represent the sugar molecules; the small circles, the water. The size of the container and the size of the particles are not to scale.

of the content of the lemon juice, and the tea itself are solutes. According to the molecular concept of a solution of sugar in water, there is a collection of sugar molecules — the solute — evenly dispersed among the water molecules — the solvent (Fig. 6-5).

Ionic and Molecular Solutions

A spoon of table salt (NaCl) in a cup of water produces a solution that is an excellent conductor of electricity (Fig. 6-6). In contrast, a similar solution of table sugar ($C_{12}H_{22}O_{11}$) in water is nonconducting. Solutes, such as salt, that render aqueous solutions electrically conducting are **electrolytes;** dissolved solutes that do not conduct are **nonelectrolytes.** Acids, bases, and salts are three major classes of electrolytes.

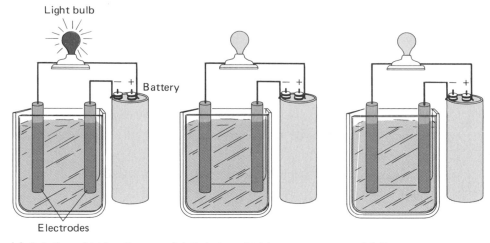

(a) Solution of table salt (an electrolytic solution)

(b) Solution of table sugar (a nonelectrolytic solution)

(c) Pure water (a nonelectrolyte)

Figure 6-6
A simple test for an electrolytic solution. In order for the light bulb to burn (a), electricity must flow from one pole of the battery and return to the battery via the other pole. To complete the circuit, the solution must conduct electricity. A solution of table salt, sodium chloride, results in a glowing light bulb. Hence, sodium chloride is an electrolyte. In (b), the light bulb does not glow. Hence, table sugar is a nonelectrolyte. In (c), it is evident that the solvent, water, does not qualify as an electrolyte, since it does not conduct electricity in this test.

The conductance of an electrolytic solution is explained by the solute particles in such solutions being ions rather than molecules. Recall from Chapter 5 that sodium chloride crystals are composed of sodium ions, which are positively charged, and chloride ions, which are negatively charged. When sodium chloride dissolves in water, **ionic dissociation** occurs (Fig. 6–7a). Ionic dissociation is the separation of ions of a solute when the substance is dissolved. Water molecules become associated with the

The two most abundant ions in ocean water are Cl^- and Na^+ ions, sufficient to recover about 27 g of NaCl per kilogram of sea water. To put it another way, there are about 129 million tons of NaCl per cubic mile of sea water.

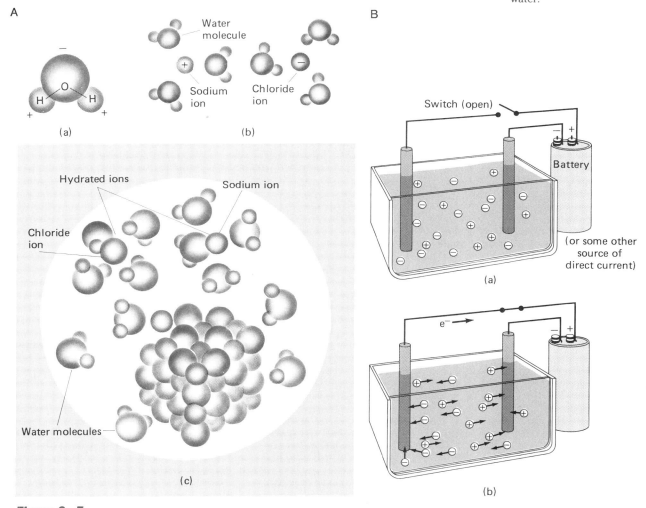

Figure 6–7
(A) Dissolution of sodium chloride in water. *(a)* Geometry of the polar water molecule. *(b)* Solvation of sodium and chloride ions due to interaction (bonding) between these ions and water molecules. *(c)* Dissolution occurs as collisions between water molecules and crystal ions result in the removal of ions from the crystal ions. In the process the ion becomes completely solvated.
(B) The hydrated ions are randomly distributed throughout the salt solution; however, the net charge on the solution is zero. The negative electrode attracts positive ions, while the positive electrode attracts negative ions. If electrons are transferred from the negative electrode to the positive ions and from the negative ions to the positive electrode, the circuit is complete, and electricity flows through it.

ions, forcing the separation. The solution then contains positive sodium ions and negative chloride ions dispersed in water (Fig. 6-7).

$$Na^+Cl^-_{(s)} \xrightarrow{water} Na^+_{(aq)} + Cl^-_{(aq)}$$

Some molecular substances react with water to produce ions and hence become electrolytes in the dissolution process. The dissolution of ammonia gas, NH_3, and the dissolution of acetic acid, $HC_2H_3O_2$, are examples. Both of these substances are molecular, but both produce a significant concentration of ions (about 1% at ordinary concentrations) when dissolved in water.

$$NH_3 + H_2O \rightleftharpoons NH_4^+ + OH^-$$
$$HC_2H_3O_2 + H_2O \rightleftharpoons H_3O^+ + C_2H_3O_2^-$$

The process that produces ions when a molecular substance dissolves in water is termed **ionization.** Electrolytic solutions, then, result from ionic solutions regardless of whether the ions come from the dissociation of an ionic compound or from the ionization of a molecular compound.

Concentrations of Solutions

Solutions may be dilute, concentrated, or anywhere in between, depending on the solubility of the solute in the solvent. Maple syrup is concentrated sugar in water; perfume is likely a dilute solution of several compounds in alcohol. As you would guess, there are a number of different ways to express solution concentrations: weight of solute per weight of solvent, weight of solute per volume of solvent, volume of solute per volume of solvent, and so on.

A solid is indicated by s and a species in aqueous (water) solution by aq.

Ammonia is the number six commercial chemical in quantity produced.

H_3O^+ is the hydronium ion.

Laboratory procedure for the preparation of a solution of known concentration.

1. Take a volumetric flask

2. Add carefully the weighed amount of solid

3. Add some water, shake, and dissolve solid

4. Fill flask to one liter mark and shake until homogeneous solution is obtained

Molar concentration is the number of moles of a substance per liter of solution. *Molarity* = M = moles of solute divided by liters of solution.

A mole of NaCl weighs 58.5 g (23 + 35.5).

A mole of any item is 6.02×10^{23} of that item.

In chemistry, concentrations of solutions are most often expressed in terms of the number of *moles* of solute *per liter* of solution. *Molar* and *molarity* denote this expression of concentration. For example, a one molar solution contains 1 mole of solute in 1 liter of solution. If a solution has a molarity of six, the solution has 6 moles of solute dissolved in 1 liter of solution. Since the mole is a number of particles (6.02×10^{23} units), molar concentrations offer an excellent way to keep up with the number of particles, as well as the weight.

Acids and Bases

Acids and **bases** are classifications of chemicals dating back to antiquity. The word *acid* comes from the Latin *acidus,* meaning "sour" or "tart," since water solutions of acids have a sour or tart taste. Acids in water react with metals such as zinc and magnesium to liberate hydrogen; neutralize bases to produce salt and water; and change the color of litmus, a vegetable dye, from blue to red. Citrus fruits offer a quick experience with natural acids because of their citric acid content, and any fruit with a high sugar content, such as apples, will readily ferment to produce a vinegar containing acetic acid. Bases in water, or alkaline solutions, have a bitter taste, feel slippery or soapy to the touch, change litmus from red to blue, and neutralize acids to form salt and water.

> Taste is not a recommended practice for distinguishing acids and bases.
>
> Litmus is but one acid-base indicator. Another, phenolphthalein, is colorless in acid and pink in base.

The variety of acids and bases in your life is illustrated by the following: Baking powder and baking soda are weak bases vital to cooking. Lye is a strong base often used as drain and toilet bowl cleaners. Lime, not as soluble in water as lye, is used to decrease the acidity of the soil. Antacids contain bases to neutralize excess acidity in the stomach. Acid skin tends to produce pimples. Your car battery depends on battery acid. Radiators corrode when antifreeze solutions acidify. Acid rain and acid mine drainage are major threats to our environment. Your digestion and body metabolism are critically dependent on narrow controls of acidity and alkalinity in body fluids and tissues. The list could go on and on.

Acids and bases are defined in terms of a particular type of chemical reaction. **Any species that will dissolve in water to produce hydronium ions, H_3O^+, is an acid.** Acids may be characterized as *strong* or *weak.* **Strong acids,** like hydrochloric acid, are completely ionized in water:

$$HCl + H_2O \longrightarrow H_3O^+ + Cl^-$$

In addition to noting the hydronium ion produced, observe that double arrows are not used because there is no evidence for a measurable back reaction; the ionization is 100%! Two other strong acids are sulfuric acid (H_2SO_4) and nitric acid (HNO_3).

Acetic acid, $HC_2H_3O_2$, is a **weak acid.** Weak acids ionize only to a slight degree in water.

$$HC_2H_3O_2 + H_2O \rightleftharpoons H_3O^+ + C_2H_3O_2^-$$

> Some eyewashes contain boric acid (H_3BO_3), a weak acid.

Notice that the reverse reaction arrow is much longer than the forward reaction arrow, in keeping with the fact that only about 1% of the acetic acid molecules is ionized. One hydrogen atom is set apart in the acetic acid formula because only one of the four can be donated to the water to produce hydronium ions.

One can test the strength of an acid with a conductivity apparatus (see Fig. 6–6). Moderate concentrations of strong acid solutions are good conductors of electricity because of the high concentration of ions. Similar concentrations of weak acid solutions, which have relatively few ions, are poor conductors.

> The organic acids, so important to the metabolism of living things, are weak acids (see Chapter 10).

The properties of an aqueous base are due to the OH^- ions in solution. **Any species that will dissolve in water to produce OH^- ions is a base.** Bases may be characterized as *strong* or *weak.* The hydroxides of the alkali metals are **strong bases.** For example, NaOH and KOH are already ionic in the solid state and dissociate readily in water solutions.

$$Na^+OH^-_{(s)} \xrightarrow{H_2O} Na^+_{(aq)} + OH^-_{(aq)}$$

> The subscripts are used to indicate solid (s) and solution (aq) species.

The hydroxides of the alkaline earth metals, such as $Ca(OH)_2$ and $Ba(OH)_2$, are also strong bases, but they are not as caustic to the skin as the alkali metal hydroxides because the alkaline earth hydroxides are not as soluble in water.

Ammonia, NH_3, is a **weak base** that ionizes only about 1% in ordinary solutions. The ionization of ammonia produces hydroxide ions:

$$NH_3 + H_2O \rightleftharpoons NH_4^+ + OH^-$$

The long reverse arrow indicates that most of the ammonia is in the molecular form (NH_3) in a solution at equilibrium.

Acids neutralize bases; the H_3O^+ ions in the acid solution readily react with OH^- ions in the basic solution to form water.

When an acid neutralizes a base, acid and base properties are suppressed.

$$H_3O^+ + OH^- \longrightarrow 2\ H_2O$$

The salt solution that is formed in an aqueous neutralization is dependent on the particular acid and base used, but the real neutralization process is always the same in water, hydronium ions combining with hydroxide ions.

The theoretical definitions of acids and bases given above were first advanced by Svante Arrhenius in the 19th century. Acids and bases can also be defined in a more general sense in terms of the giving and receiving of protons and the giving and receiving of electron pairs in the formation of covalent bonds. The proton definitions were advanced by J. N. Brønsted and T. M. Lowry in 1923.

If hydrogen chloride gas is mixed with ammonia gas, a white solid, ammonium chloride, results.

$$NH_{3(g)} + HCl_{(g)} \longrightarrow NH_4^+Cl^-_{(s)}$$

Although this reaction does not fit the Arrhenius definition of an acid-base reaction, it is a reaction of a basic material, ammonia gas, with an acidic material, hydrogen chloride gas. Note that the acid (HCl) lost a proton (hydrogen ion) to the base (NH_3). In the sense of proton transfer, the ammonia–hydrogen chloride reaction is akin to the ionization of either hydrogen chloride or ammonia in water in that all three of these reactions are proton transfer reactions.

$$HCl_{(g)} + H_2O_{(l)} \longrightarrow H_3O^+_{(aq)} + Cl^-_{(aq)}$$
$$NH_{3(g)} + H_2O_{(l)} \rightleftharpoons NH_4^+_{(aq)} + OH^-_{(aq)}$$

Brønsted and Lowry defined acids to be species that donate protons to bases and bases to be species that accept protons from acids. This approach provides a significant dimension of relativity to the acid-base concept, since it allows us to consider acid-base phenomena in nonaqueous systems, such as in the oceans of ammonia on Jupiter, and since it reduces all acid-base questions to the competition for protons by chemical species. Also according to this approach, water's acidity or alkalinity is relative, as can be seen in the two equations above. A species such as water can be **amphiprotic;** it can be a base if it receives a proton from a strong acid such as HCl to form H_3O^+ (as in the first equation) or an acid if the water molecule loses protons to a base such as ammonia and is left as the OH^- ion (as in the second equation).

The pH Scale of Acidity and Alkalinity

Pure water is a very weak conductor of electricity. Distilled water will not conduct electricity in the conductivity apparatus shown in Figure 6–6; however, very sensitive equipment shows that water does conduct electricity slightly. Thus, water must be slightly ionized.

$$H_2O + H_2O \rightleftharpoons H_3O^+ + OH^-$$

If the forward and reverse arrows were drawn to scale in length, the reverse arrow would have to be 550,000,000 times longer than the forward arrow, since water is only 0.00000018% ionized at room temperature.

In pure water at room temperature, the concentration of the H_3O^+ ion is 0.0000001 moles per liter (10^{-7} moles per liter). The concentration of the OH^- ion is the same as the concentration of the H_3O^+ ion, since the two are produced in equal amounts as the water ionizes. When the concentrations of H_3O^+ and OH^- are equal, the solution is *neutral*. Chemical equilibrium is established when the product of the concentration of the hydronium ion and the concentration of the hydroxide ion is 1.00×10^{-14}.

A mole of water is 18 g.

At pH = 7, the number of H_3O^+ ions equals the number of OH^- ions.

Pure water is neutral. Acids form H_3O^+ ions in water; bases form OH^- ions in water.

$$[H_3O^+][OH^-] = (1.00 \times 10^{-7} \text{ moles/L})(1.00 \times 10^{-7} \text{ moles/L})$$
$$= 1.00 \times 10^{-14} \text{ moles}^2/L^2$$

If we add acid to water, the concentration of H_3O^+ will become greater than 1.00×10^{-7} moles/L; the concentration of the OH^- ion will have to go down to a smaller number so that the product of the two concentrations will still be 1.00×10^{-14}. One variable goes down as the other goes up such that their product is always the same constant.

Brackets, [ ], are used as chemical shorthand to signify moles per liter.

Now look at two extreme cases:

1. What are the $[H_3O^+]$ and $[OH^-]$ concentrations in 0.1 molar hydrochloric acid?
   Answer: The strong acid is 100% ionized, so the H_3O^+ concentration is the same as the acid, 0.1, or 10^{-1} molar. The OH^- concentration is then the constant divided by the H_3O^+ concentration, or

$$1.00 \times 10^{-14}/1.00 \times 10^{-1} = 1.00 \times 10^{-13} \text{ M}$$

2. What are the $[H_3O^+]$ and $[OH^-]$ concentrations in 0.1 molar NaOH?
   Answer: The OH^- concentration from the strong base is 0.1, or 10^{-1} moles/L. The H_3O^+ concentration is then the constant divided by the OH^- concentration, or

$$1.00 \times 10^{-14}/1.00 \times 10^{-1} = 1.00 \times 10^{-13} \text{ M}$$

In summary, the concentrations of the H_3O^+ ion in the three cases studied are:

0.1 molar hydrochloric acid	1.00×10^{-1}
Pure water	1.00×10^{-7}
0.1 molar sodium hydroxide	1.00×10^{-13}

Chemists, like others, look for concise expressions. Note that the exponent goes from -1 in a strong acid solution to -7 in a neutral solution and then to -13 in a strong basic solution. The exponent tells the strength of the acidity or alkalinity. Since there is no need to distinguish between negative and positive (all the exponents are negative), we might as well communicate in the positive. Thus, the definition of pH was obtained:

$$\text{pH} = -\log [H^+]$$

The pH is the negative log (exponent of ten) of the hydrogen ion concentration.

Pure water has a pH of 7 and is a neutral solution. If the pH is below 7, the solution is acidic, and each drop of one pH unit represents a tenfold increase in acidity, or hydronium ion concentration. A pH above 7 is alkaline, with each unit of increase on the exponent scale representing a decrease in the hydronium ion concentration by a factor of one-tenth (Table 6–1).

Consumers are frequently asked to deal with pH. Figure 6–8 gives the pH of some common materials with which you are probably familiar.

Table 6–1
Relationship of pH to the Concentrations of Hydronium (H_3O^+) and Hydroxide (OH^-) ions in Water

pH	$[H_3O^+]$	$[OH^-]$	$[H_3O^+] \times [OH^-]$
1	10^{-1}	10^{-13}	10^{-14}
2	10^{-2}	10^{-12}	10^{-14}
3	10^{-3}	10^{-11}	10^{-14}
4	10^{-4}	10^{-10}	10^{-14}
5	10^{-5}	10^{-9}	10^{-14}
6	10^{-6}	10^{-8}	10^{-14}
7	10^{-7}	10^{-7}	10^{-14} Neutrality in H_2O
8	10^{-8}	10^{-6}	10^{-14}
9	10^{-9}	10^{-5}	10^{-14}
10	10^{-10}	10^{-4}	10^{-14}
11	10^{-11}	10^{-3}	10^{-14}
12	10^{-12}	10^{-2}	10^{-14}
13	10^{-13}	10^{-1}	10^{-14}

Acid-Base Buffers

Some automobiles really have bumpers that are buffers rather than just decoration.

The general idea of a buffer is that of a shock absorber—something to absorb a disturbance while retaining the original conditions or structure. The control of pH involves maintaining a steady level of acidity even when sudden "shocks" of acid or

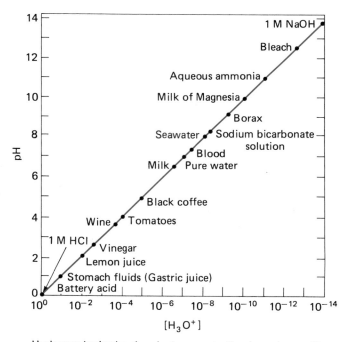

Figure 6–8
A plot of pH versus hydrogen ion concentration $[H_3O^+]$. Note that the pH increases as the $[H_3O^+]$ decreases. The pH values of some common solutions are given for reference. (A solution in which $[H_3O^+] = 1$ molar has a pH of 0, since $1 = 10^0$.)

Hydrogen (or hydronium ion) concentration in moles per liter base are added. **Buffer solutions have chemical species that can react with added acid or base and maintain a pH very close to the original value.** The pH of a buffer is dependent on the chemicals used to make the buffer.

The control of pH is necessary in many industrial and natural processes. It is a critical matter in your blood. The pH of blood is 7.4 ± 0.1, slightly alkaline, and life is in danger if the pH goes outside of this range. Blood has a relatively high concentration of the hydrogen phosphate ions, HPO_4^{2-}, and dihydrogen phosphate ions, $H_2PO_4^-$. The HPO_4^{2-} ion has a relatively strong attraction for additional protons to form the $H_2PO_4^-$ ion. Consequently, HPO_4^{2-} reacts with any acid added and thereby keeps the acidity from going up (the pH down). If a relatively strong base is added to the mixture, the $H_2PO_4^-$ gives up protons to keep the alkalinity from increasing. As long as this buffer pair is present in appreciable concentration, additions of acids and bases in small amounts will not significantly change the pH of the solution. Of course, even the best of buffer solutions can be overwhelmed.

Buffers control pH.

A buffer is made of a conjugate (kin to) pair of species. The species differ by only one proton.

SELF-TEST 6-B

1. A homogeneous mixture is a _____.
2. In salt water, is the water the solute or the solvent? _____
3. Is sugar an electrolyte or a nonelectrolyte? _____
4. A one molar solution contains how many moles in 2 liters of the solution? _____
5. When it dissolves in water, sodium chloride ( ) ionizes or ( ) dissociates.
6. The molecular weight of sulfuric acid (H_2SO_4) is 98. If 49 g of this acid are dissolved in 1 liter of water, the molarity is _____
7. Acids produce _____ ions and aqueous bases produce _____ ions in aqueous solution.
8. Give examples of the following:
    a. A strong acid
    b. A weak acid
    c. A strong base
    d. A weak base
9. The pH of pure water at room temperature is _____.
10. The hydronium ion concentration at a pH of 5 is how many times its value at a pH of 3? _____
11. In water, if the hydronium ion concentration goes up, the _____ ion concentration must go down such that the product of these two concentrations is a _____ value.
12. Two of the ions that are used to buffer the pH of your blood are: _____ and _____.

Oxidation and Reduction

Oxygen combines with more elements than any other single element. The resulting binary compounds are known as **oxides,** and the combination process is called **oxidation.** The oxides of the metals are mostly ionic solids with high melting points.

$$2\,Ca + O_2 \longrightarrow 2\,CaO$$

Calcium oxide is a metal oxide, an ionic solid with a high melting point.

The oxides of the nonmetals are molecular compounds, many of which have molecules small enough to be gases at ordinary temperatures and pressures:

$$C + O_2 \longrightarrow CO_2$$

Carbon dioxide is a nonmetal oxide, a gaseous material in which the molecules have relatively little attraction for each other.

Most compounds, as well as most elements, can be oxidized if they are heated in the presence of oxygen. An example is the burning of natural gas, composed mostly of methane, CH_4:

$$CH_4 + 2\,O_2 \longrightarrow CO_2 + 2\,H_2O$$

The complete oxidation of most compounds produces the oxides of the elements.

When oxygen combines with other elements or compounds, heat is almost always given off in the process. If heat and light are produced, the process is known as **combustion,** or **burning** (Fig. 6–9).

Historically, **reduction** was a term applied to the winning of metals from their ores (Chapter 7). It was determined before the Bronze Age that copper ores in a wood fire

Dry ice, solid carbon dioxide, is evidence that CO_2 molecules have some attraction for each other and hold together in a solid crystal if the molecules do not have too much kinetic energy.

Combustion is always accompanied by heat and light.

Oxidation is always accompanied by reduction, and vice versa. One cannot occur without the other.

Figure 6–9
Photograph of a large brush fire, common in the arid western United States. (The Bettmann Archive, Inc.)

yielded a glob of copper metal at the base of the fire. The metal went down in the fire; it was *reduced*, the reaction being:

$$CuO + C + heat \longrightarrow Cu + CO$$

In sufficient air carbon monoxide would be further oxidized to carbon dioxide, CO_2. The Iron Age was made possible by the discovery that iron oxides could be reduced to iron with carbon if the fire was made hotter with a blast of air.

$$Fe_3O_4 + 4\,C + heat \longrightarrow 4\,CO + 3\,Fe$$

Some metals, like gold, are so inactive chemically that they can be reduced from the compound to the element by simply heating the compound.

$$2\,Au_2O_3 + heat \longrightarrow 2\,Au + 3\,O_2$$

More active metals, such as sodium, are hard to reduce and were not isolated until the 19th century.

Theory Broadens Redox Concepts

Oxygen is second only to fluorine in electronegativity. As we learned in Chapter 5, oxygen tends to take electrons away from metals to form ionic bonds and to share electron pairs with the nonmetals, with the polarization of these bonds toward the oxygen, fluorine excepted. But the other nonmetals also have some electronegativity, and theory would predict that the differences in reactions between nonmetals and oxygen would be more in degree than in kind. Consider the following four reactions:

(a) $2\,H_2 + O_2 \longrightarrow 2\,H_2O$

(b) $8\,H_2 + S_8 \longrightarrow 8\,H_2S$

(c) $2\,Fe + O_2 \longrightarrow 2\,FeO$

(d) $8\,Fe + S_8 \longrightarrow 8\,FeS$

We could call (a) and (c) oxidation and (b) and (d) "sulfadation," but chemists have theorized that the same process that is happening in (a) and (c) with oxygen is also happening in (b) and (d) with sulfur. Oxygen cannot take away the electrons from hydrogen in the formation of water, but it does form polar covalent bonds with the electron pair shifted toward the oxygen. In the formation of hydrogen sulfide, sulfur does the same thing; the difference is that the bonds in H_2S are less polarized than the bonds in H_2O. In reactions (c) and (d), the electrons are removed from the iron (ionization occurs), and the results are two ionic solids. One term should suffice for what happens to iron and hydrogen in these reactions; the term is **oxidation.** There is a theoretical definition of oxidation that turns out to be very useful: **Oxidation is the loss of electrons.** The element oxygen need not be involved, and the loss of electrons need not be complete; a shifting away of the electron from an atom of one element in a polar covalent bond is sufficient.

In a **redox** reaction, reduction always occurs simultaneously with oxidation. Why? If one atom or group of atoms loses (or tends to lose) electrons, there must be another atom or group of atoms to receive the electrons. Consider again the reduction of copper oxide with carbon:

$$CuO + C \longrightarrow CO + Cu$$

Chemists often shorten the term *oxidation-reduction* to "redox."

The carbon loses full control of its electrons; it ends up in a compound with oxygen where the bonds are polarized toward the oxygen. Hence the carbon is oxidized. In contrast, the copper is reduced by gaining electrons that it had lost to oxygen in the ionic copper oxide. Consequently, copper becomes a free element. **Reduction is the gain of electrons.** It is also proper to refer to the compound copper oxide as being reduced even though we believe that the copper ions (Cu^{2+}) in the compound are actually getting the electrons.

In redox reactions the **oxidizing agent** causes the electron loss and is reduced in the process of getting electrons. Conversely, the **reducing agent** is the source of the moved electrons and is oxidized in the process. In the reaction described above, CuO is the oxidizing agent and is reduced, and carbon is the reducing agent and is oxidized.

How important is oxidation-reduction in chemistry? If, as bonding theories maintain, all atoms are held together through electrical forces, and if these electrical forces are manifest through the sharing and transfer of electrons, we shall have to return to this concept over and over again to explain atomic bonding and chemical reactions in nature around us.

> The wide applicability of electrolysis and batteries attests to the utility of redox systems for the purification of materials and the storage of energy.

Electrolysis

The suffix *-lysis* means "splitting" or "decomposition"; **electrolysis** is decomposition by electricity. The voltage that drives electrons from one atom to another in chemical reactions is only a few tenths of a volt in most ordinary chemical reactions, but the driving voltage can vary up to a maximum of about 6 volts for the most extreme cases. When we learned to make batteries and connect them in series, we were able to apply large voltages to chemical systems and provoke many previously unknown redox reactions. For example, early in the 19th century, Michael Faraday was able to isolate metallic sodium and other active metals, never before seen, by driving the reactions with battery power.

> Voltage measures the electrical pressure to move electric charge. A 6-volt source poses little threat to your body, but look out for a 110-volt socket.

$$2\ NaCl_{(molten)} \xrightarrow{\text{electric energy}} 2\ Na + Cl_2$$

A chemical reaction driven by a flow of electrons from an external voltage source is electrolysis. The principal parts of an electrolysis apparatus are shown in Figures 6–10 and 6–11. Electrical contact between the external circuit and the solution is obtained by means of electrodes, which are often made of graphite or metal. The electrode at which electrons enter an electrolysis cell is termed the **cathode;** this is the electrode at which reduction takes place. The electrode at which the electrons leave the cell is the **anode;** this is the electrode at which oxidation occurs.

> The electrode at which reduction takes place is the *cathode*. Oxidation occurs at the *anode*.

The battery or generator produces a current of electrons, which flow toward one electrode and make it negatively charged and flow away from the other electrode and make it positively charged. When the switch is closed, the positive ions in the solution migrate toward the cathode. Soon a chemical reaction is evidenced at the electrodes. Depending on the substances present in the solution, gases may be evolved, metals deposited, or ionic species changed at the electrodes. The ions that migrate to the electrodes are not necessarily the species undergoing reaction at the electrodes, because sometimes the solvent undergoes reaction more easily. Whatever happens, the chemical reactions occurring at the cathode and anode are due to electrons going into and coming out of the solution. The chemical reaction at the cathode furnishes electrons to solution species (reduction). At the anode, electrons are taken from species in solution, so the chemical reaction at the anode gives up electrons (oxidation).

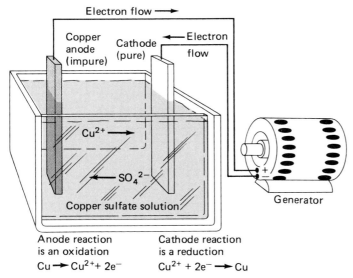

Figure 6–10
Electroplating from a copper sulfate solution.

The electroplating of copper is illustrated in Figure 6–10. Such an electrolysis can be used either to plate an object with a layer of pure copper or to purify an impure sample of copper metal; copper is transferred from the positive electrode into the solution and eventually to the negative electrode. If the positive electrode is impure copper to be purified, electrolysis deposits the copper as very pure copper on the negative electrode.

Now let us examine how the electrolysis transfers the copper from the positive electrode to the negative electrode. Electrons flow out of the negative terminal of the generator through the wire and into the negative electrode. Somehow this negative charge must be used up at the surface of the electrode.

Consider what happens when the electrons build up on the negative electrode. The positive copper ions nearby are attracted to the surface and take the electrons. Thus, the Cu^{2+} ions are reduced:

$$Cu^{2+} + 2\ e^- \longrightarrow Cu$$
Cathode Reaction

The negative sulfate ions migrate to the positive electrode (anode) in a similar way. However, it is easier to get electrons from the copper metal of the electrode than from the sulfate ions. As each copper atom gives up two electrons, the copper ion passes into solution:

$$Cu \longrightarrow Cu^{2+} + 2\ e^-$$
Anode Reaction

In effect, then, the copper of the positive electrode is oxidized (the anode reaction) and passes into solution; the copper ions in solution migrate to the negative electrode, are reduced (the cathode reaction), and plate out as copper metal. Large amounts of copper are purified in this way each year. Silver and gold can be purified in a similar fashion.

If we desire to plate an object with copper, we have only to render the surface conducting and make the object the negative electrode in a solution of copper sulfate.

Copper can be plated onto an object by making that object the negative electrode in a cell containing dissolved copper salts.

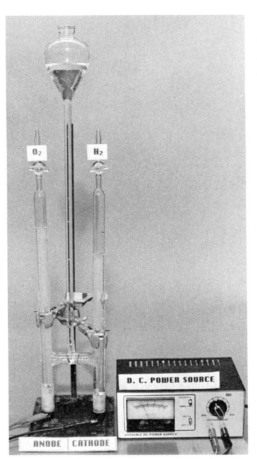

Figure 6-11
Apparatus for the electrolysis of water. The cathode, where electrons enter the cell and reduce water to hydrogen, is on the right. Electrons are removed through the wire on the left side, the anode, oxidizing the water to oxygen. (Photograph by Robert Hayes.)

The object will become coated with copper, with the copper coating growing thicker as the electrolysis is continued. If the object is a metal, it will conduct electricity by itself. If the object is a nonmetal, its surface can be lightly dusted with graphite powder to render it conducting.

A potentially very important electrolysis reaction is the electrolysis of water (Fig. 6-11). When electricity is passed into graphite electrodes immersed in a dilute salt solution, water is reduced to hydrogen and hydroxide ions at the cathode:

$$2\ H_2O + 2\ e^- \longrightarrow H_2\ (gas) + 2\ OH^-$$
Cathode Reaction

At the anode, water is oxidized to oxygen and hydrogen ions:

$$2\ H_2O \longrightarrow O_2 + 4\ H^+ + 4\ e^-$$
Anode Reaction

The OH^- and H^+ ions combine to re-form water. The overall, or net, cell reaction is:

$$2\ H_2O \xrightarrow{\text{electricity}} 2\ H_2\ (gas) + O_2\ (gas)$$

The hydrogen produced by the reduction of water can be stored and used as a fuel — for example, to power rockets into space.

Batteries

One of the most useful applications of oxidation-reduction reactions is the production of electrical energy. A device that produces an electron flow (current) is called an **electrochemical cell.** Although a series of such cells is a **battery,** the term *battery* is commonly used even for single cells such as those we shall describe.

Consider the reaction between zinc atoms and copper ions. If zinc is placed in a solution containing Cu^{2+} ions, electron transfer takes place between the zinc metal and the copper ions to produce zinc ions and copper atoms, and the energy liberated simply causes a slight heating of the solution and the zinc strip. If the zinc could be separated from the copper solution, and the two connected in such a way as to allow current flow, the reaction could proceed with the electrons transferred through the connecting wires. Figure 6-12 shows a battery that can be constructed to make use of the oxidation-reduction involved in the reaction of Zn with Cu^{2+}.

The anode reaction is the oxidation of zinc to Zn^{2+} ions.

$$Zn \longrightarrow Zn^{2+} + 2\ e^-$$

The electrons flow from the Zn electrode through the connecting wire, light the lamp in the circuit, and then flow into the copper cathode, where reduction of Cu^{2+} ions occurs:

$$Cu^{2+} + 2\ e^- \longrightarrow Cu$$

The copper is deposited on the copper cathode.

This flow of electrons (negative charge) from the anode to the cathode compartment in the battery must be neutralized electrically. This is done by use of a **salt bridge** provided to connect the two compartments. The salt bridge contains a solution of a salt

> The active metals lose electrons more easily; hence, these free metals are not found in nature.
>
> A copper ion has a greater attraction for electrons than does a zinc ion.
>
> In commercial batteries, the salt bridge is often replaced by a porous membrane.

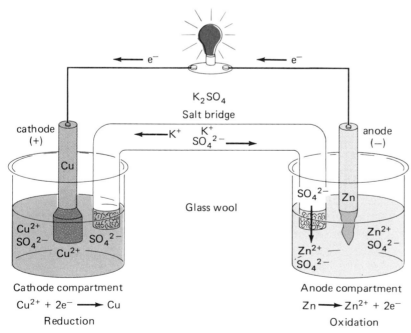

Figure 6-12
A simple battery involving the oxidation of zinc metal and the reduction of Cu^{2+} ions.

such as K_2SO_4. Its purpose is to keep the two solutions neutral. Around the cathode the deposition of positive copper ions (Cu^{2+}) would tend to cause the solution to become negative owing to the presence of excess negative sulfate ions (SO_4^{2-}). Two actions can keep the solution around the cathode neutral: positive potassium ions (K^+) can pass into the solution, and negative sulfate ions can pass out of the solution and into the salt bridge. Actually, both processes occur. Similarly, around the anode, the solution would tend to become positive because of an increase in positive zinc ions (Zn^{2+}). Two actions can keep the solution around the anode neutral: negative sulfate ions can pass into the solution, and positive zinc ions can pass out of the solution and into the salt bridge. In all this exchange, it is necessary for the solution to maintain the same number of positive charges as negative charges. The reaction between zinc atoms and copper ions continues until one or the other is consumed.

Many different oxidation-reduction combinations are used in commercial batteries to produce a flow of electrons. A few popular ones are listed in Table 6-2.

Batteries in which the stored chemical energy is simply used up are called **primary batteries.** In such batteries, the oxidation products produced at the anode are allowed to mingle with the reduction products formed at the cathode. Because of this mixing, the battery may be used only once and then discarded (or recycled). Many of the less expensive batteries used to power flashlights, toys, radios, watches, cameras, and hand-held calculators are primary batteries.

> The mercury batteries that are so popular for small electronic devices must be discarded carefully. If heated, these hermetically sealed batteries will rupture explosively because of expanding vapors within the package.

Some batteries can be recharged. These are called **secondary batteries.** In these batteries, the oxidation products stay at the anode, and the reduction products remain at the cathode. Under favorable conditions, these secondary batteries may be discharged and recharged many times over. One of the most widely used secondary batteries is the lead storage battery. As this battery is discharged, metallic lead is oxidized to lead sulfate at the anode, and lead dioxide is reduced at the cathode.

Anode: $Pb + SO_4^{2-} \longrightarrow \underline{PbSO_4} + 2\ e^-$

Cathode: $PbO_2 + 4\ H^+ + SO_4^{2-} + 2\ e^- \longrightarrow \underline{PbSO_4} + 2\ H_2O$

> Underlining indicates that the substance is insoluble.

The lead sulfate formed at both electrodes is an insoluble compound, so it stays on the elctrode surface. Since sulfuric acid is used in both the anode and the cathode reactions, the concentration of the sulfuric acid electrolyte decreases as the battery discharges. A measurement of the density of this battery acid gives a measure of the state of charge of the battery. The lower the density of the battery acid, the lower the state of charge.

Table 6-2
Characteristics of Some Batteries

System	Anode (Oxidation)	Cathode (Reduction)	Electrolyte	Typical Operating Voltage Per Cell
Dry cell	Zn	MnO_2	NH_4Cl-$ZnCl_2$	0.9–1.4
Edison storage	Fe	Ni oxides	KOH	1.2–1.4
Nickel-cadmium — NiCad	Cd	Ni oxides	KOH	1.1–1.3
Silver cell	Cd	Ag_2O	KOH	1.0–1.1
Lead storage	Pb	PbO_2	H_2SO_4	1.95–2.05
Mercury cell	Zn(Hg)	HgO	KOH-ZnO	1.30
Alkaline cell	Zn(Hg)	MnO_2	KOH	0.9–1.2

Recharging a secondary battery requires reversing the electrical current flow through the battery. When this occurs, the anode and cathode reactions are reversed.

At the negative electrode: $\text{Pb} + \text{SO}_4^{2-} \underset{\text{charge}}{\overset{\text{discharge}}{\rightleftarrows}} \text{PbSO}_4 + 2\text{ e}^-$

At the positive electrode: $\text{PbO}_2 + \text{SO}_4^{2-} + 4\text{ H}^+ + 2\text{ e}^- \underset{\text{charge}}{\overset{\text{discharge}}{\rightleftarrows}} \text{PbSO}_4 + 2\text{ H}_2\text{O}$

Normal charging of an automobile lead storage battery occurs during driving. The voltage regulator senses the output from the alternator, and when the alternator voltage exceeds that of the battery, the battery is charged. During the charging cycle in most batteries, some water is reduced at the cathode, while water is oxidized at the anode.

Oxidation of water: $2\text{ H}_2\text{O} \longrightarrow \text{O}_2 + 4\text{ H}^+ + 4\text{ e}^-$ (anode)

Reduction of water: $4\text{ H}_2\text{O} + 4\text{ e}^- \longrightarrow 2\text{ H}_2 + 4\text{ OH}^-$ (cathode)

Oxidation cannot occur without reduction.

These reactions produce a mixture of hydrogen and oxygen in the atmosphere in the top of the battery. If this mixture is accidentally sparked, an explosion results. It is a good idea always to open a battery carefully and not introduce any sparks or open flames near a lead storage battery.

During starts, especially during extremely cold weather, a car battery works very hard, causing it to need recharging. Recharging often causes elongated crystals to grow on the electrode surfaces as the lead and lead oxide are redeposited on the negative and positive electrodes. Often these crystals of lead and lead oxide grow between the electrodes, causing internal short circuits. When this happens, the battery is usually "dead" and must be replaced. If electrolyte fluid runs low, the electrode surfaces dry, tending to make the surfaces not recharge properly.

All in all, the lead storage battery is relatively inexpensive, reliable, and relatively simple, and it has an adequate life. Its high weight is its major fault. Newer secondary batteries have found use in some applications, such as electronics, but none of these newer batteries can perform as well as the lead storage battery does for its cost.

Corrosion: Unwanted Oxidation-Reduction

In the United States alone, more than $10 billion is lost each year to corrosion. Much of it is the rusting of iron and steel, although other metals may oxidize as well. The problem with iron is that its oxide, rust, does not adhere strongly to the metal's surface once the rust is formed. Because the rust flakes off or is rubbed off easily, the metal surface becomes pitted. The continuing loss of surface iron by rust formation eventually causes structural weakness.

The corrosion of metals involves oxidation and reduction. The driving forces behind corrosion are the activity of the metal as a reducing agent and the strength of the oxidizing agent. Whenever a strong reducing agent (the metal) and a strong oxidizing agent (like oxygen) are together, a reaction between the two substances is likely. Factors governing the rates of chemical reaction such as temperature and concentration will affect the rate of corrosion as well. Consider the corrosion of an iron spike (Fig. 6–13). The surface of the iron is far from perfect. There are tiny microcrystals composed of loosely bound iron atoms on the surface of the metal. The iron can readily ionize into any water present on the surface of the metal.

$$\text{Fe} \longrightarrow \text{Fe}^{+2} + 2\text{ e}^-$$
Oxidation

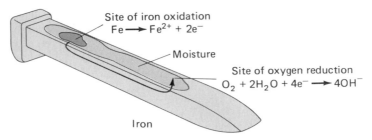

Figure 6–13
The site of iron oxidation may be different from the point of oxygen reduction owing to the ability of electrons to flow through iron. The point of oxygen reduction can be located with an acid-base indicator because of the OH⁻ ions produced.

Corrosion.

The ionization of iron atoms into Fe^{+2} ions is an oxidation process. Iron is a fairly active metal; that is, it tends to give its electrons up rather easily. Since iron is a good conductor of electricity, the electrons produced at this site can migrate to some point where they can reduce something. If these electrons did not migrate, the corrosion of iron would come to an abrupt halt as a result of a buildup of excessive negative charge. One location on the surface of the iron where electrons can be used would be any tiny drop of water containing dissolved oxygen. Here, the oxygen gains the electrons, forming hydroxide ions.

$$O_2 + 2\ H_2O + 4\ e^- \longrightarrow 4\ OH^-$$
$$\text{Reduction}$$

This reduction of oxygen occurs so readily that when Fe^{+2} ions are encountered, they are further oxidized to Fe^{+3} ions. This happens to the dissolved Fe^{+2} ions in the water on the surface of the metal. The reaction is shown below:

$$4\ Fe^{+2} + O_2 + 2\ H_2O \longrightarrow 4\ Fe^{+3} + 4\ OH^-$$

Finally, the Fe^{+3} ions combine with hydroxide ions to form the iron oxide we call rust:

$$2\ Fe^{+3} + 6\ OH^- \longrightarrow Fe_2O_3 \cdot 3\ H_2O$$

The rate of rusting is enhanced by salts, which dissolve in the water on the surface of the iron and act like tiny salt bridges of an electrochemical cell (discussed in the preceding section). The hydroxide ions and Fe^{+2} and Fe^{+3} ions migrate more easily in ionic solutions produced by the presence of the dissolved salts. Automobiles rust more quickly when exposed to road salts in wintery climates. If road salts are used in your driving area, it's a good idea after the snowy season to wash the undersides of your automobiles to remove the accumulated salts.

Rusting can be prevented by protective coatings, such as paint, grease, oil, enamel, and some corrosion-resistant metals like chromium. Some metals are more active than iron, but when these metals corrode, they form adherent oxide coatings. Coatings with these metals provide corrosion protection. One of these metals is zinc. Zinc coating of iron and steel is called **galvanizing** and may be done by dipping the object into a molten bath of zinc metal or by electroplating zinc onto the surface of an iron or steel object. In galvanized objects in which the zinc coating is exposed to air and water, a thin film of zinc oxide forms, which protects the zinc from further oxidation. Galvanizing is a

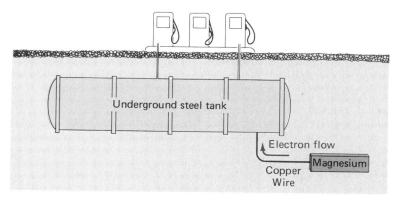

Figure 6-14
Cathodic protection. If magnesium is connected to the steel tank to be protected, the magnesium is more easily oxidized than the iron or copper connecting wire. The magnesium thus serves as a sacrificial anode, protecting the cathode, on the surface of which no oxidation occurs. The anode is the electrode at which oxidation occurs; reduction occurs at the cathode. When the magnesium is used up, it is replaced by another block. This replacement is much easier and cheaper than replacement of the tank.

type of **cathodic protection.** As the name implies, a cathode is protected by use of a more active metal in good electrical contact with the metal to be protected. The electrons for the reduction of oxygen,

$$O_2 + 2\ H_2O + 4\ e^- \longrightarrow 4\ OH^-$$

are supplied by the more active metal. Thus, a more active metal, such as magnesium, electrically connected to a piece of iron, would be oxidized before the iron is oxidized.

An important application is the cathodic protection of underground steel storage tanks (Fig. 6-14) that hold gasoline and other hazardous liquids. These tanks must be protected, since leakage would contaminate groundwater supplies. Beginning in 1986, these tanks must be cathodically protected under new federal regulations designed to protect groundwater.

Fuel Cells

Like batteries, fuel cells have a cathode and anode separated by an electrolyte. Unlike batteries, fuel cells do not store the chemicals involved in the oxidation reduction reaction. Instead, fuel cells receive a continuous supply of gaseous reactants and convert the energy of their oxidation-reduction reaction directly into electricity. The most popularized application of fuel cells has been in the space program on board the Gemini, Apollo, and space shuttle missions.

Consider the reaction between hydrogen and oxygen to produce water and energy:

$$2\ H_2 + O_2 \longrightarrow 2\ H_2O + energy$$

If a mixture of hydrogen and oxygen is sparked, the energy is released suddenly in the form of a violent explosion. In the presence of a platinum gauze, these gases will react at room temperature, slowly heating the catalytic surface to incandescence. In a fuel cell (Fig. 6-15), the oxidation of hydrogen by oxygen takes place in a controlled manner, with the electrons lost by the hydrogen molecules flowing out of the fuel cell and back in

Fuel cells convert the chemical energy of combustion directly into electrical energy.

again at the electrode at which the oxygen is reduced. This electron flow powers the electrical needs of whatever is connected to the fuel cell. The water produced in the fuel cell can be purified for drinking purposes.

Because of their light weight and their high efficiencies compared with batteries, fuel cells like the one shown in Figure 6–15 have proved valuable in the space program. Beginning with Gemini 5, alkaline fuel cells have logged over 10,000 hours of operation in space. The fuel cells used aboard the space shuttle deliver the same power that batteries weighing ten times as much would provide. On a typical seven-day mission, the shuttle fuel cells consume 1500 pounds of hydrogen and generate 190 gallons of potable water.

Potable is a term used to describe drinkable water.

Other types of fuel cells that have been developed use air as the oxidizer and less pure hydrogen or carbon monoxide as the fuel. It is hoped that fuel cells capable of direct air oxidation of cheap gaseous fuels such as natural gas will eventually be developed. These fuel cells might compete with gasoline engines, and they would not produce air pollutants such as carbon monoxide and nitric oxide.

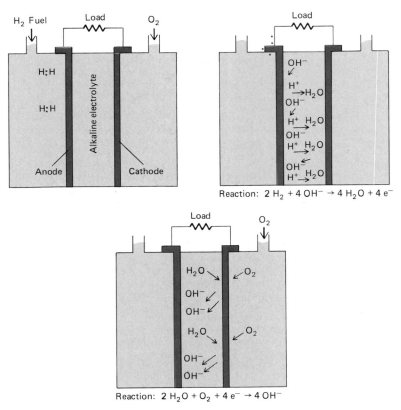

Figure 6–15
How an alkaline fuel cell works. The net reaction is $2H_2 + O_2 \rightarrow 2H_2O$. Hydrogen is oxidized at the anode, and oxygen is reduced at the cathode. The water molecules produced are discharged into the alkaline electrolyte solution, which consists of potassium hydroxide (KOH) and water.

SELF-TEST 6-C

1. A chemical reaction driven by a flow of electrons from an external voltage source is _____.
2. The electrode through which electrons enter an electrolysis cell is the _____.
3. The electrolysis of water produces what two gases? _____ and _____
4. A device that produces an electrical current through an external circuit is a/an _____. Such devices are activated by _____ reactions.
5. A battery that can be recharged by the forcing of electrons through it in reverse is a _____. A commonly used battery that cannot be recharged in this manner is a _____.
6. Name three methods that are partially effective in preventing the corrosion of iron alloys. _____, _____, and _____
7. Which is more likely to run out of reacting chemicals in its ordinary use: a battery or a fuel cell? _____ Why?
8. If an automobile lead storage battery is "overcharged," an explosive mixture of what two gases may exist within the battery? _____ and _____
9. The battery commonly used in the early telegraph systems was composed of Zn-Cu cells. Give the cathode and anode reactions for this cell.
    Cathode: _____
    Anode: _____

MATCHING SET

_____ 1. Acid-base   a. Chlorine
_____ 2. Oxidation-reduction   b. Moles of solute per liter
_____ 3. Conservation of matter   c. Oxidized form of hemoglobin
_____ 4. Chemical energy   d. Alkaline solution
_____ 5. Catalyst   e. Table salt
_____ 6. Oxyhemoglobin   f. Speeds up reaction
_____ 7. Affects reaction rate   g. Concentration
_____ 8. Solute   h. From rearranged atomic bonds
_____ 9. Electrolyte   i. Aqueous base
_____ 10. Molarity   j. Sour-bitter
_____ 11. Hydroxide ions   k. Preservation of atoms
_____ 12. Sulfuric acid   l. Disperses in solvent
_____ 13. pH = 10   m. Electron loss-gain
_____ 14. Oxidizing agent   n. Strong acid

QUESTIONS

1. The combination of carbohydrates from plants and oxygen from the air constitutes a system packed with chemical energy. Compare the energy of such a system with another one of equal mass composed of carbon dioxide and water.
2. If energy is always required to break chemical bonds between atoms, how is it possible that energy is released when two molecules of hydrogen and one molecule of oxygen are destroyed to form two molecules of water?

   $$2 H_2 + O_2 \longrightarrow 2 H_2O$$

3. If 2 g of hydrogen form 18 g of water, how much water can be formed from 1 g of hydrogen?
4. Copper sulfide is composed of just the two elements and has the formula CuS. If 100 kg of his ore yield 66.4 kg of copper, how much sulfur must have been separated from the copper? How does this illustrate the Law of Conservation of Matter in chemical change?
5. The same amount of heat is liberated in the oxidation of carbon regardless of whether it is oxidized in one step to CO_2 or in two steps to CO and then to CO_2. How does this argue for fixed amounts of energies associated with atomic combinations?
6. Name two ways to slow down the rate of a chemical reaction and one different way to speed it up.
7. Some chemical bonds are strong and some are quite weak. What do you think this has to do with the reversibility of chemical reactions?
8. If at 10 PM 500 students are at the big school party and 25 are at Joe's party down the street, and if at 11 PM the numbers are the same but the people are different (except for Joe), how could it be said that the two parties are in equilibrium? Draw an analogy between the parties and chemical equilibrium.
9. Select two different solutions and illustrate the following terms: *solute, solvent, electrolyte, nonelectrolyte, ionization, dissociation,* and *molarity*.
10. Make a list of the acids and bases that you can find in domestic products sold commercially.
11. All Arrhenius acids are Brønsted-Lowry acids, but not all Brønsted-Lowry acids are Arrhenius acids. Explain.
12. Explain how pH is kind of a thinking person's simplification of a fairly complicated expression.
13. A first-magnitude star is ten times brighter than a second-magnitude star, and an earthquake that measures 6 on the Richter scale is ten times more powerful than one that measures 5. How are these scales related to pH?
14. Using a piece of regular graph paper, plot the following numbers on a linear scale: 1, 10, 100, 1000, 10000, 100000, 1000000. Using a piece of log paper, plot the logarithm of these same seven numbers on a line. Draw any conclusion relative to pH you wish from the experience.
15. Ammonia (NH_3) and the ammonium ion (NH_4^+) differ by one proton. Ammonia can act as a base in receiving a proton to form the ammonium ion, and the ammonium ion can act as an acid in donating protons. Relate the capabilities of this pair to the definition of a buffered solution.
16. Since oxygen is not the most electronegative element (fluorine is), explain how it is possible for oxygen to be oxidized.
17. Explain the purification of copper by electrolysis.
18. A battery emplacement of guns during the Civil War is related to the cells in an electrochemical battery. The lead–lead dioxide cell in sulfuric acid produces about 2 volts. How do you think it would be possible to make a 12-volt battery out of these cells? Why was battery an apt word for this electrochemical device?
19. Cathodic protection is a sacrifice, while painting is a shield against corrosion. Explain.
20. A fuel cell can be made lighter than a battery. Explain. However, there is a catch to this savings. What has to be carried along as well? In a fuel cell, what is analogous to recharging the battery?

CHAPTER 7

Chemical Raw Materials and Products from the Earth, Sea, and Air

The crust of the Earth is well suited for the production and support of life forms. The chemical composition of the Earth's crust is dramatically different from that of the universe and even from that of the Earth as a whole. As shown in Figure 7-1, oxygen — a major part of the air, water, and rocks — constitutes about 50% of the Earth's crust. Silicon, which constitutes about 25% of the Earth's crust, is a major part of silicate rocks, clays, and sand. Then come the major metals — aluminum, iron, calcium, sodium, potassium, and magnesium — found mostly in mineral deposits and to some extent dissolved in the sea. Hydrogen, which constitutes less than 1% of the weight of the crust, is ninth in abundance, and carbon, the central element in all life forms, is present in little more than trace amounts. It is apparent, then, that life forms are a very small part of the whole and are clinging to the edge of the Earth.

Our environment is quite heterogeneous. Elements and compounds are almost lost in the complicated array of mixtures that have resulted from natural forces over very long periods of time.

Until recent history, humans had not developed the power to alter the Earth's crust significantly; wooden hammers, plows, and other instruments could produce physical, but not chemical, alterations. Then came the chemical reduction of copper from its ores, followed by the reduction of iron. Today, each passing year sees the production of a flood of new chemicals, some of which have the potential to alter radically the Earth's crust. Chemists have now developed the knowledge to change the chemical mixtures that surround us. This chapter looks at how this knowledge is applied to produce some of the major chemicals affecting human life.

Making chemicals in industrial quantities poses a threat to the natural environment. However, with the single exception of radioactive nuclear wastes, we can, if we wish to pay the cost, cycle and recycle our natural resources through countless uses. We are learning that we do not have to use our iron and aluminum just one time and then haul them away to the dump. Perhaps the most striking thing about natural chemistry is the cycling of the elements through a countless number of like uses.

Let's begin with the study of the chemistry of the major metals used in our society.

The top chemicals of commerce are listed inside the front cover of this text.

Recycling decreases our demands on many natural resources.

Metals and Their Preparation

Metals occur mostly as compounds in the crust of the Earth, although some of the less active metals, such as copper, silver, and gold, can also be found as free elements. Fortunately, the distribution of elements in the crust is not uniform. Some elements that are not particularly abundant are familiar to us because they tend to occur in very concentrated, localized deposits, called **ores,** from which they can be extracted eco-

A continual search is under way for new ore deposits.

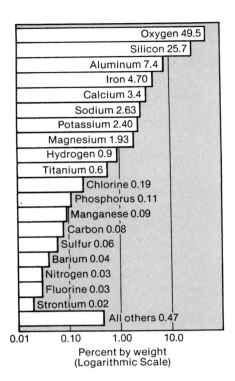

Figure 7-1
Abundance of the elements in the crust of the Earth.

nomically; examples of these are lead, copper, and tin (Fig. 7-1). Other elements that form a much larger percentage of the crust are almost unknown to us because concentrated deposits of their ores are less commonly found or the metals are difficult to extract from their ores. An example is titanium, the tenth most abundant element in the crust of the Earth. Although the ore of titanium, rutile (mostly TiO_2), is common, the use of this metal is rare because it is difficult to reclaim from the ore.

Some common minerals are listed in Table 7-1.

Table 7-1
Some Common Metals and Their Minerals

Metal	Chemical Formula of Compound of the Element	Name of Mineral
Aluminum	$Al_2O_3 \cdot xH_2O$	Bauxite
Calcium	$CaCO_3$	Limestone
Chromium	$FeO \cdot Cr_2O_3$	Chromite
Copper	Cu_2S	Chalcocite
Iron	Fe_2O_3	Hematite
	Fe_3O_4	Magnetite
Lead	PbS	Galena
Manganese	MnO_2	Pyrolusite
Tin	SnO_2	Cassiterite
Zinc	ZnS	Sphalerite
	$ZnCO_3$	Smithsonite

Metals and Their Preparation

The preparation of metals from their ores, **metallurgy,** involves chemical reduction (Chapter 6). Indeed, the concept of oxidation and reduction developed from metallurgical operations. Iron in the ore iron oxide (Fe_2O_3) is in the form Fe^{3+}. To reduce Fe^{3+} ions to Fe atoms, we must add electrons. (Recall that oxidation is the loss of electrons and that reduction is the gain of electrons.) Sometimes the desired metal is in solution (e.g., magnesium in the sea), where it exists in its oxidized form (Mg^{2+} ions). To obtain the free metal magnesium, we must add electrons to magnesium ions (reduction) to produce neutral atoms.

Reduction of magnesium:
$Mg^{2+} + 2e^- \longrightarrow Mg$

Iron and Steel

The sources of most of the world's iron are large deposits of iron oxides in Minnesota, Sweden, France, Venezuela, Russia, Australia, and England. In nature these oxides are frequently mixed with impurities, which the production process usually removes. Iron ores are then reduced to iron by use of carbon, in the form of coke, as the reducing agent.

Iron ores are iron compounds. For iron to be obtained from the ores, it must be reduced.

The reduction of iron ore is carried out in a blast furnace (Fig. 7–2). The solid material fed into the top of the blast furnace consists of a mixture of an oxide of iron (Fe_2O_3), coke (C), and limestone ($CaCO_3$). A blast of heated air is forced into the furnace near the bottom. Much heat is liberated as the coke burns, and the heat speeds up the reaction, which is important in making the process economical. These reactions occur within the blast furnace:

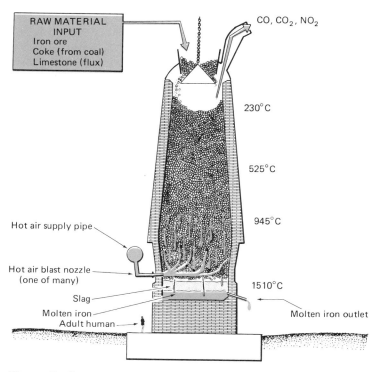

Figure 7–2
Diagram of a blast furnace.

$$2\,C + O_2 \longrightarrow 2\,CO + \text{heat}$$
$$\text{Carbon} \quad \text{Oxygen} \quad \text{Carbon Monoxide}$$

$$Fe_2O_3 + 3\,CO \longrightarrow 2\,Fe + 3\,CO_2 + \text{heat}$$
$$\text{Iron Oxide} \quad \text{Carbon Monoxide} \quad \text{Iron} \quad \text{Carbon Dioxide}$$

Limestone (calcium carbonate) is added to remove the silica (SiO_2) impurity:

$$CaCO_3 \xrightarrow{\text{heat}} CaO + CO_2$$
$$\text{Calcium Carbonate} \quad \text{Calcium Oxide} \quad \text{Carbon Dioxide}$$

$$CaO + SiO_2 \longrightarrow CaSiO_3$$
$$\text{Calcium Oxide} \quad \text{Silicon Dioxide} \quad \text{Calcium Silicate}$$

The calcium silicate, or **slag,** exists as a liquid in the furnace. Consequently, as the blast furnace operates, two molten layers collect in the bottom. The lower, denser layer is mostly liquid iron, which contains a fair amount of dissolved carbon and often smaller amounts of other impurities. The upper, lighter layer is primarily molten calcium silicate with some impurities. From time to time the furnace is tapped at the bottom and the molten iron is drawn off. Another outlet somewhat higher in the blast furnace base can be opened for removal of the liquid slag.

As it comes from the blast furnace, the iron contains too much carbon for most uses. If some of the carbon is removed, the mixture becomes structurally stronger and is known as **steel.** Steel is an **alloy** of iron with a relatively small amount of carbon (less than 1.5%); it may also contain other metals. For iron to be converted to steel, the excess carbon is burned out with oxygen.

An *alloy* is a metal consisting of two or more elements.

Most of the steel manufactured today uses the basic oxygen process furnace (Fig. 7-3) to reduce the carbon content to a suitable level. The carbon is oxidized by oxygen to carbon monoxide and carbon dioxide.

Aluminum

Before Charles Hall invented his electrolytic process in 1885, aluminum was very expensive and rare. A bar of aluminum was displayed next to the Crown Jewels at the Paris Exposition in 1855.

Aluminum, in the form of Al^{3+} ions, constitutes 7.4% of the crust of the Earth. However, because of the difficulty of reducing Al^{3+} to Al, only recently have we learned to isolate and use this abundant element. Aluminum metal is soft and has a low density. Many of its alloys, however, are quite strong. Hence, it is an excellent choice when a lightweight, strong metal is required. In structural aluminum, the high chemical reactivity of the element is offset by the fact that a transparent, hard film of aluminum oxide, Al_2O_3, forms over the surface, protecting it from further oxidation:

$$4\,Al + 3\,O_2 \longrightarrow 2\,Al_2O_3$$

The principal ore of aluminum contains the mineral bauxite, a hydrated aluminum oxide, $Al_2O_3 \cdot xH_2O$. Because impurities such as iron oxides in the ore have undesirable effects on the properties of aluminum, these must be removed, generally by purification of the ore. This is accomplished by the Bayer process, which is based on the reaction of aluminum oxide or aluminum hydroxide with strong bases. In the Bayer process the mixture of oxides is treated with a sodium hydroxide solution that dissolves aluminum oxide and leaves iron oxide, which is insoluble in the solution.

$$Al_2O_3 \cdot xH_2O + Fe_2O_3 \xrightarrow[\text{solution}]{\text{NaOH}} Al(OH)_4^- + Na^+ + Fe_2O_3$$
$$\text{Solid} \quad\quad\quad \text{Solid} \quad\quad\quad \text{Solution} \quad\quad \text{Solid}$$

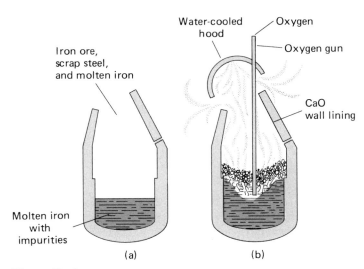

Figure 7-3
Basic oxygen process furnace. Much of the steel manufactured today is refined by the blowing of oxygen through a furnace charged with ore, scrap, and molten iron.

The mixture is filtered; $Al(OH)_3$ is then carefully precipitated out of the clear solution by the addition of carbon dioxide (an acid), which makes the solution less basic:

$$CO_2 + Al(OH)_4^- \longrightarrow Al(OH)_3\downarrow + HCO_3^-$$

The aluminum hydroxide is then heated to transform it into pure anhydrous aluminum oxide:

$$2\ Al(OH)_3 \xrightarrow{heat} Al_2O_3 + 3\ H_2O$$

Aluminum metal is obtained from the purified oxide by the Hall process, an electrolytic process that uses molten cryolite (Fig. 7-4). Cryolite, Na_3AlF_6, has a melting point of 1000°C; the molten compound dissolves considerable amounts of aluminum oxide,

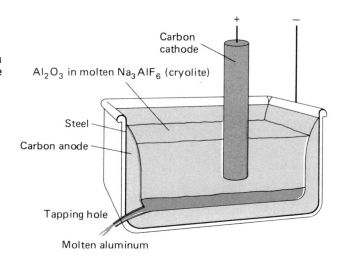

Figure 7-4
Schematic drawing of a furnace for producing aluminum by electrolysis of a melt of Al_2O_3 in Na_2AlF_6. The molten aluminum collects at the bottom of the carbon cathode container.

The top of the Washington Monument is a casting of aluminum made in 1884.

About ten times more energy is needed to produce a ton of aluminum than a ton of steel.

which in turn lowers the melting point of the cryolite solution. This mixture of cryolite and aluminum oxide is electrolyzed in a cell with carbon anodes and a carbon cell lining that serves as the cathode on which the aluminum is deposited. As the operation of the cell proceeds, the molten aluminum sinks to the bottom of the cell. From time to time the cell is tapped and the molten aluminum is run off into molds.

Aluminum is used both as a structural metal and as an electrical conductor in high-voltage transmission lines. It competes with copper as an electrical conductor because of its lower cost, although larger diameter aluminum wires must be used to offset the lower electrical conductivity of aluminum compared with that of copper.

Copper

Although copper metal occurs in the free state in some parts of the world, the supply available from such sources is quite insufficient for the world's needs. The majority of the copper obtained today is from various copper sulfide ores, most of which must be concentrated before undergoing the chemical processes that produce the metal. These minerals include $CuFeS_2$ (chalcopyrite), Cu_2S (chalcocite), and CuS (covellite). Because the copper content of these ores is only around 1% to 2%, the powdered ore is first concentrated by the flotation process (Fig. 7-5).

The preparation of copper metal from a copper sulfide ore involves roasting in air to oxidize some of the copper sulfide and any iron sulfide present:

$$2\ Cu_2S + 3\ O_2 \longrightarrow 2\ Cu_2O + 2\ SO_2\uparrow$$

$$2\ FeS + 3\ O_2 \longrightarrow 2\ FeO + 2\ SO_2\uparrow$$

Subsequently the mixture is heated to a higher temperature, and some copper is produced by the reaction:

$$Cu_2S + 2\ Cu_2O \longrightarrow 6\ Cu + SO_2\uparrow$$

The remaining copper sulfide is then converted to copper metal:

$$2\ Cu_2S + 3\ O_2 \longrightarrow 2\ Cu_2O + 2\ SO_2$$

$$Cu_2S + 2\ Cu_2O \longrightarrow 6\ Cu + SO_2$$

The copper produced in this manner is crude, or "blister," copper and is purified electrolytically.

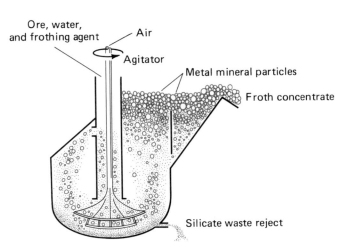

Figure 7-5
Apparatus for flotation concentration. A frothing agent, such as pine oil, is used to float the mineral particles away from the sand, rock, and clay.

In the electrolytic purification of copper, the crude copper is first cast into anodes, which are placed in a water solution of copper sulfate and sulfuric acid. The cathodes are made of pure copper. As electrolysis proceeds, copper is oxidized at the anode, moves through the solution as Cu^{2+} ions, and is deposited on the cathode (Chapter 6). The voltage of the cell is regulated so that relatively active impurities (such as iron) are left in the solution and relatively inactive ones are not oxidized at all. These less active impurities, which include gold and silver, collect as "anode slime," an insoluble residue. This anode slime is subsequently worked up for recovery of these rare metals.

The copper produced by the electrolytic cell is 99.95% pure and is suitable for use as an electrical conductor. Copper used for this purpose must be pure; even very small amounts of impurities, such as arsenic, considerably reduce the electrical conductivity of copper.

Magnesium

Magnesium, which has a density of 1.74 g/mL, is the lightest structural metal in common use. For this reason it is most often used in alloys designed for light weight and great strength. It is a relatively active metal chemically because it loses electrons easily. Magnesium "ores" include sea water, which has a magnesium concentration of 0.13%, and dolomite, a mineral with the composition $CaCO_3 \cdot MgCO_3$. Because there are 6 million tons of magnesium present as Mg^{2+} salts in every cubic mile of sea water, the sea can furnish an almost limitless amount of this element.

There are about 328 million cubic miles of sea water.

To make magnesium one uses sea water, lime from oyster shells, methane from natural gas, and electricity. The recovery of magnesium from sea water is outlined in Figure 7-6. The total world production of magnesium is only about 250,000 tons per year, although the metal is potentially available on a larger scale.

SELF-TEST 7-A

1. The most abundant element in the Earth's crust is _____.
2. The most abundant metal in the Earth's crust is _____.
3. In the United States, the largest iron ore deposits are found in the state of _____.
4. A natural material that is almost pure calcium carbonate is _____.
5. Which of the following metals may occur in the free or metallic state in mineral deposits? iron, copper, aluminum, magnesium _____
6. Which of the following metals are either produced or purified using electricity? iron, copper, aluminum, magnesium _____
7. Another name for calcium silicate as it applies to production of iron is _____.
8. For most metals to be prepared from their ores, they must be ( ) oxidized or ( ) reduced.
9. In an electrical refining process for metals, the purest metal will always be found at the ( ) anode or ( ) cathode.
10. Which of the following metals is sufficiently concentrated in the ocean to be extracted commercially? magnesium, aluminum, copper, iron _____.
11. Most metals are found in the earth as ( ) neutral atoms, ( ) positive ions, or ( ) negative ions.

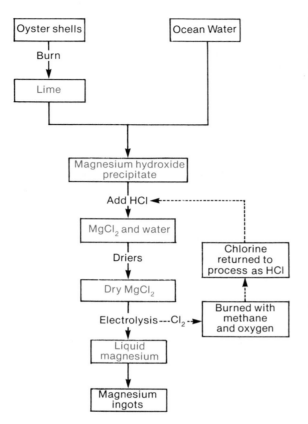

Figure 7-6
Flow diagram showing how magnesium metal is produced from sea water.

The Fractionation of Air

The atmosphere of the Earth is a fantastically large source of the elements nitrogen and oxygen, as well as of certain of the noble gases, including argon, neon, and xenon (Table 7-2).

For pure oxygen and nitrogen to be obtained from the air, water vapor and carbon dioxide must be removed first. This is usually done by precooling the air by refrigeration

Table 7-2
Composition of the Earth's Atmosphere at Sea Level

Components	Percentage by Volume
Nitrogen	78.084
Oxygen	20.948
Argon	0.934
Carbon dioxide	0.033
Neon	0.00182
Helium	0.00052
Methane	0.0002

or by using silica gel to absorb water and lime to absorb carbon dioxide. Afterward, the air is compressed to a pressure exceeding 100 times normal atmospheric pressure, cooled to room temperature, and allowed to expand into a chamber. This expansion produces a cooling effect (the Joule-Thompson effect) owing to the breaking of weak attractive London forces between the gaseous molecules. Because breaking bonds requires energy, the expanding gas absorbs energy from the surroundings, thus cooling the surroundings and the gas itself. If this expansion is repeated and controlled properly, the expanding air actually cools to the point of liquefaction (Fig. 7–7). The temperature of the *liquid air* is usually well below the boiling points of nitrogen ($-195.8°C$), oxygen ($-183°C$) and argon ($-189°C$). This liquid air is then allowed to vaporize partially again. Since nitrogen is more volatile than oxygen or argon (nitrogen has a lower boiling point), the liquid becomes more concentrated in oxygen and argon. This process, known as the Linde process, produces high-purity nitrogen (99.5+%) and oxygen with a purity of 99.5%. Further processing produces pure argon, neon (boiling point, $-246°C$), and even helium (boiling point, $-268.9°C$), although most helium used in the United States is produced from natural gas wells.

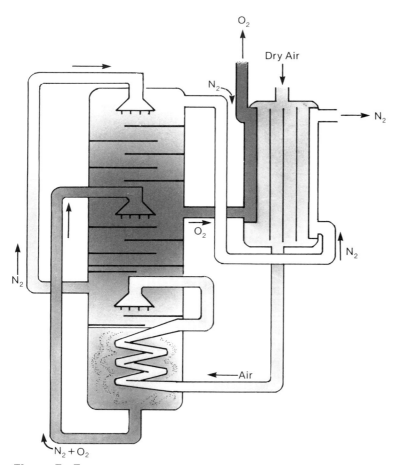

Figure 7–7
Diagram of a fractionating column for separating oxygen and nitrogen in an air supply.

Most of the oxygen produced by the fractionation of liquid air is used in steel-making, although some finds use in rocket propulsion (to oxidize hydrogen) and in controlled oxidation reactions of other types. Liquid oxygen (LOX) can be shipped and stored at its boiling temperature of $-183°C$ at atmospheric pressure. Substances this cold are called **cryogens** (from Greek, *kryos*, meaning "icy cold"). They represent special hazards, since skin contact with them produces instantaneous frostbite and structural materials such as plastics, rubber gaskets, and some metals become brittle and fracture easily at these temperatures. Liquid oxygen can accelerate oxidation reactions to the point of explosion owing to the high oxygen concentration. For this reason liquid oxygen must be prevented from contacting substances that will ignite and burn in air.

Special cryogenic containers holding liquid oxygen are actually huge vacuum-walled bottles much like those used to carry hot soup or hot coffee. These containers can be seen outside hospitals or industrial complexes, on highways and railroads, and even aboard ocean-going vessels (Fig. 7–8).

Liquid nitrogen is also a cryogen. It has uses in medicine (cryosurgery), for example in cooling a localized area of skin before removal of a wart or other unwanted or pathogenic tissue. Since nitrogen is so chemically unreactive, it is used as an inert atmosphere for certain applications such as welding, and liquid nitrogen is a convenient source of high volumes of the gas. Because of its low temperature and inertness, liquid nitrogen has found wide use in frozen food preparation and preservation during transit.

Figure 7–8
Photograph of a cargo tanker capable of carrying several thousand gallons of liquefied cryogenic oxygen, which is at a temperature of $-183°C$. Although it is extremely cold, its high concentration in the liquid state makes liquid oxygen exceptionally reactive with anything that can burn.

Containers of nitrogen atmospheres, such as railroad boxcars and truck vans, are health hazards, since they contain little (if any) oxygen to support life; workers have died when they have entered such areas without proper breathing apparatus.

Silicon Materials Old and New

Silicon and oxygen make up 75% of the crust of the Earth, and it is the bonding between these two elements in clay and rock that holds together the skin of the earth. The chemical structures involved are many and complex. However, let's examine a few molecular structures of some of the most important materials in our society, from glass to the computer chip.

Glass

Silicon dioxide, SiO_2 (also called silica), occurs naturally in large amounts in rocks and sand and more rarely in much larger crystals (quartz) (Fig. 7–9). Silica has a melting point of 1710°C. If the melted material is cooled rapidly, a noncrystalline solid is obtained. Crystalline quartz consists of an extended structure in which each silicon atom is bonded tetrahedrally to four oxygen atoms (Fig. 7–10a) and each oxygen atom is bonded to two silicon atoms. The bonding thus extends throughout the crystal (Fig. 7–10b). When silica is melted, some of the bonds are broken, and the units move with respect to each other. When the liquid is cooled, re-formation of the original solid requires a reorganization that is hard to achieve because the groups have difficulty moving. The very viscous liquid structure is thus partially preserved on cooling to give the characteristic feature of **glass,** which is an apparently solid material (pseudosolid) with some of the randomness in structure that is characteristic of a liquid. This random structure accounts for one of the typical properties of a glass: it breaks irregularly rather than splitting along a plane like a crystal.

By the addition of metal oxides to silica, the melting temperature of the mixture can be reduced from 1710°C to about 700°C. The oxides most often added are sodium oxide (added as Na_2CO_3, soda ash) and calcium oxide (added as $CaCO_3$). The metal ions form ionic bonds, which are nondirectional, with oxygen atoms that previously had been bonded rigidly to specific Si atoms. As a result, the so-called soda-lime glass has a

Glass flows slowly, like a liquid. Many colonial homes contain glass window panes that are thicker at the bottom than at the top.

Sodium carbonate (Na_2CO_3) is the number 11 commercial chemical. Calcium oxide is number 5. See inside the front cover.

Figure 7–9
Quartz. (Courtesy of McGraw-Hill).

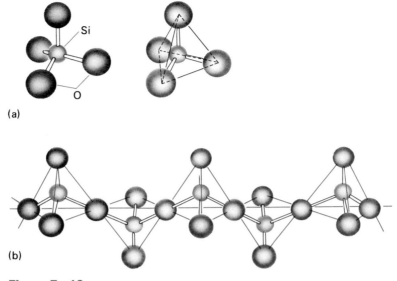

Figure 7–10
(a) Tetrahedral structure of silicon and oxygen in silicates. *(b)* Chain of tetrahedra showing that an oxygen is common at each point of contact between the tetrahedra.

Viscosity is resistance to flow.

Most glasses are made from the oxides of Si, Na, and Ca, which are melted together. Colored glass is produced by addition of other metal oxides.

lower melting temperature and viscosity than pure SiO_2 and can be produced and fabricated more easily.

Soda-lime glass is clear and colorless only if the purity of the ingredients has been controlled carefully. If, for example, too much iron oxide is present, the glass will be green. Other metal oxides produce other colors (Table 7–3). To some extent, one color can counteract another.

The substances are melted together in a gas- or oil-fired furnace. As they react, bubbles of CO_2 gas are evolved.

$$CaCO_3 + SiO_2 \longrightarrow CaSiO_3 + CO_2\uparrow$$
$$Na_2CO_3 + SiO_2 \longrightarrow Na_2SiO_3 + CO_2\uparrow$$

Table 7–3
Substances Used in Colored Glasses

Substance	Color
Copper (I) oxide	Red, green, or blue
Tin (IV) oxide	Opaque
Calcium fluoride	Milky white
Manganese (IV) oxide	Violet
Cobalt (II) oxide	Blue
Finely divided gold	Red, purple, or blue
Uranium compounds	Yellow, green
Iron (II) compounds	Green
Iron (III) compounds	Yellow

The mixture is heated to about 1500°C to remove the bubbles of CO_2. At this temperature the viscosity is low, and the bubbles of entrapped gas escape easily. The mixture is cooled somewhat and then is blown into bottles by machines, or is drawn into sheets or molded into other forms (Fig. 7-11).

It is possible to incorporate a wide variety of materials into glass for special purposes. Some examples are given in Table 7-4.

Ceramics

Ceramic materials have been made since before the dawn of recorded history (Fig. 7-12). They are generally materials fashioned at room temperature from clay or other natural earths and then permanently hardened by heat. Clays with a wide variety of properties are found in a considerable range of ceramic materials, from bricks to table china. The techniques developed with natural clay have been applied to a wide range of other inorganic materials in recent years. The result has been a considerable increase in the kinds of ceramic materials available. One can now obtain ceramic magnets as well as

Figure 7-11
Craftsman working with molten glass. (Courtesy of Corning Glass Company.)

Table 7-4
Special Glasses

Special Addition or Composition	Desired Property
Large amounts of PbO with SiO_2 and Na_2CO_3	Brilliance, clarity, suitable for optical structures: crystal or flint glass
SiO_2, B_2O_3, and small amounts of Al_2O_3	Small coefficient of thermal expansion: borosilicate glass, "Pyrex," "Kimax," etc.
One part SiO_2 and four parts PbO	Ability to stop (absorb) large amounts of X rays and gamma rays: lead glass
Large concentrations of CdO	Ability to absorb neutrons
Large concentrations of As_2O_3	Transparency to infrared radiation
Suspended Se particles	Red color

ceramics suitable for rocket nozzles; both were developed from mixtures of inorganic oxides by the use of ceramic technology, which includes heating of the materials to make them hard and resistant to wear.

The three basic ingredients of common pottery are silicate minerals: clay, sand, and feldspar. The term *clay* includes materials with a wide range of chemical compositions that are produced from the weathering of granite and gneiss rocks. Feldspars are aluminosilicates containing potassium, sodium, and other ions in addition to silicon and oxygen. An approximation of the weathering process can be shown if we write feldspar as a mixture of oxides: $K_2O \cdot Al_2O_3 \cdot 6\, SiO_2$; this includes, among other reactions, the reaction of the mineral with water containing dissolved carbon dioxide to form clay.

$$K_2O \cdot Al_2O_3 \cdot 6\, SiO_2 + 2\, H_2O + CO_2 \longrightarrow \underbrace{Al_2O_3 \cdot 2\, SiO_2 \cdot 2\, H_2O}_{\text{A Clay}} + 4\, SiO_2 + 2\, K^+ + CO_3^{2-}$$

The essential feature of the clay mineral is that it occurs in the form of extremely minute platelets which, when wet, are plastic and can easily be shaped. When dry, they are rigid; if heated to an elevated temperature, they become permanently rigid and are no longer subject to easy dispersion in water. When these clays are mixed with feldspars

Figure 7-12
Fired pottery. Different colors and surface textures can be achieved by fusion of the coating material with the clay itself. The coating ceases to be a "coat" and becomes a part of the whole, the entire structure being held together by covalent bonds. (Courtesy of the Robinson-Ransbottom Pottery Company.)

and silica, heating them produces a mixture of crystals held together by a matrix of glasslike material. The clays can be used by themselves to make bricks, flowerpots, and clay pipe, but finer quality ceramic materials contain purified clays and other ingredients in carefully controlled proportions.

Natural clays are generally extremely complex mixtures. If they are used in ceramics without treatment, the finished materials have a color and physical properties characteristic of the impurities present. The first pieces of fine oriental chinaware arrived in Europe during the late Middle Ages, and European potters envied and admired the obviously superior product. This led to the beginning of systematic studies on the effect of composition on the nature of the ceramic produced and to a keen appreciation of the role of the purity of clay in determining the color and potential decorative development of the piece.

Alchemists made notable achievements in this area. One of these, Johann Friedrich Bottger, worked from about 1705 to 1719 for King Augustus of Saxony, who kept him almost as a prisoner. The king hoped to gain power from the alchemist's discoveries. Bottger succeeded in developing several novel ceramic materials, of which the most important was the first white glazed porcelain made in Europe (in 1709). Bottger devoted the rest of his life to the perfection of the manufacture and decoration of this material, in which he enjoyed considerable success. The china was made in Meissen and was both glazed and **vitrified.** The glazing was accomplished by coating of the pieces with a material that melted and produced an impermeable layer on the surface. Vitrification was carried out by firing of the clay at a temperature sufficient to melt a portion of the material and in effect produce an impermeable glass that would hold the remaining particles together.

Vitrify means "convert into glass."

In the past few decades new ceramic materials have been developed and used on an increasingly wide scale. Nearly pure alumina (Al_2O_3) and zirconia (ZrO_2) are now used as bases for ceramic materials that are excellent electrical or thermal insulators. Magnetic ceramics, which contain iron compounds, are used as memory elements in computers.

In recent years a new class of materials, the glass ceramics, has been discovered; these have unusual but very valuable properties. Normally glass breaks because, once a crack starts, there is nothing to stop it from spreading. It was discovered that if glass is treated by heating until a very large number of tiny crystals has developed in it, the resulting material, when cooled, is much more resistant to breaking than normal glass. The process has to be controlled carefully for the desired properties to be obtained. The materials produced in this way are generally opaque and are used as cooking utensils and kitchenware; they include products marketed under the name Pyroceram (Fig. 7–13). The initial manufacturing process is similar to that of other glass objects, but once these materials have been formed into their final shapes, they are then heat-treated.

Portland Cement and Concrete

A cement is a material used to bind other materials together. Portland cement contains calcium, iron, aluminum, silicon, and oxygen in varying proportions. It has a structure somewhat similar to that described earlier for glass, except that in cement some of the silicon atoms have been replaced by aluminum atoms. Cement reacts in the presence of water to form a hydrated colloid of large surface area that subsequently undergoes recrystallization and reaction to bond to itself and to bricks or stone. Cement is made by roasting a powdered mixture of calcium carbonate (limestone or chalk), silica (sand), an aluminosilicate mineral (kaolin, clay, or shale) and iron oxide. The roasting is carried

Figure 7-13
Cookware made of Pyroceram. (Courtesy of Corning Glass Company.)

out at a temperature of up to 870°C in a rotating kiln (Fig. 7-14). As the materials pass through the kiln, they lose water and carbon dioxide and ultimately form a "clinker," in which the materials are partially fused together. The "clinker" is then ground to a very fine powder after the addition of a small amount of calcium sulfate (gypsum). The composition of Portland cement is 60% to 67% CaO, 17% to 25% SiO_2, 3% to 8% Al_2O_3, and up to 6% Fe_2O_3; Portland cement also contains small amounts of magnesium oxide, magnesium sulfate, and potassium and sodium oxides.

The reactions that occur during the setting of cement are quite complex, involving the reaction of the various constituents with water and, subsequently, at the surface, with carbon dioxide in air. The initial reaction of cement with water produces a sticky gel resulting from hydrolysis of the calcium silicates. This gel sticks to itself and to the other particles (sand, crushed stone, or gravel). It has a very large surface area and is responsible for the strength of concrete. The setting process also involves the formation of small, densely interlocked crystals after the initial solidification of the wet mass. This continues for a long time after the initial setting and increases the compressive strength of the cement. Water is required, since the setting reactions involve hydration. For this reason, freshly poured concrete is kept moist for several days.

Over 400,000,000 tons of cement are manufactured each year, most of which is used to make concrete. Concrete, like many other materials containing Si—O bonds, is highly noncompressible but lacks tensile strength. If concrete is to be used where it is subject to tension, it must be reinforced with steel.

Silicon Materials Old and New   153

(a)

(b)

Figure 7–14
(a) A cement kiln. Note the rollers on the supports, which allow the giant cylinder to rotate. Because the cylinder is at a slight angle, the powder moves down it as the kiln turns. Intense heat is produced by the combustion of gaseous fuels. The powder loses volatile materials as it moves along; the finished product is discharged from the lower end. *(b)* The first kiln used for making cement in the United States was constructed by David Saylor over a century ago in Coplay, Pennsylvania. The kiln still stands as pictured here. (Courtesy of the Portland Cement Northwestern States Company, Mason City, Iowa.)

"Pure" Silicon and "The Chip"

Silicon of about 98% purity can be obtained by the heating of silica together with a form of carbon called coke at 3000°C in an electric arc furnace.

$$SiO_2 + 2\ C \longrightarrow Si + 2\ CO$$

Silicon of this purity is alloyed with aluminum and magnesium to increase their hardness and durability; it is also used in making silicone polymers.

High-purity silicon can be prepared by the reduction of $SiCl_4$ with magnesium.

$$SiCl_4 + 2\ Mg \longrightarrow Si + 2\ MgCl_2$$

Magnesium chloride, which is water-soluble, is then washed from the silicon. Final purification of the silicon takes place by a melting process called **zone-refining** (Fig. 7–15), which produces silicon containing less than 1 part per *billion* of impurities such as boron, aluminum, or arsenic.

One outstanding property of silicon in a high state of purity is its electrical conductivity. Unlike a metal, which conducts electricity easily, and unlike a nonmetal, which fails to conduct electricity, silicon is a **semiconductor.** That is, it fails to conduct until a certain electrical voltage is applied, but beyond that it conducts moderately. By placing other atoms in a crystal of pure silicon, a process known as **doping,** experimenters have found that the conductivity properties of silicon can be changed. Doping a silicon crystal with a Group V element such as arsenic produces a crystal with extra electrons. (Arsenic has five valence electrons, whereas silicon has four.) This is known as an **n-doped semiconductor.** Doping silicon with a Group III element such as gallium produces a **p-doped semiconductor,** since gallium has only three valence electrons and looks positive in the silicon lattice.

> The extra electrons simulate a negatively charged material, hence the name *n-doped semiconductor.*

In 1947 an electrical device called the **transistor** was invented. The simplest device used layers of n-p-n– or p-n-p–doped silicon. Germanium, a Group IV element just below silicon in the periodic table, was also used. Later, scientists used electrical fields to control conductivity in silicon transistors. These **field-effect transistors** (FETs) have been put to good use by engineers designing low-noise amplifiers, receivers, and other forms of electronic equipment.

The most revolutionary application of silicon's semiconductor properties has been

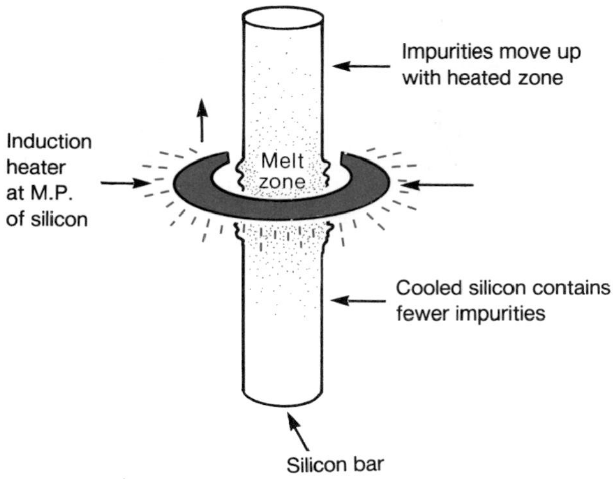

Figure 7–15
Zone refining. The hot zone moves upward on the silicon bar. As the silicon melts, impurities become mobile and move with the molten zone. Repeated passes of the heater produce a crystalline silicon bar with fewer than one part impurities per billion parts of silicon (1 ppb).

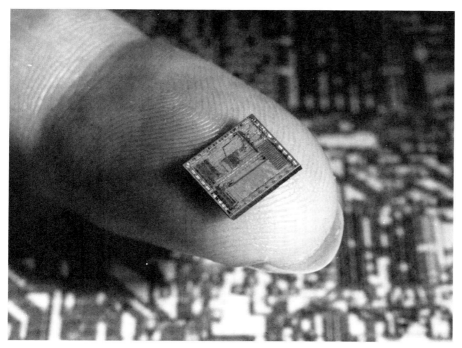

Figure 7-16
A tiny microcomputer (Intel 8748) fabricated from a single piece of highly purified silicon. Such computers are capable of many millions of computations per second. Their speed and small size have revolutionized computers and their applications. (Courtesy of Intel Corp.)

the design of **integrated electrical circuits** (ICs), computer memories, and even whole computers called **microprocessors** on tiny chips of silicon scarcely larger than a millimeter or so in diameter (Fig. 7-16). These devices can be found in calculators, cameras, watches, toys, coin changers, cardiac pacemaker devices, and many other products. Truly, silicon is both the world we walk on and our constant companion in communications and electronic controls.

SELF-TEST 7-B

1. Name the two elements that are commercially prepared by the fractionation of the air.
   _____ and _____
2. LOX is the industrial abbreviation for _____.
3. What is the meaning of the word *cryogen*? _____
4. Give four uses of the products of liquid air, and identify each with a particular element.

5. The two principal nonmetals in glass are _____ and _____.
6. Flint or crystal glass contains a large amount of a compound of what metal?
   _____
7. What element is associated with the miniaturization of electronic components?
   _____
8. What three silicate minerals are used to make common pottery materials?
   _____, _____, and _____
9. Roasting a mixture of powdered limestone, clay, sand, and iron oxide produces what important commercial building material? _____

Sulfur and Sulfuric Acid

Sulfur in underground mineral deposits is brought to the surface by the Frasch process (Fig. 7–17), in which sulfur is melted by superheated steam. The molten sulfur is raised to the surface of the earth by means of compressed air and is then allowed to cool in large vats.

Burning sulfur-containing fuels produces acid rain (see Chapter 15).

The conversion of sulfur to sulfuric acid is carried out by means of a four-step process called the contact process. In the first step the sulfur is burned in air to yield mostly sulfur dioxide.

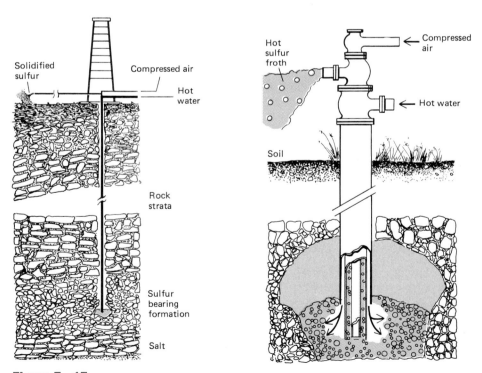

Figure 7–17
Diagram of the mining of sulfur by the Frasch process, which uses melting and pressure. Three concentric pipes are directed into the sulfur deposit. Compressed air and superheated water are sent down two of the pipes, and molten sulfur is forced up the third pipe.

$$S + O_2 \longrightarrow SO_2(g)$$

The conversion of the gaseous SO_2 to SO_3 is then achieved by passage over a hot, catalytically active surface, such as platinum or vanadium pentoxide:

$$2\ SO_2 + O_2 \xrightarrow{catalyst} 2\ SO_3(g)$$

Although SO_3 can be converted directly into H_2SO_4 by passing into water, the enormous amount of heat released in the reaction causes the formation of a stable fog of H_2SO_4. This is avoided by passage of the SO_3 into H_2SO_4

$$\underset{\text{Sulfuric Acid}}{SO_3 +\ H_2SO_4} \longrightarrow \underset{\text{Pyrosulfuric Acid}}{H_2S_2O_7}$$

and then dilution of the $H_2S_2O_7$ with water:

$$H_2S_2O_7 + H_2O \longrightarrow 2\ H_2SO_4$$

Several million tons of sulfuric acid are prepared by this method each year.

Sulfuric acid is used in huge quantities in the manufacture of fertilizers, in the petroleum industry, and in the production of steel. It also plays an important role in the manufacture of organic dyes, plastics, and drugs, among other products. The cost of sulfuric acid, about 1¢ per pound, has not changed much in 300 years, a tribute to improving technology.

Sodium Hydroxide, Chlorine, and Hydrogen Chloride

Sodium hydroxide, hydrogen, and chlorine are prepared simultaneously by the electrolysis of a concentrated solution of sodium chloride in water (Fig. 7–18). Several variations in the basic process improve its efficiency and the purity of the products. The basic reactions, which occur at nonreactive solid electrodes, are shown below:

Anode: $\quad 2\ Cl^- \longrightarrow Cl_2(g) + 2\ e^- \qquad$ *(Oxidation)*

Cathode: $\ 2\ H_2O + 2\ e^- \longrightarrow 2\ OH^- + H_2 \quad$ *(Reduction)*

The reaction replaces the Cl^- of the NaCl with OH^-.

Purer sodium hydroxide is produced when the process is carried out with a mercury cathode. In this case Na^+ is reduced to Na metal, which dissolves in the mercury electrode:

Cathode: $\quad Na^+ + e^- \xrightarrow{Hg} Na(Hg)$

The major uses of NaOH, Cl_2, and HCl are presented inside the front cover.

The sodium-mercury amalgam* is a liquid. It is removed from the cell continuously and reacted with pure water:

$$2\ Na(Hg) + 2\ H_2O \longrightarrow 2\ Na^+ + 2\ OH^- + H_2(g)$$

The chlorine gas is collected from the anode compartment, compressed into tanks, and sold for further use. The hydrogen gas is also collected and sold. An important reaction between hydrogen and chlorine is the production of hydrogen chloride gas and hydrochloric acid:

Large amounts of mercury have been discharged carelessly into the environment.

$$H_2(g) + Cl_2(g) \longrightarrow 2\ HCl(g)$$

This process can be used to prepare hydrogen chloride, which is quite pure. Hydrochlo-

---

* An amalgam is an alloy of metal with mercury; it can be either liquid or solid.

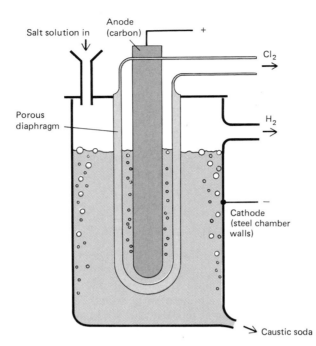

Figure 7-18
Cross section of a cell for the electrolysis of brine to yield chlorine, hydrogen, and caustic soda (NaOH).

ric acid is a solution of hydrogen chloride in water. The use of mercury in this process is a potential hazard, since mercury contaminates streams when the waste solutions from the process are thrown out.

Because electrical energy is quite expensive, an electrochemical process is economical only if all the products are sold. In recent years the demand for chlorine has grown very rapidly while the demand for sodium hydroxide has not. Consequently it has become necessary to devise some way of disposing of relatively large amounts of sodium hydroxide. This has been accomplished in part by the transformation of sodium hydroxide into sodium hydrogen carbonate:

$$Na^+ + OH^- + CO_2(g) \longrightarrow Na^+ + HCO_3^- \longrightarrow NaHCO_3$$

Because of the surplus of NaOH, this process has largely replaced older processes for the manufacture of $NaHCO_3$.

Phosphoric Acid

Phosphoric acid or its salts are used in the manufacture of a large number of materials encountered in our daily lives. These include baking powder, carbonated beverages, detergents, fertilizers, and fire-resistant textiles. Several processes are available for the manufacture of phosphoric acid, but the basic raw material is usually calcium phosphate, which occurs naturally in apatite minerals of the general formula $Ca_5X(PO_4)_3$, where X may be F^-, Cl^-, or OH^-.

An electric furnace is used to heat a mixture of the phosphate ore $[Ca_5F(PO_4)_3]$, silica (SiO_2), and coke (C). At an elevated temperature a reaction occurs in which elemental phosphorus vapor (P_4) is produced:

$$4\ Ca_5F(PO_4)_3 + 18\ SiO_2 + 15\ C \longrightarrow 18\ CaSiO_3 + 2\ CaF_2 + 15\ CO_2\uparrow + 3\ P_4\uparrow$$

The phosphorus is then condensed from the gaseous vapor and purified. Because of the low melting point of phosphorus (44.1 °C), it can easily be stored and handled as a liquid if it is protected from air, in which it ignites spontaneously.

Elemental phosphorus is transformed into phosphoric acid by oxidation with air to give P_4O_{10}:

$$P_4(gas) + 5\ O_2 \longrightarrow P_4O_{10}(gas)$$

This is then hydrated by absorption into hot phosphoric acid containing about 10% water:

$$P_4O_{10} + 6\ H_2O \xrightarrow{H_3PO_4} 4\ H_3PO_4$$

Arsenic, when present in the original ore, is carried through to the phosphoric acid produced, at which point it can be precipitated by treatment with H_2S. This is important if the phosphoric acid is to be used in the manufacture of food products.

Sodium Carbonate

As noted earlier, sodium carbonate, Na_2CO_3, is a major ingredient in making glass. Sodium carbonate is also important as a cheap alkaline material in the production of numerous chemicals and in the paper industry. It is especially useful as a water softener, since the carbonate ion precipitates the metal ions of iron, calcium, and magnesium that cause water hardness.

Trona ore, an impure form of sodium carbonate, is insufficient as a source of this important chemical. Several chemical methods have been employed in the production of sodium carbonate; the Solvay process dominates. In this process, ammonia from the Haber process (see Chapter 12), and carbon dioxide from the lime kiln are dissolved in a concentrated brine (NaCl) solution. Sodium hydrogen carbonate ($NaHCO_3$) precipitates from this solution:

$$NH_3 + CO_2 + H_2O \rightleftharpoons NH_4^+ + HCO_3^-$$
$$NH_4^+HCO_3^- + Na^+Cl^- \rightleftharpoons NaHCO_3\downarrow + NH_4^+Cl^-$$

On gentle heating, the dry sodium hydrogen carbonate yields the anhydrous sodium carbonate:

$$2\ NaHCO_3 \longrightarrow Na_2CO_3 + H_2O\uparrow + CO_2\uparrow$$

Anhydrous sodium carbonate is known industrially as soda ash; washing soda is the decahydrate, $Na_2CO_3 \cdot 10\ H_2O$, which is prepared by crystallization directly from water solution.

Sources of New Materials

For most of history, we have been limited to the use of materials found on the surface of the Earth and to relatively few chemical transformations. Advances in mining techniques due to technological advances and the control of chemical change through basic research have radically changed the materials available for human use.

Four sources of new materials are promised by recent advances:

1. Thousands of new compounds should be produced, described, and catalogued for years to come as a continuation of ongoing basic and applied research efforts.

2. The production of new stable elements appears possible. If elements are made with atomic numbers up to 150, as predicted, some of them should be stable, resulting in a multitude of new pure substances and mixtures.
3. Our reach through space programs comes ever closer to the exploitation of extraterrestrial materials from the moon, meteors, and nearby planets. Indications now are that the elements of space materials are the same, but that the mineral configurations are often different.
4. Recent experiments make it clear that chemicals formed and purified in a gravity-free environment are significantly different from the "same" chemicals formed and purified in the presence of Earth's gravity.

> K is Kelvin temperature; K = 273 + °C.
>
> Light fibers have successfully carried 20 billion laser pulses per second, equivalent to 300,000 simultaneous conversations over a 42-mile distance. Under-ocean and intercity systems are presently being developed.

Having noted the enhanced effects of purified copper in electrical conduction and ultrapure silicon in the control of electrical circuits and in the processing of information, one has to wonder what new materials and applications are in the offing. For example, consider again the making of glass. One of the biggest problems in controlling the purity of glass is contamination from the container in which it is made. In space, no container is needed! New applications in making glass for improved optical fibers are exciting to researchers seeking still another breakthrough in the communications industry.

A recent discovery of materials that exhibit superconductivity at temperatures as high as 98 K illustrates the impact that research on new materials can have on society. In February 1987, researchers reported that a complex oxide of lanthanum, barium, and copper becomes superconducting at much higher temperatures than the previously known superconductors made of niobium alloys, which require cooling to 23 K. What is the potential significance of this discovery?

A superconductor is a material that has the ability to conduct electric current with no apparent resistance. Superconducting materials are used to build more powerful electromagnets such as those used in nuclear particle accelerators or in magnetic resonance imaging (MRI) machines, which are used in medical diagnosis. The main drawback from wider application is the necessity to cool the magnet with liquid helium, which costs $4 per liter, in order to keep the material below the temperature where superconductivity occurs. The discovery of materials that superconduct as high as 98 K permits the use of liquid nitrogen as a coolant, at a cost of only $0.40 per liter.

Many scientists are saying that this discovery is more important than the discovery of the transistor because of its potential effect on electrical and electronic technology. For example, the use of superconducting materials for transmission of electric power could save as much as 30% of the energy now lost because of the resistance of the wire. Superchips for computers could be up to 1000 times faster than existing conventional silicon chips. Electromagnets could be both more powerful and smaller, which could hasten the day of a practical nuclear fusion reactor.

Although the discovery is a significant one, translating the research into practical applications such as those described above will take years. The technology is just beginning to be developed, but there is hope that problems relating to applying the discovery to a variety of applications will be overcome. In addition, there is now hope that sometime in the future, new materials that are superconductors at room temperature might be discovered.

SELF-TEST 7-C

1. Which element is pumped from underground deposits as a molten material mixed with hot water? _____

2. Give two major uses for sulfuric acid.
   _____
   _____
3. What three chemicals are produced in the electrolysis of a concentrated solution of sodium chloride? _____, _____, and _____
4. Phosphorus can be moved about as a liquid because of its low melting point, provided it is kept out of contact with the _____.
5. What is the chemical name of washing soda? _____ What related chemical is likely to be on the shelf in the kitchen? _____
6. Give two reasons why it is quite likely that many new substances will be available for technological development.
   _____
   _____

MATCHING SET

_____ 1. Copper
_____ 2. Aluminum
_____ 3. Milk glass
_____ 4. Magnesium
_____ 5. Oyster shells
_____ 6. Sodium bicarbonate
_____ 7. Sulfur
_____ 8. Iron
_____ 9. Phosphate
_____ 10. Silicon

a. Reduced in a blast furnace
b. Calcium fluoride added to glass
c. Contains phosphorus and oxygen
d. Used in making transistors
e. $NaHCO_3$
f. Mined by superheated water
g. Present in unlimited supply in sea water
h. Supply calcium hydroxide for magnesium production
i. Purified electrolytically
j. The most abundant metal in the Earth's crust
k. Sodium carbonate
l. An element distilled from air

QUESTIONS

1. Name three metals you would expect to find free in nature and three that you would expect to find only in compound form.
2. What is the primary reducing agent in the production of iron from its ore?
3. Why is CaO necessary for the production of iron in a blast furnace?
4. What is the chemical difference between iron and steel?
5. Are natural materials most likely to be elements, compounds, or mixtures? Explain.
6. What metal is most used in industry?
7. What metal is recovered from sea water in industrial quantities?
8. Describe the solution used in a commercial cell for the electrolytic reduction of aluminum.
9. Why is it so important to purify industrial quantities of copper electrolytically to a level above 99.9% pure?
10. What chemical is obtained from oyster shells in the production of magnesium from sea water? What is the role of this chemical in the production of magnesium?

11. Explain how the structures of glass and a liquid are similar.
12. What oxide is the main ingredient in glass?
13. Give reactions involved in the preparation of (a) calcium hydroxide from calcium carbonate, (b) iron from iron oxide, (c) sulfuric acid from sulfur, and (d) phosphoric acid from phosphorus.
14. Hard water can be rendered soft with washing soda because the alkaline carbonate solution precipitates what three metal ions from the water?
15. The computer chip, or microprocessor chip, is dependent on a high level of purity of what element?
16. Which will precipitate at the lower pH, sodium carbonate or sodium bicarbonate? Explain.
17. What is the purpose of using sodium carbonate in the family wash?
18. What is the purpose of oxidizing phosphorus in air after it has been reduced from phosphate rock in a furnace?
19. Two elements and a compound are produced in the electrolysis of brine. Name them, and write the reactions involved in their production.
20. Why is it necessary to have two oxidation steps in the production of sulfuric acid from elemental sulfur?
21. Some argue that atomic energy is the most significant scientific and technical event of the 20th century. Others say that the silicon chip will have the greatest impact on our society. Do you think genetic engineering should also be mentioned in this context? Give reasons for your answer.
22. What chemicals are in Portland cement? Give the source of each.
23. When clay is fired, a rigid glasslike framework is set up that is not attacked by water. Where do you think the "glass" comes from in the fired clay?
24. Is glass a mixture or a compound? Explain.
25. Is the recovery of oxygen from the air a chemical or a physical process? Give a reason for your answer.
26. Name two commercial sources for lime, CaO, and give the equations involved. Also, name two uses of this important chemical.
27. Two elements discussed in this chapter have special electrical properties only when they are in a high state of purity. What are they, and what are the applications for the pure materials?

CHAPTER 8

What Every Consumer Should Know About Energy

In our industrialized, high-tech, appliance-oriented society, the average use of energy per individual is at its highest point in the history of the world (Fig. 8-1). In the United States alone, with only 5% of the world's population, we consume 30% of the daily supply of energy. We are highly dependent on a huge supply of energy.

Who is using energy?

What is it we use? **Energy**—defined as the ability to do work, which is accomplished only by moving things—is involved every time anything moves or changes. Types of energy involved with matter in motion include heat (molecules in motion), electricity (electrons in motion), sound (compression and expansion of the space between molecules), and mechanical energy (macroscopic objects in motion). All matter in motion involves **kinetic energy,** the energy of motion.

What is energy?

Energy can be stored. Examples include energy stored in chemical bonds (as in wood and food), in the nuclei of atoms (atomic energy), and in gravitational systems (rocks on the top of a hill). Stored energy is **potential energy.**

What are the practical sources of energy? In one way or another, most of our daily energy needs are supplied by the fossil fuels petroleum, coal, and natural gas (Fig. 8-2). Smaller amounts are supplied by nuclear energy and food, and still smaller amounts are provided in the form of direct solar energy, geothermal energy, wind currents, and ocean currents. These so-called **primary sources** of energy may be converted into electricity (a secondary source), the form of energy we may find more useful.

Table 8-1 is a chart of energy units based on equivalences of coal, oil, natural gas, and electricity. A **British thermal unit** (Btu) is the amount of heat required to raise the temperature of 1 pound of water 1°F. A smaller unit of energy, the **calorie,** is the amount of heat required to raise the temperature of 1 gram of water 1°C. A kilocalorie (kcal) is 1000 calories. From the table you can see that 252 kcal (or 252,000 calories) are the same amount of heat as 1000 Btu. Thus, 252 calories are the same amount of heat as 1 Btu. Equivalent values for other units of energy can also be deduced. For example, 4.18 **joules** are the same as 1 calorie, and 860 kcal are the same as one kilowatt-hour of energy. (The kilowatt-hour will be defined in the later discussion of electricity.) A convenient unit to use when discussing large amounts of energy is the **quad,** which is 1 quadrillion (10^{15}) Btu.

Our ultimate source of energy is the fusion of nuclei in our Sun. In a major way, the energy of the Sun has been and is being made available to us through photosynthesis. It is thought that photosynthesis of long ago produced the plants that were converted into coal, petroleum, and natural gas, though the processes by which this occurred are not completely understood. During every moment of daylight today, yesterday, and tomorrow, photosynthesis produces our food supply either directly or through animal chains. The products of photosynthesis store chemical energy as only a holding form for the

Our dependence on the Sun is immense and multifaceted.

Photosynthesis is discussed in Chapter 11.

Figure 8–1
Energy use per capita by various types of societies. (From Turk, A., et al.: *Environmental Science,* 2nd ed. Philadelphia, Saunders College Publishing, 1978.)

Table 8–1
A Chart of Energy Units*

Cubic Feet of Natural Gas	Barrels of Oil	Tons of Bituminous Coal	British Thermal Units (Btu)	Kilowatt Hours of Electricity	Joules	Kilo-Calories†
1	0.00018	0.00004	1000	0.293	1.055×10^6	252
1000	0.18	0.04	1×10^6	293	1.055×10^9	0.25×10^6
5556	1	0.22	5.6×10^6	1628	5.9×10^9	1.40×10^6
25,000	4.50	1	25×10^6	7326	26.4×10^9	6.30×10^6
1×10^6	180	40	1×10^9	293,000	1.055×10^{12}	0.25×10^9
3.41×10^6	614	137	3.41×10^9	1×10^6	3.6×10^{12}	0.86×10^9
1×10^9	180,000	40,000	1×10^{12}	293×10^6	1.055×10^{15}	0.25×10^{12}
1×10^{12}	180×10^6	40×10^6	1×10^{15}	293×10^9	1.055×10^{15}	0.25×10^{15}

* Based on normal fuel heating values. 10^6 = 1 million. 10^9 = 1 billion, 10^{12} = 1 trillion, 10^{15} = 1 quadrillion (quad).
† 1 food calorie = 1000 calories = 1.000 kcal.

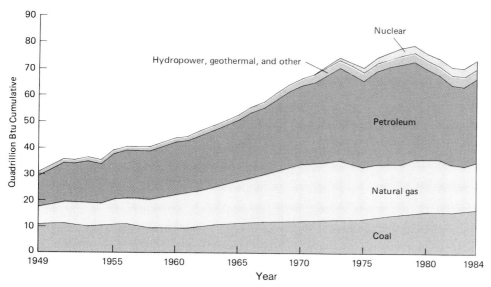

Figure 8-2
Energy consumption in the United States—the immediate view. Note that the burning of fossil fuels (coal, petroleum, and natural gas) furnishes nearly all of our present energy in spite of all the talk about hydroelectric and nuclear energy. *Btu* = British thermal unit (see Table 8-1). (From *Annual Energy Review*, 1984. Department of Energy, Energy Information Administration.)

almost unlimited amount of nuclear energy in our Sun. In addition to carrying out photosynthesis, the Sun's energy controls wind patterns, precipitation, water power, and seasons. And of course the Sun also provides us with heat, sunsets, suntanning, and the energy for solar batteries. Knowing how enormous, complex, and absolutely necessary are the Sun's contributions to life on Earth, we may be concerned about how long the Sun will last; with a mass of primarily highly concentrated hydrogen 330,000 times greater than the mass of Earth, the Sun is expected to last an estimated 4 to 5 billion years.

The amount of energy that enters the Earth's atmosphere from the Sun each day is enormous (about 2×10^{15} kcal/min, or 2.0 cal/cm^2/min), although it is only about three ten-millionths (0.0000003) of the total energy emitted by the Sun. About half (about 1 cal/cm^2/min) of this energy reaches the surface of Earth; the rest goes to reradiation from the atmosphere and to the absorption and scattering of radiant energy by the lower portion of the atmosphere. The actual amount that reaches the surface depends on location, season, and weather conditions. However, even 1 cal/cm^2/min is a large amount of energy. For example, at this rate, the roof of an average-sized house receives about 10^8 cal/day, equivalent to the heat energy derived from the burning of about 32 pounds of coal a day or to 120 kilowatt-hours of electrical energy a day—more than enough to heat an average American home in the winter.

The purposes of this chapter are to understand some consumer interest principles about energy, to describe how energy is extracted from chemicals, and to summarize our energy situation.

Purposes of this chapter

Fundamental Principles of Energy

The Distinction Between Energy and Power

Energy is the ability to move matter and may be expressed in units of calories, Btu, or joules; in contrast, **power** is the rate at which energy is used. Power is expressed in units of energy used per time, such as calories per second or joules per second (watts). Some consumers confuse energy and power when looking at their electrical bills; consumers pay for the amount of energy they use (kilowatt-hours), not for how fast they use it (kilowatts, or kilojoules per second). Some industries that consume huge amounts of electricity are given restrictions on the rate at which they can use energy.

Power is the rate at which energy is used.

A joule has fundamental units of $kg \cdot m^2 s^{-2}$.

When Shopping for Energy, Choices Are Limited

By far most substances around us are low-energy compounds. Most mixtures, such as rocks, dirt, earth, and water, are chemically oxygen-saturated. Only a few products of photosynthesis (food, fossil fuels, wood) are in the chemical position of providing energy through oxidation, the principal way in which we obtain energy from chemicals. Our choices of sources of energy from chemical reactions are limited and dwindling.

Where are dirt and water on the chemical energy supply scale?

Energy Extraction from Chemical Change is Net—Not All The Energy in a Chemical

Energy is absorbed (is endothermic) when a chemical bond is broken to yield isolated atoms. The energy required to break 1 mole of a particular kind of bond is the **bond energy** (Table 8-2). The same amount of energy is released (is exothermic) when 1 mole of bonds is formed from isolated atoms.

A given chemical reaction is exothermic if the formation of new bonds liberates more energy than is required to break the bonds in the reaction; a reaction is endothermic if the bonds in the reactants are stronger than those in the products. In summary, we receive energy from a chemical reaction only when more energy is produced by bond-making than is required for bond-breaking.

For example, consider the oxidation of methane, the principal component of natural gas:

$$H-\underset{\underset{H}{|}}{\overset{\overset{H}{|}}{C}}-H + 2\,\dot{O}-\dot{O} \longrightarrow O=C=O + 2\,H\overset{O}{\diagup \diagdown}H + 192 \text{ kcal/mole } CH_4$$

The energy released in this exothermic reaction can be used to heat houses, drive gas turbines, and generate electricity. The fact that energy is released means that it takes less energy to break the C—H bonds and Ȯ—Ȯ bonds than is produced when the C=O bonds and O—H bonds are formed. The bond energies in Table 8-2 bear this out. It takes only 632 kcal to break 2 moles of Ȯ—Ȯ bonds and 4 moles of C—H bonds [2(118) + 4(99) = 632]. The making of bonds produces 828 kcal generated by the formation of 2 moles of C=O bonds and 4 moles of O—H bonds [2(192) + 4(111) = 828]. This gives a net release of 196 kcal (828 − 632 = 196), which closely agrees with the experimental value of 192 kcal/mole of CH_4 burned.

Combustion is oxidation that produces heat and light. Burning is an example.

If the supply of oxygen is sufficient, **combustion** of fossil fuels and wood produces principally carbon dioxide and water. The net energy received from the burning of the fuel is the difference between the energy given off in making bonds in CO_2 and H_2O and the energy required to break bonds in O_2 and the fuel. Water and carbon dioxide have

Table 8-2
Approximate Bond Energies in kcal/mole

Bond	Energy	Bond	Energy	Bond	Energy
H—H	104	I—I	36	I—Cl	50
C—C	83	H—F	135	O—Cl	49
C=C	146	H—Cl	103	S—Cl	60
C≡C	200	H—Br	88	N—Cl	48
N—N	40	H—I	71	P—Cl	79
N=N	100	H—O	111	C—Cl	79
N≡N	225	H—Se	66	Si—Cl	86
O—O	33	H—S	81	C—O	84
Ȯ—Ȯ in O_2	118	H—N	93	C=O	173
S—S	51	H—P	76	C=O in CO_2	192
F—F	37	H—C	99		
Cl—Cl	58	H—Si	70		
Br—Br	46	Br—Cl	52		

The O_2 molecule has two unpaired electrons, as indicated by the dots in Ȯ—Ȯ.

their capacities for oxygen satisfied and therefore, cannot be burned to extract more energy. The amount of energy derived from the combustion of some of nature's storehouses of energy are given in Table 8-3.

The Law of Conservation of Energy

Also known as the first law of thermodynamics, the law of conservation of energy asserts that energy is neither lost nor gained in all energy processes. When a beaker of water is heated on a burner, all of the energy given off by the flame can be accounted for in the increased energy of the water and its surroundings; no energy is lost or gained in the

Thermodynamics is the movement of energy.

Table 8-3
Heat Produced by the Combustion of Some Organic Materials

Substance	Heat (kcal/g)
Methane (principal component of natural gas)	13.2
Gasoline, kerosene, crude petroleum, tallow	9.5–11.5
Lipids	9.0–9.5
Carbon (coal)	7.8
Ethyl alcohol	7.1
Proteins	4.4–5.6
Carbohydrates (sugars and starches)	3.6–4.2

These figures correspond to laboratory combustion to yield CO_2, H_2O, and oxides of nitrogen. In the body, proteins are oxidized to CO_2, H_2O, and urea. For the latter process, the heat yield is less than the value indicated above. Thus, proteins and carbohydrates yield (per gram) about the same energy in the body.

First statement of the first law of thermodynamics: "A force [translated: energy] once in existence cannot be annihilated"—Julius Robert Mayer, a ship's doctor, 1840.

The law of conservation of energy was extended by Albert Einstein, who showed the interrelationship between mass and energy by $E = mc^2$.

The more general law is: The total amount of matter and energy in the universe is constant.

Usable energy is not conserved.

No matter how we try, we can never convert all of the stored energy in a system into usable energy.

transformation of chemical energy to heat energy. Furthermore, the transformation is quantitative, in that a certain amount of gas burned produces a certain amount of energy (see Table 8–3); when one kind of energy is changed into another, the exchange rate is definite, reliable, and reproducible.

The law of conservation of energy also implies that the total amount of energy in the universe is constant. Energy is transformed regularly from one kind to another, but the total remains the same. This means that the Sun and the energy stored in chemicals on the Earth are what we have to use—that is all! There is no creation of new energy.

One last implication of the law applicable to this study concerns perpetual motion. The law recognizes that a machine cannot produce enough energy to run itself, much less create enough energy to be used elsewhere. At a minimum, the machine would have to create enough energy to move its parts and to overcome friction, but this is creation, not transformation, and is therefore impossible.

Energy Is Conserved in Quantity but Not in Quality

What does this second law of thermodynamics mean: that energy is conserved in quantity but not in quality? Perhaps two examples will clarify the concept. Consider first the release of some energy by the burning of coal, petroleum, or wood. Recall from our previous discussion that the main products of these combustion reactions, carbon dioxide and water, will not burn and release more energy. In the burning process, both matter and energy are conserved, as required by the laws of conservation of matter and energy, respectively. However, the reactants and their stored energy are more useful in energetic terms than the products and their spent energy.

As a second example, consider an electrical motor. The electricity that runs the motor is more useful than the heat that comes from the warm motor. Again, energy is conserved in the process of running an electrical motor, but the usable energy is not conserved.

In concept, the energy relationships in the second law of thermodynamics can be compared to the relationships among gross income, deductions, and net pay (or realizable income) in a paycheck. A certain amount of energy is available for the process considered; this is analogous to gross income. Some of the energy is not usable owing to frictional losses, electrical shorts and drains, retention of some energy in the chemical products, or some other factor affecting efficiency; this is represented by paycheck deductions. Finally, some of the energy is usable; it is analogous to net pay. The energy, like the money, is accounted for as required by the first law of thermodynamics.

In all processes, then, some energy is wasted—not lost—by conversion into energy that is not usable in doing work. The wasted (or unusable) energy is represented by **entropy,** a measure of the disorder in a physical system. Entropy is not energy per se, but it is a function of energy with units of energy per degree, such as calories/degree.

Entropy means disorder, and measures nonuseful energy.

Another statement of the second law of thermodynamics is based on entropy: In all natural processes, entropy is increased. Taken to its extreme, this means that the entropy of the whole universe is increasing at the expense of stars running down in usable energy at a tremendous rate. This is not a reason for worry, because the universe is so vast that enough usable energy is there for all conceivable purposes for many billions of years. However, sources of usable energy that are not limitless are the so-called fossil fuels (coal, petroleum, and natural gas), which when gone are not easily restored. It would take eons for photosynthesis to regenerate the material for new fossil fuel deposits.

Why is the energy used to increase entropy instead of usable energy? The derivation of the word *entropy,* meaning "disorder," explains. The ultimate fate of any change in energy is a form of heat energy caused by the random, disordered motion of molecules. Have you ever thought about what happens to the light energy coming from a light bulb, or what happens to electrical energy once it is used to run an electrical motor, or what happens to the sometimes large amounts of energy that result from an explosion? All forms of energy, including sound, are converted eventually into random molecular motion, in which the molecules move faster and (or) further apart. Molecules moving in all directions are not as useful in bringing about controlled change as are electrons, photons of light, or molecules when they are moving from one point to another in organized fashion.

The end of the line for energy is the random motion of molecules.

Let us summarize this brief discussion of the second law of thermodynamics by describing the energy coming from a burning match. Some of the energy can be used to ignite another object, or to heat an object, or to provide light; this is the organized energy. However, while the usable energy is being used, some of the total energy emanating simply heats molecules in the vicinity and increases the entropy of the molecules. Eventually, all of the heat and light coming from the match becomes increased random motion of the molecules.

What does the second law of thermodynamics mean to the informed citizen? Simply stated, when usable energy-rich chemicals such as coal and petroleum are consumed, the usable energy is less than the total energy produced.

When fossil fuels are gone, then what?

The Efficiency of Energy Use Is Low

In every energy process, the efficiency of energy use is less than 100%, usually far less. Automobiles are about 20% to 25% efficient; that is, about 80% of the useful energy available to do work is lost and not applied to the turning of the wheels. Some fuel cells are about 70% efficient. The human body is about 45% efficient in converting the energy of glucose metabolism to muscle movement. Photosynthesis is 2% to 10% efficient, steam turbines for producing electricity are about 38% efficient, heating homes with electricity is about 38% efficient, and heating homes with natural gas is about 70% efficient. The efficiency is usually greater when a **primary source** of energy is used (gas) than when a **secondary source** is used (electricity). For example, it takes about 10,000 Btu's to produce 1 kilowatt-hour of electricity. If this 1 kilowatt-hour is then used for heating, only 3800 Btu's of heat are produced. Natural gas burned on site would be more efficient than natural gas burned in a steam generator plant to produce electricity.

Efficiency is used energy ÷ available energy.

Primary source of energy: one transformation on site (e.g., chemical → heat via combustion). Secondary source: usually more than one transformation, plus long-distance transport (e.g., chemical → heat via combustion → steam → mechanical → electricity).

Energy Not Lost Is "Energy Gained"

Energy can be transported through wires (electricity), stored in chemicals (batteries), and carried through space (radio waves). On the other hand, energy can be prevented from moving by means of insulators. The insulation of houses has popularized the **R value** for heat insulators. The R value (the resistance) is inversely proportional to the conductivity of heat through a slab of material; a common unit of R value is (ft^2) $(°F)$ (hr/Btu). An R value of 30 is typically recommended for the ceilings of single-family dwellings; an average square foot of such a ceiling would lose heat by conduction at a rate of $(\frac{1}{30})$ Btu/hr for every 1°F difference in temperature. The higher the R value, the fewer Btu's escape per hour per square foot of ceiling. Some R values for 1-inch slabs of material (in units of $ft^2 °F\ hr/Btu$) are air, 5.9; polyurethane foam, 5.9; rock wool, 3.3; fiberglass, 3.0; white pine, 1.3; and window glass, 0.14. Dry, still air has an insulating value (R value) as great as almost any building material. In fact, many commercial materials owe their heat-insulating ability to trapped, isolated pockets of air.

The R factor for heat insulators.

Some Materials Have a Higher Energy Cost Than Others

It costs more in energy terms to produce a ton of some substances than to produce a ton of other substances (Table 8–4). Certain applications now using plastics or metals

It costs less energy to make a ton of steel than to make a ton of aluminum.

Table 8–4
Energy Requirements to Produce Some Common Products*

Product	Millions of Btu/Ton
Titanium	482
Aluminum	244
Copper	112
Polyethylene	100
Polystyrene	64
Polyvinylchloride	49
Plate glass	25
Steel slabs	24
Paper	22
Portland cement	8
Brick	4

* From Peter R. Payne, "Which material uses the least energy," *Chem. Tech,* September 1980, p. 550.

might more efficiently use ceramics or brick to conserve energy. Of course, other factors such as labor costs also influence the economic decisions involved.

SELF-TEST 8–A

1. In 1984, what energy source provided the most energy for U.S. consumption (see Fig. 8–2)? _____
2. Which furnishes the most heat energy per gram: coal, petroleum, or natural gas? _____
3. The typical efficiency of an electrical generating plant is about ( ) 100%, ( ) 50%, ( ) 33%, or ( ) 10%.
4. Examples of fossil fuels are _____, _____, and _____.
5. Natural gas and petroleum react with _____ to produce CO_2 and _____.
6. All combustions of fossil fuels give off energy.   True ( ) or False ( )
7. Energy is the ability to do _____.
8. One type of energy (for example, light) is always transformed into another type of energy (for example, heat) ( ) quantitatively, ( ) not quantitatively, or ( ) sometimes quantitatively, sometimes not quantitatively.
9. The ultimate fate of all types of energy is an increase in _____.
10. Although the quantity of energy is conserved, the _____ of energy is not conserved.
11. Three units of energy are _____, _____, and _____.

12. Two units of power are _____ and _____.
13. Which costs more energy to produce, a ton of aluminum or a ton of brick? _____ Which costs more money to buy? _____

Fossil Fuels

The oxidation of coal, petroleum, and natural gas provides most of the energy for our nation as well as for the world. For each material, the products of complete combustion are carbon dioxide and water. When coal or petroleum is combusted incompletely or contains certain impurities, major air pollution is possible. For example, the sulfur in coal or petroleum burns to form sulfur dioxide, SO_2, a major cause of acid rain.

Acid rain is discussed in Chapter 15.

No ongoing coal or petroleum production underground has been detected. The fossil fuels that have already been found may be all there is. We must use these limited fossil substances wisely.

Petroleum

Petroleum was first discovered in the United States (in Pennsylvania) in 1859 and in the Middle East (in Iran) in 1908. Today petroleum is pumped from the ground in many parts of the world. As an energy source, petroleum was first used as kerosene for lighting, then as gasoline and aircraft fuel for transportation, and most recently as fuel oil to produce electricity. Of the 75 quads of energy used per year in the United States in recent years, 35 quads (47%) were supplied by petroleum (see Fig. 8-2).

One quad is 10^{15} Btu.

Petroleum is a mixture of many hydrocarbons. By refining (separation of the components of the mixture by distillation) and subsequent conversions, much crude oil is turned into gasoline (42.6%, or 6.7 million barrels per day in the United States), while lesser amounts are turned into fuel oil (26.8%), jet fuel (7.4%), and other miscellaneous fuels. From the small fraction of petroleum that is not burned comes thousands of chemicals known as the petrochemicals; they will be discussed in Chapter 10. Other uses of petroleum include the production of edible fats, which was done in Germany during World War II. Glycerol is now made from petroleum in Germany on a commercial scale, and the process for making sugar from oil has also been developed. The present and next few generations must decide whether to burn our limited supply of petroleum for energy production or to save oil for petrochemical products.

See Figure 9-5 for a distillation tower for petroleum and for a discussion of chains of carbon atoms. There are 42 gallons of oil per barrel.

How limited is the supply of petroleum? Figure 8-3 shows how much oil the world has used since 1900 and how much the world has available for about the next 100 years. New oil wells are being drilled every day; other wells come in every day. These new wells raise the projected curve only slightly. As Third World countries become more industrialized, their energy requirements will increase, lowering the projected curve. Some new wells will be required just to keep oil use in the advanced nations at a status quo.

If we are willing to pay the price, there is more oil in the ground that is hard to remove. When an oil well comes in, only the **recoverable oil** is removed. Usually about 30% is obtained, with 70% left in the ground. As supplies decrease and costs go up, it may prove economically feasible to "recover" more oil.

A huge, mostly untapped source of petroleum is **oil shale rock.** Three immense deposits of oil shale rock in the United States are the most-developed deposit in Utah and Colorado, a giant U-shaped formation from Michigan and Pennsylvania to Alabama (believed to hold a trillion barrels of oil), and a vast deposit in north-central Alaska. The Alaskan oil shale yields only a few gallons of oil per ton of rock, whereas saturated ores test at 102 gallons of oil per ton of rock.

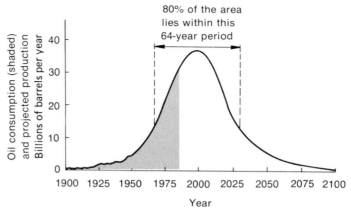

Figure 8–3
World oil use: past and projected.

Most oil shale lies near the surface of the Earth, and in its economic favor is the fact that it's not possible to hit a dry hole. The oil is obtained by heating of the rock in the absence of air. One technique involves blasting the rock, burning part of the oil underground and using the heat to force the oil out of the rock into underground holding reservoirs, and then pumping the collected oil to the surface.

Since the energy costs of obtaining petroleum from oil shale rock are high and the process of recovery leaves large amounts of waste rock, the production of petroleum from oil shale is not attractive as long as other sources of petroleum provide oil for $30 or less per barrel.

Coal

Coal is a mixture of hydrocarbons with a relatively small amount of sulfur. By way of contrast with petroleum, coal has more rings of carbon atoms (some rings are bonded to each other—fused) and more bonding of chains of carbon atoms to other chains of carbon atoms.

See Chapter 9 for typical chains and rings of carbon atoms.

Minable coal is defined as 50% of all coal in a seam at least 12 inches thick and within 4000 feet of the Earth's surface. In the United States the minable coal reserves are divided among anthracite (2%), lignite (8%), subbituminous coal (38%), and bituminous coal (52%). Some properties of the different kinds of coal are listed in Table 8–5.

Table 8–5
Some Properties and Characteristics of Types of Coal

Characteristics	Anthracite	Bituminous Coal	Subbituminous Coal	Lignite
Heat content	High	High	Medium	Low
Sulfur content	Low	High	Low	Low
Hydrogen/carbon mole ratio	0.5	0.6	0.9	1.0
Major deposits	New York, Pennsylvania	Appalachian Mts., Midwest, Utah	Rocky Mts.	Montana

From the information given in Figure 8–2, it can be seen that in recent years coal provided about 18 quads (23%) of the nation's energy requirements per year. This required the mining of about 900 million short tons of coal each year.

The largest portion of mined coal (about 75%) is burned to produce electricity. Only about 1% is used for residential and commercial heating. Coal's decline as a heating fuel was caused by its being a relatively dirty fuel, bulky to handle, and a major cause of air pollution (because of its sulfur content). The dangers of deep coal mining and the environmental disruption caused by strip mining contributed to the decline in the use of coal.

Some of the problems associated with the use of coal as a fuel can be alleviated by converting coal to combustible gases (coal gasification; see Chapter 10) or to liquid fuels (coal liquefaction). The liquid fuels are made from coal by reacting the coal with hydrogen under high pressure in the presence of catalysts (hydrogenating the coal). The process produces more straight chains of carbon atoms, like those found in petroleum. The products of coal gasification and liquefaction burn more cleanly, pollute less, and can be more easily transported than coal.

Like petroleum, coal supplies chemicals to the chemical industry. Most of the useful compounds in coal contain rings of carbon atoms (these rings are discussed in Chapter 9).

Given our great dependence on coal for the production of electricity and our lesser but still significant dependence on coal for the production of industrial chemicals, just how much coal do we have and how long is it likely to last? Geologists believe that all of the world's coal has now been discovered. The world's reserves are estimated to be about 990 billion short tons, of which about 29% is in the United States. How much coal has been used and how long coal is expected to last are summarized in Figure 8–4. Coal is expected to last several hundred years longer than petroleum. New mining techniques that make more of the deposited coal minable can extend the usable life of coal. However, as is the case with petroleum, as more Third World countries advance industrially and technologically, more coal will be used and the total supply will be depleted more quickly.

A short ton is 2000 pounds. A long ton is 2200 pounds.

Most coal is burned to make electricity.

Problems with coal.

The largest supply of fossil fuel is in the form of coal.

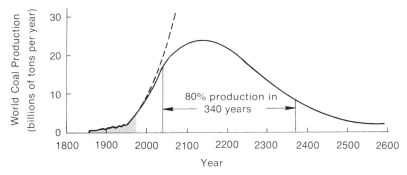

Figure 8–4
The coal mined to date (shaded area) represents only a small fraction of minable coal. The rate of increase in coal consumption (dashed line) is 4% per year. It is obvious that such an exponential rise cannot continue long after the year 2000. At the present usage (held constant), coal would last for many hundreds of years.

Natural Gas

> Natural gas is mostly methane, CH_4.

Natural gas burns with a high heat output (see Table 8-3), produces little or no residue or pollution from burning, and is transported easily. Practically the only pollution produced by the combustion of natural gas is CO_2, which can cause the greenhouse effect discussed in Chapter 15.

The production of natural gas peaked in the United States in 1973 at 22 trillion cubic feet per year. The world's reserves of natural gas are estimated to be about 3400 trillion cubic feet, most of them in the Middle East, Eastern Europe, and the Soviet Union. Even if significant new deposits of natural gas are discovered in such locations as the outer continental shelves or deep below the Earth's mantle, the North American natural gas deposits are already about 60% depleted, and not much time is available for tapping new sources. Importation is complicated by the danger of transporting and storing concentrated, condensed, volatile natural gas.

Even with higher prices and an impending depletion of natural gas, more homes are heated by the burning of natural gas than by any other means. In 1983 55% of the homes in the United States were heated by natural gas, followed by electricity (18.5%), fuel oil (14.9%), wood (4.8%), and liquefied gas such as butane and propane (4.6%). Coal and kerosene came in at a low 0.5%, and solar heating of homes was even lower.

Novel, Nonmainstream Sources of Energy

Some energy sources offer more potential than they now supply, though most of these sources are available only in certain areas. For example, in some locales (e.g., Boise, Idaho) it is possible to heat homes and make electricity from hot springs and geysers. Other areas near oceans derive energy from ocean currents and temperature differences between warm surface water and colder, deeper water. For many years wind currents and windmills have provided energy for pumping water and making electricity in some places.

> Everybody has it; nobody wants it. Why not use it to provide energy? Great idea!

One novel source of energy is available in every populated area but is currently tapped in only a few locations. Garbage is everywhere and can be used to produce energy by burning or fermenting. Burning the garbage is a solution to an energy problem as well as to the problem of what to do with tons and tons of trash.

Plants for burning garbage to extract energy are in operation in several countries, including France and the United States. The Nashville Thermal Transfer Corporation in downtown Nashville, Tennessee, began operation in February 1974 (Fig. 8-5). The plant supplies steam and (or) cold water to 28 buildings and about 30,000 people in the downtown area. The energy comes from burning 400 tons of residential garbage each day. Pipes laid under the city carry the steam and (or) cold water to the various buildings. The cold water used to cool the buildings is cooled by electricity produced at the plant by steam-powered turbines. The electricity cools the water in a manner similar to the action of a very large refrigerator. The ash from the burning of the garbage has to go to a landfill, but this ash is less than 10% of the volume and 30% of the weight of the original garbage. Oil and gas are available as backup energy sources if needed, but they are used rarely.

A product of the fermentation of garbage is extracted from the world's largest garbage dump at Fresh Kills on Staten Island, New York. Underneath the huge mounds of garbage, bacteria turn some of the old buried garbage into methane (natural gas). The Brooklyn Union Gas Company has tapped this gas, which provides enough methane to fuel 16,000 homes on Staten Island.

Figure 8-5
The Nashville, Tennessee, thermal energy plant.

Controlled Nuclear Reactions as a Source of Energy

When basic concepts of nuclear reactions were discussed in Chapter 4, fission and fusion of nuclei were described as sources of vast amounts of energy from a relatively small amount of fuel. The emitted energy comes from the conversion of some mass into energy according to Einstein's equation, $E = mc^2$. Interest in controlling nuclear energy has led to the production of controlled nuclear fission reactors and to intense research on sustaining and controlling nuclear fusion. Nuclear fission reactors are a practical reality; nuclear fusion reactors are only in the early experimental stage.

The fission of a $^{235}_{92}U$ nucleus by a slow-moving neutron to produce smaller nuclei, extra neutrons, and large amounts of energy suggested to Enrico Fermi and others that the reaction could proceed at a moderate rate if the number of neutrons could be controlled. If a neutron control could be found, the concentration of neutrons could be maintained at a level sufficient to keep the fission process going but not high enough to allow an uncontrolled explosion. It would then be possible to drain the heat away from such a reactor on a continuing basis to do useful work. In 1942 Fermi, working at the University of Chicago, was successful in building the first atomic reactor, called an **atomic pile**.

An atomic reactor has a number of essential components. The charge material (fuel) must be fissionable or contain significant concentrations of a fissionable isotope such as $^{235}_{92}U$, $^{239}_{94}Pu$, or $^{233}_{92}U$. Ordinary uranium, which is mostly the nonfissionable $^{238}_{92}U$, cannot be used because of its small concentration of the $^{235}_{92}U$ isotope. A moderator is required to slow the speed of the neutrons produced in the reactions without absorbing them. Graphite, water, and other substances have been used successfully as moderators. A substance that absorbs neutrons, such as cadmium or boron steel, must be present as a control over the neutron concentration. Shielding to protect workers from dangerous radiation is an absolute necessity. Such shielding (thick concrete or heavy water composed of deuterium hydrogen, $^{2}_{1}H$) makes reactors heavy and bulky. A heat-transfer fluid (water or liquid sodium) provides a large and even flow of heat away from the reaction center. The heat extracted from the nuclear reactor can be used to generate electricity or to operate any device that uses heat energy. A system for the nuclear production of electricity is illustrated schematically in Figure 8-6.

Fission is the breakup of heavy nuclei.

Review the discussion of $E = mc^2$ in Chapter 4.

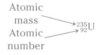

Atomic pile:
1. Carefully diluted fissionable material
2. Moderator to control fission reaction
3. Coolant to control heat
4. Shielding to limit radiation

The first one was piled together at the University of Chicago in 1942.

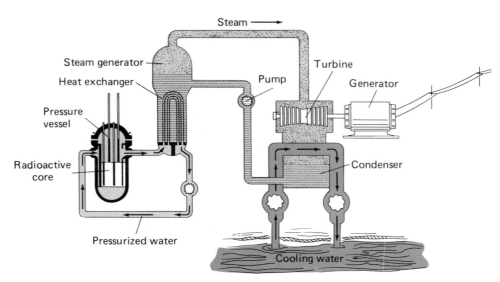

Figure 8-6
Schematic illustration of a nuclear power plant used for the production of electricity. No shielding is shown on the reactor.

When 1 g of ^{235}U undergoes fission, it provides the same energy as does burning about 6 tons of coal.

One gram of $^{235}_{92}$U undergoing fission produces the same amount of energy as the combustion of either 5.95 tons of coal or 13.7 barrels of oil. For a 1000-megawatt electric power plant, about 3 kg of $^{235}_{92}$U would be required per day. Since the fuel contains only about 3% $^{235}_{92}$U (the remainder being nonfissionable $^{238}_{92}$U), about 300 kg of packaged fuel would be required per day, compared with 18,000 tons of coal or 42,000 barrels of oil.

In 1965 when controlled nuclear energy usage began, and until about 1974, more energy was produced in the United States from the burning of firewood than from nuclear energy. In 1984 4.8% of the energy used in the United States was supplied by nuclear sources; by 1990 almost 10% of our energy needs are expected to be met by nuclear sources. On January 1, 1986, there were 98 operable nuclear reactors in the United States, with 32 on order but not yet in operation (Fig. 8-7).

See Chapter 1 for a more complete description of the Chernobyl accident and its effects.

The Chernobyl nuclear reactor accident in 1986 was caused by an experiment gone awry. The experiment was designed to determine how long it would take the reactor to reach the danger point when the control rods were out. Some automatic shutdown systems did not work, others had been cut off for reasons that are unknown, and the experiment got out of hand. The reactor did not have containment or as much shielding as is found on more modern plants. Some of the consequences of a nuclear meltdown such as the one that occurred at Chernobyl are illustrated in Figure 8-8.

$^{235}_{92}$U is in very short supply.

Other than the problems of possible human error (which may cause a shower of radioactive debris from a hydrogen-oxygen explosion — not a nuclear explosion) and of what to do with the long half-lived products of nuclear reactors (Chapter 4), there is the problem of a possibly limited supply of uranium. The United States is producing progressively less uranium, down from a high of 21,850 short tons of U_3O_8 in 1980 to 7500 short tons in 1984. Reserves in 1983 were 885,000 short tons of U_3O_8 reasonably assured and 4,394,000 short tons speculated and estimated. This should be enough uranium to last for several decades at the present rate of consumption.

There is a way to make the uranium last longer. **Breeder reactors** convert nonfissionable $^{238}_{92}$U and $^{232}_{90}$Th into fissionable fuels. In a breeder reactor, a blanket of nonfissionable material is placed outside the fissioning $^{235}_{92}$U fuel, which serves as the source of neutrons for the breeder reactions (Fig. 8–9). The two breeder reactions are

$$^{238}_{92}\text{U} + ^{1}_{0}\text{n} \longrightarrow ^{239}_{92}\text{U} \xrightarrow{\beta} ^{239}_{93}\text{Np} \xrightarrow{\beta} ^{239}_{94}\text{Pu}$$

$$^{232}_{90}\text{Th} + ^{1}_{0}\text{n} \longrightarrow ^{233}_{90}\text{Th} \xrightarrow{\beta} ^{233}_{91}\text{Pa} \xrightarrow{\beta} ^{233}_{92}\text{U}$$

The products of the breeder reactions, $^{233}_{92}$U and $^{239}_{94}$Pu, are both fissionable by slow neutrons, and neither is found in the Earth's crust.

Breeder reactors present many technological problems, not the least of which is the potential of a disaster caused by mishandling of the $^{239}_{94}$Pu isotope, which is extremely toxic and can also be fabricated into a fission bomb. Nevertheless, the expected benefit from the breeder program is massive amounts of energy. If breeder reactors were used to produce electricity, the breeder-produced fuels would supply the electrical requirements for the United States for more than 2000 years at present rates of electricity use. If a breeder program does not become more widespread, the $^{235}_{92}$U will be gone someday and another, as yet undiscovered, source of neutrons will be needed to convert $^{238}_{92}$U and $^{232}_{90}$Th into fissionable fuels.

The breeder concept came from Walter Zinn, first director of Argonne National Laboratory outside Chicago. The first nuclear reactor of any kind to produce electricity was a research breeder reactor at Idaho Falls, Idaho, which began operating in 1951. An updated version (EBR-2, experimental breeder reactor) has been in operation for more

$^{239}_{94}$Pu is toxic from a radiation as well as from a chemical point of view.

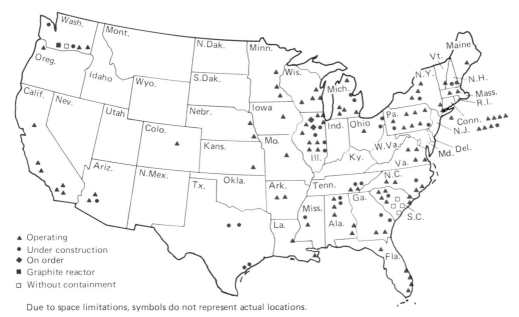

▲ Operating
● Under construction
◆ On order
■ Graphite reactor
□ Without containment

Due to space limitations, symbols do not represent actual locations.

Figure 8–7
Status of nuclear reactors in the United States on January 1, 1986.

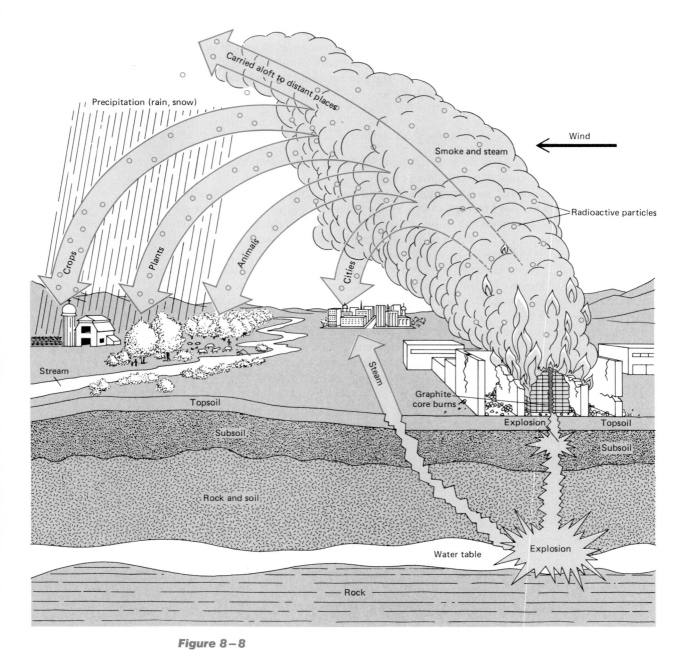

Figure 8–8
After a nuclear reactor meltdown, fire, and explosion, radioactive particles are broadcast from the site by explosive forces, rising hot air, and steam. Such an explosion is chemical, *not nuclear*. With current reactors, the steam in contact with zirconium or other active metals produces hydrogen gas, which explodes. If the explosion and/or fire produces cracks to the water table, groundwater carries radioactive particles throughout the region. Radioactivity incorporated into plants or absorbed by streams on the surface finds its way up the food chain into human beings. Either through ingested radioactive food or through inhaled radioactive particles, human beings and animals become susceptible to cancer.

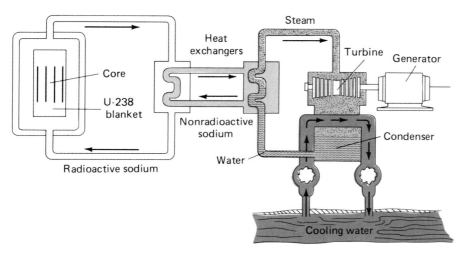

Figure 8-9
Schematic diagram of a fast breeder reactor and steam-turbine power generator.

than 20 years. Research on breeder reactors continues at Argonne as well as in other places such as France, England, and Japan.

Electricity: The Major Secondary Source of Energy

A secondary source of energy is made from a primary source on the way to the end user. Most of the nuclear energy, a primary source, generated by nuclear reactors at the locations shown in Figure 8-7 is used to produce electricity. At present other primary sources produce more electricity than nuclear energy (Fig. 8-10). Coal is used to produce more electricity than all of the other primary sources combined.

About 35% of all the energy consumed in the United States is used in the production of electricity. In 1984 the 26 quads of energy put into the production of electricity yielded about 8 quads of electricity in the home or factory using the electricity. The 18 quads difference was lost in the production and the transmission of electricity. At least part of this loss is expected because of the second law of thermodynamics (discussed earlier in this chapter), which states that a natural process loses some nonuseful energy as entropy (disorder) is increased.

A specific example of energy loss in the production of electricity in a coal-burning power plant is shown in Figure 8-11 and summarized below.

For a 1000-megawatt coal-burning plant, one hour of operation might look like this:

Coal consumed	696 tons producing 2.270 billion kcal
Smokestack heat loss	0.227 billion kcal
Heat loss in plant	0.106 billion kcal
Heat loss in evaporator to cool condenser	1.080 billion kcal
Electrical energy delivered to power lines	0.857 billion kcal
Percentage of energy delivered as electricity before transmission losses	$\dfrac{0.857}{2.27} \times 100\% = 37.8\%$

A few nuclear reactors, such as some at Oak Ridge, Tennessee, are used to produce radioactive isotopes for medicine and industry, not to produce electricity.

Electrical generating plants yield about one third of the fuel energy in the form of electrical energy.

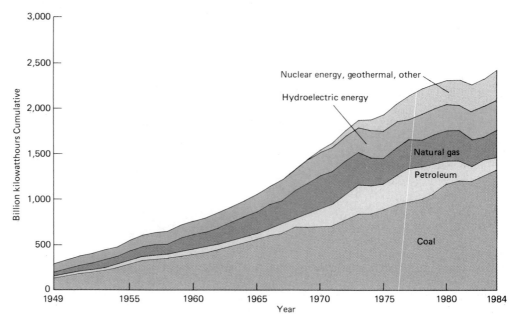

Figure 8–10
The generation of electricity by type of energy source. (From *Annual Energy Review*, 1984. Department of Energy, Energy Information Administration.)

> Research on superconductors, described in Chapter 7, may lead to a dramatic reduction in energy loss from transmission of electricity.

There is a further energy loss in the power lines and the transformers, lowering the useful output of the plant to 30% of the energy consumed. This is the **efficiency** figure for the overall operation. It is important to note that we pay for 300 kcal of heat energy in the form of coal or fuel oil but receive less than 100 kcal of energy in the form of electricity. Obviously, it requires much less fuel to heat homes with the fuel itself than with electricity made from the fuel.

> When we pay the electricity bill, we pay for energy (kilowatt-hours), not for power (how fast we use the energy, or kilowatts).

Controlled Nuclear Fusion

Three critical requirements must be met for controlled fusion. First, the temperature must be high enough for fusion to occur. The fusion of deuterium (^2_1H) and tritium (^3_1H) requires a temperature of 100 million degrees or more.

$$^3_1H + {}^2_1H \longrightarrow {}^4_2He + {}^1_0n + 4.1 \times 10^8 \text{ kcal/mol } {}^2_1H$$

> A plasma is a gaseous state composed of ions.

Second, the plasma must be confined long enough to release a net output of energy. Third, the energy must be recoverable in some usable form.

Attractive features that encourage research in controlled nuclear fusion are the rather limited production of dangerous radioactivity and the great abundance of hydrogen (in water), a most abundant resource. Any radioisotopes produced by fusion have short half-lives and therefore are a serious hazard for only a short period of time.

> Containment is one of the biggest problems in developing controlled fusion.
>
> Tokamaks and Stellarators are magnetic bottle research devices.

Fusion reactions have not been "controlled." No physical container can contain the plasma without cooling it below the critical fusion temperature. Magnetic "bottles," enclosures in space bounded by a magnetic field, have confined the plasma, but not for long enough periods of time. Recent developments suggest that these "bottles" may hold the plasma long enough for a fusion reaction to occur.

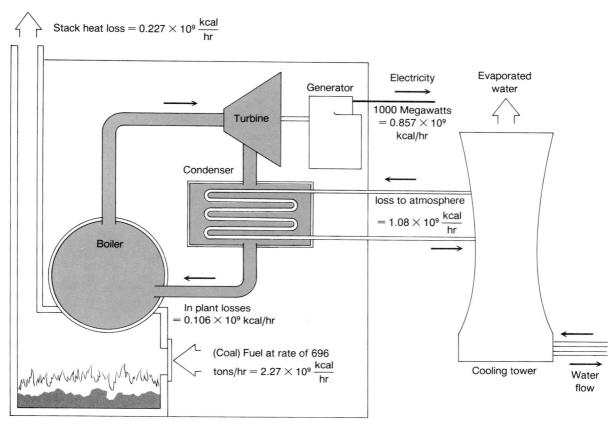

Figure 8-11
The heat balance of a 1000-megawatt coal-burning electrical generating plant. Note that the 969 tons of coal burned per hour furnish 2.27×10^9 kcal of heat energy, but that only 0.857×10^9 kcal of energy, or 38%, is converted to electricity. Note also the large amounts of heat energy lost to the cooling water and atmosphere.

A newer confinement method is based on a laser system that simultaneously strikes tiny hollow glass spheres called **microballoons;** these microballoons enclose the fuel, which consists of equal parts of deuterium and tritium gas at high pressures (Fig. 8-12).

It is hoped that controlled fusion will be demonstrated during the next decade, but it is unlikely that fusion will furnish any significant fraction of the world's energy needs before the turn of the century.

Solar Energy

Although the Sun is our ultimate source of energy, we use only a small fraction of what the Earth receives from it; efforts are now being made in several directions to use more of the Sun's gift to us. One technique is to use algae to produce hydrogen, which then can be used in fuel cells to produce electricity. Another device is the solar collector, which uses the warmth of sunlight to heat water and air to heat homes. A third device uses photosensitive materials to make a solar electrical cell, such as the type that is commonly used to power hand calculators.

Solar energy is transmitted nuclear energy.

See Chapter 4 for some equations on the fusion that produces the Sun's energy.

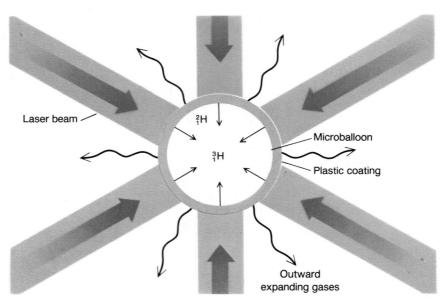

Figure 8–12
Focused laser light strikes the microballoon filled with deuterium and tritium, causing a plastic outer layer to burn off (or ablate). The outwardly expanding gases from the plastic material drive the glass sphere and its fuel contents inward. The high density and high temperatures produced might result in fusion.

Solar Energy and Algae Produce Hydrogen

Certain blue-green algae, *Anabaena cylindrica*, convert sunlight and water into hydrogen and oxygen. A colony of such algae, coupled with a fuel cell utilizing hydrogen and oxygen, could be a source of electricity during sunlight hours (Fig. 8–13).

Some catalysts reported in late 1982 for the decomposition of water in the presence of sunlight are indium phosphide (InP), phosphorus-doped silicon coated with Pt or Ni, and p-type iron oxide semiconductor. All are presently too expensive and (or) inefficient (12%, 12%, and 0.05%, respectively) to compete with hydrogen produced by the reaction of coal with steam.

$$C + H_2O \longrightarrow CO + H_2$$

Solar Energy and the Solar Battery

The goal is to capture and use the solar energy "on the run" without upsetting the energy flow into (light side) and away from (dark side) the Earth.

Another approach to the direct utilization of solar energy is the **solar battery,** known as a photovoltaic device. The solar battery converts energy from the sun into electron flow. Solar batteries are about 13% to 14% efficient and are capable of generating electrical power from sunlight at the rate of at least 90 watts per square yard of illuminated surface. They are now used in space flight applications and communication satellites and in Israel, India, Pakistan, South Africa, and Azerbaijan SSR to obtain electrical power.

Doped silicon for transistors is discussed in Chapter 7.

One type of solar battery consists of two layers of almost pure silicon. The lower, thicker layer contains a trace of boron, and the upper, thinner layer a trace of arsenic. As pointed out in Chapter 7, the arsenic-enriched layer is an n-type semiconductor, and the boron-enriched layer is a p-type semiconductor. Recall that silicon has four valence electrons and is covalently bonded to four other silicon atoms (Fig. 8–14). Arsenic has

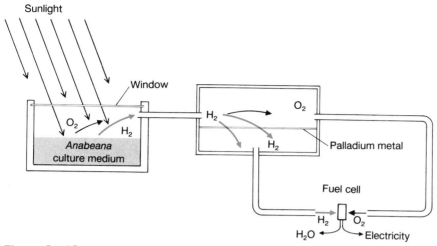

Figure 8–13
Schematic diagram of an electricity-producing photosynthesis process. H_2 and O_2 produced by the *Anabeana* are separated by palladium metal, which is permeable to H_2 but not to O_2. The H_2 and O_2 are then combined in the fuel cell to produce electricity.

five valence electrons. When arsenic is included in the silicon structure, only four of the five valence electrons of arsenic are used for bonding with four silicon atoms; one electron is relatively free to move. Boron has three valence electrons. When boron atoms are included in the silicon structure, there is a deficiency of one electron around the boron atom; this creates "holes" in the boron-enriched layer.

There is a strong tendency for the "free" electrons in the arsenic layer to pair with the unpaired silicon electrons in the "holes" in the boron layer. If the two layers are

Figure 8–14
Schematic drawing of semiconductor crystal layers derived from silicon. (From W. L. Masterton and E. J. Slowinski: *Chemical Principles*. Philadelphia, Saunders College Publishing, 1977.)

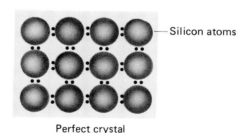

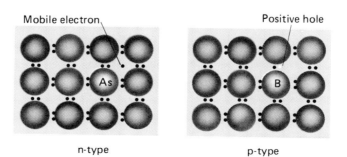

connected by an external circuit and quanta of light of sufficient energy strike the surface, excited electrons can leave the arsenic layer and flow to the boron layer. As the boron layer becomes more negative because of added electrons, electrons are repelled *internally* back into the arsenic layer, which is now positive and attracts the electrons from the boron layer. The process can continue indefinitely as long as the cell is exposed to sunlight.

Advantages of the solar battery are that it has no moving parts, no liquids, and no corrosive chemicals. The drawbacks of the solar battery are the large area required for large amounts of power, the high costs of the pure materials needed, and the fact that solar cells work only when the sun is shining. Since the first practical use of solar batteries in 1955 to power eight rural telephones in Georgia, solar cells have undergone a great deal of development, and much more is expected because of their great potential use. For example, if solar batteries become cheap enough, the electricity they supply could be used to electrolyze water to yield hydrogen and oxygen. The hydrogen could be piped to where the energy is needed and burned to heat water to steam, which in turn could generate electricity. Such an arrangement could give rise to a hydrogen economy and release us from our dependence on fossil fuels and nuclear energy.

> Solar batteries are still very expensive.

> Hydrogen can be burned in most devices that now burn natural gas.

> Some hydrogen-powered buses and cars are now operating on an experimental basis.

Solar Energy and Solar Collectors

The four major types of solar devices for heating water or air in buildings are illustrated in Figure 8–15. A total of 16.83 million square feet of solar collectors was sold in 1983.

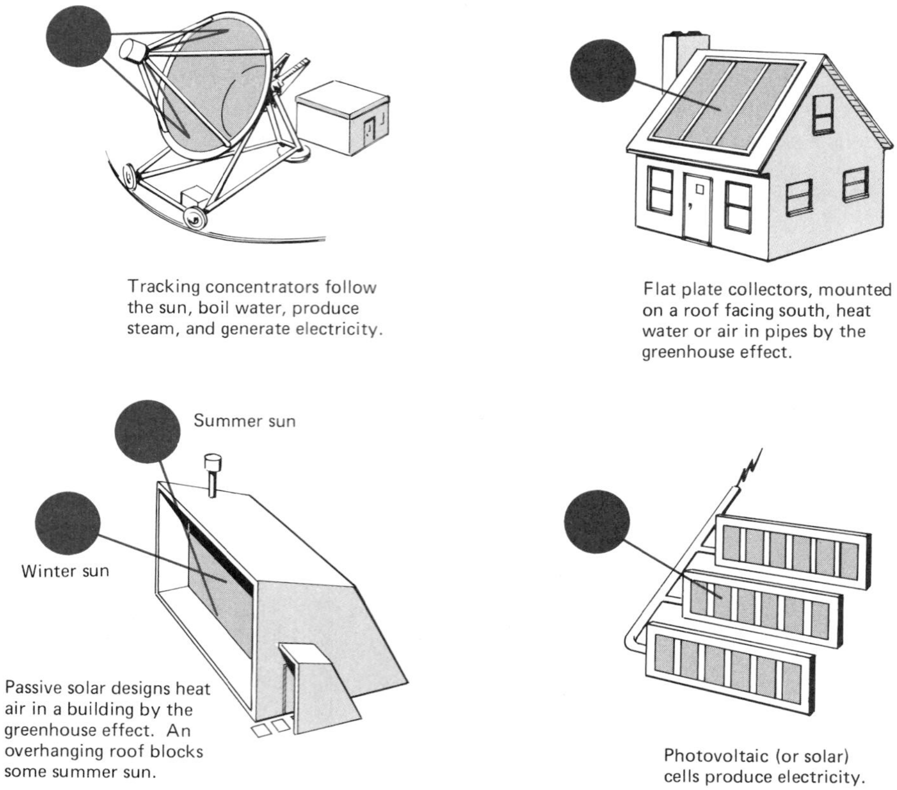

Tracking concentrators follow the sun, boil water, produce steam, and generate electricity.

Flat plate collectors, mounted on a roof facing south, heat water or air in pipes by the greenhouse effect.

Passive solar designs heat air in a building by the greenhouse effect. An overhanging roof blocks some summer sun.

Photovoltaic (or solar) cells produce electricity.

Figure 8–15
Solar devices for heating water and air in buildings and for producing electricity.

Energy from these and all other solar systems depends on changes in the cloud cover and the seasons. Cloudy and cold days considered, solar energy is expected to meet the official government goal of 20% of our energy needs by the year 2000.

Energy Prognosis

More than perhaps any other single physical factor, the availability of energy determines our way of life. What, then, can we expect in the way of energy availability (and a way of life) in the year 2000?

We can expect to be less vulnerable to oil supply disruption than we were in the 1970s. There will be a slower growth of energy demand, a trend toward more efficient use of energy, and a shift from oil to other sources. Hence, a reduced dependence on OPEC is expected. The total oil demand is predicted to be 16.3 million barrels per day, with 10.5 million barrels supplied by the United States and 5.8 million barrels imported.

By the year 2000, use of coal (primarily for electricity generation) is expected to increase from 900 million short tons (1985) to 1700 million short tons, and use of natural gas is expected to decrease from 20 trillion cubic feet (1980) to 17.5 trillion cubic feet. Nuclear power may go from producing 16% of our electricity to producing 25%.

Synthetic fuels probably will not contribute much until the late 1990s. By the year 2000, the production of coal gas is expected to be 255 billion cubic feet per year. Shale oil may contribute 515 thousand barrels per day. Energy from biomass is estimated to be equivalent to 65 thousand barrels of oil per day. This will be energy primarily from alcohol fuels produced from crops, wood, and waste.

Jot these numbers down, and check them when you are about 30 to 35 years old. We hope the energy situation provides a good lifestyle for us all in the year 2000.

SELF-TEST 8–B

1. The splitting of an unstable nucleus to produce energy is termed ( ) fission or ( ) fusion.
2. When light nuclei combine to form heavy nuclei and energy, the process is ( ) fission or ( ) fusion.
3. The major problem with obtaining fusion energy is ( ) containment of reactants at high temperature, ( ) enough fuel, or ( ) costs.
4. A radioactive isotope of hydrogen used as a fuel in experimental fusion reactions is _____.
5. How many gallons of oil are in one barrel? _____
6. When fission is used to produce energy in a breeder reaction, the fuel produced will be _____ and _____.
7. Flat-plate solar collectors on roofs of houses use solar energy to heat _____ or _____ in pipes by the _____ effect.
8. Which is more efficient, heating a home by a primary or by a secondary source of energy? _____
9. What is the major secondary source of energy? _____
10. Which primary source of energy is used most for producing electricity? _____

11. The cleanest burning fossil fuel is _____.
12. The solar battery (cell) contains a small amount of _____ in the n-type silicon layer and a small amount of _____ in the p-type silicon layer.

MATCHING SET

____ 1. User of 35% of the world's energy
____ 2. Fossil fuels
____ 3. Combustion products of fossil fuels
____ 4. Minable coal
____ 5. Fissionable isotope
____ 6. Product of a fission breeder reactor
____ 7. Deuterium
____ 8. Date of petroleum discovery in United States
____ 9. Tritium
____ 10. Source of deuterium
____ 11. Used to confine fusion fuel
____ 12. One use of solar radiation
____ 13. Approximate efficiency of a solar battery

a. 1859
b. CO_2 and H_2O
c. Uranium-235 ($^{235}_{92}U$)
d. n-type transistor
e. Plutonium-239 ($^{239}_{94}Pu$)
f. United States
g. $^{2}_{1}H$
h. Sea water
i. 90%
j. Within 4000 feet of the Earth's surface
k. Microballoons
l. Coal, petroleum, natural gas
m. Photosynthesis
n. CO and H_2
o. 10–14%
p. $^{3}_{1}H$
q. China
r. 1740

QUESTIONS

1. What is your attitude toward using up the fossil fuels within a few decades? Do we owe future generations a supply of these resources? Would you agree to give up air-conditioning, private cars, and power tools, to mention a few examples, and to limit heating and cooking if necessary to share these fuels with your grandchildren?
2. Which theoretically yields the greatest energy per mole?
    a. The burning of gasoline
    b. The fission of uranium-235
    c. The burning of methane (natural gas)
3. Which is the more efficient use of energy: burning coal in a house to heat it, or heating the house electrically with energy produced in a coal-burning power plant?
4. Give three examples of systems with stored chemical energy that can be used as a source of heat energy.
5. Is the electrical energy where you live produced by burning fossil fuels? If not, what is the energy source? Are there pollution problems associated with the generation of the electrical energy?
6. What was the original source of energy that is tied up in fossil fuels?
7. Define *power*. Give a unit (label) for power.
8. What is meant by an insulator R value of 30?
9. Do you pay for electricity as electrical power (kilowatts) or electrical energy (kilowatt-hours)? What is the difference?
10. What major problem is associated with harnessing the energy from a fusion reaction?

11. Suggest several ways in which solar energy might be harnessed.
12. Name two sources of energy not specifically mentioned in this chapter.
13. Explain how useful energy might be obtained from garbage.
14. Which is more fundamental: a supply of energy or a supply of food? Explain.
15. The energy consumption of the United States in 1970 was 2×10^{13} kilowatt-hours. What is this amount of energy expressed in kilocalories? In Btu's?
16. Assume the world population to be 5 billion and calculate the Earth's energy needs if everyone used as much energy as is used in the United States.
17. Which fuel has the greatest energy content per gram of fuel burned: coal, natural gas, or petroleum? Is this factor more important to you, the consumer, than is economics or pollution?
18. Define *energy*. Give a unit (label) for energy.
19. List three so-called fossil fuels. Why are they called fossil fuels?
20. Which fuel—coal, petroleum, or natural gas—burns naturally with the least amount of pollution?
21. What are two dangerous properties of plutonium-239?
22. Which is the more efficient transport of energy: gas through pipes, or electricity through wires?
23. If solar energy is so clean, why are we so slow in moving to its use?
24. Do you think the United States should go forward with the use of the breeder reactor? Why?
25. How much has the price of oil changed during your lifetime?
26. Is an energy crisis a crisis of quality or quantity of energy?
27. Is electricity a primary or secondary source of energy? Explain your answer.
28. As a project, update the energy situation in the United States and the world by consulting the most recent edition of the *Annual Energy Review* (published by the Department of Energy, Energy Information Administration), a copy of which is probably in your library.

CHAPTER 9

Introduction to Organic Chemistry: The Ubiquitous Carbon Atom

Friedrich Wöhler (1800–1882) was Professor of Chemistry at the University of Berlin and later at Göttingen. His preparation of the organic compound urea from the inorganic compound ammonium cyanate did much to overturn the theory that organic compounds must be prepared in living organisms.

Carbon compounds hold the key to life on Earth. Consider what the world would be like if all carbon compounds were removed; the result would be much like the barren surface of the moon. If carbon compounds were removed from the human body, there would be nothing left except water and a small residue of minerals. The same would be true for all forms of living matter. Carbon compounds are also an integral part of our lifestyle. Fossil fuels, foods, and most drugs are made of carbon compounds. Since we live in an age of plastics, our clothes, appliances, and most other consumer goods contain a significant portion of carbon compounds.

The very large and important branch of chemistry devoted to the study of carbon compounds is **organic chemistry**. The name *organic* is actually a relic of the past, when chemical compounds produced from once-living matter were called "organic" and all other compounds were called "inorganic." Before 1828 chemists believed that chemical compounds synthesized by living matter could not be made without living matter, that a "vital force" was necessary for the synthesis. In 1828 a young German chemist, Friedrich Wöhler, destroyed the vital force myth and opened the door to the synthesis of organic compounds in the laboratory. Wöhler heated a solution of silver cyanate and ammonium chloride, neither of which had been derived from any living substance.

$$\text{AgOCN} + \text{NH}_4\text{Cl} \longrightarrow \text{AgCl} + \text{NH}_4\text{OCN}$$

Silver Cyanate   Ammonium Chloride   Silver Chloride (Precipitate)   Ammonium Cyanate

$$\text{NH}_4\text{OCN} \xrightarrow{\text{heat}} \underset{\text{Urea}}{\text{H}_2\text{NCNH}_2}\overset{\overset{\displaystyle O}{\|}}{}$$

Ammonium Cyanate

A solution of ammonium cyanate is obtained by filtering the mixture to remove the silver chloride precipitate. During one of his experiments, Wöhler evaporated the solution to obtain solid ammonium cyanate. When he heated this salt, he discovered that another substance had formed; this he analyzed and found it to be urea, a compound found in urine.

The notion of a mysterious vital force declined as other chemists began to synthesize more and more "organic" chemicals without the aid of a living system. Organic chemistry became an active area of research; now over 5 million different carbon compounds have been synthesized and described in the chemical literature, and thousands of new ones are reported every year.

Since organic chemistry is central to the chemistry of life and to the quality of that life, an understanding of the basic principles of organic chemistry is an important part of our study of the impact of chemistry on society.

Why Are There So Many Organic Compounds?

Carbon can form more compounds than any other element because of its unique ability to form covalent bonds with other carbon atoms in a seemingly endless array of possible combinations. The simplest organic compounds are compounds of carbon and hydrogen, **hydrocarbons.** Carbon also forms stable bonds with many other elements, such as oxygen, nitrogen, sulfur, phosphorus, chlorine, fluorine, bromine, iodine, silicon, and even many metals. For classification purposes, all organic compounds can be considered as hydrocarbons or derivatives of hydrocarbons in which one or more hydrogen atoms have been replaced by other atoms or groups of atoms.

Another reason for the large number of organic compounds is the ability of a given number of atoms to combine in more than one molecular pattern and hence produce more than one compound. Such compounds, each of which has molecules containing the same number and kinds of atoms, but arranged differently relative to each other, are called **isomers.** For example, the molecular structure represented by A—B—C is different from the molecular structure A—C—B, as is C—A—B; these three species are isomers. If we consider the number of possible ways in which the digits 1 through 9 can be ordered to make nine-digit numbers, we can begin to imagine how a single group of atoms can form hundreds of different molecules. Carbon, with its ability to bond to other carbon atoms, is especially well suited to form isomers.

Isomers are two or more different compounds with the same number of each kind of atom per molecule.

Hydrocarbons

Carbon atoms can join to form long chains up to thousands of carbon atoms long. The chains can also form branches or rings. Only two elements, hydrogen and carbon, are needed to form stable molecules with these straight-chain, branched-chain, or ring structures.

The hydrocarbon compounds produced by carbon atoms forming single, double, or triple bonds with other carbon atoms fall into four classes: the **alkanes,** which contain C—C bonds; the **alkenes,** which contain one or more C=C bonds; the **alkynes,** which contain one or more C≡C bonds; and the **aromatics,** which consist of benzene, benzene derivatives, and fused benzene rings.

Alkanes

Alkanes are **saturated hydrocarbons,** that is, hydrocarbons in which all carbon atoms are bonded to the maximum number of hydrogen atoms. The simplest alkane is methane, the principal component of natural gas. Four ways of representing the methane molecule are shown below.

$$CH_4 \qquad H:\overset{\overset{\displaystyle H}{\cdot\cdot}}{\underset{\underset{\displaystyle H}{\cdot\cdot}}{C}}:H \qquad H-\underset{\underset{\displaystyle H}{|}}{\overset{\overset{\displaystyle H}{|}}{C}}-H \qquad \overset{\displaystyle H}{\underset{\displaystyle H\quad H}{\overset{|}{C}}}\;H$$

(a)   (b)   (c)   (d)

The condensed formula *(a)* conveys the composition of the substance. The Lewis structure *(b)* shows the presence of four bonding pairs, and the structural formula *(c)*

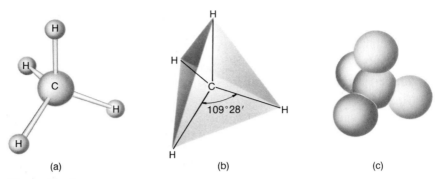

Figure 9-1
Methane. *(a)* Ball-and-stick model showing tetrahedral structure. *(b)* Geometry of regular tetrahedron. *(c)* Model of methane, CH_4, showing the relative size of atoms in relation to interatomic distances.

uses a dash for each bonding electron pair. The last representation *(d)* depicts in a perspective drawing the geometrical shape of CH_4; the solid wedge represents a bond extending in front of the page, and the dashed line represents a bond pointing behind the page. Recall from Chapter 5 that a tetrahedral structure is predicted when the central atom has four bonding pairs. Methane molecules are tetrahedral and have four C—H bonds, as shown in Figure 9-1.

The next member of the alkane family is ethane, C_2H_6. The bonding in ethane is illustrated by the following structural formula:

$$\begin{array}{c} \text{H} \quad \text{H} \\ | \quad \; | \\ \text{H}-\text{C}-\text{C}-\text{H} \\ | \quad \; | \\ \text{H} \quad \text{H} \end{array}$$

We can apply what we have learned to extend the concept of carbon-carbon bonding to the three-carbon molecule of propane, C_3H_8, and to the four-carbon molecule of butane, C_4H_{10}. Ten straight-chain alkanes, or saturated hydrocarbons, are listed in Table 9-1. Note that each succeeding formula is obtained simply by the addition of CH_2 to the previous formula. Alkanes are an example of a **homologous series,** a series of compounds of the same chemical type that differ only by a fixed increment. In this case the fixed increment is CH_2. The alkane homologous series can be represented by the general formula C_nH_{2n+2}, where n is the number of carbon atoms for a member of the series.

Note in Figure 9-2 that the carbon atoms in propane do not lie in a straight line. Although the hydrocarbon compounds in Table 9-1 are referred to as straight-chain hydrocarbons, each carbon has tetrahedral geometry, so that the molecules are not really a straight chain of carbon atoms. Structural formula representations such as those shown in Table 9-1 are used for convenience. Actually, as shown in Figure 9-2, the carbon chain is bent (109.5°) at each carbon atom, and "straight chain" means that the carbon atoms are bonded together in succession, not in a straight line.

Structural Isomers

Butane is the first alkane in which carbon atoms can be arranged in more than one way. Both a straight-chain structure and a branched-chain structure can be drawn for a composition of C_4H_{10}.

Table 9–1
The First Ten Straight-Chain Saturated Hydrocarbons

Name	Formula	Boiling Point, °C	Structural Formula	Use
Methane	CH_4	−162		Principal component in natural gas
Ethane	C_2H_6	−88.5		Minor component in natural gas
Propane	C_3H_8	−42		Bottled gas for fuel
n-Butane	C_4H_{10}	0		Bottled gas for fuel
n-Pentane	C_5H_{12}	36		Some of the components of gasoline
n-Hexane	C_6H_{14}	69		Some of the components of gasoline
n-Heptane	C_7H_{16}	98		Some of the components of gasoline
n-Octane	C_8H_{18}	126		Some of the components of gasoline
n-Nonane	C_9H_{20}	151		Some of the components of gasoline
n-Decane	$C_{10}H_{22}$	174		Found in kerosene

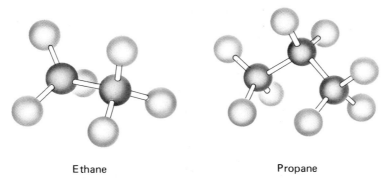

Figure 9–2
Ball-and-stick models of ethane and propane. Ball-and-stick models are not to scale on atom sizes and bond distances. They only give spatial arrangements of atoms in a molecule.

Structural isomers have the same molecular formulas but a different pattern of bonds.

	n-Butane	Methylpropane (Isobutane)
Melting Point	−138.3°C	−160°C
Boiling Point	−0.5°C	−12°C
Density (at 20°C)	0.579 g/mL	0.557 g/mL

In all branched-chain structures, at least one carbon atom is bonded to three or four other carbon atoms.

The two formulas represent two distinctly different compounds, each with its own particular set of properties. Since these are different compounds, they have different names. The straight-chain compound is normal butane, or n-butane, and the branched-chain compound is isobutane. Isobutane and n-butane are called **structural isomers** because the molecules have the same number and kinds of atoms, C_4H_{10}, but some atom-to-atom connections are different. Structural isomerism can be compared to the results you might expect from a child building many different structures with the same collection of linking blocks, using all of the blocks in each structure.

The number of possible structural isomers for the homologous series C_nH_{2n+2} increases rapidly with n. For example, C_5H_{12} has three structural isomers, C_6H_{14} has five, $C_{10}H_{22}$ has 75, and $C_{30}H_{62}$ is predicted to have over 4 billion! Although all the isomers for the alkanes with n above 10 have not been isolated, structural isomerism in hydrocarbons certainly helps explain the vast number of carbon compounds.

Alkenes

Molecules of alkenes have one or more carbon-carbon double bonds. The general formula for alkenes with one double bond is C_nH_{2n}. The first two members of the homologous alkene series are ethene, C_2H_4, and propene, C_3H_6, the structural formulas of which are shown below:

Ethene (or Ethylene)   Propene (or Propylene)

The structural formulas illustrate why alkenes are said to be **unsaturated hydrocarbons.** They contain fewer hydrogen atoms than the corresponding alkanes and can be made to react with hydrogen to form alkanes.

$$\underset{\text{Unsaturated Ethene}}{\overset{H}{\underset{H}{>}}C=C\overset{H}{\underset{H}{<}}} + H_2 \longrightarrow \underset{\text{Saturated Ethane}}{H-\underset{\underset{H}{|}}{\overset{\overset{H}{|}}{C}}-\underset{\underset{H}{|}}{\overset{\overset{H}{|}}{C}}-H}$$

Ethene, the most important raw material used in the organic chemical industry, ranks fourth in the top 25 chemicals (see inside front cover) behind the inorganic chemicals sulfuric acid, nitrogen, and oxygen. Over 30 billion pounds were produced in 1986 for use in making polyethylene, antifreeze (ethylene glycol), ethanol, and other chemicals.

A double bond between carbon atoms does not allow free rotation, and this lack of rotation provides a structural basis for **geometric isomerism.** If two chlorine atoms replace two hydrogen atoms in ethene, one on each carbon atom, the result is $CHCl=CHCl$. Since the double bond does not allow free rotation, there are two possible geometric arrangements: the chlorine atoms can be on the same side of the molecule *(cis)* or on opposite sides of the molecule *(trans)*. Both compounds are called 1,2-dichloroethene (the *1* and *2* indicate that the two chlorine atoms are attached to different carbon atoms). Note that the two isomeric compounds have significant differences in properties.

	Cis-1,2-Dichloroethene	Trans-1,2-Dichloroethene
Melting Point	−80.5°C	−50°C
Boiling Point	60.3°C	47.5°C
Density (at 20°C)	1.284 g/mL	1.265 g/mL

The third possible isomer, 1,1-dichloroethene, is a structural isomer and does not have *cis* and *trans* structures, but it does have different properties.

	1,1-Dichloroethene
Melting Point	−122.1°C
Boiling Point	37°C
Density (at 20°C)	1.218 g/mL

Alkynes

The alkynes have one or more carbon-carbon triple bonds and are very reactive unsaturated hydrocarbons. Alkynes with one triple bond have the general formula C_nH_{2n-2}. The simplest alkyne is ethyne, commonly known as acetylene, C_2H_2:

$$H-C\equiv C-H$$

A mixture of acetylene and oxygen burns with a flame hot enough to cut steel (3000°C).

Alkenes and alkynes can also have an additional type of structural isomerism. Beginning with butene or butyne, the position of the double or triple bond can be between either the two end carbons or the two middle carbons. This is illustrated for butyne along with some physical properties to illustrate that the two isomers have different properties.

	1-Butyne	2-Butyne
Melting Point	−125.8°C	−32.2°C
Boiling Point	8.1°C	27°C
Density	0.678 g/mL (0°C)	0.691 g/mL (20°C)

> No number is necessary for ethene or propene because there is only one possible position for the double bond.

Again, the number of possible isomers increases with an increase in the number of carbon atoms.

Hydrocarbon Nomenclature

There are too many organic compounds for a system of common names to work; therefore, a system of nomenclature has been developed that makes use of numbers as well as of names. Much attention has been given to the problem of naming organic compounds, and several international conventions have been held to work out a satisfactory system that can be used throughout the world. The International Union of Pure and Applied Chemistry has given its approval to a very elaborate nomenclature system (*IUPAC* system), which is now in general use.

A few IUPAC names will be needed for our discussion, and an appreciation of the basic simplicity of the IUPAC approach is desirable. Inscrutable names, such as some of those encountered on medicine bottles, have been replaced by systematic names that are descriptive of the molecules involved.

The names for the first ten alkanes are given in Table 9–1. The name of each of the members of the hydrocarbon groups consists of two parts. The first part — *meth-*, *eth-*, *prop-*, *but-*, and so on — reflects the number of carbon atoms. When more than four carbons are present, the Greek number prefixes are used: *pent-*, *hex-*, *hept-*, *oct-*, *non-*, and *dec-*. The second part of the name, or the suffix, tells the class of the hydrocarbon. Alkanes are saturated, alkenes have carbon-carbon double bonds, and alkynes have carbon-carbon triple bonds.

Branched-Chain Hydrocarbons

For branched-chain hydrocarbons, it becomes necessary to name submolecular groups. The —CH_3 group is called the methyl group; this name is derived from methane by the deletion of the *-ane* and the addition of *-yl*. Any of the nine other hydrocarbons listed in Table 9–1 can give rise to a similar group. For example, the propyl group would be —C_3H_7. As an illustration of the use of the group names, consider this formula:

> —CH_3 is the same as
> —C—H with H above and below
>
> The *-yl* ending indicates an attached group such as —CH_3, methyl.

The *longest* carbon chain in the molecule is five carbon atoms long; hence, the root name is *pentane*. Furthermore, the molecule is a methylpentane (written as one word) because a methyl group is attached to the pentane structure. In addition, a number is needed because the methyl group could be bonded to either the second or the third carbon atom.

Position of group on chain
↓
2-*methylpentane*
↑   ↖
group   longest
attached chain of
to chain C atoms

3-Methylpentane   2-Methylpentane

Note that 2-methylpentane is the same as 4-methylpentane, just turned around; the rule requires numbering from that end of the carbon chain that will yield the smallest numbers. Therefore, *2-methylpentane* is the correct name.

Positional numbers of groups should have the smallest possible sum.

Any number of substituted groups can be handled in the same fashion. Consider the name and the formula for 3,3,4,6-tetramethyl-5,5-diethyloctane.

$$CH_3CH_2-\underset{\underset{CH_3}{|}}{\overset{\overset{CH_3}{|}}{C}}-\underset{\underset{CH_3}{|}}{\overset{\overset{H}{|}}{C}}-\underset{\underset{C_2H_5}{|}}{\overset{\overset{C_2H_5}{|}}{C}}-\underset{\underset{CH_3}{|}}{\overset{\overset{H}{|}}{C}}-CH_2CH_3$$

If both a branched chain and a double or triple bond are present, numbers are used in the name to designate the positions of both components. An example is 2-methyl-2-pentene:

$$CH_3-\underset{}{\overset{\overset{CH_3}{|}}{C}}=CHCH_2CH_3$$

2-Methyl-2-Pentene

Cyclic Hydrocarbons

Hydrocarbons can form rings as well as straight chains and branched chains. The cyclic hydrocarbons include cycloalkanes (all single bonds), cycloalkenes (one carbon-carbon double bond), cycloalkynes (one carbon-carbon triple bond), and the aromatics (a unique combination of single bonds and electron delocalization around the ring).

Cycloalkanes.
The simplest **cycloalkane** is cyclopropane, a highly strained ring compound:

Cycloalkanes have the same general formula as alkenes: C_nH_{2n}.

or △ C_3H_6

The bonds are strained because of the 60° angles in the ring; angles above 90° show a much greater stability. Cyclopropane is an anesthetic. However, cyclopropane is highly flammable, and extreme caution must be taken when it is used in surgery. Cyclopropane is an isomer of C_3H_6, an alkene. Although isomers of alkenes, cyclic isomers are called cycloalkanes because all the bonds in the molecules are single bonds.

The cycloalkanes are commonly represented by a polygon. Each corner represents a carbon atom, and the lines represent C—C bonds. The C—H bonds are not shown but are understood. Other common homologous cycloalkanes include cyclobutane, cyclopentane, and cyclohexane. These are represented as:

cyclobutane   cyclopentane   cyclohexane

Cycloalkenes and Cycloalkynes. Examples of the unsaturated cyclic hydrocarbons include cyclohexene, which is used as a stabilizer in high-octane gasoline, and cyclooctyne, a compound with no known practical use.

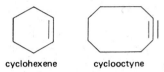

cyclohexene   cyclooctyne

Aromatic Compounds. Hydrocarbons containing one or more benzene rings are called **aromatic hydrocarbons.** The word *aromatic* was derived from "aroma," which describes the rather strong and often pleasant odor of these compounds. However, benzene and some other aromatic compounds, such as benzo(*a*)pyrene, are both toxic and carcinogenic. The main structural feature, which is responsible for the distinctive chemical properties of the aromatic compounds, is the benzene ring.

Carcinogenic means cancer-causing. The type of cancer caused may vary from one carcinogen to another. Benzene causes a form of leukemia.

The molecular structure of benzene is a ring of carbon atoms in a plane, with one hydrogen atom bonded to each carbon atom. The bonds between these carbon atoms are shorter than single bonds but longer than double bonds. The measured bond angles are 120°. The benzene structure is sometimes written with alternating double bonds:

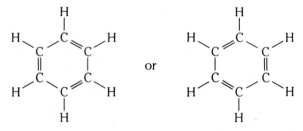

The six electrons in the three double bonds are not localized between atoms like the

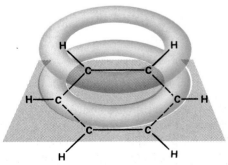

Figure 9–3
Bonding in benzene.

Delocalized electrons above and below ring

double bonds in alkenes, but are spread over the ring. Figure 9-3 illustrates the delocalization of six electrons above and below the plane of the ring. In other words, all six C—C bonds are equivalent, so that a better representation of benzene is:

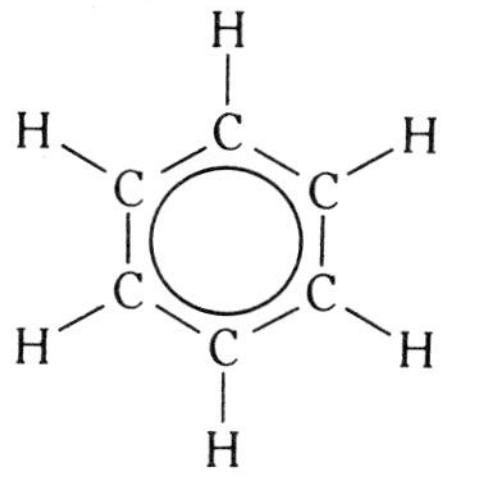

When hydrogen and carbon atoms are not shown, benzene is represented by a circle in a hexagon.

The delocalization of electrons around the ring accounts for the greater stability of aromatic compounds relative to that of unsaturated compounds, which contain double or triple bonds.

Examples of some aromatic hydrocarbons are shown in Figure 9-4. Many aromatic compounds, such as benzene, toluene, and the xylenes, are on the list of top 50 chemicals (see Table 10-11) because of their use in the manufacture of plastics, detergents, pesticides, drugs, and other organic chemicals.

Nomenclature in aromatic compounds:
Ortho- groups are on adjacent carbon atoms (1,2 positions).
Meta- groups have one carbon atom between them (1,3 positions).
Para- groups have two carbon atoms between them (1,4 positions).

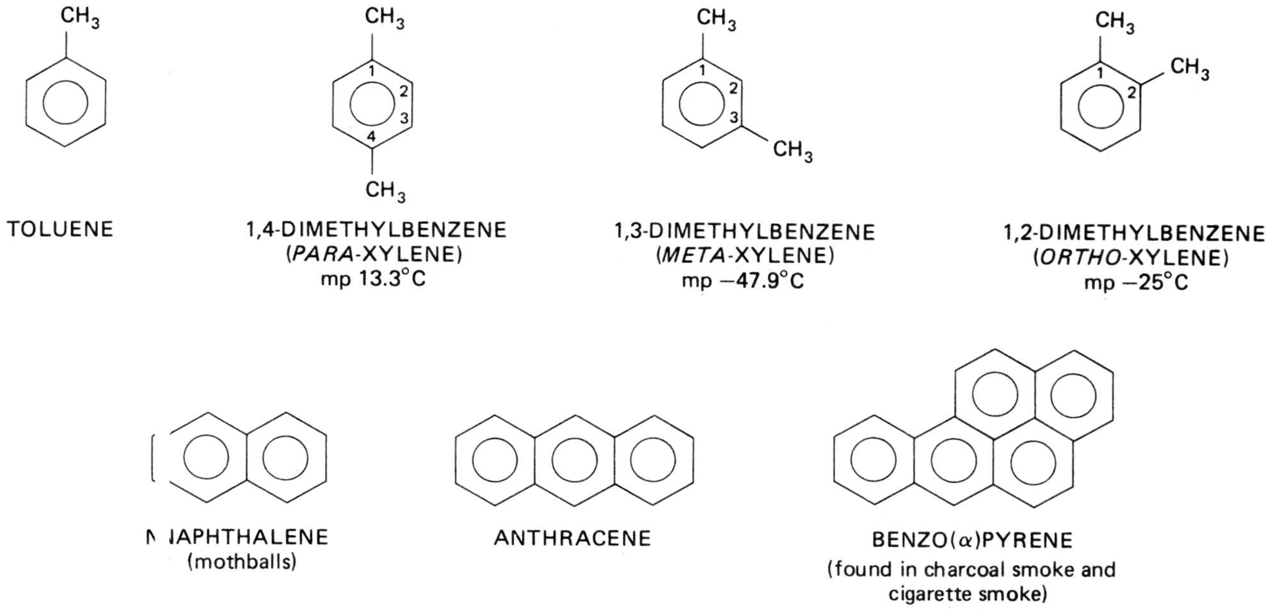

TOLUENE

1,4-DIMETHYLBENZENE
(*PARA*-XYLENE)
mp 13.3°C

1,3-DIMETHYLBENZENE
(*META*-XYLENE)
mp −47.9°C

1,2-DIMETHYLBENZENE
(*ORTHO*-XYLENE)
mp −25°C

NAPHTHALENE
(mothballs)

ANTHRACENE

BENZO(α)PYRENE
(found in charcoal smoke and cigarette smoke)

Figure 9-4
Examples of aromatic hydrocarbons

SELF-TEST 9-A

1. Each carbon in a saturated hydrocarbon has _____ geometry.
2. _____ was the first organic compound synthesized in the laboratory.
3. The number-one organic chemical produced in the United States is _____ _____.
4. A carcinogenic aromatic hydrocarbon found in smoke is _____.
5. Butane and isobutane are examples of _____ isomers.
6. _____ is the first member of the alkyne series of hydrocarbons.
7. ⬡ is chemical shorthand for _____.
8. ▢ is chemical shorthand for _____.
9. A saturated hydrocarbon contains the maximum number of _____ __ atoms per _____ atom.
10. The formula for the ethyl group is _____.
11. When the name of a compound ends in -ene (for example, *butene*), what structural feature is indicated? _____
12. The rigidity of the double and triple carbon-carbon bonds allows for the possibility of _____ isomers.
13. Name the compound shown below:

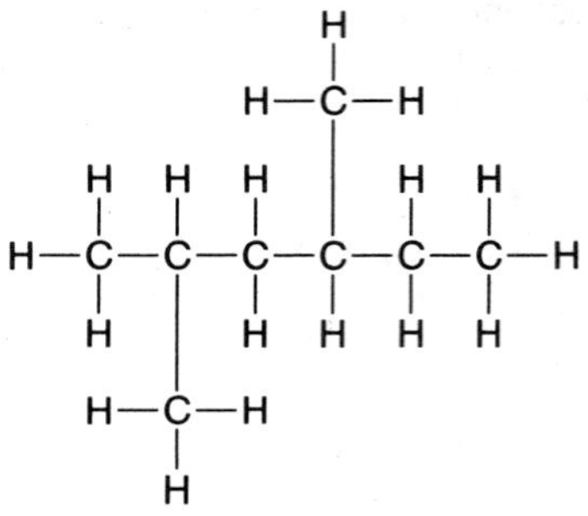

```
              H
              |
            H-C-H
  H  H  H   |   H  H
  |  |  |   |   |  |
H-C--C--C---C---C--C-H
  |  |  |   |   |  |
  H  |  H   H   H  H
     H-C-H
       |
       H
```

Where Do Hydrocarbons Come From?

Complex mixtures of hydrocarbons occur in enormous quantities in nature as natural gas, petroleum, and coal. These materials were formed from organisms that lived millions of years ago. After their death, the organisms became covered with layers of sediment and ultimately were subjected to high temperatures and pressures in the depths of the Earth's crust. In the absence of free oxygen, these conditions converted this once-living tissue into petroleum and coal. Hence, they are known as *fossil fuels*.

Natural Gas

Natural gas is a mixture of gases trapped with petroleum in the Earth's crust and is recoverable from oil wells or gas wells where the gases have migrated through the rock. The composition of natural gas varies widely, but methane is always the major constituent. The natural gas found in North America is a mixture of C_1 to C_4 alkanes, methane

(60–90%), ethane (5–9%), propane (3–18%), and butane (1–2%), with a number of other gases, such as CO_2, N_2, H_2S, and the noble gases present in varying amounts. In Europe and Japan the natural gas is essentially all methane. Most natural gas is used as a fuel, but it is also an important source of raw materials for the organic chemical industry (see Chapter 10).

Energy is obtained by burning the constituents of natural gas in air:

$$CH_4 + 2\ O_2 \longrightarrow CO_2 + 2\ H_2O + 213 \text{ kcal/mole } CH_4$$

$$2\ C_2H_6 + 7\ O_2 \longrightarrow 4\ CO_2 + 6\ H_2O + 372.8 \text{ kcal/mole } C_2H_6$$

Petroleum

Petroleum is a complex mixture of alkanes, cycloalkanes, and aromatic hydrocarbons. Thousands of compounds are present in crude petroleum, and the actual composition of petroleum varies with location. For example, Pennsylvania crude oils are primarily chain hydrocarbons, whereas California crude is composed of a larger portion of aromatics. The composition of petroleum can be classified according to boiling point ranges (Fig. 9–5a). Each of these ranges has important uses. The refining of petroleum is the separation of fractions with a certain boiling point range by a process called *fractional distillation*.

Refining Petroleum

Figure 9–5a is a schematic drawing of a fractional distillation tower used in the petroleum refining process; a picture of a fractionating tower is shown in Figure 9–5b. The crude oil is heated to about 400°C to produce a hot vapor and fluid mixture that enters the fractionating tower. The vapor rises and condenses at various points along the tower. The lower boiling fractions (those that are more volatile) will remain in the vapor stage longer than the higher boiling fractions. This difference in boiling point ranges allows the separation of fractions in the same way that simple distillation allows the partial separation of water and ethanol. Some of the gases do not condense and are drawn off the top of the tower. The unvaporized residual oil is collected at the bottom of the tower. Typical products of the fractionation of petroleum are listed in Table 9–2.

The "straight-run" gasoline fraction obtained from the fractional distillation of petroleum contains primarily straight-chain hydrocarbons that burn too rapidly to be suitable for use as a fuel in vehicles. Rapid ignition causes a "knocking" or "pinging" sound in the engine that reduces engine power and may damage the engine.

Knocking in gasoline engines is a sign of improper combustion.

Table 9–2
Hydrocarbon Fractions from Petroleum

Fraction	Size Range of Molecules	Boiling Point Range (°C)	Uses
Gas	C_1–C_4	−160 to 30	Gaseous fuel, production of H_2
Straight-run gasoline	C_5–C_{12}	30 to 200	Motor fuel
Kerosene, fuel oil	C_{12}–C_{18}	180 to 400	Diesel fuel, furnace fuel, cracking
Lubricants	C_{17} and up	350 and up	Lubricants
Paraffins	C_{20} and up	Low-melting solids	Candles, matches
Asphalt	C_{36} and up	Gummy residues	Surfacing roads, fuel

Figure 9-5
(a) Diagram of a fractionating column for distilling petroleum. The boiling range and composition of each fraction are given to the right of the column. Notice that the higher boiling substances condense at the lower levels, whereas the lower boiling substances do not condense until the higher, cooler levels. (From J. I. Routh, D. P. Eyman, and D. J. Burton: *A Brief Introduction to General, Organic and Biochemistry.* Philadelphia, Saunders College Publishing, 1971.) (b) The fractional distillation of petroleum is carried out in columns such as the one shown on the right. The middle unit is a catalytic cracking unit. (Courtesy of Ashland Oil Company.)

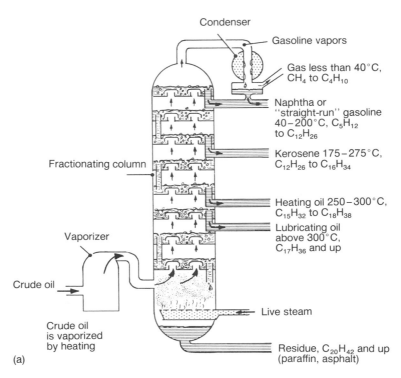

Condenser
Gasoline vapors
Gas less than 40°C, CH_4 to C_4H_{10}
Naphtha or "straight-run" gasoline 40–200°C, C_5H_{12} to $C_{12}H_{26}$
Kerosene 175–275°C, $C_{12}H_{26}$ to $C_{16}H_{34}$
Fractionating column
Heating oil 250–300°C, $C_{15}H_{32}$ to $C_{18}H_{38}$
Lubricating oil above 300°C, $C_{17}H_{36}$ and up
Vaporizer
Crude oil
Crude oil is vaporized by heating
Live steam
Residue, $C_{20}H_{42}$ and up (paraffin, asphalt)
(a)

(b)

An arbitrary scale for rating the relative knocking properties of gasolines has been developed. Normal heptane, typical of straight-run gasoline, knocks considerably and is assigned an octane rating of 0:

$$CH_3CH_2CH_2CH_2CH_2CH_2CH_3 \quad \text{(octane rating} = 0)$$
n-Heptane

2,2,4-Trimethylpentane (isooctane) is far superior in this respect and is assigned an octane rating of 100:

$$CH_3-\underset{\underset{CH_3}{|}}{\overset{\overset{CH_3}{|}}{C}}-CH_2-\underset{\underset{H}{|}}{\overset{\overset{CH_3}{|}}{C}}-CH_3 \quad \text{(octane rating} = 100)$$
2,2,4-Trimethylpentane

The octane rating of a gasoline is determined by use of the gasoline in a standard engine and recording of its knocking properties. This is compared with the behavior of mixtures of n-heptane and isooctane; the percentage of isooctane in the mixture with identical knocking properties is called the octane rating of the gasoline. Thus, if a gasoline has the same knocking characteristics as a mixture of 9% n-heptane and 91% isooctane, it is assigned an octane rating of 91. This corresponds to a regular grade of gasoline. Since the octane rating scale was established, fuels superior to isooctane have been developed, so the scale has been extended well above 100.

The octane rating of "straight-run" gasoline is increased by **catalytic reforming** and **octane enhancers.** Table 9–3 lists octane ratings for some hydrocarbons and octane enhancers.

The **catalytic reforming** process is used to produce branched-chain and aromatic hydrocarbons. Under the influence of certain catalysts, such as finely divided platinum, straight-chain hydrocarbons with low octane numbers can be re-formed into their branched-chain isomers, which have higher octane numbers.

The octane scale measures the ability of a mixture to burn without knocking in a gasoline engine.

Table 9–3
Octane Numbers of Some Hydrocarbons and Gasoline Additives

Name	Octane Number
n-Heptane	0
n-Pentane	62
2,4-Dimethylhexane	65
1-Pentene	91
Tertiary-butyl alcohol	98
2,2,4-Trimethylpentane (isooctane)	100
Benzene, technical grade	106
Ethanol	112
Methanol	116
Para-xylene	116
Toluene, technical grade	118

$$\text{CH}_3\text{CH}_2\text{CH}_2\text{CH}_2\text{CH}_3 \xrightarrow[\text{heat}]{\text{platinum}} \text{CH}_3\text{CH}_2\underset{\underset{\text{CH}_3}{|}}{\text{CH}}\text{CH}_3$$

n-Pentane
Octane Number 62

2-Methylbutane
94

Catalytic reforming is also used to produce aromatic hydrocarbons such as benzene, toluene, and xylenes by use of different catalysts and petroleum mixtures. For example, when the vapors of naphtha, kerosene, and light oil fractions are passed over a copper catalyst at 650°C, a high percentage of the original material is converted into a mixture of aromatic hydrocarbons from which benzene, toluene, xylenes, and similar compounds may be separated by fractional distillation. For example, n-hexane is converted into benzene:

$$\text{CH}_3\text{CH}_2\text{CH}_2\text{CH}_2\text{CH}_2\text{CH}_3 \xrightarrow{\text{catalyst}} \text{C}_6\text{H}_6 + 4\text{H}_2$$

n-Hexane

Benzene

The hydrogen produced here can be used in the synthesis of ammonia by the Haber process. (See Chapter 12.)

n-Heptane is changed into toluene:

$$\text{CH}_3\text{CH}_2\text{CH}_2\text{CH}_2\text{CH}_2\text{CH}_2\text{CH}_3 \longrightarrow \text{C}_6\text{H}_5\text{CH}_3 + 4\text{H}_2$$

n-Heptane

Toluene

The octane number of a given blend of gasoline can also be increased by the addition of "antiknock" agents or **octane enhancers.** Before 1975 the most widely used antiknock agent was tetraethyllead, $(C_2H_5)_4Pb$. The addition of 3 g of $(C_2H_5)_4Pb$ per gallon increases the octane rating by 10 to 15, and before the Environmental Protection Agency (EPA) required reductions in lead content, both regular and premium gasoline contained an average of 3 g of $(C_2H_5)_4Pb$ or $(CH_3)_4Pb$ per gallon.

Lead compounds are extremely toxic (see Chapter 14), and the low boiling point of tetraethyllead constitutes an additional hazard. However, levels of lead in the environment have been reduced drastically as a result of two actions on the part of the EPA. First, the decision to use a platinum-based catalytic converter to reduce emissions of carbon monoxide and nitrogen oxides required lead-free gasolines, since lead deactivates the platinum catalyst. Beginning in 1975, new automobiles were required to use lead-free gasoline to protect the catalytic converter and to decrease the amount of airborne lead. Scheduled reductions in the lead content of gasoline have led to as little as 0.1 g of lead per gallon, and the goal is completely lead-free gasoline.

Tertiary-*Butyl* Alcohol

MTBE

Alcohols and ethers are discussed in Chapter 10.

Methanol is more soluble in water (recall hydrogen bonding) than in hydrocarbons. Hence, water will extract the methanol into a two-layered system.

With the decreased use of tetraethyllead, other octane enhancers are being added to gasoline to increase its octane rating. These include toluene, 2-methyl-2-propanol (also called tertiary-butyl alcohol), methyl-tertiary-butyl ether (MTBE), methanol, and ethanol. In 1984, the most popular octane enhancer was MTBE, which joined the top-50 chemical list for the first time (number 41 in 1986).

Gasoline blends that contain methanol and ethanol are also being used as fuels. The EPA and all U.S. car manufacturers have approved the use of ethanol-gasoline blends up to 10% ethanol (known as gasohol when introduced in the 1970s). However, methanol is receiving much attention because it offers several advantages as an octane enhancer. When properly blended, methanol is more economical, has a higher octane rating, and can reduce emission levels of particulates, hydrocarbons, carbon monoxide, and nitrogen oxides. However, the biggest disadvantage of methanol relates to moisture. Small amounts of moisture destabilize the methanol-gasoline mixture, and metal corrosion of the engine becomes a serious problem.

The methanol moisture problem is solved by use of another alcohol (ethanol, propanols, butanols) as a co-solvent in methanol blends. The EPA has approved several methanol blends that meet the vehicle emission standards and provide a high-octane gasoline. Most methanol blends contain about 2.5% methanol, 2.5% tertiary-butyl alcohol, 95% gasoline, and a corrosion inhibitor. Although 309,000 tons of methanol were blended into gasoline in 1984, auto makers still disagree about the use of methanol-gasoline blends. AMC/Renault, Ford, and General Motors have approved the use of methanol blends. However, a recent owner's manual for Chrysler cars warns that use of gasolines containing methanol could void the car's warranty. About half of the major foreign car makers do not recommend methanol blends either.

SELF-TEST 9-B

1. The fractions of petroleum are separated by _____ .
2. The principal component in natural gas is _____ .
3. A widely used octane enhancer is _____ .
4. The _____ process is used to produce branched-chain and aromatic hydrocarbons from straight-chain hydrocarbons.
5. Which of the following hydrocarbons would be expected to have the highest octane rating? _____

   a. $CH_3CH_2CH_2CH_2CH_2CH_2CH_3$

   b. $CH_3CH_2-\overset{\overset{\displaystyle CH_3}{|}}{CH}-CH_2CH_2CH_3$

   c. $CH_3-\overset{\overset{\displaystyle CH_3}{|}}{\underset{\underset{\displaystyle CH_3}{|}}{C}}-\overset{\overset{\displaystyle CH_3}{|}}{\underset{\underset{\displaystyle H}{|}}{C}}-CH_3$

MATCHING SET

____	1. Organic chemistry	a. Breaks hydrocarbons into smaller molecules
____	2. Isomers	b. Found in benzene
____	3. Cracking	c. Compound such as C_3H_8
____	4. Hydrocarbon	d. Chemistry of nonmineral carbon compounds
____	5. Methyl group	e. Same number and kinds of atoms arranged differently
____	6. Synthesized from NH_4OCN by Wöhler	f. $-CH_3$
____	7. Delocalized electrons	g. Urea
____	8. Octane rating	h. Measures knocking behavior in gasoline

QUESTIONS

1. *Saturated hydrocarbons* are so named because they have the maximum amount of hydrogen present for a given amount of carbon. The saturated hydrocarbons have the general formula C_nH_{2n+2}, where n is a whole number. What are the names and formulas of the first four members of this series of compounds?
2. What is the simplest aromatic compound?
3. Using the periodic table and electron dot formulas, illustrate the bonding in the compound cyclopropane, C_3H_6.
4. Write the structural formulas for the following:
   a. 2-Methylbutane
   b. 4,4-Dimethyl-5-ethyloctane
   c. 2-Methyl-2-hexene
5. Give the names of the following:

   a.
   ```
        H       H       H
        |       |       |
   H — C ——— C ——— C — H
        |       |       |
        H   H — C — H   H
                |
                H
   ```

   b.
   ```
        H   H       H           H   H
        |   |       |           |   |
   H — C — C ——— C ——— C — C — H
        |   |       |           |   |
        H   H   H — C — H   H   H
                    |
                    H
   ```

   c.
   ```
              H       H
              |       |
          H — C — H   H — C — H   H
        H     |       |           |
        |     |       |           |
   H — C ——— C ═══ C ——— C — H
        |                           |
        H                           H
   ```

   d.
   ```
        H           H
        |           |
   H — C — C ≡ C — C — H
        |           |
        H           H
   ```

6. What unique type of bond is present in an alkyne hydrocarbon?
7. Distinguish between the classical and modern use of the word *organic* in chemistry.
8. What structural feature characterizes aromatic compounds?
9. List four gasoline additives that will increase the octane rating of "straight-run" gasoline.
10. What is the difference between natural gas found in North America and that found in Europe?
11. What is the difference between benzene and cyclohexane?
12. Make a drawing that illustrates the petroleum refinement process and label the fractions that are separated.
13. Why and how are methanol blends used in gasoline?
14. What is catalytic reforming?
15. Explain the octane rating scale.
16. Use 1,2-dichloroethene to explain what geometric isomers are. Why doesn't 1,2-dichloroethane have geometric isomers?
17. a. Carbon and hydrogen have almost the same electronegativities. On the basis of this information and the tetrahedral structure around each carbon atom in a hydrocarbon such as octane (C_8H_{18}), would you say that octane is polar or nonpolar?
    b. Is octane likely to dissolve in water?
18. Define *ubiquitous*, and explain how this word is descriptive of the carbon atom.

CHAPTER 10

Organic Chemicals of Major Importance

The prospect of studying the organic chemistry of over 5 million compounds is discouraging until you realize that this myriad of chemicals can be organized into only a few classes of compounds. Acquaintance with group characteristics, such as those of hydrocarbons, along with a few representative compounds will allow you to have a feel for the subject as a whole. Even though not all the classes of organic chemicals are included in this brief discussion, the approach used here will allow you to understand those organic chemicals that are important in your life.

Functional Groups

The millions of organic compounds include classes of compounds that are obtained by replacing hydrogen atoms of hydrocarbons with atoms or groups of atoms known as **functional groups.** The important classes of compounds that result from attaching functional groups to a hydrocarbon framework are shown in Table 10–1. The "R" attached to the functional group represents the hydrocarbon framework with one hydrogen atom removed for each functional group added. The name for this R group is obtained by removing -ane from the parent hydrocarbon and adding -yl.

Parent Compound	R Group	Functional Group	Substituted Hydrocarbon
H H │ │ H—C—C—H │ │ H H	H H │ │ H—C—C— │ │ H H	—Cl	H H │ │ H—C—C—Cl │ │ H H
Ethane	Ethyl	Chloride	Ethyl Chloride

Substituted alkanes that contain both fluorine and chlorine are called chlorofluorohydrocarbons (CFCs). Chlorofluorohydrocarbons are relatively nontoxic, nonflammable, noncorrosive, odorless gases or liquids. These properties led to the use of CFC-11 and CFC-12 as

$$\begin{array}{ccc} \text{F} & \text{F} & \text{H} \\ | & | & | \\ \text{Cl—C—Cl} & \text{F—C—Cl} & \text{F—C—Cl} \\ | & | & | \\ \text{Cl} & \text{Cl} & \text{F} \\ \text{CFC-11} & \text{CFC-12} & \text{CFC-22} \end{array}$$

propellants in aerosol spray cans and as refrigerants in refrigerators and air conditioning

CFCs are perhaps better known as Freons, the duPont trade name for a variety of CFCs used as refrigerants in refrigerators and air conditioning units.

The numbers in CFC-11, CFC-12, and CFC-22 are industrial code numbers.

Organic Chemicals of Major Importance

Table 10-1
Classes of Organic Compounds Based on Functional Groups*

General Formulas of Class Members	Class Name	Typical Compound	Compound Name	Common Use of Sample Compound
R—X	Halide	H—C(H)(Cl)—Cl	Dichloromethane (methylene chloride)	Solvent
R—OH	Alcohol	H—C(H)(H)—OH	Methanol (wood alcohol)	Solvent
R—C(=O)—H	Aldehyde	H—C(=O)—H	Methanal (formaldehyde)	Preservative
R—C(=O)—OH	Carboxylic acid	H—C(H)(H)—C(=O)—OH	Ethanoic acid (acetic acid)	Vinegar
R—C(=O)—R'	Ketone	H—C(H)(H)—C(=O)—C(H)(H)—H	Propanone (acetone)	Solvent
R—O—R'	Ether	C_2H_5—O—C_2H_5	Diethyl ether (ethyl ether)	Anesthetic
R—O—C(=O)—R'	Ester	CH_3—CH_2—O—C(=O)—CH_3	Ethyl ethanoate (ethyl acetate)	Solvent in fingernail polish
R—N(H)(H)	Amine	H—C(H)(H)—N(H)(H)	Methylamine	Tanning
R—C(=O)—N(H)—R'	Amide	CH_3—C(=O)—N(H)(H)	Acetamide	Plasticizer

* R stands for an H or a hydrocarbon group such as —CH_3 or —C_2H_5. R' could be a different group from R.

The reaction of ozone with CFCs is discussed in Chapter 15.

units for buildings and vehicles. The widespread use of CFCs led to the production of over 1 billion pounds in 1974. However, in 1974 scientists pointed out that the release of CFCs in the atmosphere was potentially harmful to the ozone layer in the stratosphere. The lack of reactivity of CFCs released from aerosol cans in the lower atmosphere would eventually allow them to reach the ozone layer, where ultraviolet radiation from the Sun could decompose them to reactive species that could react with ozone in the stratosphere and decrease the ozone concentration. Any decrease in the ozone layer would cause an increase in the ultraviolet radiation reaching the Earth's surface, and this would result in an increase in the incidence of skin cancer. As a result of these studies, in 1978 the Environmental Protection Agency imposed a ban on the use of CFCs in aerosol cans.

Alkyl Halides

The alkyl halides, the first class of compounds listed in Table 10–1, are molecules in which one or more hydrogen atoms are replaced by halogen atoms. In the IUPAC system for naming these compounds, the halogen is specified as fluoro, chloro, bromo, or iodo. Structural formulas and names of some common alkyl halides are

The halogens are Group VIIA elements (F, Cl, Br, I).

Trichloromethane (chloroform)

Tetrachloromethane (carbon tetrachloride)

1,1,1-Trichloroethane

Dichloromethane (methylene chloride)

1,2-Dibromoethane (ethylene dibromide)

At one time chloroform and carbon tetrachloride were widely used as solvents in the laboratory and in industrial cleaning, but their toxicity and carcinogenicity have led them to be removed from use. Many chlorinated hydrocarbons are on either the carcinogen or the suspect carcinogen list of the Environmental Protection Agency. Only 1,1,1-trichloroethane appears to be safe, and it has replaced the others in many solvent applications.

A compound that has shown up in trace amounts in flour because of its use as a fumigant in grain is ethylene dibromide, or EDB. EDB is also a suspected carcinogen.

Although CFCs are no longer used in aerosol cans in the United States, CFC-11 and CFC-12 are still the major chemicals used in refrigeration. In 1986 the annual value of the CFCs produced was $26.9 billion. Concern about the increasing atmospheric levels of CFC-11 and CFC-12 has led to a search for environmentally safer substances, such as CFC-22. The lower chlorine content of CFC-22 reduces its tendency to form reactive species in the presence of ultraviolet radiation.

Several other halogenated hydrocarbons have also caused environmental problems because of their toxicity or carcinogenicity (Table 10–2). Dichlorodiphenyltrichloroethane (DDT), used on a worldwide basis after World War II, was very effective in the eradication of mosquitoes that carried malaria. The World Health Organization estimated that about 25 million lives were saved and hundreds of millions of illnesses were prevented by the use of DDT and other chlorinated hydrocarbons. However, DDT is not metabolized very rapidly by animals, and since it is soluble in fatty tissues, the concentration of DDT increases with time. For many animals that is not a problem, but for some predators, such as eagles and ospreys that feed on other animals and fish, the buildup of DDT interfered with normal calcium metabolism. This led to a reduction in the thickness of their eggshells, which in turn led to the nearly complete extinction of eagles and ospreys in some parts of the United States. DDT was outlawed for use in the United States

Table 10-2
Some Organic Chlorine Compounds

Name	Formula	Uses
Dichlorodiphenyltrichloroethane (DDT)	Cl—C₆H₄—CH(CCl₃)—C₆H₄—Cl	Insecticide
2,4-Dichlorophenoxyacetic acid (2,4-D)	2,4-Cl₂C₆H₃—O—CH₂—COOH	Herbicide
3,4,3',4',5'-Pentachlorobiphenyl (a typical polychlorinated biphenyl —PCB)	(structure: pentachlorobiphenyl)	Plasticizer, solvent, coolant
Vinyl chloride	$CH_2=CHCl$	Plastics
2,3,7,8-Tetrachlorodibenzo-p-dioxin	(structure: tetrachlorodibenzo-p-dioxin)	Herbicides

Plasticizers are substances added to polymers to make them flexible.

When the —OH is attached directly to the aromatic ring, the compounds are known as phenols. These are considered a separate class because they are much more acidic than alcohols owing to the electron-withdrawing properties of the aromatic ring.

in 1973, although it is still manufactured for export. Many Third World countries use DDT because it is still one of the most effective insecticides of low toxicity for controlling malaria.

Polychlorinated biphenyls (PCBs) similar to those in Table 10-2 are also no longer used in the United States because they have a physiological effect similar to that of DDT. PCBs were once used as insulating materials in transformers and as plasticizers in plastics.

Dioxins similar in structure to the one shown in Table 10-2, an impurity in herbicides and defoliants such as Agent Orange, are among the most toxic substances known. In 1983 residents of Times Beach, Missouri, were evacuated permanently because of high dioxin concentrations in the soil caused by dioxin-contaminated oil. As little as 10 ppb dioxin (10 parts dioxin per billion parts soil) is considered dangerous, and samples of soil from Times Beach contained from 100 ppb to 2000 ppb. The source of the dioxin in the Times Beach area was a result of the use of waste oil contaminated with trace amounts of dioxin as a spray to keep down the dust on soil in horse arenas. The soil was later removed and used as fill dirt for road construction in Times Beach. The company using the waste oil was not aware that dioxin was a contaminant that had been inadvertently introduced from wastes picked up from a chemical company that was manufacturing herbicides.

Alcohols

Alcohols contain the hydroxyl group —OH. Alcohols are classified as primary, secondary, or tertiary based on the number of other carbons bonded to the —C—OH carbon.

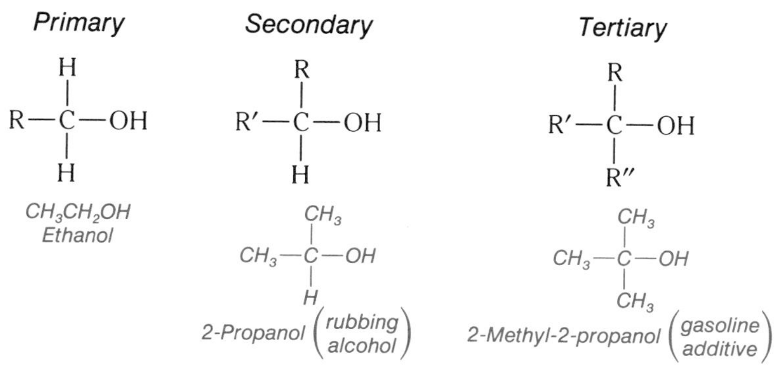

Functional Groups

Primary

R—C(H)(H)—OH

CH_3CH_2OH
Ethanol

Secondary

R'—C(R)(H)—OH

CH_3—C(CH$_3$)(H)—OH

2-Propanol (rubbing alcohol)

Tertiary

R'—C(R)(R")—OH

CH_3—C(CH$_3$)(CH$_3$)—OH

2-Methyl-2-propanol (gasoline additive)

Some of the most important alcohols are listed in Table 10–3. Note that all classes of alcohols are represented, along with two that have more than one hydroxyl group.

Table 10–3
Some Important Alcohols

Formula	IUPAC Name	Common Name	Typical Use
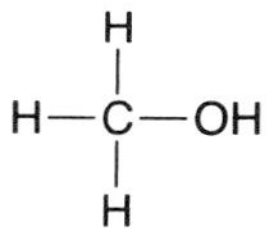 H—C(H)(H)—OH	Methanol	Methyl alcohol (wood alcohol)	Industrial solvent
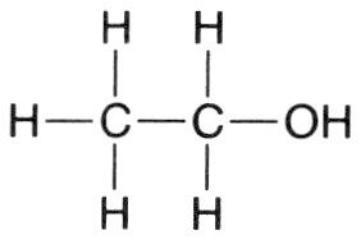 H—C(H)(H)—C(H)(H)—OH	Ethanol	Ethyl alcohol (grain alcohol)	Beverage
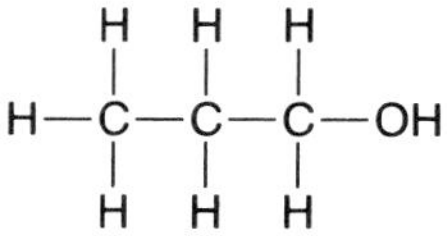 H—C(H)(H)—C(H)(H)—C(H)(H)—OH	1-Propanol	n-Propyl alcohol	Chemical intermediate
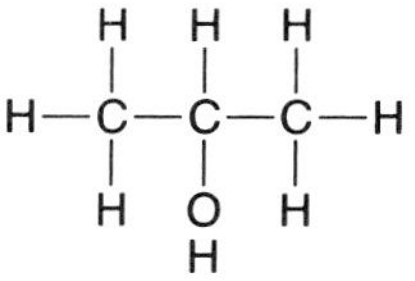 H—C(H)(H)—C(H)(OH)—C(H)(H)—H	2-Propanol	Isopropyl alcohol	Rubbing alcohol
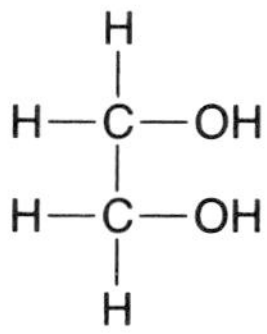 H—C(H)(OH)—C(H)(H)—OH (1,2-Ethanediol structure)	1,2-Ethanediol	Ethylene glycol	Permanent antifreeze
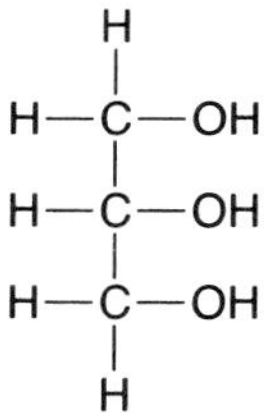 H—C(H)(OH)—C(H)(OH)—C(H)(H)—OH (glycerol structure)	1,2,3-Propanetriol	Glycerol	Manufacture of drugs and cosmetics

Alcohols among the top 50 chemicals produced in the United States (see Table 10–11) include methanol (22), ethylene glycol (28), and isopropyl alcohol (49).

Methanol. Methanol, which has the formula CH_3OH, is the simplest of all alcohols. Over 7 billion pounds of methanol are produced each year from **synthesis gas,** a mixture of carbon monoxide and hydrogen. High pressure, high temperature, and a mixture of catalysts are used to increase the yield. The uses of synthesis gas are discussed in more detail later in this chapter.

$$\underset{\text{Coal}}{C} + \underset{\text{Steam}}{H_2O} \longrightarrow \underset{\text{Synthesis gas}}{CO + H_2} \xrightarrow[300°C]{ZnO,\ Cr_2O_3} CH_3OH$$

An old method of producing methanol involved heating of a hardwood such as beech, hickory, maple, or birch in a retort in the absence of air. For this reason methanol is sometimes called wood alcohol.

About 50% of the methanol produced in United States is used in the production of formaldehyde (used in plastics, embalming fluid, germicides, and fungicides); 30% is used in the production of other chemicals, and the remaining 20% is used for jet fuels, antifreeze mixtures, and solvents, as a gasoline additive, and as a denaturant (a poison added to make ethanol unfit for beverages). Methanol is a deadly poison that causes blindness in less than lethal doses. Many deaths and injuries have resulted from the accidental substitution of methanol for ethanol in beverages.

Methanol is the main ingredient in many windshield washer fluids.

Ethanol. Ethanol, also called ethyl alcohol or grain alcohol, can be obtained by the fermentation of carbohydrates (starch, sugars). For example, glucose is converted into ethanol and carbon dioxide by the action of yeast in the absence of oxygen.

Ethanol can be prepared by the fermentation of grains.

$$\underset{\text{Glucose}}{C_6H_{12}O_6} \xrightarrow{\text{yeast}} 2\ \underset{\text{Ethanol}}{C_2H_5OH} + 2\ CO_2$$

A mixture of 95% ethanol and 5% water can be recovered from the fermentation products by distillation. Ethanol is the active ingredient of alcoholic beverages. Some of the most commonly encountered alcoholic beverages and their characteristics are presented in Table 10–4. The "proof" of an alcoholic beverage is twice the volume percent of ethanol; 80 proof vodka, for example, contains 40% ethanol.

95% ethanol (190 proof) is a strong dehydrating agent. Never drink it straight.

Table 10–4
Common Alcoholic Beverages

Name	Source of Fermented Carbohydrate	Amount of Ethyl Alcohol	Proof
Beer	Barley, wheat	5%	10
Wine	Grapes or other fruit	12% maximum, unless fortified*	20–24
Brandy	Distilled wine	40–45%	80–90
Whiskey	Barley, rye, corn, etc.	45–55%	90–110
Rum	Molasses	~45%	90
Vodka	Potatoes	40–50%	80–100

* The growth of yeast is inhibited at alcohol concentrations over 12%, and fermentation comes to a stop. Beverages with a higher concentration are prepared either by distillation or by fortification with alcohol that has been obtained by the distillation of another fermentation product.

Table 10-5
Alcohol Blood Level and Effect

Blood Alcohol Level (Percentage by Volume)	Effect
0.05–0.15	Lack of coordination
0.15–0.20	Intoxication
0.30–0.40	Unconsciousness
0.50	Possible death

Although ethanol is not as toxic as methanol, 1 pint of pure ethanol, rapidly ingested, would kill most people. Ethanol is a depressant for nonalcoholics. The effects of different blood levels of alcohol are shown in Table 10-5. Rapid consumption of two 1-ounce "shots" of 90-proof whiskey or of two 12-ounce beers can cause one's blood alcohol level to reach 0.05%.

The breathalyzer test used to detect drunken drivers is based on the color change that occurs when ethanol is oxidized to acetic acid by dichromate anion ($Cr_2O_7^{2-}$) in acidic solution.

$$16\ H^+ + 2\ Cr_2O_7^{2-} + 3\ CH_3CH_2OH \longrightarrow 3\ CH_3COOH + 4\ Cr^{3+} + 11\ H_2O$$

Yellow-orange → Green

Ethanol is quickly absorbed by the blood and metabolized by enzymes produced in the liver. The rate of detoxification is about 1 ounce of pure alcohol per hour. The ethanol is oxidized to acetaldehyde, which is further oxidized to acetic acid; eventually CO_2 and H_2O are produced and eliminated through the lungs and kidneys.

$$\underset{\text{Ethanol}}{H-\underset{\underset{H}{|}}{\overset{\overset{H}{|}}{C}}-\underset{\underset{H}{|}}{\overset{\overset{H}{|}}{C}}-OH} \xrightarrow[\text{enzymes}]{\text{liver}} \underset{\text{Acetaldehyde}}{H-\underset{\underset{H}{|}}{\overset{\overset{H}{|}}{C}}-C\overset{\displaystyle O}{\underset{\displaystyle H}{\diagup\!\!\!\!\diagdown}}}$$

Alcoholism. Alcoholism is one of the largest health problems in the United States, where there are at least 10 million alcoholics and an estimated 200,000 deaths per year are attributed to alcohol abuse. A metabolic change that accompanies detoxification of ethanol in the liver is the synthesis of fat, which is deposited in liver tissue. Excessive drinking causes deterioration of the liver, known as cirrhosis. Cirrhosis of the liver is eight times more common among alcoholics than among nonalcoholics. Since 1974, cirrhosis of the liver has surpassed arteriosclerosis, influenza, and pneumonia to become the seventh leading cause of death. Alcoholics also tend to suffer from malnutrition and cardiovascular disease.

Genetic factors appear to play an important role in alcoholism. Research indicates that alcoholics metabolize acetaldehyde less effectively than nonalcoholics, probably because of a deficiency in the enzyme alcohol dehydrogenase. As a result, blood acetaldehyde levels are higher in alcoholics than in nonalcoholics for the same amount of alcohol intake. Although the biological effects of high acetaldehyde levels have not been explained fully, there are indications that higher than normal acetaldehyde levels enhance organ damage and influence brain chemistry, possibly causing the production of small amounts of compounds more addictive than morphine.

Industrial Use of Ethanol.
The federal tax on alcoholic beverages is about $20 per gallon. Since the cost of producing ethanol is only about $1 per gallon, ethanol intended for industrial use must be **denatured** to avoid the beverage tax. **Denatured alcohol** contains small amounts of a toxic substance, such as methanol or gasoline, that cannot be removed easily by chemical or physical means.

Apart from being used in the alcoholic beverage industry, ethanol is used widely in solvents and in the preparation of many other organic compounds. Over 1 billion pounds of ethanol are used for these purposes each year, and this is produced synthetically rather than by a fermentation process. The reaction involves the addition of water vapor to ethylene under high pressure in the presence of a catalyst.

Where does the ethylene come from?

$$CH_2=CH_2 + HOH \xrightarrow[300°C]{70 \text{ atm.}} CH_3-CH_2-OH$$

Ethylene → Ethanol

Other important alcohols listed in Table 10–3 include 2-propanol, commonly known as isopropyl alcohol or rubbing alcohol; ethylene glycol, used widely as an automotive antifreeze and in the manufacture of plastics; and glycerol, which is found in fats and is also used in antifreeze applications and in the manufacture of drugs and cosmetics.

Ethers

Dehydration reactions involve the formation and removal of water molecules.

$CH_3CH_2-O-CH_2CH_3$

Diethyl Ether

Ethers, which contain the R—O—R' linkage, are formed by the dehydration of alcohols. The most common ether is diethyl ether, which for many years was used as an anesthetic. It produces unconsciousness by depressing the activity of the central nervous system. It is no longer used because it irritated the respiratory system and caused postanesthetic nausea and vomiting. Methyl propyl ether, $CH_3OCH_2CH_2CH_3$, known as neothyl, is currently used as an anesthetic because it is relatively free of side effects.

Aldehydes and Ketones

Aldehydes and ketones contain a **carbonyl group**, $>C=O$, and are generally obtained by the oxidation of alcohols. In aldehydes the carbonyl group is on an end carbon, whereas in ketones the carbonyl group is bonded to two carbon atoms.

$$\underset{\text{Aldehyde}}{R-\overset{\overset{O}{\|}}{C}-H} \qquad \underset{\text{Ketone}}{R-\overset{\overset{O}{\|}}{C}-R}$$

$H-\overset{\overset{O}{\|}}{C}-H$

Formaldehyde

Formaldehyde is a suspected carcinogen.

Formaldehyde, the simplest aldehyde, has a foul odor. It is the starting material in the production of several plastics and is used in the laboratory as a preservative for dead animals. Aldehydes with an aromatic ring have pleasant odors, and some are used in food flavors and perfumes.

Cinnamaldehyde is a cis isomer.

Benzaldehyde (bitter almonds) Vanillin (vanilla bean) Cinnamaldehyde (cinnamon)

The simplest ketone is acetone, an important commercial solvent. Methyl ethyl ketone is a solvent in model airplane glue.

$$\underset{\text{Acetone}}{H_3C-\underset{\underset{O}{\|}}{C}-CH_3} \qquad \underset{\text{Methyl ethyl ketone}}{H_3C-\underset{\underset{O}{\|}}{C}-CH_2-CH_3}$$

Carboxylic Acids

Carboxylic acids, which contain the **carboxyl group,** —COOH, can be prepared by the oxidation of alcohols or aldehydes. These reactions occur quite readily, as evidenced by the souring of wine, which is the oxidation of ethanol to acetic acid in the presence of oxygen from the air.

Carboxylic acids are found in both the plant and animal kingdoms. The first six carboxylic acids, with their sources, common names, and odors, are given in Table 10–6. Longer chain carboxylic acids do not smell as bad, in part because they are less volatile. Some of the other common carboxylic acids found in nature are given in Table 10–7. As can be seen in the table, some organic acids have more than one carboxyl group as well as other groups, usually hydroxyl groups.

A fatty acid is a naturally occurring organic acid with a long hydrocarbon chain. The chain often contains only C—C single bonds, but several fatty acids also contain some C=C double bonds. Stearic acid (saturated) and oleic acid (unsaturated), shown in Table 10–7, are examples of these two types of fatty acids.

Recall from Chapter 9 that saturated *compounds contain the maximum number of hydrogen atoms per carbon atom.*

Formic Acid. The simplest organic acid has the carboxyl group attached directly to a hydrogen atom:

$$\underset{\text{Formic Acid}}{H-\underset{\underset{O}{\|}}{C}-OH}$$

This acid is found in ants and other insects as part of the irritant that produces itching and swelling after a bite.

Table 10–6
First Six Carboxylic Acids

Formula	Source	Common Name	Odor
HCOOH	Ants (Latin, *formica*)	Formic acid	Sharp
CH₃COOH	Vinegar (Latin, *acetum*)	Acetic acid	Sharp
CH₃CH₂COOH	Milk (Greek, *protos pion*, "first fat")	Propionic acid	Swiss cheese
CH₃(CH₂)₂COOH	Butter (Latin, *butyrum*)	Butyric acid	Rancid butter
CH₃(CH₂)₃COOH	Valerian root (Latin, *valere*, "to be strong")	Valeric acid	Manure
CH₃(CH₂)₄COOH	Goats (Latin, *caper*)	Caproic acid	Goat

Condensed formulas are often used to save space. For example, CH₃(CH₂)₃COOH is the same as CH₃CH₂CH₂CH₂COOH.

Table 10-7
Some Other Naturally Occurring Carboxylic Acids

Name	Structure	Natural Source
Citric acid	HOOC—CH$_2$—C(OH)(COOH)—CH$_2$—COOH	Citrus fruits
Lactic acid	CH$_3$—CH(OH)—COOH	Sour milk
Malic acid	HOOC—CH$_2$—CH(OH)—COOH	Apples
Oleic acid	CH$_3$(CH$_2$)$_7$—CH=CH—(CH$_2$)$_7$—COOH	Vegetable oils
Oxalic acid	HOOC—COOH	Rhubarb, spinach, cabbage, tomatoes
Stearic acid	CH$_3$(CH$_2$)$_{16}$—COOH	Animal fats
Tartaric acid	HOOC—CH(OH)—CH(OH)—COOH	Grape juice, wine

Acetic Acid. The most widely used of the organic acids is found in vinegar, an aqueous solution containing 4% to 5% acetic acid.

Flavor and colors are imparted to vinegars by the constituents of the alcoholic solutions from which they are made. Ethanol in the presence of certain bacteria and air is oxidized to acetic acid:

$$\underset{\text{Ethanol}}{CH_3CH_2OH} + \underset{\text{Oxygen}}{O_2} \xrightarrow{\text{bacteria}} \underset{\text{Acetic Acid}}{CH_3COOH} + \underset{\text{Water}}{H_2O}$$

The bacteria, called **mother of vinegar,** form a slimy growth in a vinegar solution. The growth of bacteria can sometimes be observed in a bottle of commercially prepared vinegar after it has been opened to the air.

Acetic acid is an important starting substance in the production of textile fibers, vinyl plastics, and other chemicals, and it is a convenient choice when a cheap organic acid is needed.

Salts. Like other salts, organic acid salts are formed by the reaction of an acid with a base. An example of an organic salt is sodium benzoate, a food preservative:

$$[C_6H_5COO^-] Na^+$$

Sodium Benzoate

Esters

The functional group in esters is $-\overset{\overset{O}{\|}}{C}-O-R$. In the presence of a strong acid such as sulfuric acid, organic acids react with alcohols to form **esters.** For example, ethyl acetate is formed by the reaction of ethanol with acetic acid in the presence of sulfuric acid. This reaction is a dehydration in which sulfuric acid acts as both a dehydrating agent and a catalyst.

$$C_2H_5O-[H + HO]-\overset{\overset{O}{\|}}{C}-CH_3 \xrightarrow{H_2SO_4} C_2H_5O-\overset{\overset{O}{\|}}{C}-CH_3$$

Ethanol   Acetic Acid   Ethyl Acetate

Organic esters are compounds of the type $R-O-\underset{\underset{O}{\|}}{C}-R'$ formed by the reaction of organic acids and alcohols.

Ethyl acetate is a common solvent for lacquers and plastics and is often used as fingernail polish remover.

Many esters have a pleasant odor. They are used in the manufacture of perfumes and as flavoring agents in the confectionery and soft-drink industries. Many fruits owe their characteristic smell and flavor to the presence of small quantities of esters. Esters are present in natural flavors and are used to make artificial flavorings. Table 10-8 illustrates the pleasant odors of several esters derived from acids with foul odors.

SELF-TEST 10-A

1. Give examples of the following:
    a. An ether _____
    b. An alcohol _____
    c. An organic acid _____
    d. A ketone _____

2. Identify the functional groups present in each of the following molecules.

    a. [benzene ring with $O-\overset{\overset{O}{\|}}{C}-CH_3$ and $\overset{\overset{O}{\|}}{C}-OH$ substituents] _____

    b. $HO-\overset{\overset{O}{\|}}{C}-\overset{\overset{H}{|}}{\underset{\underset{H}{|}}{C}}-\overset{\overset{H}{|}}{\underset{\underset{H}{|}}{C}}-\overset{\overset{O}{\|}}{C}-OH$ _____

    c. $H-\overset{\overset{H}{|}}{\underset{\underset{H}{|}}{C}}-\overset{\overset{H}{|}}{\underset{\underset{NH_2}{|}}{C}}-\overset{\overset{O}{\|}}{C}-OH$ _____

3. Gin that is 84 proof contains what percentage of alcohol? _____
4. Ethanol is quickly absorbed by the blood and oxidized to _____ in the liver.
5. Ethanol intended for industrial use is _____ by the addition of small amounts of a toxic substance.
6. 2-Propanol is commonly known as _____ alcohol.
7. A long-chain organic acid found in nature is called a _____ acid.
8. The organic acid found in vinegar is _____.

More on Functional Groups in Nature

Fats and Oils

Fats and oils are esters of glycerol (glycerin) and a fatty acid. R, R′, and R″ stand for the hydrocarbon chains of the acids in the following equation:

Fats and oils are esters of fatty acids and glycerol. Fats are solids, and oils are liquids.

Table 10–8
Some Alcohols, Acids, and Their Esters

Alcohol	Acid	Ester	Odor of the Ester
CH₃CHCH₂CH₂OH | CH₃ *Isopentyl Alcohol*	CH₃COOH *Acetic Acid*	CH₃CHCH₂CH₂—O—C—CH₃ | ‖ CH₃ O *Isopentyl Acetate*	Banana
CH₃CHCH₂CH₂OH | CH₃ *Isopentyl Alcohol*	CH₃CH₂CH₂CH₂COOH *Pentanoic Acid*	CH₃CHCH₂CH₂—O—C—CH₂CH₂CH₂CH₃ | ‖ CH₃ O *Isopentyl Pentanoate*	Apple

CH₃CH₂CH₂CH₂OH
n-Butyl Alcohol
	CH₃CH₂CH₂COOH
	Butanoic Acid
		CH₃CH₂CH₂CH₂—O—C—CH₂CH₂CH₃
		                  ‖
		                  O
		Butyl Butanoate
			Pineapple

CH₃CHCH₂OH | CH₃ *Isobutyl Alcohol*	CH₃CH₂COOH *Propionic Acid*	CH₃CHCH₂—O—C—CH₂CH₃ | ‖ CH₃ O *Isobutyl Propionate*	Rum
CH₃CHCH₂OH | CH₃ *Isobutyl Alcohol*	HCOOH *Formic Acid*	CH₃CHCH₂—O—C—H | ‖ CH₃ O *Isobutyl Formate*	Raspberry

C₆H₅—CH₂—OH
Benzyl Alcohol
	CH₃CH₂CH₂COOH
	Butanoic Acid
		C₆H₅—CH₂—O—C—CH₂CH₂CH₃
		              ‖
		              O
		Benzyl Butanoate
			Rose

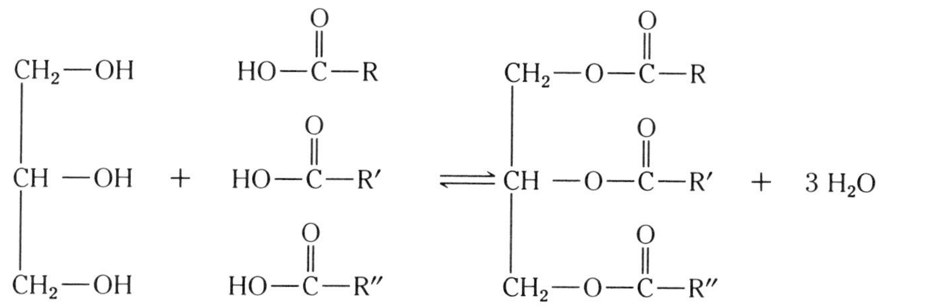

$$
\begin{array}{c}
\text{CH}_2\text{—OH} \\
| \\
\text{CH —OH} \\
| \\
\text{CH}_2\text{—OH}
\end{array}
+
\begin{array}{c}
\text{HO—C(=O)—R} \\
\text{HO—C(=O)—R}' \\
\text{HO—C(=O)—R}''
\end{array}
\rightleftharpoons
\begin{array}{c}
\text{CH}_2\text{—O—C(=O)—R} \\
| \\
\text{CH —O—C(=O)—R}' \\
| \\
\text{CH}_2\text{—O—C(=O)—R}''
\end{array}
+ 3\,\text{H}_2\text{O}
$$

Glycerol (one molecule) + Fatty Acid (three molecules that may or may not be the same) ⇌ Fat or Oil (one molecule) + Water (three molecules)

The term *fat* is usually reserved for solid glycerol esters (butter, lard, tallow) and *oil* for liquid esters (castor, olive, linseed, tung, and so forth). The term *lipid* includes fats, oils, and fat-soluble compounds.

Saturated fatty acids (which contain all single bonds with maximum hydrogen content) are usually found in solid or semisolid fats, whereas **unsaturated** fatty acids (containing one or more double bonds) are usually found in oils. Hydrogen can be catalytically added to the double bonds of an oil to convert it into a semisolid fat. For example, liquid soybean and other vegetable oils are hydrogenated to produce cooking fats and margarine.

Consumers in Europe and North America have historically valued butter as a source of fat. As the population of these parts of the world increased, the advantages of a substitute for butter became apparent, and efforts to prepare such a product began about 100 years ago. One initial problem was that common fats are almost all *animal* products with very pronounced tastes of their own. Analogous compounds from vegetable oils, which are bland or have mixed flavors, were generally *unsaturated* and consequently *oils*. A solid fat could be made from the much cheaper vegetable oils if an inexpensive way could be discovered to add hydrogen across the double bonds. After extensive experiments, many catalysts were found, of which finely divided nickel is among the most effective. The nature of the process can be illustrated by the following reaction:

> Lipids are soluble in fats and oils.

> Catalytic hydrogenation can convert a liquid oil into a solid fat.

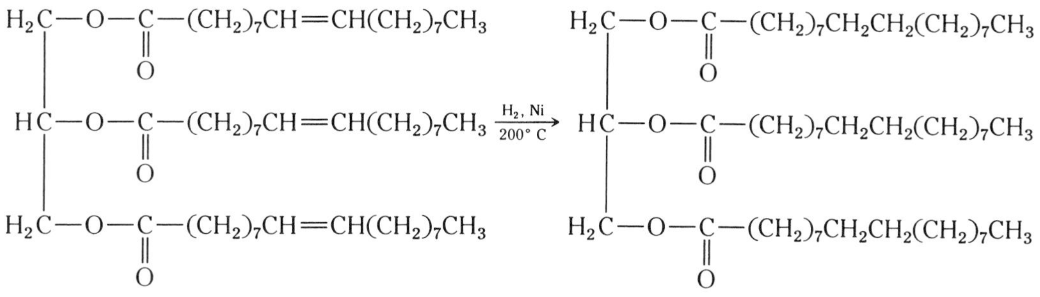

$$
\begin{array}{c}
\text{H}_2\text{C—O—C(=O)—(CH}_2\text{)}_7\text{CH=CH(CH}_2\text{)}_7\text{CH}_3 \\
| \\
\text{HC—O—C(=O)—(CH}_2\text{)}_7\text{CH=CH(CH}_2\text{)}_7\text{CH}_3 \\
| \\
\text{H}_2\text{C—O—C(=O)—(CH}_2\text{)}_7\text{CH=CH(CH}_2\text{)}_7\text{CH}_3
\end{array}
\xrightarrow{\text{H}_2,\ \text{Ni},\ 200°\text{C}}
\begin{array}{c}
\text{H}_2\text{C—O—C(=O)—(CH}_2\text{)}_7\text{CH}_2\text{CH}_2\text{(CH}_2\text{)}_7\text{CH}_3 \\
| \\
\text{HC—O—C(=O)—(CH}_2\text{)}_7\text{CH}_2\text{CH}_2\text{(CH}_2\text{)}_7\text{CH}_3 \\
| \\
\text{H}_2\text{C—O—C(=O)—(CH}_2\text{)}_7\text{CH}_2\text{CH}_2\text{(CH}_2\text{)}_7\text{CH}_3
\end{array}
$$

Triolein, A Liquid Oil → Tristearin, A Solid Fat

Oils commonly subjected to this process include those from cottonseed, peanuts, corn germ, soybeans, coconuts, and safflower seeds. In recent years, as it has become apparent that saturated fats may encourage diseases of the heart and arteries, soft margarines and cooking oils (which still contain some of the unhydrogenated fatty acid) have been placed on the market.

Dietary Fats and Essential Fatty Acids

Forty to fifty percent of the calories in most diets in the United States are obtained from fats or oils. This is rather high when compared with diets in most other parts of the world. Natural fats and oils are generally mixtures of various esters of glycerol with more than one kind of fatty acid. In our diets, most fatty acids are **saturated** fatty acids (Table 10–9). Such fatty acids can be (1) used as a source of energy if the body converts them to CO_2 and H_2O, (2) stored for possible future use in fat cells, or (3) used as starting materials for the synthesis of other compounds needed by the body. Fats are the most concentrated source of food energy in our diets; they furnish about 9000 cal/g when burned for energy, compared with glucose, which furnishes about 3800 cal/g. The human body can make some fats from carbohydrates and carries out such processes to store excess energy furnished in the diet.

> An ordinary scientific calorie is the amount of heat required to raise 1 g of water 1°C. A food calorie is 1000 scientific calories (or 1 kcal).

A high intake of dietary fat has been implicated as a factor in the development of **atherosclerosis,** a complex process in which the walls of the arteries suffer damage and ultimately develop scar tissue and fatty deposits. Atherosclerosis is generally considered to be a precursor to certain types of heart disease and strokes. Atherosclerosis may also be related to the amount of cholesterol in the diet, but the relationship of both dietary fat and cholesterol intake to atherosclerosis does not appear to be a simple one.

It has been known for about 60 years that the human body has a small requirement for certain types of fatty acids (called **essential fatty acids**), and in recent years the basis for this need has been determined. Essential fatty acids were thought to be **linoleic, linolenic,** and **arachidonic** acids, until it was learned that the human body can produce the latter two from the former:

$$CH_3\underset{18}{C}H_2\underset{17}{C}H_2\underset{16}{C}H_2\underset{15}{C}H_2\underset{14}{C}H=\underset{13}{C}H\underset{12}{C}H_2\underset{11}{C}H=\underset{10}{C}H\underset{9}{C}H_2\underset{8}{C}H_2\underset{7}{C}H_2\underset{6}{C}H_2\underset{5}{C}H_2\underset{4}{C}H_2\underset{3}{C}H_2\underset{2}{C}\underset{1}{\overset{O}{\underset{OH}{\diagup\!\!\!\diagdown}}}$$

Linoleic Acid ($C_{18}\Delta_{9,12}$)

> Δ indicates the positions of the double bonds.

Table 10–9
Ratios of Saturated and Unsaturated Fatty Acids from Common Fats and Oils*

	Percentage of Total Fatty Acids by Weight		
Oil or Fat	Saturated	Monounsaturated	Polyunsaturated
Coconut oil	93	6	1
Corn oil	14	29	57
Cottonseed oil	26	22	52
Lard	44	46	10
Olive oil	15	73	12
Palm oil	57	36	7
Peanut oil	21	49	30
Safflower oil	10	14	76
Soybean oil	14	24	62
Sunflower oil	11	19	70

* *Saturated* means full complement of hydrogen (only C—C single bonds); *monounsaturated* means one C=C double bond per fatty acid molecule; *polyunsaturated* means two or more C=C double bonds per molecule of fatty acid. The chief unsaturated fatty acid is linoleic acid. Although derived from vegetable rather than animal fats, both coconut oil and peanut oil have been associated recently with hardening of the arteries when combined with a high cholesterol intake.

$$\text{CH}_3\text{CH}_2\underset{18\quad17\quad16}{\text{CH}}=\underset{15\quad14}{\text{CHCH}_2}\underset{13}{\text{CH}}=\underset{12\quad11}{\text{CHCH}_2}\underset{10}{\text{CH}}=\underset{9\quad8}{\text{CHCH}_2}\underset{7}{\text{CH}_2}\underset{6}{\text{CH}_2}\underset{5}{\text{CH}_2}\underset{4}{\text{CH}_2}\underset{3}{\text{CH}_2}\underset{2}{\text{CH}_2}\underset{1}{\text{C}}\diagdown_{\text{OH}}^{\text{O}}$$

Linolenic Acid ($C_{18}\Delta_{9,12,15}$)

$$\text{CH}_3\text{CH}_2\text{CH}_2\text{CH}_2\text{CH}_2\underset{15}{\text{CH}}=\underset{14\quad13}{\text{CHCH}_2}\underset{12}{\text{CH}}=\underset{11\quad10}{\text{CHCH}_2}\underset{9}{\text{CH}}=\underset{8\quad7}{\text{CHCH}_2}\underset{6}{\text{CH}}=\underset{5\quad4}{\text{CHCH}_2}\underset{3}{\text{CH}_2}\underset{2}{\text{C}}\underset{1}{\diagdown_{\text{OH}}^{\text{O}}}$$

Arachidonic Acid ($C_{20}\Delta_{5,8,11,14}$)

The presence of fatty acids in the diet permits the body to synthesize a very important group of compounds, the prostaglandins. The key compound here again is linoleic acid.

Prostaglandins are a group of more than a dozen related compounds with potent effects on physiological activities such as blood pressure, relaxation and contraction of smooth muscle, gastric acid secretion, body temperature, food intake, and blood platelet aggregation. Their potential use as drugs is currently under widespread investigation. Two of the prostaglandins that have been characterized are prostaglandin E_1 (used to induce labor to terminate pregnancy) and prostaglandin E_2.

Medical evidence links the decrease in peptic ulcer disease in the United States and Britain to an increase in the consumption of polyunsaturated fats, which provide linoleic acid for the synthesis of prostaglandins. The prostaglandins protect the stomach and the intestinal tract from ulcers.

Prostaglandin E_1 ($C_{20}H_{34}O_5$)

Prostaglandin E_2 ($C_{20}H_{32}O_5$)

Note that both of these prostaglandins contain exactly the same number of carbon atoms as arachidonic acid.

Amines and Amides

Organic amines can be considered derivatives of ammonia (NH_3). The nitrogen in the amine may be attached to R groups or may be part of a ring.

CH₃—NH₂

Methyl Amine
(fish odor)

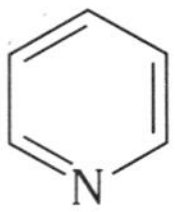

Pyridine

Most amines have unpleasant odors. The stench of decaying protein, for example, is due to some of the compounds listed below.

H₂N—CH₂—CH₂—CH₂—CH₂—NH₂   *Putrescine*

H₂N—CH₂—CH₂—CH₂—CH₂—CH₂—NH₂   *Cadaverine*

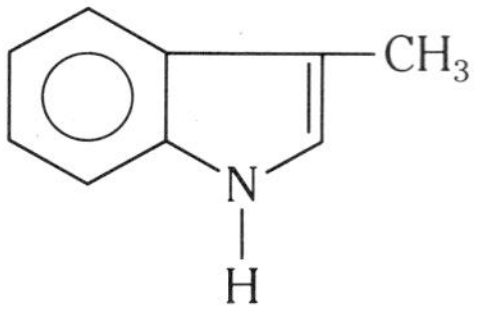

Skatole (in feces)

Alkaloids (alkali-like) are amines derived from plants. The amine nitrogen is usually part of a ring. Caffeine, nicotine, morphine, and coniine are examples of alkaloids (Fig. 10–1).

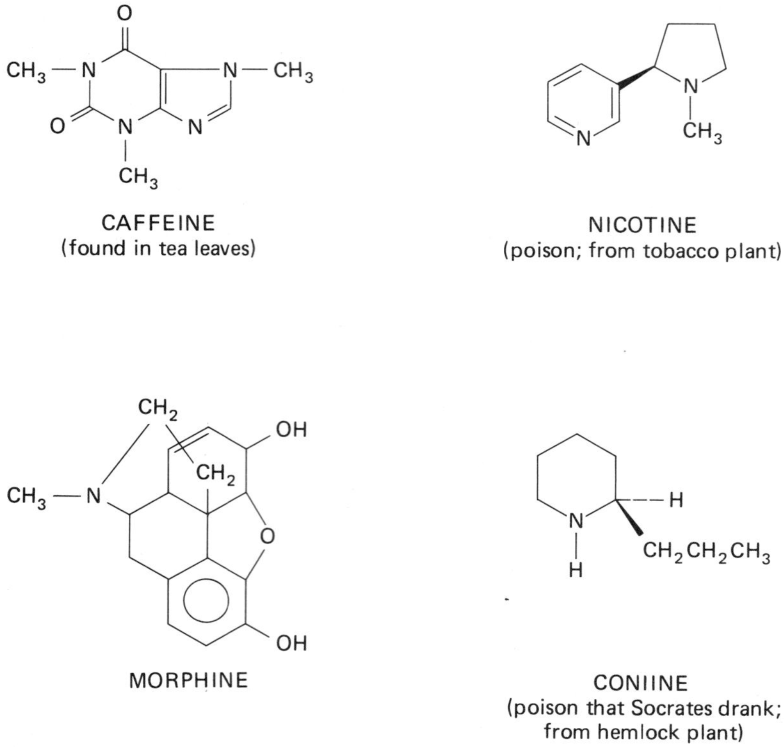

CAFFEINE
(found in tea leaves)

NICOTINE
(poison; from tobacco plant)

MORPHINE

CONIINE
(poison that Socrates drank;
from hemlock plant)

Figure 10–1
Some common alkaloids.

The functional group in amides is —C(=O)—NH$_2$. The functional groups of amines and amides are found in many important biological compounds (discussed in Chapter 11).

Urea

Nicotinamide

Optical Isomers

Structural and geometrical isomers were described in Chapter 9. A third type of isomerism that is important to an understanding of chemical reactions in the body and the specificity of certain drugs is **optical isomerism**.

Optical isomerism is possible when a molecular structure is **asymmetric** (without symmetry). One common example of an asymmetric molecule is one containing a tetrahedral carbon atom bonded to four *different* atoms or groups of atoms. Such a carbon atom is called an asymmetric carbon atom; an example is the carbon atom in the molecule CBrClIH.

Figure 10–2 shows the two ways to arrange four different atoms in the tetrahedral positions about the central carbon atom. These result in two nonsuperimposable, mirror-image molecules that are optical isomers.

Figure 10–2
Optical isomers. Four different atoms, or groups of atoms, are bonded to tetrahedral center atoms so that the upper isomeric form cannot be turned in any way to match exactly the lower structure. The upper structure and the lower structure are nonsuperimposable mirror images. (See also Fig. 10–3.)

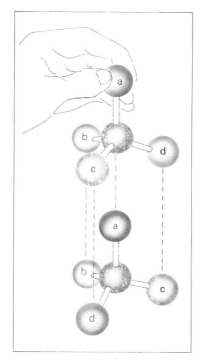
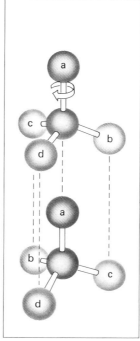

There are many examples of nonsuperimposable mirror images in the macroscopic world. Consider your hands or right- and left-hand gloves, for instance. They are mirror images of one another and are nonsuperimposable.

All amino acids except glycine can exist as one of two optical isomers. In Figure 10–3 the mirror-image relationship is shown for optical isomers of alanine, an amino acid with a tetrahedral carbon atom surrounded by an amino group ($-NH_2$), a methyl group ($-CH_3$), an acid group ($-COOH$), and a hydrogen atom. Note that the carbon atoms in the methyl and acid groups are not asymmetric, since they are not bonded to four different groups.

The properties of some optical isomers are almost identical. Different compounds whose molecules are mirror images of one another have the same melting point, the same boiling point, the same density, and many other identical physical and chemical properties. However, they always differ with respect to one physical property: they rotate the plane of **polarized** light in opposite directions. According to the wave theory of light, a light wave traveling through space vibrates at right angles to its path (Fig. 10–4). If a group of waves is passed through a polarizing crystal, such as Iceland spar (a form of $CaCO_3$), or through a sheet of Polaroid material, the light is split into two rays, and the waves emerging along the incoming axis vibrate in only one plane perpendicular to the light path. Such light is said to be **plane-polarized.** When plane-polarized light is passed through a solution of D-lactic acid, the light is still polarized, but the plane of vibration is rotated counterclockwise. If the other lactic acid isomer is substituted (L-lactic acid), the light rotates in a clockwise direction.

Optical isomers can also differ with respect to biological properties. An example is the hormone adrenalin (or epinephrine). Adrenalin is one of a pair of optical isomers, the L-form. C* designates the asymmetric carbon atom. Only the L-isomer is effective in starting a heart that has stopped beating momentarily, or in giving a person unusual strength during times of great emotional stress. The other isomer is inactive.

All optically active amino acids in proteins are left-handed (L-isomers). Nature's preference for L-amino acids has provoked much discussion and speculation among scientists since Pasteur's discovery of optical activity in 1848 from studies of crystals of

---

levo, L, left
dextro, D, right

D- and L- simply indicate that two structures are possible around an asymmetric C atom. The D- and L- notations do not indicate which way the substance will rotate the plane-polarized light.

Adrenalin (Epinephrine)

Is there another set of life forms in another setting that may be "right-handed"?

All amino acids have an amine group ($-NH_2$) and an acid group ($-COOH$).

The formula of glycine is H_2NCH_2COOH. Why doesn't glycine have optical isomers?

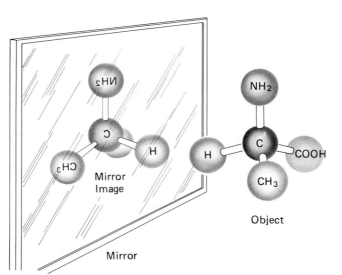

Figure 10–3
Optical isomers of the amino acid alanine, 2-aminopropionic acid.

The D-form is the nonsuperimposable mirror image of the L-form.

tartaric acid salts. However, no satisfactory explanation has been found for this "handedness" of life.

Enzymes, the catalysts for biochemical reactions, also have a handedness and, like a glove, bind to only one of the optical isomers. For example, during contraction of muscles the body produces only the L-form of lactic acid and not the D-form.

Large organic molecules may have many asymmetric carbon atoms within the same molecule. At each such carbon atom there exists the possibility of *two* arrangements of

D-Lactic Acid

L-Lactic Acid

The concentration of lactic acid in the blood is associated with the feeling of tiredness, and a period of rest is necessary to reduce the concentration of this chemical by oxidation.

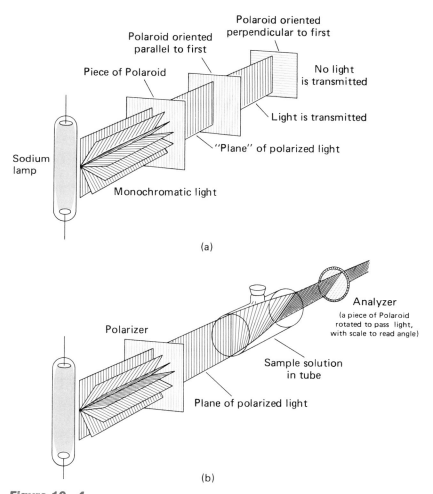

Figure 10–4
Rotation of plane-polarized light by an optical isomer. *(a)* A sodium lamp provides a monochromatic yellow light. The original beam is nonpolarized; it vibrates in all directions at right angles to its path. After passing through a Polaroid filter, the light vibrates in only one direction. This polarized light will pass through another Polaroid filter if the filter is lined up properly but will not pass through the third Polaroid filter if it is at right angles to the other two. The direction of the Polaroid filters determines the direction of the polarization. *(b)* The plane of polarized light is rotated by a solution of an optically active isomer. The analyzer can be a second Polaroid filter that can be rotated to find the angle for maximum transmission of light. If the solution rotates the plane of polarized light, the analyzer will not be at the same angle as the polarizer for maximum transmission.

the molecule. The total number of possible molecules, then, increases exponentially with the number of asymmetric centers. With two asymmetric carbon atoms there are 2^2, or four, possible structures; for three, there are 2^3, or eight, possible structures. It should be emphasized that each of the eight isomers can be made from the *same* set of atoms with the *same* set of chemical bonds. Glucose, a simple blood sugar also known as dextrose, contains four asymmetric carbon atoms per molecule. Thus, there are 2^4 (16) isomers in the family of stereoisomers to which D-glucose belongs. However, of the 16 possible isomers, only 3 are important. These are D-glucose, D-mannose, and D-galactose. Of these, D-glucose is by far the most common.

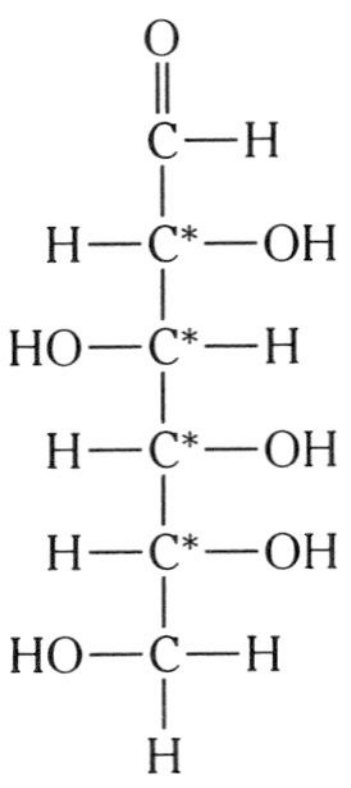

D-Glucose
(C* = Asymmetric Carbon Atom)

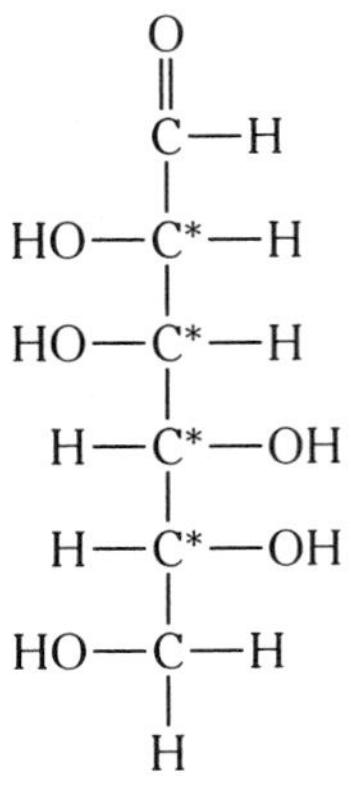

D-Mannose

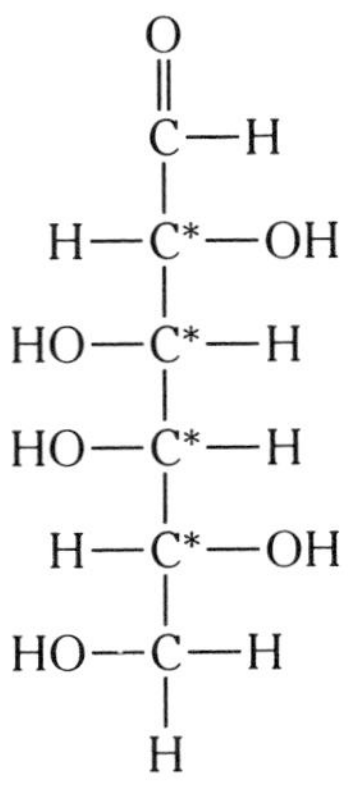

D-Galactose

SELF-TEST 10-B

1. In order to have optical isomers in carbon compounds, a carbon atom must have _____ different groups attached.
2. In what physical property do optical isomers that are mirror images differ? _____
3. a. When referring to edible lipids, what is the difference between a fat and an oil? ____
   b. How can the melting points of most edible oils be increased? _____
4. A high intake of dietary fat is one of the factors that lead to _____.
5. Amines derived from plants are known as _____.

Organic Chemicals: Energy and Materials for Society

Although fuels are the principal products of the petroleum industry, **petrochemicals** (chemicals produced from petroleum) are essential to modern society. About 10% of the refined petroleum is the source of 90% of the organic chemicals used to make plastics, synthetic rubber, synthetic fibers, fertilizers, and thousands of other consumer products. Figure 10–5 summarizes the organic chemicals obtained from fossil fuels and their uses as raw materials in the commercial production of a wide range of products.

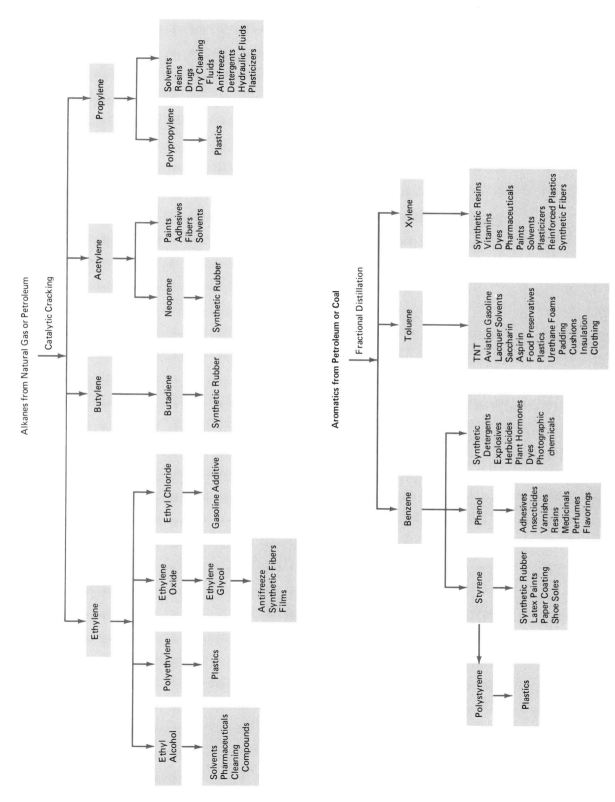

Figure 10–5
Some uses of organic compounds obtained from fossil fuels. (Data from the American Petroleum Institute.)

Should petroleum be reserved by law as a source of chemicals rather than being burned for its energy content?

World reserves of fossil fuels are discussed in Chapter 8.

Petroleum and natural gas are abundant now, but their cost will escalate significantly by the end of the century. Current projections are that 80% of all the known and estimated world reserves of petroleum and natural gas will be consumed by the year 2050 (see Chapter 8). In other words, alternative sources for energy and raw materials need to be developed to replace petroleum and natural gas if the 8 billion people expected to be living on this planet in 2025 are to have a decent standard of living. For this reason, there is increased interest in coal as a source of energy and raw materials, since the Earth's reserves of coal are expected to last about 500 years. Coal, the most abundant fossil fuel, currently constitutes 80% of the U.S. fossil fuel reserves and 90% of the world reserves of fossil fuel. The discussion here will center on the use of coal as a source of organic chemicals.

Coal

Heating coal at high temperatures in the absence of air produces a mixture of coke, coal tar, and coal gas. The process, called **pyrolysis,** is represented by

$$\text{Coal} \longrightarrow \text{Coke} + \text{Coal tar} + \text{Coal gas}$$

One ton of bituminous (soft) coal yields about 1500 pounds of coke, 8 gallons of coal tar, and 10,000 cubic feet of coal gas. Coal gas is a mixture of H_2, CH_4, CO, C_2H_6, NH_3, CO_2, H_2S, and other gases. At one time coal gas was used as a fuel. Coal tar can be distilled to yield the fractions listed in Table 10–10. Note the predominance of aromatic compounds.

Coal Gasification

When coal is pulverized and treated with superheated steam, a mixture of CO and H_2, **synthesis gas,** is obtained. Synthesis gas is used as a fuel:

$$C + H_2O \longrightarrow CO + H_2$$

Recall that synthesis gas is used to make methanol.

It is also used as a starting material for the production of organic chemicals. As a fuel, synthesis gas produces about one third of the heat produced by an equal volume of methane (natural gas).

$$2\ CO + O_2 \longrightarrow 2\ CO_2 + 135.3 \text{ kcal } (67.6 \text{ kcal/mole CO})$$

$$2\ H_2 + O_2 \longrightarrow 2\ H_2O + 115.6 \text{ kcal } (57.8 \text{ kcal/mole } H_2)$$

Table 10–10
Fractions from Distillation of Coal Tar

Boiling Range	Name	Tar, Mass %	Primary Constituents
Below 200°C	Light oil	5	Benzene, toluene, xylenes
200–250	Middle oil (carbolic oil)	17	Naphthalene, phenol, pyridine
250–300	Heavy oil (creosote oil)	7	Naphthalenes and methylnaphthalenes, cresols, quinoline
300–350	Green oil	9	Anthracene, carbazole
Residue		62	Pitch or tar

In a newer coal gasification process, methane is the end product. In the process, crushed coal is mixed with an aqueous catalyst; the mixture is dried and CO and H_2 are added. The resulting mixture is then heated to 700°C to produce methane and carbon dioxide. The overall reaction is the following:

$$2\,C + 2\,H_2O \longrightarrow CH_4 + CO_2 - 2\text{ kcal/mole }CH_4$$

Since the combustion of methane releases 213 kcal/mole, coal gasification is an energy-efficient process.

Synthesis gas from coal is receiving increased attention as a starting material for the production of organic chemicals that are among the top 50 produced in the United States (Table 10–11). For example, the first complete "chemicals from coal" plant, built by Eastman Kodak in Kingsport, Tennessee, started production in 1983. Figure 10–6a is a schematic drawing of the various components of the plant, which is pictured in Figure 10–6b. The basic reactions are to produce synthesis gas from coal, to use the synthesis gas to make methanol, and to use the methanol in the synthesis of acetic anhydride. Acetic anhydride is used by Eastman Kodak to make cellulose acetate, a polymer used in the manufacture of photographic film base, synthetic fibers, plastics, and other products.

Within the complex pictured in Figure 10–6b are nine separate plants, four related to the gasification of coal, two for synthesis gas preparation, and three for the synthesis of methanol, methyl acetate, and acetic anhydride.

The main chemical reactions used in the process are shown below:

1. $C + H_2O \longrightarrow CO + H_2$
   Coal Steam   Synthesis Gas

2. $CO + 2\,H_2 \longrightarrow CH_3OH$
   Methanol

Table 10–11
Top 50 Chemicals Produced in the United States in 1986*

1. Sulfuric acid	18. Carbon dioxide	35. Carbon black
2. Nitrogen	19. Vinyl chloride	36. Potash
3. Oxygen	20. Styrene	37. Aluminum sulfate
4. Ethylene	21. Terephthalic acid	38. Propylene oxide
5. Lime	22. Methanol	39. Acrylonitrile
6. Ammonia	23. Hydrochloric acid	40. Vinyl acetate
7. Sodium hydroxide	24. Ethylene oxide	41. Methyl t-butyl ether
8. Chlorine	25. Formaldehyde	42. Cyclohexane
9. Phosphoric acid	26. Toluene	43. Acetone
10. Propylene	27. Xylene	44. Titanium dioxide
11. Sodium carbonate	28. Ethylene glycol	45. Sodium silicate
12. Ethylene dichloride	29. p-xylene	46. Calcium chloride
13. Nitric acid	30. Ammonium sulfate	47. Sodium sulfate
14. Urea	31. Cumene	48. Adipic acid
15. Ammonium nitrate	32. Acetic acid	49. Isopropyl alcohol
16. Benzene	33. Phenol	50. Sodium tripolyphosphate
17. Ethylbenzene	34. Butadiene	

* Data from *Chemical and Engineering News*, April 13, 1987, p. 21.

228 *Organic Chemicals of Major Importance*

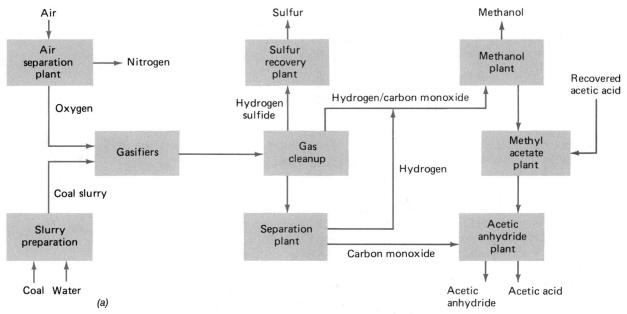

Figure 10-6
(a) Schematic drawing showing the production of acetic anhydride from coal. (b) Eastman Kodak's "chemicals-from-coal" plant in Kingsport, Tennessee. Numbers on the photograph represent different parts of the plant: 1, coal unloading; 2, coal silos; 3, steam plant; 4, slurry preparation; 5, coal gasification plant; 6, gas cleanup and separation; 7, sulfur recovery plant; 8, gas flare stack; 9, chemical storage; 10, methanol plant; 11, methyl acetate plant; 12, acetic anhydride plant. (Photo courtesy of Tennessee Eastman.)

3. $\underset{\text{Acetic Acid}}{CH_3OH + CH_3\overset{\overset{\displaystyle O}{\|}}{C}OH} \longrightarrow \underset{\text{Methyl Acetate}}{CH_3\overset{\overset{\displaystyle O}{\|}}{C}OCH_3}$

4. $CH_3\overset{\overset{\displaystyle O}{\|}}{C}OCH_3 + CO \longrightarrow \underset{\text{Acetic Anhydride}}{CH_3-\overset{\overset{\displaystyle O}{\|}}{C}-O-\overset{\overset{\displaystyle O}{\|}}{C}-CH_3}$

Acetic anhydride reacts with water to give acetic acid, number 32 in the top 50 chemical list.

About 900 tons per day of high-sulfur coal from nearby Appalachian coal mines are ground in water to form a slurry of 55% to 65% by weight of coal in water. The slurry is fed into two gasifiers to make synthesis gas. To produce the same amounts of these chemicals by conventional means would require the annual equivalent of 1 million barrels of oil.

Plant design uses the latest environmental control technologies to protect the environment. For example, the sulfur recovery unit converts the hydrogen sulfide gas that was removed during the gasification of coal into free sulfur. This process removes over 99% of the sulfur from the coal, and this sulfur is sold to chemical companies.

Synthesis gas will become increasingly important as a raw material, since it can be used to make a variety of hydrocarbons that can then be converted into other commercially important organic chemicals.

Industrial Catalysts

Many of the reactions listed in the discussions of petroleum cracking, coal gasification, and other industrial processes involve the use of a catalyst. Recall from Chapter 6 that a catalyst increases the rate of a reaction without being consumed in the reaction. In many of the industrial processes, the catalyst is one of the noble metals, such as platinum, rhodium, or gold. A metal surface is often used to catalyze a gas-phase reaction. For example, the platinum catalyst in the catalytic reforming process described in Chapter 9 adsorbs low-octane alkanes such as *n*-hexane and converts them to compounds with higher octane numbers, such as benzene and branched or cyclic alkanes. The product depends on the pattern of platinum atoms (crystal face) at the surface where adsorption takes place. Figure 10–7 illustrates how products depend on the particular characteristics of the exposed surface. Note in Figure 10–7b that undesirable alkanes are produced at the "kinked" surface. The catalytic converter found in American automobiles produced since 1975 is also a good example of a catalytic process that uses platinum as a surface catalyst.

Key Chemicals

The top 50 chemicals produced in the United States in 1986, listed in Table 10–11, can serve as a basis for a summary of the importance of organic chemistry to society. Of the top 50, 28 are organic compounds. Many of these compounds have already been discussed, but one that merits further discussion is methanol, number 22.

Urea, the first organic compound synthesized in the laboratory, ranks fourteenth.

Methanol-to-Gasoline

Methanol will likely move upward in the ranking as petroleum and natural gas become too expensive as sources of both energy and chemicals. Although most of the world's methanol currently comes from synthesis gas made from natural gas, coal gasification will become a more important source of methanol as the natural gas reserves are used

Details about the top 25 chemicals produced in the United States are listed on the inside front cover.

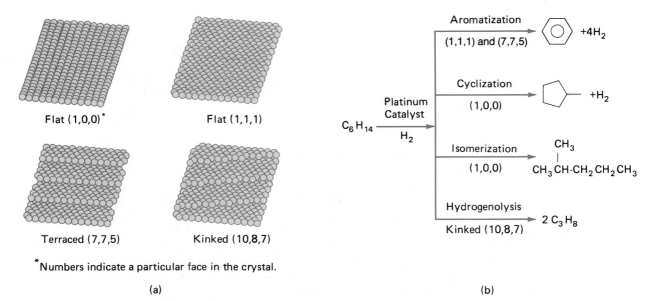

Figure 10-7
(a) Chemistry on a platinum surface depends on which surface is exposed. (b) Different surfaces favor different products. (Redrawn from George C. Pimentel, "Surfaces and Condensed Phases," *Chem Tech*, September 1986, p. 537.)

up. Since methanol is relatively cheap, its potential as a fuel and as a starting material for the synthesis of other chemicals is receiving more attention.

In Chapter 9, the use of methanol as a gasoline additive was described. In addition, methanol is being considered as a fuel to replace gasoline. There are a number of methanol-fueled test vehicles on the highways, and although there is little likelihood that consumers will have the option of buying methanol-fueled vehicles during the next 20 years, the technology is available for using methanol to make gasoline.

The production of 92-octane gasoline from methanol is taking place in New Zealand. New Zealand Synthetic Fuels Company is operating a plant based on a methanol-to-gasoline process developed by Mobil Oil Company. The process starts with the production of synthesis gas from natural gas and then uses the synthesis gas to make methanol. The key reaction for the production of gasoline from methanol is the dehydration of methanol with a clay catalyst developed by Mobil; this catalyst is known as the ZSM-5 zeolite catalyst. The catalyst aids the dehydration to yield short-chain alkenes, which then cyclize and polymerize to give a mixture of C_5 to C_{12} hydrocarbons made up of branched chains, straight chains, and aromatics, similar to the 92-octane gasoline currently obtained by the refinement of "straight-run" gasoline from petroleum refining. The dehydration and subsequent polymerization can be represented by the following reactions:

Zeolites are porous clays.

$$2\ CH_3OH \xrightarrow[\text{catalyst}]{\text{ZSM-5}} (CH_3)_2O + H_2O$$
Dimethyl ether

$$2\ (CH_3)_2O \xrightarrow[\text{catalyst}]{\text{ZSM-5}} 2\ C_2H_4 + 2\ H_2O$$
Ethylene

$$C_2H_4 \xrightarrow[\text{catalyst}]{\text{ZSM-5}} \text{Hydrocarbon Mixture in the } C_5-C_{12} \text{ Range}$$
Gasoline

The New Zealand plant is currently producing 14,000 barrels per day of gasoline with an octane rating of 92 to 94. This is about one-third the amount of gasoline used in New Zealand.

Plastics

The prominence of plastics in consumer products is indicated by the fact that 24 of the top 50 chemicals are used in the production of plastics and synthetic fibers.

Polymers

Polymers are large molecules made by the joining together of many small molecules called **monomers.** Polymers can be classified as natural or synthetic. Examples of natural polymers include proteins, nucleic acids, starch, cellulose, and rubber. Most synthetic polymers are organic compounds used in the production of plastics. Synthetic polymers usually contain only one or two different types of monomers, joined in chains that may be thousands of units long. We shall see in the sections that follow that synthetic polymers can be **addition polymers,** in which monomer units are joined directly, or **condensation polymers,** in which monomer units combine by splitting out a small molecule, usually water.

Nature made the first plastics: tar, resins, and rubber are examples. The chemist learned to "copy" nature.

Addition Polymers. The monomer for addition polymers normally contains a double bond. The polymerization process involves conversion of the double bond to a single bond with the formation of a reactive species that has unpaired electrons at either end. The process is shown below for ethylene:

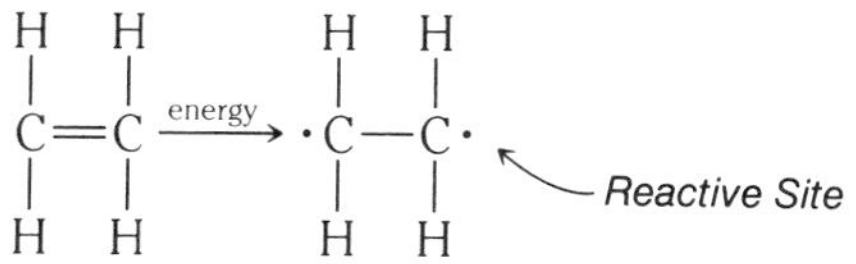

$$\begin{array}{c} H\ \ H \\ |\ \ \ | \\ C=C \\ |\ \ \ | \\ H\ \ H \end{array} \xrightarrow{\text{energy}} \begin{array}{c} H\ \ H \\ |\ \ \ | \\ \cdot C-C\cdot \\ |\ \ \ | \\ H\ \ H \end{array} \longleftarrow \text{Reactive Site}$$

The reactive species join to form long chains:

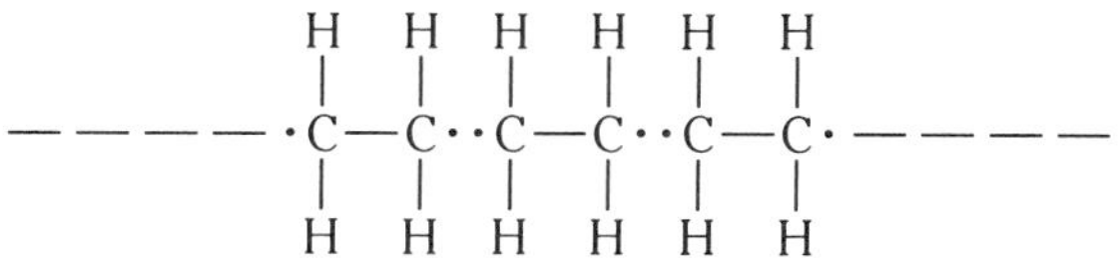

$$----\cdot \overset{H}{\underset{H}{C}}-\overset{H}{\underset{H}{C}}\cdot\cdot\overset{H}{\underset{H}{C}}-\overset{H}{\underset{H}{C}}\cdot\cdot\overset{H}{\underset{H}{C}}-\overset{H}{\underset{H}{C}}\cdot----$$

For polyethylene this chain can be represented as

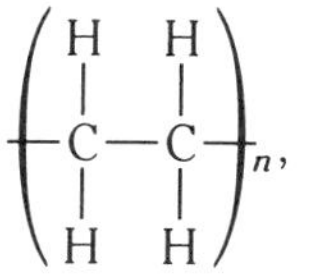

$$\left(\begin{array}{c} H\ \ H \\ |\ \ \ | \\ C-C \\ |\ \ \ | \\ H\ \ H \end{array} \right)_n,$$

where n is 500 to 5000, depending on the type of polymer.

Polyethylenes formed under various pressures and catalytic conditions have different molecular structures and hence different physical properties. For example, chromium oxide as a catalyst yields almost exclusively the linear polyethylene shown in the margin. Actually, a methyl group is attached to about every eighth or tenth carbon in the chain. If ethylene is heated to 230°C at a pressure of 200 atmospheres, irregular branches result.

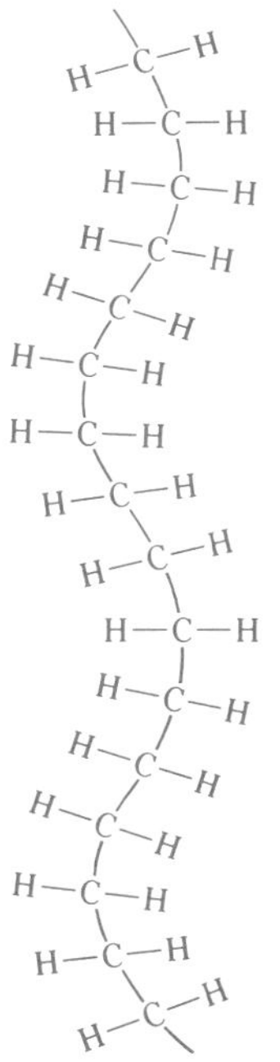

The molecules in linear polyethylene can line up with one another very easily, yielding a tough, high-density crystalline compound that is useful in making toys, bottles, and structural parts. The polyethylene with irregular branches is less dense, more flexible, and not nearly as tough as the linear polymer, since the molecules are generally farther apart and their arrangement is not as precisely ordered. This material is used for trash bags, squeeze bottles, and other similar applications.

Other monomers used to make addition polymers can be regarded as derivatives of ethylene. Table 10–12 lists several common addition polymers, as well as two condensation polymers.

Condensation Polymers. Condensation polymers form by the elimination of a small molecule such as water. The reaction of an alcohol with an organic acid to form an ester can be used as the basis for a condensation reaction. By using alcohols and acids with two functional groups per molecule, a "polyester" such as **Dacron** is formed.

$$\text{HO-C(=O)-C}_6\text{H}_4\text{-C(=O)-OH} + \text{HO-CH}_2\text{-CH}_2\text{-OH} \longrightarrow$$

Terephthalic Acid   Ethylene Glycol

$$\text{HO-C(=O)-C}_6\text{H}_4\text{-C(=O)-OCH}_2\text{CH}_2\text{O}\left(-\text{C(=O)-C}_6\text{H}_4\text{-C(=O)-OCH}_2\text{CH}_2\text{O}-\right)_n$$

$$-\text{C(=O)-C}_6\text{H}_4\text{-C(=O)-OCH}_2\text{CH}_2\text{OH}$$

Poly(ethylene Glycol Terephthalate)

Another common condensation polymer is **nylon,** which is formed by condensing a bifunctional amine with a bifunctional acid. These could be called polyamides because they are held together by the $-\overset{\overset{\text{O}}{\|}}{\text{C}}-\overset{|}{\underset{\text{H}}{\text{N}}}-$ linkage; in proteins this linkage is called a **peptide** bond.

Proteins are condensation polymers of amino acid monomers bonded together by peptide bonds (Chapter 11).

$$\text{HO-C(=O)-(CH}_2)_4\text{-C(=O)-OH} + \text{H}_2\text{N-(CH}_2)_6\text{-NH}_2 \longrightarrow$$

Adipic Acid   Hexamethylenediamine

$$-\text{C(=O)-(CH}_2)_4-\boxed{\text{C(=O)-N(H)}}-(\text{CH}_2)_6-\boxed{\text{N(H)-C(=O)}}-(\text{CH}_2)_4-\boxed{\text{C(=O)-N(H)}}-(\text{CH}_2)_6- + x\text{H}_2\text{O}.$$

Nylon 66
(The amide groups are outlined for emphasis.)

Table 10-12
Some Common Polymers

Monomer	Name (Top 50 Rank)	Polymer	Uses
Addition Polymers			
H₂C=CH₂	Ethylene (4)	Polyethylene	Bags, coatings, toys
H₂C=CHCH₃	Propylene (10)	Polypropylene	Beakers, milk cartons
H₂C=CHCl	Vinyl chloride (19)	Polyvinyl chloride (PVC)	Floor tile, raincoats, pipe, phonograph records
H₂C=CHCN	Acrylonitrile (39)	Polyacrylonitrile (PAN)	Rugs; Orlon and Acrilan are copolymers with other monomers
H₂C=CH(C₆H₅)	Styrene (20)	Polystyrene	Cast articles using a transparent plastic
H₂C=C(CH₃)C(=O)OCH₃	Methyl methacrylate	Plexiglas, Lucite, acrylic resins	High-quality transparent objects, latex paints
F₂C=CF₂	Tetrafluoroethylene	Teflon	Gaskets, insulation, bearings, pan coatings
Condensation Polymers			
	Ethylene glycol (28) and terephthalic acid (21)	Dacron	Synthetic fabrics
	Adipic acid (48) and hexamethylenediamine	Nylon-66	Synthetic fabrics

Synthetic "Natural" Rubber: A Tailor-Made Addition Polymer

Natural rubber, a product of the *Hevea brasiliensis* tree, is a hydrocarbon with the composition C_5H_8; when it is decomposed in the absence of oxygen it yields the monomer isoprene:

$$CH_2=C(CH_3)-CH=CH_2$$

Isoprene

Natural rubber occurs as latex (an emulsion of rubber particles in water) that oozes from rubber trees when they are cut. Precipitation of the rubber particles yields a gummy mass that is not only elastic and water-repellent but also very sticky, especially when warm. In 1839, after ten years' work on this material, Charles Goodyear (1800–1860) discovered that the heating of gum rubber with sulfur produced a material that was no longer sticky but was still elastic, water-repellent, and resilient.

Vulcanized rubber, as Goodyear called his product, contains short chains of sulfur atoms that bond together the polymer chains of the natural rubber and reduce its unsaturation. The sulfur chains help align the polymer chains, so the material does not undergo a permanent change when stretched but springs back to its original shape and size when the stress is removed. Substances that behave this way are called **elastomers.**

In later years chemists searched for ways to make a synthetic rubber so we would not be completely dependent on imported natural rubber during emergencies, such as during the first years of World War II. In the mid-1920s German chemists polymerized butadiene (obtained from petroleum and structurally similar to isoprene, but without the methyl group side chain). The product was buna rubber, so named because it was made from butadiene (Bu-) and catalyzed by sodium (-Na).

The behavior of natural rubber (polyisoprene), it was learned later, is due to the specific arrangement within the polymer chain. We can write the formula for polyisoprene with the CH_2 groups on opposite sides of the double bond (the *trans* arrangement):

Poly-trans-isoprene (the $—CH_2—CH_2—$ groups are trans*)*

The formula can also be written with the CH_2 groups on the same side of the double bond (the *cis* arrangement, from Latin meaning "on this side").

Poly-cis-isoprene (the $—CH_2—CH_2—$ groups are cis*)*

Natural rubber is poly-*cis*-isoprene. However, the *trans* material also occurs in nature in the leaves and bark of the sapotacea tree and is known as *gutta-percha*. It is used as a thermoplastic for golf ball covers, electrical insulation, and other such applications. Without an appropriate catalyst, polymerization of isoprene yields a solid that is like neither rubber nor gutta-percha. Neither the *trans* polymer nor the randomly arranged material is as good as natural rubber *(cis)* for making automobile tires.

In 1955 chemists at the Goodyear and Firestone companies almost simultaneously discovered how to use stereoregulation catalysts to prepare synthetic poly-*cis*-isoprene. This material is structurally identical to natural rubber. Today, synthetic poly-*cis*-isoprene can be manufactured cheaply and is used almost equally well (there is still an increased cost) when natural rubber is in short supply. More than 2.4 million tons of synthetic rubber are produced in the United States yearly. Table 10–13 gives a typical rubber formulation as it might be used in a tire.

Table 10-13
Rubber Formulation

Ingredient	Name	Percentage	Function
Rubber	Poly-*cis*-isoprene	62.0	Elastomer
Activators	Zinc oxide	2.7	Activates vulcanizing agents; stearic acid acts as a lubricant in processing
	Stearic acid	0.6	
Vulcanizing agent	Sulfur	1.5	Cross-links polymer chains
Filler	Carbon black	30.5	Provides strength and abrasion resistance
Accelerator	Dibenzthiozole disulfide	1.1	Catalyzes vulcanization
Antioxidant	Alkylated diphenylamine	1.1	Inhibits attack by oxygen or ozone in the air
Processing oil	Hydrocarbon oil	0.5	Plasticizer

SELF-TEST 10-C

1. The process of heating coal at high temperatures in the absence of air is called _____.
2. _____ (a class of hydrocarbons) are the principal components of coal tar.
3. Two fossil fuels that will become scarce during your lifetime are _____ and _____.
4. The individual molecules from which polymers are made are called _____.
5. Draw the formulas of the monomers used to prepare the polymers listed below. For example, $CH_2\!=\!CH_2$ is used to prepare polyethylene.
    a. Polypropylene
    b. Polystyrene
    c. Teflon
6. Natural rubber is a polymer of _____.
7. Nylon is an example of a _____ polymer.
8. Polyamides are formed when _____ is split out from the reaction of many organic acid groups and many amine groups.
9. Polyesters are formed by ( ) addition or ( ) condensation.

MATCHING SET

_____ 1. Asymmetric carbon atom   a. Made from essential fatty acids
_____ 2. RCOOH   b. Unsaturated fatty acid
_____ 3. Nylon   c. Amine
                   O
                   ‖
_____ 4. R—O—C—R'   d. Ester of saturated fatty acid and glycerol
                                           e. Carboxylic acid

_____ 5. Prostaglandins
_____ 6. Fat
_____ 7. Linoleic acid
_____ 8. RNH$_2$
_____ 9. Alkaloid
_____ 10. Poly-*cis*-isoprene
_____ 11. Monomer
_____ 12. Vulcanization

f. Caffeine
g. Has four different groups attached to it
h. Alcohol
i. Ester
j. Cross-linking via reaction with sulfur
k. Polyamide
l. Building unit for polymer
m. Natural rubber

QUESTIONS

1. What is meant by the following terms?
   a. Proof rating of an alcohol
   b. Denatured alcohol
2. Wood alcohol is a deadly poison that can be made from what deadly gas?
3. Draw a structural formula for each of the following:
   a. An alcohol
   b. An organic acid
   c. An ester
   d. Glycerol
4. How do primary, secondary, and tertiary alcohols differ?
5. Explain the common names of *wood alcohol* for methanol and *grain alcohol* for ethanol.
6. Why is a fatty acid so named?
7. Is pure synthetic ethanol different from pure grain alcohol? Explain.
8. Which propanol is used as rubbing alcohol?
9. Pure ethyl alcohol is what proof?
10. Prostaglandins belong to what group of organic compounds?
11. What structural features and properties make ethylene glycol a desirable antifreeze agent?
12. What is the acid in vinegar?
13. After reading this chapter and the previous one, what new thoughts do you have when you view a lump of coal or a drop of petroleum?
14. Consult a medical dictionary to determine the difference between atherosclerosis and arteriosclerosis.
15. What functional groups are found in glucose?
16. Indicate the functional groups present in the following molecules:
    a. CH$_3$CH$_2$CH$_2$COOH
    b. CH$_3$CH$_2$NH$_2$
    c. CH$_3$CHCH$_2$CH$_2$COOH
          |
         NH$_2$
    d. CH$_3$CHCH$_2$COOH
          |
         OH
    e. CH$_3$CCH$_2$CH$_2$COOH
          ||
          O
    f. CH$_3$CHCH$_2$OH
          |
         NH$_2$
17. What feature do all condensation polymerization reactions have in common?
18. Which do you think is the source of most polymers used today, green plants or petroleum? Do you think this will ever change? Explain.
    a. Should we stop burning petroleum? What are the problems involved?
    b. Should we start to develop research on how to change wood and straw into plastics? What are a few of the problems involved?
19. What properties of plastics make them superior to metals? What properties of plastics make them inferior to metals?
20. A tiny sample of rubber, held in the flame of a match, burns with a *white* flame, in contrast to the black smoke of burning tires. Explain.
21. Given an example of
    a. An alkane
    b. An amine
    c. A carboxylic acid
    d. An ether
    e. An ester
    f. An alkene
    g. An alkyne
    h. An alcohol
    i. A ketone
22. Explain how gasoline can be made from methanol.
23. What single property must a molecule possess in order to be a monomer (a) for addition polymerization? (b) for condensation polymerization?
24. What is synthesis gas? How can it be used to produce chemicals?

25. Biphenyl, 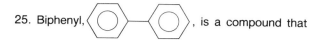, is a compound that can be chlorinated to make chlorinated biphenyl. The family of chlorinated biphenyls is called polychlorinated biphenyls, or PCBs. Draw the structures of several polychlorinated biphenyls. There are 209 possible structures.
26. What ester has the smell of bananas?
27. In what ways is a railroad train like polystyrene? Where do you suppose the first chemist who prepared a polymer got the idea for giant molecules?
28. What is the origin of the word *polymer*?
29. What is meant by the term *macromolecule*?
30. How can optical isomers be distinguished from each other experimentally?
    a. Which arrangement has a mirror image that is nonsuperimposable?

    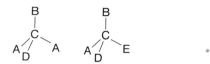

    b. Use the two structures to explain the term *asymmetric carbon atom*.

CHAPTER 11

Chemistry of Living Systems

Biochemistry embodies the relationships between chemicals and life forms.

Have you ever marveled at the vastness of the universe or the intricacies of the atom? Have you wondered how a huge airplane flies or a computer works? The most marvelous and intricate wonderment of them all is a living being. One cannot help but be amazed at the complicated chemical sequences that occur when, for example, we eat, digest, use, and eliminate a burger, french fries, and a milkshake. **Biochemistry,** the science of life, has helped us make great strides in medicine and health care, and thus in our general well-being.

The purposes of this chapter are to acquaint you with some of the biochemicals in our bodies, to discuss how biochemicals give us energy, and to help you understand from a chemical point of view how like begets like.

Biochemicals common to all living systems are fats and oils, carbohydrates (sugars and starches), proteins, enzymes, vitamins, hormones, nucleic acids, and compounds for the storage and exchange of energy, such as adenosine triphosphate (ATP).

Starch, glycogen, cellulose, and proteins are condensation polymers (Chapter 10).

Some biochemicals are polymers. Starches are condensation polymers of simple sugars (the monomers); sucrose (table sugar) is composed of only two simple sugars. Proteins are condensation polymers of amino acids (the monomers). Nucleic acids are condensation polymers of simple sugars, nitrogenous bases, and phosphoric acid species.

A nitrogenous base is basic because hydrogen ions are attracted to the nonbonding pairs of electrons on nitrogen atoms:

$$-\overset{\cdot\cdot}{\underset{|}{N}}-$$

Other biochemicals are composed of two or more smaller molecular structures. Recall from Chapter 10 that a fat molecule is composed of one glycerol and three fatty acid molecules bonded by ester linkages. Enzymes are constructed of a protein alone or a protein bonded to a metal ion or a vitamin.

Carbohydrates contain the elements carbon, hydrogen, and oxygen, with hydrogen atoms and oxygen atoms generally in the ratio of 2 to 1.

Carbohydrates

Carbohydrates have the three elements carbon, hydrogen, and oxygen arranged primarily into three structural groups: alcohol (—OH), aldehyde ($-\overset{\overset{O}{\|}}{C}H$), and ketone

mono—one
oligo—few
poly—many

($-\overset{\overset{O}{\|}}{C}-$). Carbohydrates are divided into three groups on the basis of degree of condensation polymerization: monosaccharides (Latin *saccharum,* "sugar"), oligosaccharides, and polysaccharides. Monosaccharides are simple sugars that cannot be dissociated into smaller units by acid hydrolysis. Hydrolysis of a molecule of an oligosaccharide yields two to six molecules of a simple sugar; complete hydrolysis of a polysaccharide molecule produces many (sometimes thousands) monosaccharide monomers.

Hydrolysis is a water-splitting reaction in which H· bonds with one fragment of the attacked molecule and ·OH bonds with the other fragment.

Glucose, $C_6H_{12}O_6$, and some of the other simple sugars are quick sources of energy for cells. Large amounts of energy are stored in polysaccharides, such as starch. The stored energy is usable by living cells only if polysaccharides are hydrolyzed into monosaccharides. Some complex polysaccharides are used by some organisms for structural purposes. Cellulose, for example, is partially responsible for plant support.

Monosaccharides

Approximately 70 monosaccharides are known; 20 occur naturally. The most common simple sugar is D-glucose (Fig. 11-1), which is found in fruit, blood, and living cells. As can be seen from the structure of D-glucose, the great solubility of monosaccharides in water is caused by the numerous —OH groups, which hydrogen-bond with water. An aqueous solution of D-glucose contains all three structures shown in the figure in dynamic equilibria involving mostly the two ring forms in the presence of a relatively small amount of the straight-chain form. The aldehyde group in the straight-chain structure of D-glucose qualifies this sugar as an aldose monosaccharide.

Also known as dextrose, grape sugar, and blood sugar, D-glucose is used in the manufacture of candy and in commercial baking. A solution of D-glucose is often given intravenously when a quick source of energy is needed to sustain life.

Because D-fructose, a monosaccharide found in many fruits and table sugar, has a ketone group in its straight-chain form (Fig. 11-2), it is classified as a ketose monosaccharide.

Oligosaccharides

The most commonly encountered oligosaccharides are the disaccharides, which have two simple sugar monomers per molecule. Examples include three widely used disaccharide sugars:

An aldose is a sugar with the aldehyde group.

D-glucose has a relative sweetness of 74.3, compared with sucrose, which has an assigned value of 100.0. The sweetness value of fructose is 173.3.

Sweetness is judged by taste testers.

A ketose is a sugar with the ketone group.

Recall from Chapter 9 that a carbon atom is at each vertex not occupied by another atom.

Refer to Chapter 9 for a discussion of cis and trans (geometric) isomers.

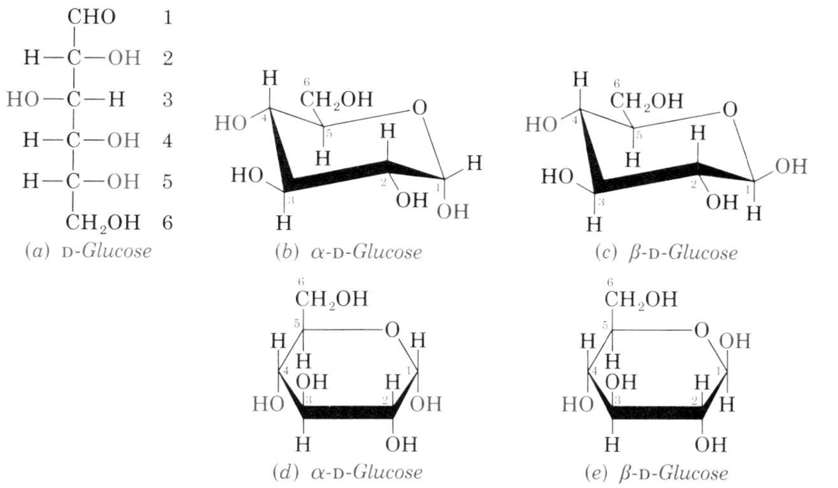

Figure 11-1
The structures of D-glucose; d and e are two-dimensional representations of b and c, respectively. Note the difference in the positions of the —OH groups *(color)* in the alpha and beta forms of glucose: the —OH groups on the 1 and 4 carbons are *trans* when the structure is beta (β), and the —OH groups are *cis* when the structure is alpha (α). In both alpha and beta glucose, the —OH group on the number-4 carbon atom must be in the same position.

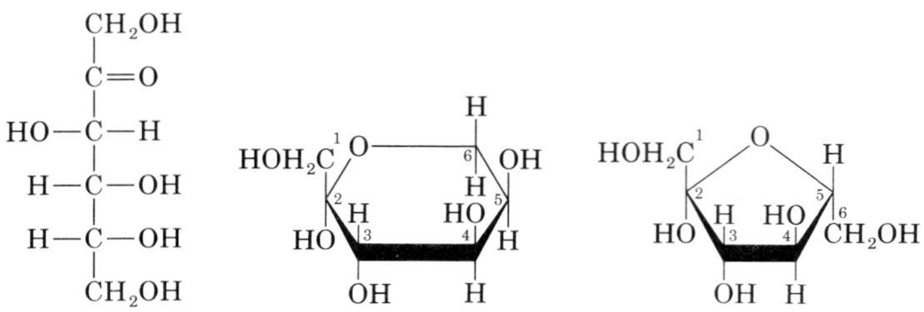

Figure 11-2
The structures of D-fructose. The alpha-ring structure (not shown) differs from the beta-ring structure in that the CH_2OH and OH groups are in reversed positions on carbon 2.

(a) Ketone structure

(b) β-Ring structure (Pyranose structure: 6-membered ring with an oxygen atom in the ring)

(c) β-Ring structure (Furanose structure: 5-membered ring with an oxygen atom in the ring)

> **Sucrose** (from sugar cane or sugar beets), which consists of a glucose monomer and a fructose monomer
> **Maltose** (from starch), which consists of two glucose monomers
> **Lactose** (from milk), which consists of a glucose monomer and a galactose (an optical isomer of glucose) monomer

The structure of galactose is shown on page 241.

Disaccharide molecules contain two simple sugars bonded together; sucrose, for example, contains a glucose and a fructose unit in each molecule. A water molecule is eliminated when the bond forms between the two simple sugars.

The formula for these disaccharides, $C_{12}H_{22}O_{11}$, is not simply the sum of two monosaccharides, $C_6H_{12}O_6 + C_6H_{12}O_6$. A water molecule is eliminated as two monosaccharides are united to form the disaccharide. The structures of sucrose, maltose, and lactose, along with their hydrolysis reactions, are shown in Figure 11-3.

Sucrose is produced in a high state of purity on an enormous scale — over 80 million tons per year. Originally produced in India and Persia, sucrose is now used universally as a sweetener. About 40% of the world sucrose production comes from sugar beets and 60% from sugar cane. Sucrose provides a high caloric value (1794 kcal/pound); it is also used as a preservative in jams, jellies, and candied fruit.

Polysaccharides

Polysaccharides are condensation polymers.

Nature's most abundant polysaccharides are the starches, glycogen, and cellulose. Molecular structures are known to combine more than 500 monosaccharide monomers into molecules with molecular weights of over 1 million. The monosaccharide most commonly used to build polysaccharides is D-glucose.

Starches and Glycogen. Plant starch is found in protein-covered granules. If these granules are ruptured by heat, they yield a starch that is soluble in hot water, **amylose**, and an insoluble starch, **amylopectin.** Amylose constitutes about 25% of most natural starches. When tested with iodine solution, amylose turns blue-black, whereas amylopectin turns red.

Starch molecules consist of many glucose monomers bonded together.

Structurally, amylose is a straight-chain condensation polymer with an average of about 200 alpha-D-glucose monomers per molecule. Each monomer is bonded to the next with the loss of a water molecule, just as the two units are bonded in maltose (Fig. 11-3). A representative portion of the structure of amylose is shown in Figure 11-4.

A typical amylopectin molecule has about 1000 alpha-D-glucose monomers arranged into branched chains (Fig. 11-5). Complete hydrolysis yields D-glucose; partial

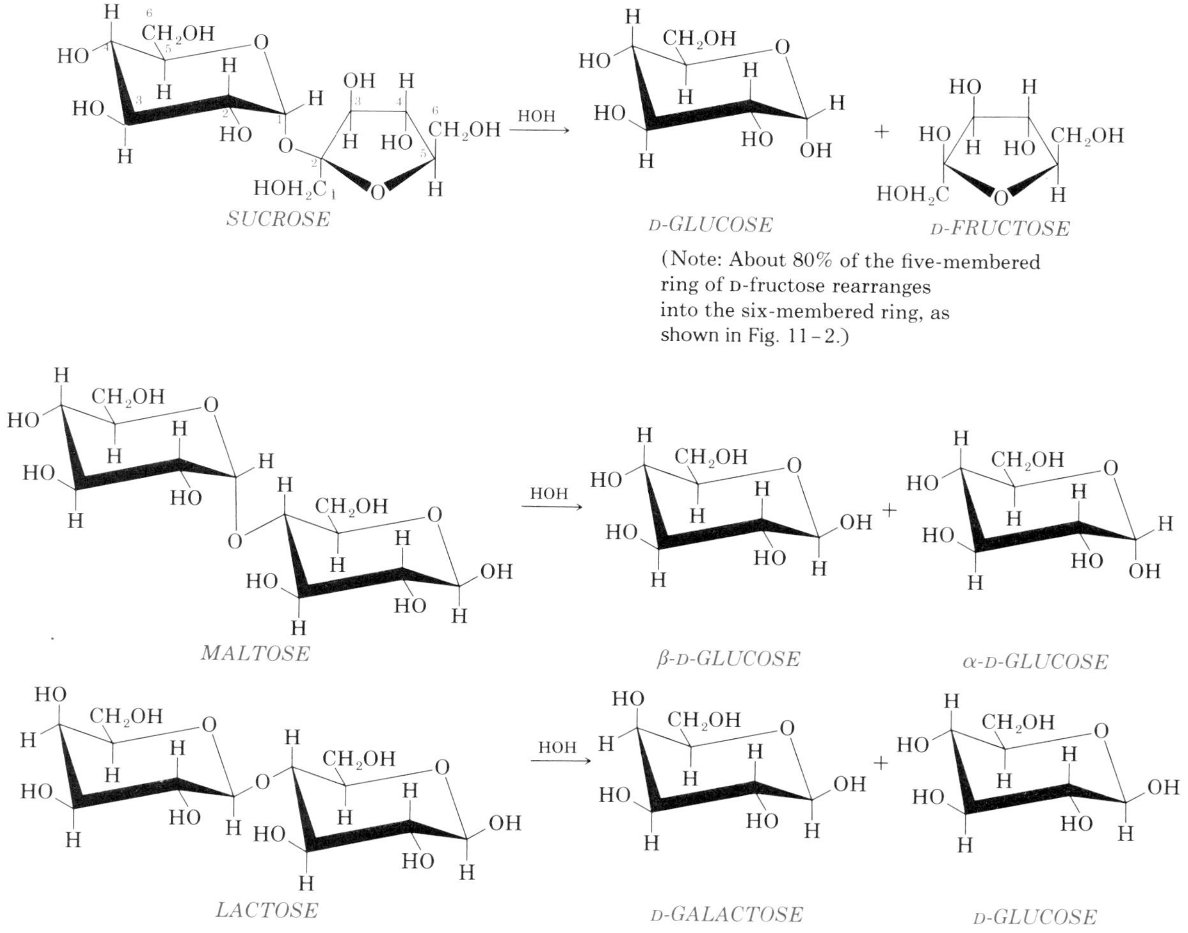
HOH
SUCROSE
D-GLUCOSE
D-FRUCTOSE
(Note: About 80% of the five-membered ring of D-fructose rearranges into the six-membered ring, as shown in Fig. 11-2.)
MALTOSE
β-D-GLUCOSE
α-D-GLUCOSE
LACTOSE
D-GALACTOSE
D-GLUCOSE

Figure 11–3
Hydrolysis of disaccharides (sucrose, maltose, and lactose).

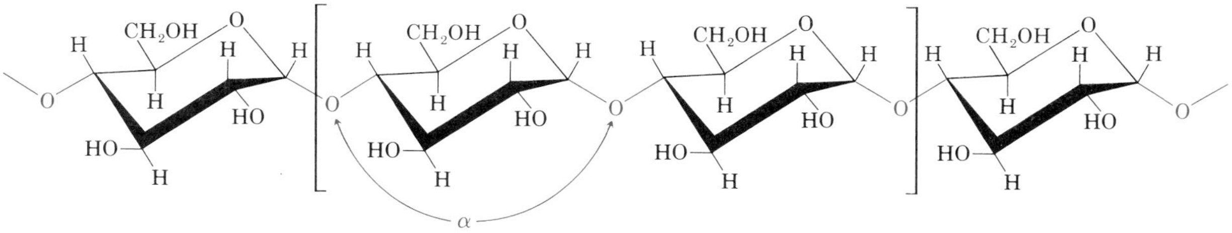
α

Figure 11–4
Amylose structure. From 60 to 300 alpha-D-glucose units are bonded together by alpha linkages to form amylose molecules. In alpha linkages, only the alpha structure of glucose is used. The bonding is between monomers at the 1 and 4 carbon atoms (see Fig. 11–1). The —OH groups on the 1 and 4 carbon atoms are *cis* in alpha glucose, leading all bonds between alpha-glucose monomers (⌒O⌒) to point in the same direction.

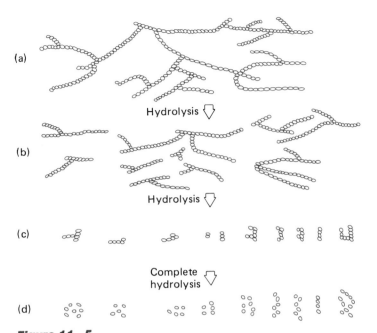

Figure 11-5
(a) Partial schematic amylopectin structure. (b) Dextrins from incomplete hydrolysis of a. (c) Oligosaccharides from hydrolysis of dextrins. (d) Final hydrolysis product: D-glucose. Each circle represents a glucose unit.

hydrolysis produces mixtures called **dextrins.** Dextrins are used as food additives and in mucilage, paste, and finishes for paper and fabrics.

Glycogen is an energy reservoir in animals, just as starch is in plants. The alpha-glucose chains in glycogen are more highly branched than the chains in amylopectin.

Cellulose. Cellulose is the most abundant polysaccharide in nature. Like amylose, it is composed of D-glucose units. The difference between the structures of cellulose and

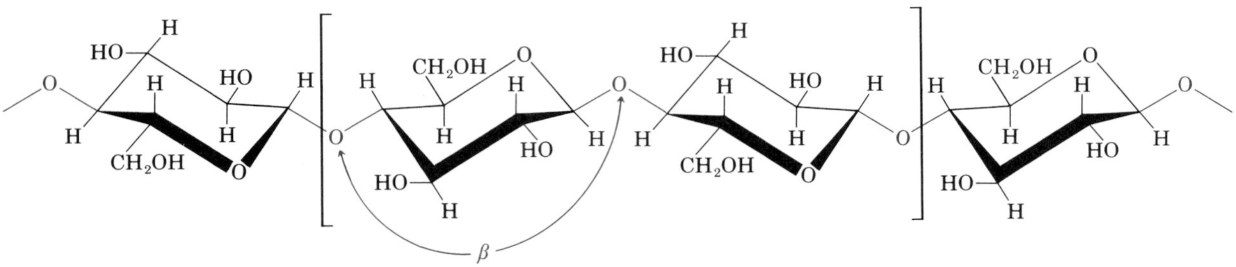

Figure 11-6
Cellulose structure. About 2800 beta-D-glucose units are bonded together by beta linkages to form an unbranched cellulose structure. Cellulose contains only the beta form of glucose. The —OH groups on the 1 and 4 carbon atoms (see Fig. 11-1) are *trans* in beta glucose, leading the bonds between beta-glucose monomers (∼O⌒ and ⌒O∼) to alternate in direction. Compare cellulose with amylose (Fig. 11-4). Note in cellulose that every other beta-glucose monomer is turned over. In amylose all alpha-glucose monomers are in the same position.

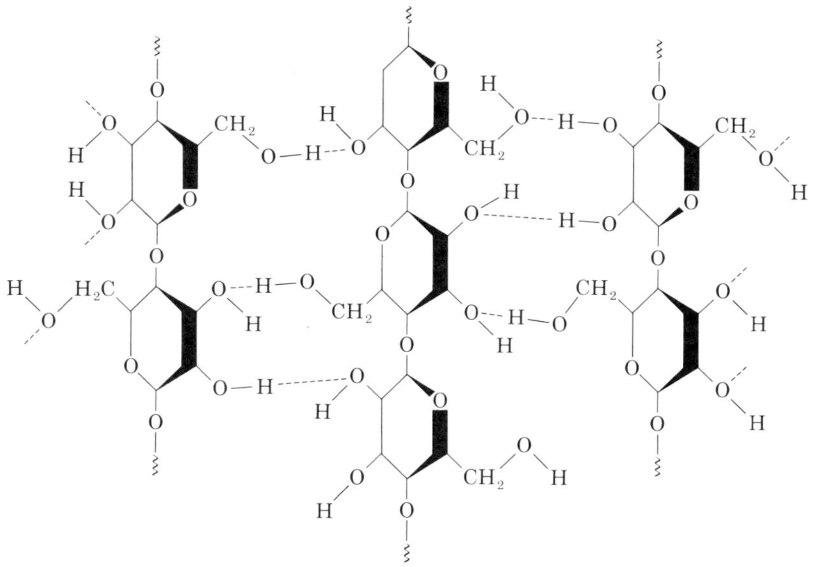

Figure 11–7
The properties of cotton, which is about 98% cellulose, can be explained in terms of this submicroscopic structure. A small group of cellulose molecules, each with 2000 to 9000 units of D-glucose, are held together in an approximately parallel fashion by hydrogen bonding (-----). When several of these *chain bundles* cling together in a relatively vast network of hydrogen bonds, a *microfibril* results; the microfibril is the smallest microscopic unit that can be seen. The macroscopic *fibril* is a collection of numerous microfibrils. The absorbent nature of cotton results from the numerous capillaries wherein the smaller water molecules are held by hydrogen bonds.

amylose lies in the bonding between the D-glucose units; in cellulose all of the glucose units are in the beta-ring form, whereas in amylose they are in the alpha-ring form. (Review the ring forms in Fig. 11–1 and compare the structures in Figs. 11–4 and 11–6). This subtle structural difference between starch and cellulose causes their differences in digestibility. Human beings and carnivorous animals do not have the necessary enzymes (biochemical catalysts) to hydrolyze cellulose, as do numerous microorganisms (such as bacteria in the digestive tracts of termites).

D-glucose can be obtained from cellulose by heating a suspension of the polysaccharide in the presence of a strong acid. At present wood cannot be hydrolyzed into food (D-glucose) economically enough to satisfy the world's growing need for an adequate food supply.

Paper, rayon, cellophane, and cotton are principally cellulose. A representative portion of the structure of cotton is shown in Figure 11–7. Note the hydrogen bonding between cellulose chains.

> Human beings do not have an enzyme to hydrolyze cellulose into its glucose monomers.

SELF-TEST 11–A

1. Carbohydrates contain the elements _____, _____, and _____.

2. The complete hydrolysis of a polysaccharide yields _____.

3. When a molecule of sucrose is hydrolyzed, the products are one molecule each of the monosaccharides _____ and _____.
4. The sugar referred to as blood sugar, grape sugar, or dextrose is actually the compound _____.
5. Starch is a condensation polymer built of _____ monomers.
6. What type of bonding holds polysaccharide chains together, side by side, in cellulose? _____

Proteins

There are about 20 common amino acids.

Proteins are high-molecular-weight compounds made up of amino acid monomers.

Amino acids are compounds that generally have the structure

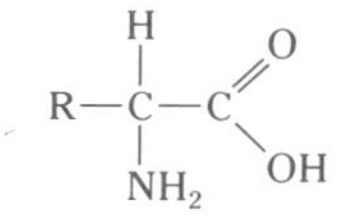

$$R-\underset{\underset{NH_2}{|}}{\overset{\overset{H}{|}}{C}}-C\underset{OH}{\overset{O}{\nwarrow}}$$

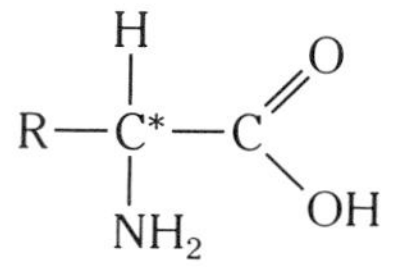

Review the discussion of optically active amino acids in Chapter 10.

Essential amino acids are amino acids that the body needs but cannot make.

For good nutrition we require all of the essential amino acids in our daily diet, but the amount required does not exceed 1.5 g per day for any of them.

Peptide bonds form polyamides like nylon 66 (Chapter 10).

Proteins are condensation polymers of **amino acids.** The twenty different amino acids that can be found in proteins are made primarily from carbon, oxygen, hydrogen, and nitrogen. Small amounts of other elements are also found in proteins, the most common one being sulfur. As the name implies, amino acids have an amino group ($-NH_2$) and an acid (carboxyl) group ($-COOH$). Most amino acids have an amine group and an acid group bonded to the same carbon atom (see Table 11–1). The general formula for an amino acid is shown below:

$$R-\underset{\underset{NH_2}{|}}{\overset{\overset{H}{|}}{C^*}}-C\underset{OH}{\overset{O}{\nwarrow}}$$

R is a characteristic group for each amino acid, and * identifies an asymmetric carbon atom. The simplest amino acid is **glycine,** in which R is a hydrogen atom. Except for glycine, the amino acids have asymmetric carbon atoms and can be optical isomers. Nature prefers the left-handed optical isomers of amino acids.

The close relationship between proteins and living organisms was first noted in 1835 by the German chemist G. T. Mulder. He named proteins from the Greek **proteios** ("first"), thinking that proteins are the starting point for a chemical understanding of life. Proteins play a role in a wide variety of functions, including motion of organisms, defense mechanisms against foreign substances, makeup of enzymes, and makeup of the all-important cell wall. Each unique kind of protein is composed of several specific amino acids arranged in a definite molecular structure. In a few proteins the major fraction is only one kind of amino acid; the protein in silk, for example, is 44% glycine.

The **essential amino acids** must be ingested from food; they are indicated by asterisks in Table 11–1. The other amino acids can be synthesized by the human body.

Amino acid monomers are bonded together by **peptide bonds.** The chemical reaction is an acid-base reaction in which two monomers bond and water is split out. For example, when two glycine molecules react, a peptide bond is formed and a water molecule is produced:

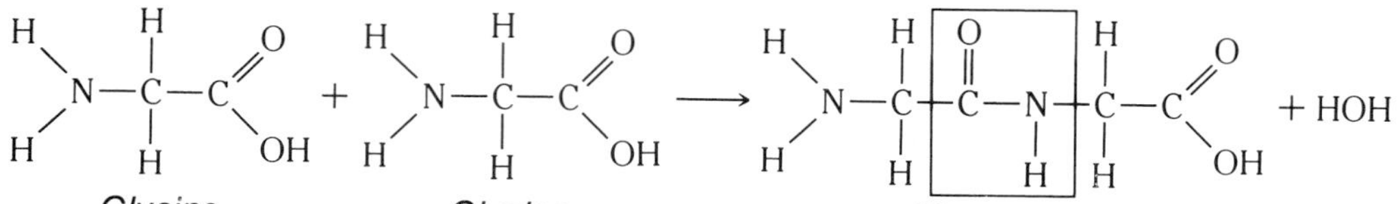

Glycine + Glycine → Glycylglycine + HOH

Table 11-1
Common Amino Acids

All of the amino acids except proline and hydroxyproline have the general formula

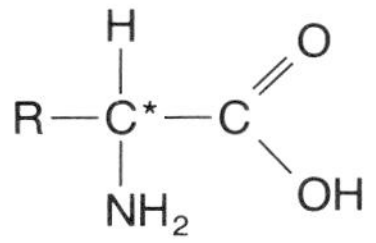

$$R-\overset{H}{\underset{NH_2}{C^*}}-C\overset{O}{\underset{OH}{\diagup}}$$

in which R is the characteristic group for each acid. The R groups are as follows.

1. Glycine —H
2. Alanine —CH_3
3. Serine —CH_2OH
4. Cysteine —CH_2SH
5. Cystine —CH_2—S—S—CH_2—
*6. Threonine —CH—CH_3
                 |
                OH
*7. Valine CH_3—CH—CH_3
*8. Leucine —CH_2—CH—CH_3
                        |
                        CH_3
*9. Isoleucine—CH with CH_3 and CH_2—CH_3

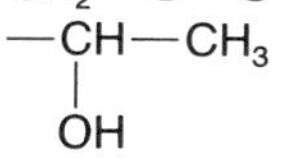

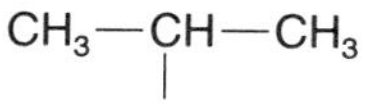

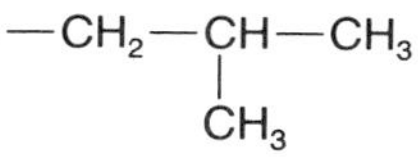

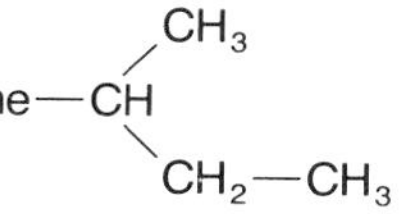

*10. Methionine —CH_2—CH_2—S—CH_3
11. Aspartic acid —CH_2CO_2H
12. Glutamic acid —CH_2—CH_2—CO_2H
*13. Lysine —CH_2—CH_2—CH_2—CH_2—NH_2
*14. Arginine —CH_2—CH_2—CH_2—$NH\overset{NH}{\overset{\|}{C}}NH_2$
*15. Phenylalanine—CH_2—⟨phenyl⟩
16. Tyrosine—CH_2—⟨phenyl⟩—OH
*17. Tryptophan—CH_2—⟨indole⟩
*18. Histidine—CH_2—⟨imidazole⟩

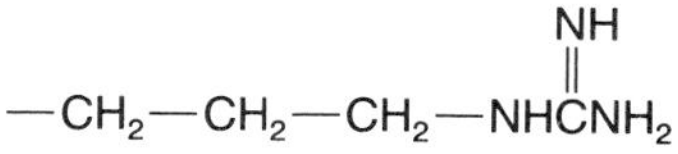

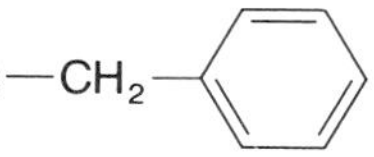

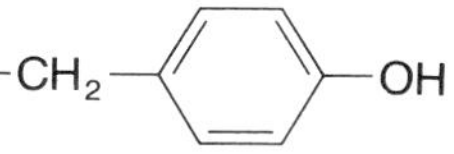

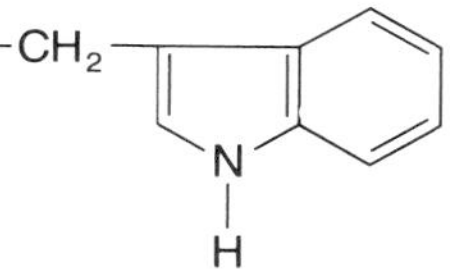

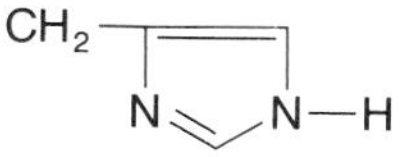

The structures for the other two are:

19. Proline
$$\begin{array}{c} H_2C\text{——}CH_2 \\ |\qquad\quad| \\ H_2C\quad CHCO_2H \\ \diagdown N \diagup \\ | \\ H \end{array}$$

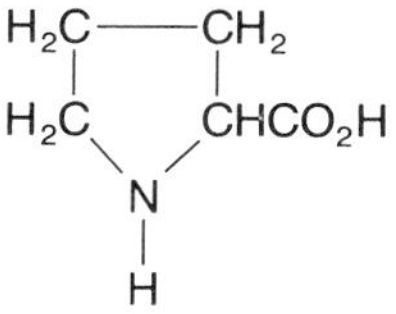

20. Hydroxyproline
$$\begin{array}{c} HOHC\text{——}CH_2 \\ |\qquad\quad| \\ H_2C\quad CHCO_2H \\ \diagdown N \diagup \\ | \\ H \end{array}$$

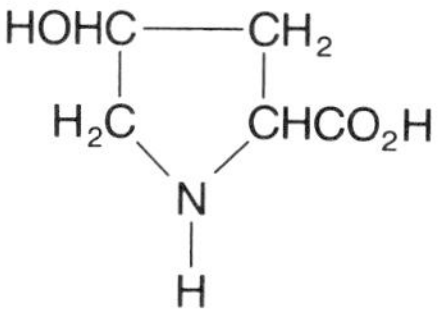

*Essential amino acids; arginine and histidine are essential for children but may not be essential for adults.

Part of the uniqueness of each human being is caused by the uniqueness of some of that person's protein structures. For billions of people, this implies a tremendously large number of protein structures.

When two different amino acids are bonded, two different combinations are possible, depending on which amine reacts with which acid group. For example, when glycine and alanine react, both glycylalanine and alanylglycine can be formed.

The peptide bond

$$-\overset{O}{\overset{\|}{C}}-\underset{H}{\overset{|}{N}}-$$

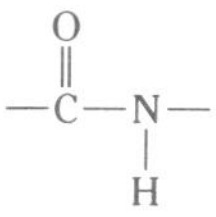

binds amino acid units together in proteins.

Chemistry of Living Systems

Peptide Bonds

GLYCYLALANINE — ALANYLGLYCINE

A very large number of different proteins can be prepared from a small number of different amino acids.

If four different amino acids are bonded in all possible combinations, 24 different molecules are formed.* If 17 different amino acids are bonded, the sequences alone

* If the amino acids are all different, the number of arrangements is $n!$ (read n factorial). For five different amino acids, the number of different arrangements is 5! (or $5 \times 4 \times 3 \times 2 \times 1 = 120$).

A coiled spring is helical in structure.

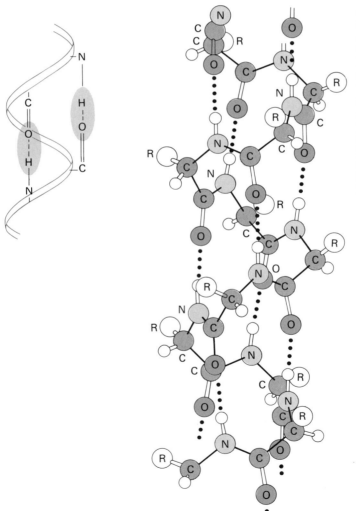

Figure 11-8
(a) Helical structure for a polypeptide in which each oxygen atom can be hydrogen bonded to a nitrogen atom in the third amino acid unit down the chain. *(b)* Alpha-helix structure of proteins. The sketch represents the actual position of the atoms and shows where intrachain hydrogen bonds occur.

(b)

make 3.56 × 10¹⁴ (356 trillion) uniquely different 17-monomer molecules. Many more combinations can be made when more than one molecule of each amino acid is taken. However, of the many different proteins that could be made from a set of amino acids, a living cell will make only the relatively small, select number it needs.

There are several distinguishing characteristics in the structures of proteins. The sequence of amino acids bonded to one another in a chain is the **primary structure.** The twisting of the amino acid chain into a helical shape is a **secondary structure** (Fig. 11–8). Hydrogen bonds hold the helices in place as a nitrogen atom hydrogen-bonds with the oxygen atom in the third amino acid down the chain.

Another secondary structure of proteins is like a sheet in which several chains of amino acids are joined side to side by hydrogen bonds (Fig. 11–9). Most of the properties of silk can be explained by its sheetlike structure.

The twisted or folded form of the helix is the third level of protein structure, the **tertiary structure.** One type of tertiary structure is found in **collagen,** a fibrous protein tissue. Three amino acid chains twisted into left-handed helices are then twisted into a right-handed superhelix to form an extremely strong fibril (Fig. 11–10a). Bundles of fibrils form the tough collagen. A second type of tertiary structure is globular protein in which the helix chain is folded and twisted into a definite geometric pattern (Fig. 11–10b). Many enzymes are globular proteins. Tertiary structures are held together by different kinds of chemical bonds; one of these is the —S—S— disulfide bond, which is

Linus Pauling (b. 1901), along with R. B. Corey, proposed the helical and sheetlike secondary structures for proteins. For his bonding theories and for his work with proteins, Pauling was awarded the Nobel Prize in 1954. For his fight against nuclear danger, he received the 1963 Nobel Peace Prize.

Figure 11–9
Sheet structure for polypeptide. *(a)* A two-dimensional drawing emphasizes that all of the oxygen and nitrogen atoms are involved in hydrogen bonds for the most stable structure. *(b)* Illustration of the bonds in perspective, showing that the sheet is not flat; rather, it is sometimes called a pleated-sheet structure.

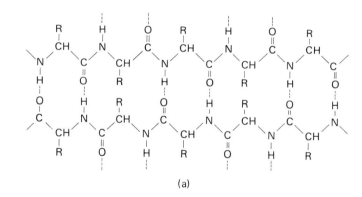

(a)

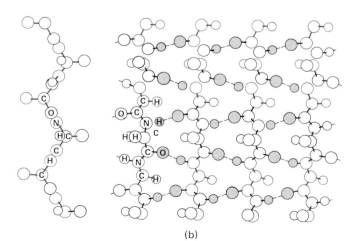

(b)

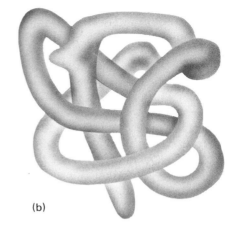

(a)

(b)

Figure 11-10
Tertiary molecular structures of proteins. *(a)* The imaginary twisted structure of collagen. *(b)* The imaginary folded structure of the helix in a globular protein.

used frequently when cysteine or cystine (Table 11-1) is part of the amino acid sequence.

The **quaternary structure** of proteins is the degree of aggregation of protein units. Human hemoglobin, a globular protein with a molecular weight of 68,000, must have its four amino acid chains properly aggregated in order to form active hemoglobin. Insulin is also composed of subunits of protein properly arranged into its quaternary structure.

How important is structure to protein? If hemoglobin, for example, has an abnormal primary, secondary, tertiary, or quaternary structure because of a wrong amino acid in a given position, it may be unable to transfer oxygen through the bloodstream. The cause of **sickle cell anemia** is the alteration of only one specific amino acid of the 146 amino acids in a single hemoglobin chain.

> Hemoglobin carries oxygen and carbon dioxide in the bloodstream and helps control pH.

Enzymes

> In 1926 at Cornell, James B. Sumner (1887-1955) separated, crystallized, identified, and characterized the first enzyme, urease. Sumner had been advised not to enter the field of chemistry because he had only one arm. In 1946 he won the Nobel prize.
>
> Activation energy is required to start a chemical reaction.
>
> Enzymes are specific catalytic molecules with a specific catalytic task.

Enzymes function as catalysts for chemical reactions in living systems. As we shall discuss later, a major part of the structure of an enzyme is globular protein. Like all catalysts, enzymes increase the rate of a reaction by weakening bonds and causing a lowering of the **energy of activation** (Fig. 11-11). The action of an enzyme on a chemical reaction is similar to the effect of a key opening a lock (Fig. 11-12). The lock can be opened without the key by use of more energy (i.e., the lock can be broken). Similarly, the reaction will occur without the enzyme, but at a much slower rate. The enzyme makes the procedure easier and faster. For example, enzyme-catalyzed action allows a single molecule of beta-amylase to catalyze the breaking of bonds between the alpha-glucose monomers in amylose at the rate of 4000 per second.

Most enzymes are very specific. The enzyme maltase hydrolyzes maltose into two molecules of D-glucose. This is the only function of maltase, and no other enzyme can substitute for it. Sucrase, another enzyme, hydrolyzes only sucrose. Some enzymes are less specific. The digestive enzyme trypsin, for example, primarily hydrolyzes peptide bonds in proteins. However, the structure and polarity of trypsin are such that it can also catalyze the hydrolysis of some esters.

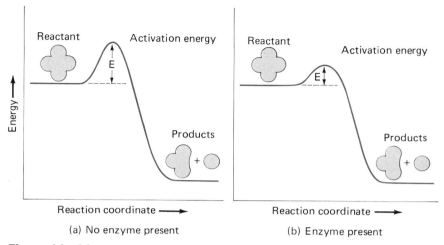

Figure 11–11
Effect of enzyme activity on activation energy. The vertical axis represents energy, and the horizontal axis represents time. For energy-producing reactions, the reactant molecules are at a higher energy than the product molecules. It is necessary for the reactant molecules to "get over" the energy barrier (acquire the activation energy, E) in going from being reactants to being products. The function of an enzyme is to lower the activation energy as illustrated in (b), and thereby to speed up the reaction.

Some enzymes are globular proteins only. Other enzymes are globular proteins plus either a metal ion (e.g., Co^{3+}, Fe^{3+}, Mg^{2+}, or another essential mineral) or a vitamin. The vitamin or the mineral is the **coenzyme,** and the protein is the **apoenzyme.** Both parts are needed for enzymatic activity, just as two keys are required to open a bank lock-box. Neither your key nor the bank's key alone will open the box; both are needed. The B vitamins are coenzymes in various oxidative processes in the human body. For example, niacin (vitamin B_3) is part of a larger molecule, **nicotinamide adenine dinucleotide (NAD$^+$);** NAD$^+$ serves as a coenzyme in concert with the apoenzyme, a globular protein. Riboflavin (vitamin B_2) is part of the coenzyme **flavin adenine dinucleotide (FAD).**

The names of most enzymes end in *-ase*.

Why must we have minerals and vitamins? Answer: In part, because vitamins and minerals serve as coenzymes.

Niacin prevents pellagra.

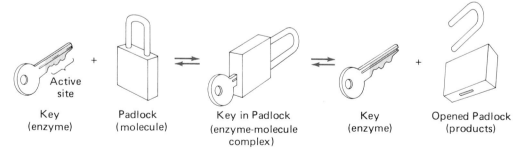

Figure 11–12
Lock-and-key theory for enzymatic catalysis. Although it is generally agreed that this analogy is an oversimplification, it does make one very important point; the enzyme makes a difficult job easy by reducing the energy required to get the job started. It also suggests that the enzyme has a particular structure at an active site that will allow it to work only for certain molecules, similar to a key that fits the shape of a particular keyhole and a particular sequence of tumblers.

250   *Chemistry of Living Systems*

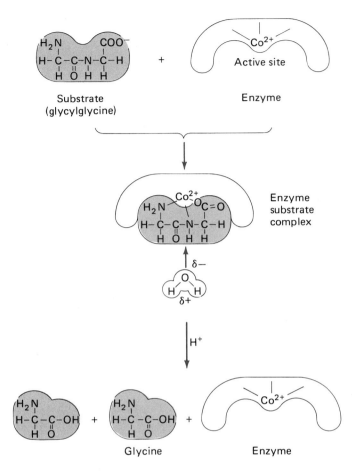

Figure 11-13
Action of an enzyme. The substrate molecule is chemically bonded to the enzyme (glycylglycine dipeptidase). The negative oxygen and the nitrogen atoms of the substrate bond to the positive cobalt ion in the enzyme. The bonding of the substrate makes it more susceptible to attack by water. Hydrolysis occurs and the glycine molecules are released by the enzyme, which is then ready to play its catalytic role again.

Riboflavin promotes growth, healthy eyes and skin, and the oxidation of foods.

Enzyme structure is the key to specific catalytic activity.

The action of an enzyme is shown in Figure 11-13. The reactant molecule is the **substrate.** Enzymes and substrates have electrically polar regions, partially charged groupings, or ionic sections that attract and guide the enzyme and substrate together; these regions of chemical activity are the **active sites.** Substrates sit down on active sites on enzymes in assembly-line fashion at a remarkably fast rate. For example, enzymes renew three million red blood cells in the human body *every second*.

SELF-TEST 11-B

1. The fundamental building units in proteins are the _____.
2. Amino acids that the body cannot synthesize from other molecules are called _____.
3. The peptide linkage that bonds amino acids together in protein chains has the structure _____.
4. The basic structure present in almost all of the amino acids can be represented as _____.

5. a. The primary structure of a protein refers to its _____;
   b. the secondary structure refers to its _____;
   c. its tertiary structure refers to _____;
   d. and its quaternary structure refers to _____.
6. a. If we have three different amino acids and can use each three times in any given tripeptide, we can make a total of _____ different tripeptides.
   b. If we can use each amino acid only once, there are still _____ possible different tripeptides.
7. Describe how hydrogen bonding is involved in the secondary structures of proteins.
   _____
8. The best term to describe the general function of enzymes is ( ) *catalyst,* ( ) *intermediate,* or ( ) *oxidant.*
9. In the lock-and-key analogy of enzyme activity, the enzyme functions as the _____ while the substrate molecule serves as the _____.
10. The activation energy of many biological reactions is decreased if a(n) _____ is present.
11. Apoenzyme + coenzyme ⟶ _____
12. That portion of the enzyme at which the reaction is catalyzed is called the _____.

Energy and Biochemical Systems

Energy for life's processes comes from the Sun. During photosynthesis, green plants absorb energy from the Sun to make glucose and oxygen from carbon dioxide and water. The energy stored in glucose is transferred eventually to the bonds in molecules such as ATP (Fig. 11–14). When needed, the ATP molecules release energy to drive other chemical reactions.

Photosynthesis

In the complex process of photosynthesis, carbon dioxide is reduced to make sugar,

$$6\ CO_2 + 24\ H^+ + 24\ e^- \longrightarrow C_6H_{12}O_6 + 6\ H_2O$$

and water is oxidized to oxygen:

$$12\ H_2O \longrightarrow 6\ O_2 + 24\ H^+ + 24\ e^-$$

The oxidation and reduction reactions added together give the overall reaction:

$$6\ CO_2 + 6\ H_2O + 688\ kcal \longrightarrow C_6H_{12}O_6 + 6\ O_2$$

Carbon Water Energy   Glucose   Oxygen
Dioxide   (Sunlight)

The oxygen produced in photosynthesis is the source (and only present source) of all of the oxygen in our atmosphere. Only this life-giving gas, given off by trees, grass, greenery, and even by algae in the sea, makes possible human life and most animal life on Earth. We are dependent on the plant life of our planet, and we must live in balance

Photosynthesis involves a number of different steps and is a very complex process.

Reduction is the gain of electrons or hydrogen. Oxidation is the loss of electrons or hydrogen.

It was discovered in 1985 that a blue-green protozoa, Stentor coeruleus, contains a light-absorbing substance, stentorin, that allows it to undergo a unique type of photosynthesis.

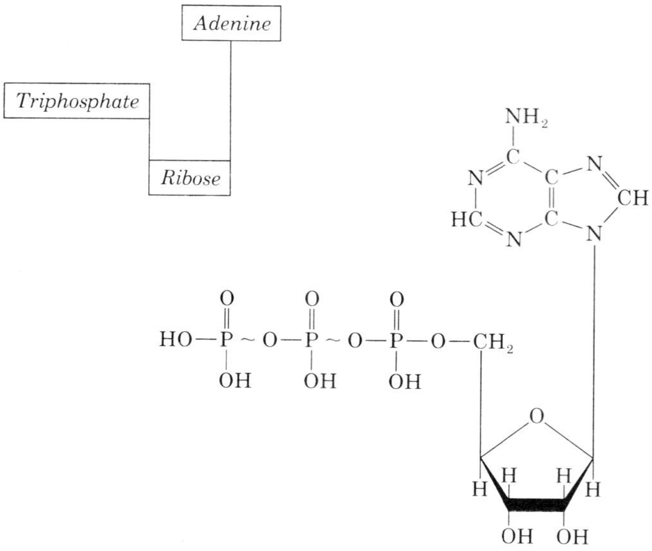

Figure 11–14
Molecular structure of adenosine triphosphate (ATP).

The energy of a photon is captured by chlorophyll when an electron is raised to a higher energy state.

Refer to atomic theory in Chapter 3.

with the oxygen output of that plant life, as well as with the food output of the same plant life. Photosynthesis is thus absolutely vital to life on Earth.

Photosynthesis is generally considered a series of **light reactions,** which occur only in the presence of light energy, and a series of **dark reactions,** which can occur in the dark. The dark reactions feed on high-energy compounds (such as ATP) produced by the light reactions. During the light reactions, green pigments such as the chlorophylls (either A or B) absorb photons of light and raise electrons within these structures to

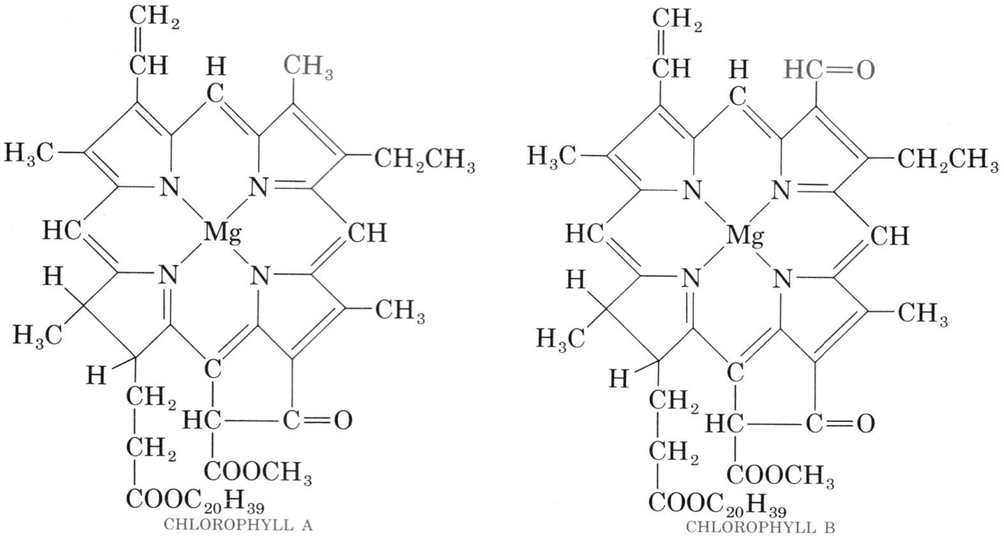

CHLOROPHYLL A

CHLOROPHYLL B

Chlorophyll is green because violet light and red light are absorbed and green light is reflected.

higher energy levels. As electrons move back to the ground state, chloroplasts absorb this energy. Through a series of reactions, water is oxidized to oxygen, and energy is stored in the bonds of energy-bank compounds such as ATP (see Fig. 11–14). ATP

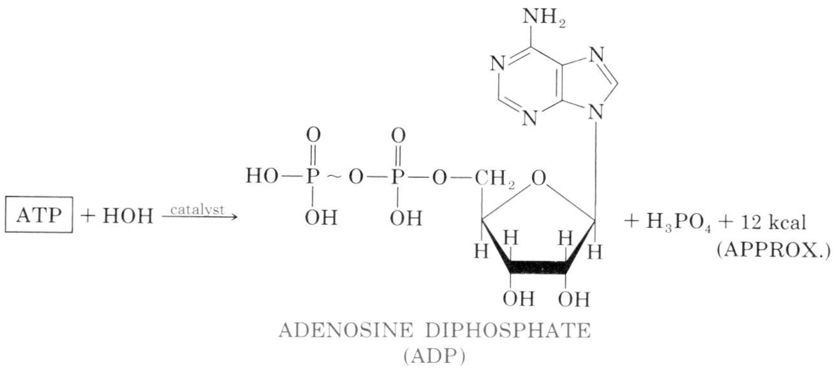

Figure 11-15
Hydrolysis of ATP to ADP.

stores energy in two high-energy phosphate bonds, shown as wiggle lines in Figure 11–14.

In the presence of a suitable catalyst, ATP undergoes a three-step hydrolysis. In the first step ATP is hydrolyzed to adenosine diphosphate (ADP) and releases about 12 kcal per mole (Fig. 11–15). The second hydrolysis step, ADP to adenosine monophosphate (AMP), also produces about 12 kcal of energy per mole. The last hydrolysis step, AMP to adenosine, releases only about 2.5 kcal per mole. The hydrolysis of ATP releases energy (is **exothermic**); the synthesis of ATP from AMP or ADP requires energy (is **endothermic**). It is the synthesis of ATP that occurs during the light reactions of photosynthesis, and it is this process that stores the Sun's energy in chemical compounds.

Exothermic means energy is released.

During the dark reactions, hydrolysis of the P—O bonds of ATP provide the energy to convert CO_2 and hydrogen (from water) into glucose through a series of chemical reactions.

Endothermic means energy is required.

After photosynthesis the living plant may convert glucose to oligosaccharides, starches, cellulose, proteins, or oils. The end-product depends on the type of plant involved and the complexity of its biochemistry.

The next steps involved in use of the energy stored in high-energy compounds are for the compounds to be eaten, digested, transported to the cells of the body, and metabolized.

Digestion

From a chemical point of view, digestion is the breakdown of ingested foods by hydrolysis. The products of digestion are relatively small molecules that can be absorbed through the intestinal walls. The hydrolytic reactions of digestion are catalyzed by enzymes, there being a specific enzyme for the hydrolysis of each type of substance. The hydrolysis of carbohydrates ultimately yields simple sugars, proteins yield amino acids, and fats and oils yield fatty acids and glycerol.

Digestion is the hydrolysis of carbohydrates, fats, and proteins to provide small molecules that can be absorbed.

In our food, carbohydrates requiring digestion are polysaccharides such as starch and disaccharides such as sucrose and lactose. The digestion process begins in the mouth with salivary amylase, or ptyalin. Starch is partially hydrolyzed into the disaccharide maltose by ptyalin, which is later rendered inactive by the high acidity of the stomach. No more digestion of carbohydrates occurs in the stomach. When the food passes from the stomach into the small intestine, the acidity is neutralized by a secretion from the pancreas. Enzymes from the pancreas complete the hydrolysis of carbohy-

Acidic solutions: pH below 7
Basic solutions: pH above 7

Human blood normally contains between 0.08% and 0.1% glucose.

Insulin is a protein.

drates into simple sugars such as glucose, fructose, and galactose. These simple sugars are then absorbed into the bloodstream. The hormone insulin (a protein) escorts simple sugars through the cell membranes and into the cells. There, in the mitochondria, these simple sugars are oxidized for their energy content.

If the sugar level in the bloodstream becomes too high, the simple sugars are converted into glycogen in the liver. If the sugar level is too low, stored glycogen is hydrolyzed to raise it. Malfunctions in these processes can lead to too much blood sugar, **hyperglycemia,** or too little blood sugar, **hypoglycemia.** Either condition, if sustained, produces a type of **diabetes.**

Types of diabetes are discussed in Chapter 13.

The digestion of fats and oils, such as the triesters of fatty acids and glycerol, occurs primarily in the small intestine. The enzyme that catalyzes the hydrolysis of fatty acid esters is water-soluble, but the fats and oils themselves are not. Bile salts, secreted by the liver, emulsify the oil by forming an interface between the nonpolar oil and the polar water and thereby making it possible for the oil to "dissolve" in water. For a molecule to be an emulsifier between polar and nonpolar molecules, it must have both polar and nonpolar structures. The sodium salt of glycocholic acid, a bile salt, contains the bulky nonpolar hydrocarbon groups, which are compatible with fat or oil, and the —OH and ionic groups, which attract water molecules:

Bile salts act chemically much like detergent and soap molecules (Chapter 17).

Sodium Salt of Glycocholic Acid

The stomach is protected from protein-splitting enzymes by a mucous lining.

The digestion of proteins begins in the stomach and is completed in the small intestine. Many enzymes are known to be involved. In the stomach pepsin catalyzes the hydrolysis of only about 10% of the bonds in a typical protein, leaving protein fragments with molecular weights of 600 to 3000. In the small intestine hydrolysis is completed to amino acids, which are absorbed through the intestinal wall.

Some protein enzymes are sold commercially as meat tenderizers and stain removers. Some are used to free the lens of the eye before cataract surgery.

The Liver: The Nutrient Bank of the Body

After digestion most food nutrients pass directly to the liver for distribution to the body. Glucose is used for energy in the liver and to prepare glucose phosphate as the first step in the preparation of glycogen (the storage carbohydrate); in addition, about one third goes on in the bloodstream to nourish the cells. From the liver a fraction of the amino acids is sent to the cells to build proteins. In the liver amino acids are used to form enzymes, and some are oxidized to obtain energy. The liver is thus the central nutrient bank, or warehouse, of the body in that it stores, converts, and classifies nutrients. In addition to carrying out these varied activities, the liver also detoxifies pollutants invading the body (see Chapter 14).

Glucose Metabolism

Aerobic reactions use elemental oxygen. Anaerobic reactions use no elemental oxygen.

The sequence of reactions by which energy is obtained from glucose and similar compounds begins with an **anaerobic** (without elemental oxygen) series of reactions followed by a series of **aerobic** (with elemental oxygen) reactions.

The anaerobic process was discovered by the German chemist Otto Fritz Meyerhof (1884–1951). More details were discovered by Gustav Embden (1884–1933). In 1918 Meyerhof showed that animal cells break down sugar in much the same way as does yeast, a plant. This work made clear for the first time that, with only minor differences, glucose metabolism follows the same sequences in all creatures. Details of the individual steps were discovered between 1932 and 1933 by Embden and between 1937 and 1941 by Carl Ferdinand Cori (b. 1896) and his wife Gerty Theresa Cori (1896–1957). The Coris shared a Nobel Prize in 1947. Meyerhof shared the Nobel Prize in physiology and medicine in 1922 with Archibald Vivian Hill (1886–1977), who had investigated muscle from the point of view of heat production. The anaerobic sequence of reactions is known as the **Embden-Meyerhof pathway.**

Gerty Cori was the third woman to receive a Nobel Prize. The first two were Marie Curie and her daughter, Irene Joliet.

The aerobic sequence of reactions was discovered by the German biochemist Sir Hans Adolf Krebs (1900–1981). In 1933 Krebs fled from Hitler's Germany to England, where he studied at Cambridge, later joined the faculty at Oxford, and was knighted in 1958. In 1953 he shared the Nobel prize in physiology and medicine with Fritz Albert Lipmann (b. 1899), who discovered the roles of ATP and other such compounds in the storing of energy in phosphorus-oxygen bonds. The aerobic sequence of reactions is known as the **Krebs cycle.**

When a muscle is used, glucose is converted anaerobically to lactic acid by a series of 11 steps in the **Embden-Meyerhof pathway.** The overall reaction can be represented by the following equation:

Muscular activity converts glucose to lactic acid, which produces fatigue in muscles as the lactic acid accumulates.

$$C_6H_{12}O_6 + 2\ ADP + 2\ H_3PO_4 \longrightarrow 2\ CH_3-\underset{OH}{\underset{|}{C}}H-\overset{O}{\overset{\|}{C}}-OH + 2\ ATP + 2\ H_2O$$

Glucose   Lactic Acid

Note that the energy of one molecule of glucose is transferred to two molecules of ATP (two P—O bonds), and the remainder resides in two molecules of lactic acid. Since the process is anaerobic, elemental oxygen (O_2) is not a reactant.

If muscle is used strenuously for a sufficiently long period of time, the lactic acid buildup produces tiredness and a painful sensation. The bloodstream carries the lactic acid away eventually, but time and oxygen are needed to convert it to carbon dioxide and water, which are excreted.

The conversion of lactic acid to CO_2 and H_2O is accomplished by the **Krebs cycle.** The equation for the aerobic conversion of lactic acid into carbon dioxide and water contains the number of ATP molecules formed in the process. Note that it includes elemental oxygen (O_2), which is required for an aerobic process:

$$C_3H_6O_3 + 18\ ADP + 18\ H_3PO_4 + 3\ O_2 \longrightarrow 3\ CO_2 + 21\ H_2O + 18\ ATP$$

Lactic Acid

The energy available to the body from one glucose molecule is now stored in 38 ATP molecules, 2 from the Embden-Meyerhof pathway and 36 from two lactic acid molecules. The complete oxidation of glucose into CO_2 and H_2O with energy stored in ATP bonds is 41% efficient. This is remarkable when you consider that the efficiency of the automobile engine is only about 20% and that the efficiency of a heat engine of any size seldom goes above 35%.

When ATP hydrolyzes back to ADP by losing a phosphate group, the energy is used to drive chemical reactions that cause muscles to move, such as the heart to pump, the diaphragm to breathe, the ear to twitch, the jaw to chew, the throat to swallow, and so on.

You now have seen some of the detailed chemistry involved in simply raising your arm, and you are now aware of what happens to some of the sugars and starches you ingest. As far as our bodies are concerned, this is the fate of the quantity of the Sun's energy absorbed by the green plant. Of course, there is much more known than is presented here, and there appears to be no end to what is left to be discovered.

SELF-TEST 11-C

1. The source of energy for photosynthesis is the _____.
2. Most of the energy obtained by the oxidation of food is used immediately to synthesize the molecule _____.
3. The hydrolysis of ATP results in the molecules _____ and _____; The other "product" is _____.
4. The reactants in the photosynthesis process are _____ and _____; _____ must also be supplied.
5. Energy from the Sun is absorbed by _____ in the green cells of a plant.
6. Digestion is the breakdown of foodstuffs by _____.
7. Bile salts act as ( ) catalysts, ( ) emulsifying agents, or ( ) enzymes.
8. The two products of the Embden-Meyerhof pathway are _____ and _____.
9. The end-products of the Krebs energy cycle are _____, _____, and _____.

Nucleic Acids

Like polysaccharides and polypeptides, **nucleic acids** are condensation polymers. The components of the monomers are one of two simple sugars, phosphoric acid, and one of a group of ringed nitrogen compounds that have basic (alkaline) properties. The structures of the two sugars are shown in Figure 11–16. The names and formulas of the basic nitrogen compounds are given in Figure 11–17.

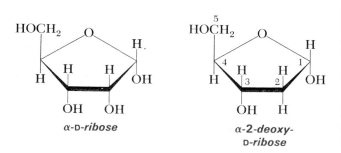

Figure 11–16
The structure of alpha-D-ribose and alpha-2-deoxy-D-ribose. In the names given, *alpha* indicates the one of two ring forms possible, *D* distinguishes the isomers that rotate plane-polarized light in opposite directions, and the 2 indicates the carbon to which no oxygen is attached in the second sugar.

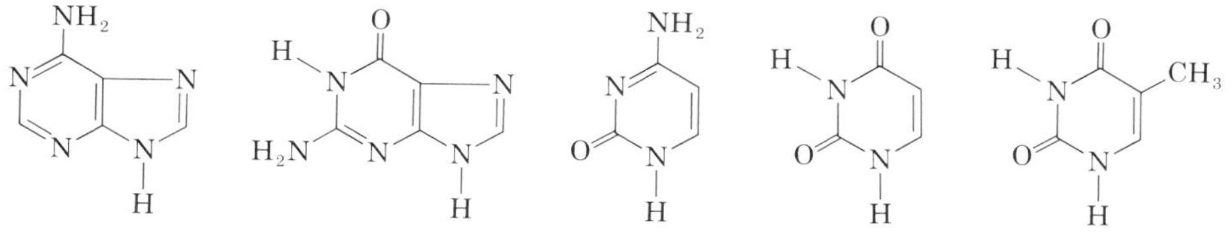

Nucleic Acids 257

Figure 11–17
Some nitrogenous bases obtained from the hydrolysis of nucleic acids.

Nucleic acids are **deoxyribonucleic acids (DNA)** if they contain the sugar **alpha-2-deoxy-D-ribose** or **ribonucleic acids (RNA)** if they contain the sugar **alpha-D-ribose**). DNA is found primarily in the nucleus of the cell (Fig. 11–18), whereas RNA is found mainly in the cytoplasm, outside of the nucleus. Nucleic acids are found in all living cells, with the exception of the red blood cells of mammals.

Three major types of RNA have been identified. They are messenger RNA (mRNA), transfer RNA (tRNA), and ribosomal RNA (rRNA). Each has a characteristic molecular weight and base composition. Messenger RNAs are generally the largest, with molecular weights between 25,000 and 1 million. They contain from 75 to 3000 mononucleotide units. Transfer RNAs have molecular weights in the range of 23,000 to 30,000 and contain 75 to 90 mononucleotide units. Ribosomal RNAs, which have molecular weights between those of mRNAs and tRNAs, make up as much as 80% of the total cell RNA.

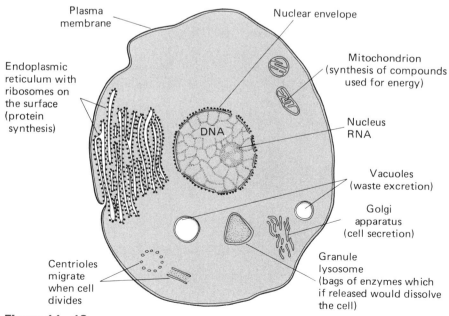

Figure 11–18
Diagrammatic generalized cell showing the relationships among the various components of the cell. Cytoplasm is the material of the cell, exclusive of the nucleus. Many of the components shown are not visible through an ordinary optical microscope.

Besides having different molecular weights, the three types of RNA differ in function. One difference in function is described in the discussion of natural protein synthesis.

The monomers of DNA and RNA contain a simple sugar, one of the nitrogenous bases, and one or two phosphoric acid units. The structure of a monomer, a **nucleotide,** is shown in Figure 11-19a. The only two structural differences between nucleotides of DNA and RNA are the sugar and one base. Three monomers condensed into an oligonucleotide can be seen in Figure 11-19b.

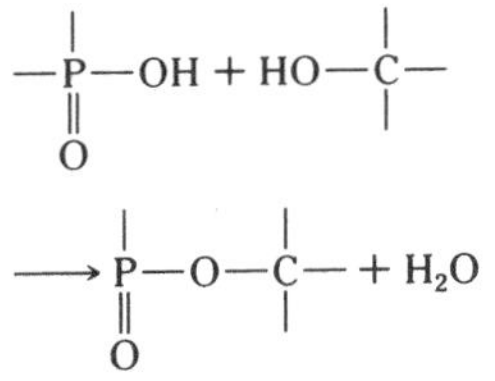

Polynucleotides with molecular weights up to several million are known. The sequence of nucleotides in the polynucleotide chain is its **primary structure.**

In 1953, James D. Watson and Francis H. C. Crick (Fig. 11-20) proposed a **secondary structure** for DNA that has since gained wide acceptance. Figure 11-21 illustrates a small portion of the structure, in which two polynucleotides are arranged in a double helix stabilized by hydrogen bonding between the base groups opposite each other in the two chains. RNA is generally a single strand of helical polynucleotide.

The function of polynucleotides is to transcribe cellular and organism information

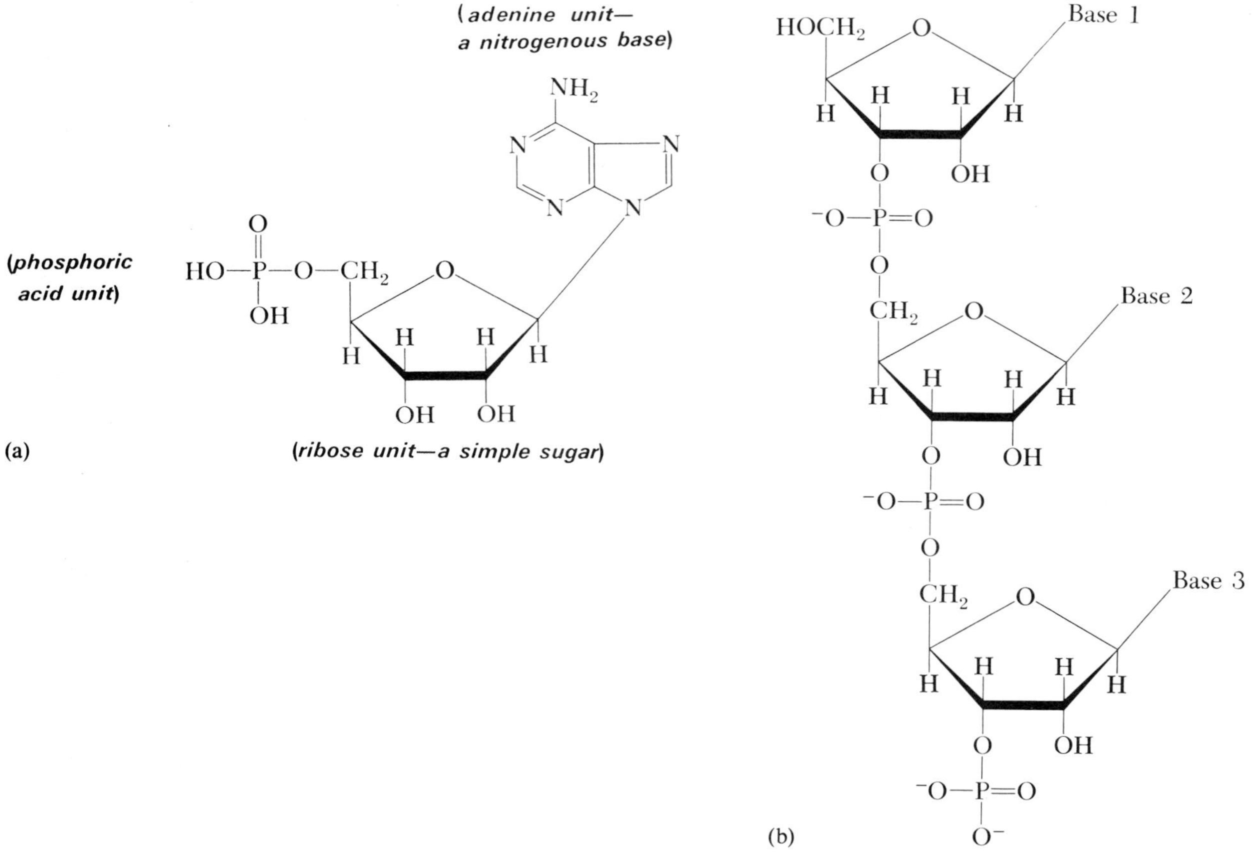

One nucleotide is joined to another by an ester-forming reaction:

$$-\overset{\underset{\parallel}{O}}{P}-OH + HO-\overset{|}{\underset{|}{C}}- \longrightarrow -\overset{\underset{\parallel}{O}}{P}-O-\overset{|}{\underset{|}{C}}- + H_2O$$

Figure 11-19
(a) A nucleotide. If other bases are substituted for adenine, several nucleotides are possible for each of the two sugars shown in Figure 11-16. *(b)* Bonding structure of a trinucleotide. Bases 1, 2, and 3 represent any of the nitrogenous bases obtained in the hydrolysis of DNA and RNA (Fig. 11-17). The primary structure of both DNA and RNA is an extension of this structure and produces molecular weights as high as a few million.

Figure 11-20
Francis H. C. Crick (b. 1916) (right) and James D. Watson (b. 1928) (left), working in the Cavendish Laboratory at Cambridge, built scale models of the double helical structure of DNA based on the X-ray data of Maurice H. F. Wilkins. Knowing distances and angles between atoms, they compared the task to the working of a three-dimensional jigsaw puzzle. Watson, Crick, and Wilkins received the Nobel Prize in 1962 for their work relating to the structure of DNA.

so that like begets like. The almost infinite variety of primary structures of polynucleotides allows an almost infinite variety of information to be recorded in the molecular structures of the strands of nucleic acids. The different arrangements of just a few different bases give the large variety of structures. In a somewhat similar fashion, the multiple arrangements of just a few language symbols convey the many ideas in this book. The coded information in the polynucleotide is believed to control the inherited characteristics of the next generation as well as most of the continuous life processes of the organism.

The inherited traits of an organism are controlled by DNA molecules.

The transfer of coded information begins wih the replication of DNA and continues with natural protein synthesis as well as with the synthesis of body tissues. In the following section we shall see how DNA replicates and how protein is synthesized naturally.

Replication of DNA: Heredity

Almost all nuclei in an organism's cells contain the same chromosomal composition. This composition remains constant regardless of whether the cell is starving or has an ample supply of food materials. Each organism begins life as a single cell with this same chromosomal composition; in sexual reproduction half of a chromosome comes from each parent. These well-known biological facts, along with recent discoveries concerning polynucleotide structures, have led scientists to the conclusion that the DNA structure is faithfully copied during normal cell division (**mitosis**—both strands) and that only half is copied in cell division producing reproductive cells (**meiosis**—one strand).

The DNA molecule is capable of causing the synthesis of its duplicate.

Chemistry of Living Systems

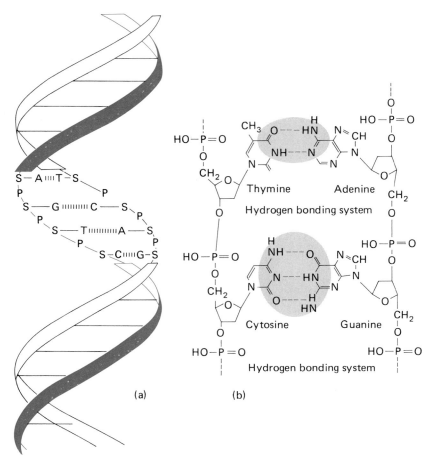

Figure 11–21
(a) Double helix structure proposed by Watson and Crick for DNA. S = sugar, P = phosphate, A = adenine, T = thymine, G = guanine, C = cytosine. *(b)* Hydrogen bonds in the thymine-adenine and cytosine-guanine pairs stabilize the double helix. Adenine also pairs with uracil in mRNA, which contains no thymine.

In replication the double helix of the DNA structure unwinds and each half of the structure serves as a template, or pattern, from which the other complementary half can be reproduced from the molecules in the cell environment (Fig. 11–22). Replication of DNA occurs in the nucleus of the cell.

Natural Protein Synthesis

The proteins of the body are continually being replaced and resynthesized from the amino acids available to the body.

The use of isotopically labeled amino acids has made possible studies of the average lifetimes of amino acids as constituents in proteins — that is, the time it takes the body to replace a protein in a tissue. For a process that must be extremely complex, replacement is very rapid. Only minutes after radioactive amino acids are injected into animals, radioactive protein can be found. Although all the proteins in the body are continually being replaced, the rates of replacement vary. Half of the proteins in the liver and plasma

The proteins in the human body are continually being replaced.

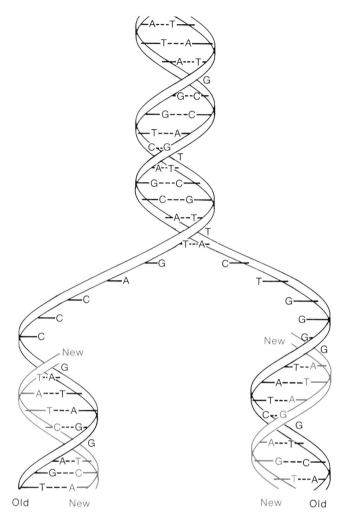

Figure 11-22
Replication of DNA structure. When the double helix of DNA (black) unwinds, each half serves as a template on which to assemble subunits (color) from the cell environment.

are replaced in 6 days; the time needed for replacement of muscle proteins is about 180 days, and replacement of protein in other tissues, such as bone collagen, takes even longer.

Recall that each organism has its own kinds of proteins. The number of possible arrangements of 20 amino acid units is 2.43×10^{18}, yet proteins characteristic of a given organism can be synthesized by the organism in a matter of a few minutes.

The DNA in the cell nucleus holds the code for protein synthesis. Messenger RNA, like all forms of RNA, is synthesized in the cell nucleus. The sequence of bases in one strand of the chromosomal DNA serves as the template from which a single strand of a messenger ribonucleotide (mRNA) is made (Figure 11-23). The bases of the mRNA strand complement those of the DNA strand. A pair of complementary bases is structured such that each one fits the other and forms one or more hydrogen bonds. Messenger RNA contains only the four bases adenine (A), guanine (G), cytosine (C), and uracil

The DNA molecule tells the cell what kind of protein to synthesize.

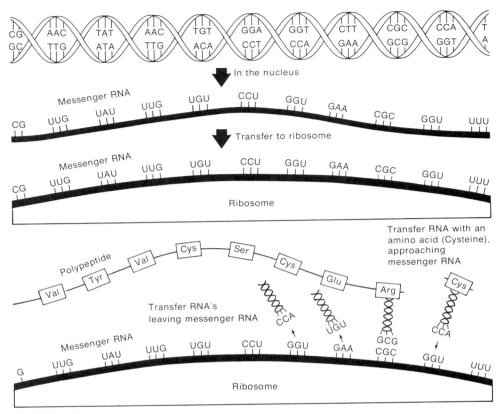

Figure 11–23
A schematic illustration of the role of DNA and RNA in protein synthesis. A, C, G, T, and U are nitrogen bases characteristic of the individual nucleotides. See Figure 11–17 for structures of the bases and Table 11–2 for abbreviations of the amino acids used.

(U). DNA contains principally the four bases adenine (A), guanine (G), cytosine (C), and thymine (T). The base pairs are as follows:

These base pairs "fit" for hydrogen bonding.

DNA	mRNA
A	U
G	C
C	G
T	A

This means that, provided the necessary enzymes and energy are present, wherever a DNA has an adenine base (A), the mRNA will transcribe a uracil base (U).

After transcription, mRNA passes from the nucleus of the cell to a ribosome, where it serves as the template for the sequential ordering of amino acids during protein synthesis. As its name implies, messenger RNA contains the sequence message, in the form of a three-base code, for ordering amino acids into proteins. Each of the thousands of different proteins synthesized by cells is coded by a specific mRNA or segment of an mRNA molecule.

Transfer RNA carries specific amino acids to the mRNA.

Transfer RNAs carry the specific amino acids to the mRNA. Each of the 20 amino acids found in proteins has at least one corresponding tRNA, and some have multiple

Table 11-2
Messenger RNA Codes for Amino Acids*

Amino Acid	Shortened Notation Used for Amino Acids in Fig. 11-23	Base Code on mRNA
Alanine	Ala	GCA, GCC, GCG, GCU
Arginine	Arg	AGA, AGG, CGA, CGG, CGC, CGU
Asparagine	Asp-NH$_2$	AAC, AAU
Aspartic acid	Asp	GAC, GAU
Cysteine	Cys	UGC, UGU
Glutamic acid	Glu	GAA, GAG
Glutamine	Glu-NH$_2$	CAG, CAA
Glycine	Gly	GGA, GGC, GGG, GGU
Histidine	His	CAC, CAU
Isoleucine	Ileu	AUA, AUC, AUU
Leucine	Leu	CUA, CUC, CUG, CUU, UUA, UUG
Lysine	Lys	AAA, AAG
Methionine	Met	AUG
Phenylalanine	Phe	UUU, UUC
Proline	Pro	CCA, CCC, CCG, CCU
Serine	Ser	AGC, AGU, UCA, UCG, UCC, UCU
Threonine	Thr	ACA, ACG, ACC, ACU
Tryptophan	Try	UGG
Tyrosine	Tyr	UAC, UAU
Valine	Val	GUA, GUG, GUC, GUU

The bases in groups of three are called codons.

* In groups of three (called codons), bases of mRNA code the order of amino acids in a polypeptide chain. A, C, G, and U represent adenine, cytosine, guanine, and uracil, respectively. Some amino acids have more than one codon, and hence more than one tRNA can bring the amino acid to mRNA. The research on this coding was initiated by Nirenberg. (Adapted from J. I. Routh, D. P. Eyman, and D. J. Burton: *Essentials of General, Organic, and Biochemistry*, 3rd ed. Philadelphia, Saunders College Publishing, 1977.)

tRNAs (Table 11-2). For example, there are five distinctly different tRNA molecules specifically for the transfer of the amino acid leucine in cells of the bacterium *Escherichia coli*. At one end of a tRNA molecule is a trinucleotide base sequence (the **anticodon**) that fits a trinucleotide base sequence on mRNA (the **codon**). At the other end of a tRNA molecule is a specific base sequence of three terminal nucleotides—CCA—with a hydroxyl group on the sugar exposed on the terminal adenine nucleotide group. With the aid of enzymes, this hydroxyl group reacts with a specific amino acid by an esterification reaction.

$$\text{(Mononucleotides)}_{75-90}\text{CCA—OH} + \text{HOCCH(NH}_2\text{)R} \longrightarrow$$
$$\text{tRNA} \qquad \text{Amino Acid}$$

$$\text{(Mononucleotides)}_{75-90}\text{CCA—OCCH(NH}_2\text{)R} + \text{H}_2\text{O}$$
$$\text{tRNA-Amino Acid}$$

The P—O bonds of ATP provide energy for the reaction between an amino acid and its tRNA. A molecule of ATP first activates an amino acid.

$$\text{ATP} + \text{Amino acid} \longrightarrow \text{ATP—Amino acid-activated species}$$

The activated complex then reacts with a specific tRNA and forms the products shown in Figure 11-24.

The tRNA and its amino acid migrate to the ribosome, where the amino acid is used in the synthesis of a protein. The tRNA is then free to migrate back to the cell cytoplasm and repeat the process.

Messenger RNA is used at most only a few times before being depolymerized. Although this may seem to be a terrible waste, it allows the cell to produce different proteins on very short notice. As conditions change, different types of mRNA come from the nucleus, different proteins are made, and the cell responds adequately to a changing environment.

The ribosome is the part of the cell in which protein synthesis takes place.

Synthetic Nucleic Acids

Slow progress has been made in the synthesis of polynucleotides, principally because of the difficulties involved in determining the proper blocking groups.

In 1959 Arthur Kornberg synthesized a DNA type of polynucleotide, for which he received a Nobel prize. He used natural enzymes as templates to arrange the nucleotides in the order of the desired polynucleotide. His product was not biologically active. In 1965 Sol Spiegelman synthesized the polynucleotide portion of an RNA virus. This polynucleotide was biologically active and reproduced itself readily when introduced into living cells. In 1967 Mehran Goulian and Kornberg synthesized a fully infectious virus of a more complicated DNA type.

In 1970 Gobind Khorana synthesized a complete, double-stranded, 77-nucleotide gene. He, too, used natural enzymes to join previously synthesized, short, single-stranded polynucleotides into the double-stranded gene.

If scientists can construct DNA, can they then control the genetic code? Genes are the submicroscopic, theoretical bodies proposed by early geneticists to explain the transmission of characteristics from parents to progeny. It was thought that genes composed the chromosomes, which are large enough to be observed through the microscope as the central figures in cell division. It is now generally believed that DNA structures carry the message of the genes; hence, DNA contains the **genetic code.** If scientists can construct DNA, they could very well alter its structure and thereby control the genetic code. The ability to alter genes has led to a new field of science known as **biogenetic engineering.** This is a very active field of research that has developed (among other accomplishments) bacteria that can clean up oil spills. In this case a patent was granted to the General Electric Company for the production of life—a unique

Biogenetic engineering is the alteration of genes for a desired purpose.

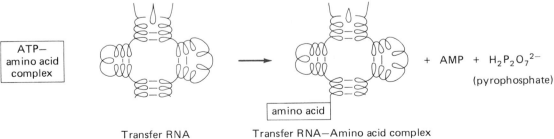

Figure 11-24
Bonding of activated amino acid to tRNA. *AMP* is adenosine monophosphate.

patent. Other bacteria have been produced that can synthesize protein, human growth hormone, and insulin. The method of producing bacteria for a particular function involves removing a gene from the bacterium, splicing in part of a gene from a human or other organism (the part that produces human insulin, for example), placing the spliced gene back into the bacterium, and letting the bacterium make millions of other insulin-producing bacteria. This process of splicing and recombining genes is referred to as **recombinant DNA technology.** The implications of gene splicing are tremendous — for both good and bad — and will demand responsible human decision-making for guidance toward the common good.

The process of forming and cloning recombinant DNA is discussed in Chapter 1. See Figure 1-7 for an outline of this process.

A **mutation** occurs whenever an individual characteristic appears that has not been inherited but is duly passed along as an inherited factor to the next generation. A mutation can readily be accounted for in terms of an alteration in the DNA genetic code; that is, some force alters the nucleotide structure in a reproductive cell. Some sources of energy, such as gamma radiation, are known to produce mutations. This is entirely reasonable because certain kinds of energy can disrupt some bonds, which can re-form in another sequence.

A mutation results when there has been an alteration of the genetic code contained within the DNA molecule.

If scientists can control the genetic code, can they control hereditary diseases such as sickle cell anemia, gout, some forms of diabetes, and mental retardation? If our understanding of detailed DNA structure and the enzymatic activity required to build these structures continues to grow, it is reasonable to believe that some detailed relationships between structure and gross properties will emerge. If this happens, it may be possible to build compounds that, when introduced into living cells, can combat or block inherited characteristics.

Ethics and risks were discussed in Chapter 1.

SELF-TEST 11-D

1. a. The basic code for the synthesis of protein is contained in the _____ molecule.
   b. The synthesis of a protein is carried out when _____ molecules bring up the required amino acids to mRNA.
2. When DNA replicates itself, each nitrogenous base in the chain is matched to another one via _____ bonds.
3. The energy for DNA replication and natural protein synthesis is supplied by substances such as _____ .
4. What nitrogenous base complements (matches through hydrogen bonding)
   a. adenine (A)? _____
   b. cytosine (C)? _____
   c. guanine (G)? _____
   d. thymine (T)? _____
   e. uracil (U)? _____
5. A gene has been synthesized from individual nucleotides in the laboratory without the aid of natural enzymes.   True ( ) or False ( )
6. Replication means the same as *duplication*.   True ( ) or False ( )
7. The sugar in RNA is _____ , whereas the one in DNA is _____ .

8. A nucleotide contains _____, _____, and _____.
9. The secondary structure of DNA is in the shape of a(n) _____.

MATCHING SET

_____ 1. Energy "cash" in the living cell
_____ 2. Mutation
_____ 3. Enzyme that splits polysaccharides in the mouth
_____ 4. Natural protein
_____ 5. Occurs under aerobic (with air) conditions
_____ 6. Product of ATP hydrolysis
_____ 7. Molecules that absorb light energy
_____ 8. D-glucose
_____ 9. Methionine
_____ 10. Enzymes
_____ 11. Carbohydrate stored in animals
_____ 12. Starch
_____ 13. Polypeptides
_____ 14. DNA
_____ 15. Fibrous protein
_____ 16. Cellulose
_____ 17. Vitamins
_____ 18. Enzyme that splits sucrose into fructose and glucose

a. Ptyalin
b. ADP + energy
c. Krebs cycle
d. Chlorophylls
e. Structure determined by DNA and RNA
f. Altered DNA
g. ATP
h. Embden-Meyerhof pathway
i. Polymer consisting of alpha-D-glucose monomers
j. Proteins
k. Sugar present in the blood
l. Amino acid
m. Sucrase
n. Polynucleotide
o. Biochemical catalysts
p. Coenzymes
q. Glycogen
r. Collagen
s. Polymer consisting of beta-D-glucose monomers

QUESTIONS

1. Show the structure of the product that would be obtained if two alanine molecules (Table 11-1) were to react to form a dipeptide.
2. What is an essential amino acid?
3. Name a polysaccharide that yields only alpha-D-glucose upon complete hydrolysis. Name a disaccharide that yields the same hydrolysis product.
4. What is the chemical difference between the starch amylopectin and the "animal starch" glycogen?
5. What is the chief function of glycogen in animal tissue?
6. Explain the basic differences between amylose and cellulose.
7. Why does cotton, a cellulose material, absorb moisture so much better than nylon 66? (The structure of nylon 66 is given in Chapter 10.)
8. What functional groups are always present in each molecule of an amino acid?
9. a. What element is necessarily present in proteins that is not present in either carbohydrates or fats?
   b. Name another element that is probably

present in proteins but is not present in either carbohydrates or fats.
10. What are the meanings of the terms *primary, secondary,* and *tertiary structures of proteins*?
11. In a protein, what type of bond holds the helical structure in place?
12. What type of proteins are enzymes?
13. a. Which of the following biochemicals are polymers: starch, cellulose, glucose, fats, proteins, DNA, and RNA?
    b. What are the monomer units for those that are polymers?
14. What is the chemical function of many vitamins? Give some examples.
15. Why are carbohydrates considered "energy-rich"?
16. Why can humans not digest cellulose?
17. The molecular structures of enzymes (particularly apoenzymes) are most closely related to which of the following structures: proteins, fats, carbohydrates, or polynucleic acids?
18. What are the two major divisions of reactions in photosynthesis? Express in words what is accomplished in each.
19. What is the basic nature of the digestion processes for large molecules?
20. What compound produces soreness in the muscles after a period of vigorous exercise?
21. If protein digestion is facilitated by enzymes and these enzymes are produced in body organs made of proteins, why do the enzymes not cause rapid digestion of the organs themselves?
22. Give the structure of ATP and point out the region of the molecule that contains bonds that are hydrolyzed to give energy.
23. What is a storehouse chemical for biochemical energy?
24. What are the end-products in the digestion of carbohydrates? Of fats? Of proteins?
25. How do living beings store and transfer energy?
26. When water reacts with ATP, is this an energy-releasing or an energy-requiring process?
27. What is the role of enzymes in digestion?
28. What is the purpose of ATP?
29. The importance of water in living systems is emphasized by incidences of hydrolysis. Cite examples.
30. What important types of chemicals can function as coenzymes?
31. What three molecular units are found in nucleotides?
32. What are the basic differences between DNA and RNA structures?
33. What stabilizing forces hold the double helix together in the secondary structure of DNA proposed by Watson and Crick?
34. What is recombinant DNA?
35. What happens if an amino acid is needed for protein synthesis and the amino acid can neither be made by the body nor obtained from the diet?
36. Does a strand of DNA actually duplicate itself base for base in the formation of a strand of mRNA? Explain.
37. A mutation can be explained in terms of a change in which chemical in the cell?
38. The replication of DNA occurs in what part of the cell?
39. What is meant by a *base pair* in protein synthesis? What type of bond holds base pairs together?
40. Why is the liver called the central nutrient bank of the body?
41. Check recent issues of *Science* or other scientific news publications to find out about recent work done on the synthesis of polynucleotides.

CHAPTER 12

Chemistry of Food Production

The British Isles recorded 201 famines between 10 A.D. and 1846, with none since. In China there were 1846 famines between 108 B.C. and 1828 A.D.

Through the 8000 to 10,000 years of recorded human history, food production techniques have developed enormously. Today's efficient farming methods are required to feed the approximately 5 billion people who are now alive.

To help grow the enormous amount of food we need, the chemical industry supplies modern scientific agriculture with a large assortment of chemicals — the agrichemicals — including fertilizers, medicine for livestock, chemicals to destroy unwanted pests and plant diseases, food supplements, and many others. It is clear that an adequate food supply for the Earth's exploding population (Fig. 12-1) can be provided only through a chemical understanding of food production and the general practice of modern food production techniques. Even with the use of new farming techniques and chemicals related to farming, an improper climate or political-economic conditions that deter the flow of food can still result in massive starvation.

Annual U.S. crop exports: $40 billion, including 38% of our corn, 43% of our soybeans, 27% of our wheat, and 64% of our rice.

A number of the ancient civilizations apparently developed good farming practices that are not directly recorded in history. In Roman times, Cato the Elder described seed selection, green manuring with legumes, testing of the soil for acidity, the use of marl, the value of alfalfa and clover, composting, the preservation and use of animal manures, pasture management, early-cut hay, and the importance of livestock in any general farming operation. Such progress based on practical experience was apparently gained and lost a number of times in the course of history.

A Statement of Philosophy

"Perfect agriculture is the true foundation of all trade and industry — it is the foundation of the riches of nations. But a rational system of agriculture cannot be formed without the application of scientific principles, for such a system must be based on an exact acquaintance with the means of vegetable nutrition. This knowledge we must seek through chemistry."

Justus von Liebig (1803–1878)

An Agricultural Law, The "Law of the Minimum"

If one of the nutritive elements is deficient or lacking, plant growth will be poor even when all other elements are abundant. If the deficient element is supplied, growth will be increased up to the point where the supply of that element is no longer the limiting factor.

Liebig

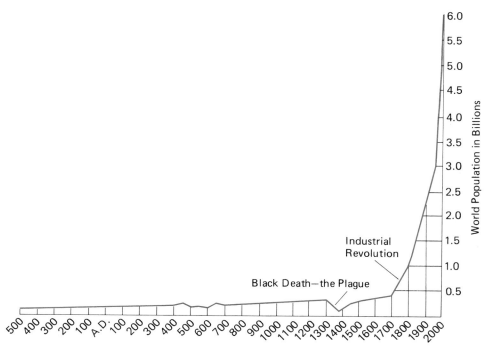

Figure 12-1
World population from 500 B.C. to 2000 A.D. The curve is shaped like a J because slow growth over a long period of time was followed by rapid growth over a short period of time. It is projected that 6 billion humans will inhabit the earth by the year 2000.

In 18th-century England, Arthur Young set the stage for modern scientific agriculture: as the first noted agricultural extension worker, he set in motion a system of dissemination of agricultural information that has now been developed on a worldwide basis. In 1840 Justus von Liebig published his *Organic Chemistry in Its Applications to Agriculture and Physiology*. Liebig began the description of the chemicals required by plants, and his work set the stage for the chemical fertilizer industry. As a result of his work, Liebig has been called the Father of Modern Soil Science. The Industrial Revolution provided better tools and power, which made the physical tasks in agriculture relatively easy. Even now, however, the best estimate is that two thirds of the world's agriculture is "backward."

The use of too much of a chemical or the use of the wrong chemical produces unwanted side effects that negate effective food production. Two notable examples are the phosphate runoff from fertilized fields, which can pollute streams, and the inclusion of DDT in bird eggs, which interferes with calcium metabolism and causes eggshells to be too thin and therefore to crack.

Our fundamental food is plants. We eat either plants or animals that eat plants. In order to grow, plants require the proper temperature, nutrients, air, water, and freedom from disease, weeds, and harmful pests. Chemistry has provided chemicals to give plants the proper nutrients and freedom from disease; these chemicals are the subject of this chapter.

Benefits and problems related to DDT are discussed in Chapter 10 and later in this chapter.

Natural Soils Supply Nutrients to Plants

Layers within the soil are **horizons.** The **topsoil** contains most of the presently living material and humus from dead organisms. It is not uncommon to find as much as 5% organic matter in topsoil. The topsoil is usually several inches thick, and in places more than 3 feet can be found. The **subsoil,** up to several feet in thickness, contains the inorganic materials from the parent rocks as well as organic matter, salts, and clay particles washed out of the topsoil. The root systems of large plants penetrate the subsoil. Under the subsoil is usually a layer of weathered parent rock on top of bedrock.

Since healthy topsoil is alive with life forms and their remains, it must contain an abundant supply of oxygen. Soil that supports a rich vegetative growth and serves as a host for insects, worms, and microbes is typically full of pores; such soil is likely to have as much as 25% of its volume occupied by air. The ability of soil to hold air depends on soil particle size and how well the particles pack and cling together to form a solid mass. The particle size groups in soils, called **separates,** vary from clays (the finest) through silt and sand to gravel (the coarsest). The particle size of a **clay** is 0.005 mm or less. The small particles in a clay deposit pack close together, eliminating essentially all air and thus supporting little or no life. A typical soil horizon is composed of several separates. A **loam,** for example, is a soil consisting of a friable mixture of varying proportions of clay, sand, and organic matter; a loam is rich in air content.

Friable material easily crumbles under slight pressure.

Air in soil has a different composition from air above. Normal dry air at sea level contains about 21% oxygen and 0.03% carbon dioxide. In soil the percentage of oxygen may drop to as low as 15%, and the percentage of carbon dioxide may rise above 5%. This results from the partial oxidation of organic matter in the closed space. The carbon in the organic material uses oxygen to form carbon dioxide. This increased concentration of carbon dioxide tends to cause the groundwater flowing through to become acidic; acidic soils are described as **sour soils** because of the sour taste of aqueous acids.

$$CO_2 + H_2O \longrightarrow H^+ + HCO_3^-$$

George Washington had marl, an alkaline mixture of limestone and clay, dug from the Potomac River bed for application to his fields.

Crushed limestone, $CaCO_3$, applied to soil combines with hydrogen ions to form bicarbonate ions.

$$H^+ + CO_3^{2-} \rightleftharpoons HCO_3^-$$

A slightly basic soil is a sweet soil.

If enough limestone is added to neutralize the acid in the soil and leave an excess of limestone, the pH of the soil is raised to slightly basic.

Water in Soil: Too Much, Too Little, or Just Right

Water can be held in soil in three ways: it can be absorbed into the structure of the particulate material, it can be adsorbed onto the surface of the soil particles, and it can occupy the pores ordinarily filled with air. **Sorption** is the general term covering absorption and adsorption; water held by either method of sorption is not readily released by soil. For example, if you take the driest soil you can find (perhaps from under an old house) and heat it in a test tube, it will release a considerable amount of moisture. "Dry soil" is not really dry in the chemical sense.

The sorption of water in soil is caused primarily by the abundance of oxygen in the chemical structures of most soil materials. Some examples are silicates (SiO_3^{2-}), phosphates (PO_4^{3-}), and carbonates (CO_3^{2-}). The high electronegativity of oxygen produces strongly polar bonds in the soil materials. Strong attractions between polar water molecules and polar centers in the soil materials cause water to wet soil.

Water is removed from soil in four ways: plants transpire it while carrying on life processes, soil surfaces evaporate it, it is carried away in plant products, and gravity pulls it to the subsoil and rock formations below. **Percolation** is the ability of a solid material to drain a liquid from the spaces between the solid particles. Soils with good percolation drain water from all but the small pores in the natural flow of the water. The flow of water through soil is a necessity; it takes several hundred pounds of water for the typical food crop to make 1 pound of food. A negative aspect of the massive flow of water through soil is the **leaching effect.** Water, known as the universal solvent because of its ability to dissolve so many different materials, dissolves away, or leaches, many of the chemicals needed to make a soil productive. If the leached material is not replaced, the soil becomes increasingly unproductive.

The percolation of a soil depends on the soil particle size and on the chemical composition of the soil material. Because of the small particle sizes involved, clays, and to a lesser degree silts, tend to pack together in an impervious mass with little or no percolation. Of course sand, gravel, and rock pass water readily. Water-logged soils that will not percolate support few crops because of their lack of air and oxygen. Rice is an exception.

Soils become acidic, or sour, not only because of the oxidation of organic matter but also because of selective leaching by the passing groundwater. Salts of the alkali and alkaline earth metals are more soluble than salts of the Group III and transition metals. For example, a soil containing calcium, magnesium, iron, and aluminum is likely to be slightly alkaline, or sweet, before leaching with water. If calcium and magnesium are removed and iron and aluminum remain, the soil becomes acidic. The iron and aluminum ions each tie up hydroxide ions from water and release an excess of hydrogen ions:

In some arid regions, calcium collects as calcium carbonate just under the solum, *or true soil.*

$$Fe^{3+} + H_2O \longrightarrow FeOH^{2+} + H^+$$

$$Al^{3+} + H_2O \longrightarrow AlOH^{2+} + H^+$$

Leaching is not bad insofar as it removes unwanted elements. In dry regions selenium is present in relatively high concentrations because of its concentration in parent rocks. The selenium would have been removed by leaching if the soil had been formed in a moist climate. Plants that grow in soil rich with selenium often take up large amounts of this element and are, as a result, poisonous to animals and humans. There are several such plants in the western United States and Mexico.

Humus

Organic matter varies in soil from the relatively fresh remains of leaves, twigs, and other plant and animal parts to peat, the precursor of coal and oil. Humus is not far removed in time from the living debris. However, it is well decomposed, dark-colored, and rather resistant to further decomposition. As a source of nutrients for plants, humus is almost like a time-release capsule, taking considerable time to release its contents while holding them in an insoluble form. Finally, humus is decomposed into minerals and inorganic oxides. In soils with poor drainage, the decomposition of humus is stopped and peat is formed. Such mucky soils are worthless for crop production unless they are drained and fertilized with phosphates and potash; then they become highly productive.

Humus releases its nutrients to plants slowly.

In addition to being a source of plant nutrients, humus is important in maintaining good soil structure, often keeping the soil friable in a soil rich in clay. Soil rich in humus may contain as much as 5% organic matter. Soils in the grasslands of North America are rich in humus to a considerable depth, in contrast to humid forest regions, where there is only a thin film on the ground surface.

Maintaining humus in the soil is of major concern to the agriculturist. Humus such as peat moss or organic fertilizer can be added. However, there is no real substitute for natural plant growth that is returned to the ground for humus formation. Clover is often grown for this purpose and plowed under at the point of its maximum growth. The compost pile of the gardener is another effort to maintain humus for a productive soil.

If large amounts of organic matter such as leaves or sawdust are added to soil to promote humus formation, it should be remembered that the oxidation of this material produces acid that will have to be counterbalanced by lime if the soil is to remain sweet.

Chemical Composition of Soil

The chemical compositions of soils reflect the Earth's crust composition, the composition of the parent rocks, and chemical and physical activity during and after soil formation. Even though it is second to oxygen in percent composition, silicon is the central element in explaining soil chemicals. Sand is silicon dioxide, clays are mixtures of silicates, and the different kinds of silicate rock fragments are numerous. The bulk of most soil horizons is composed of silicate materials. A wide variety of other elements is present in any given soil sample.

Black soils are usually rich in organic matter and consequently contain the elements required for plant life. However, the same amount of organic matter that makes a soil black in a temperate region will make it only brown in a tropical region. Red soils are likely to be rich in iron, and soils that are nearly white have been heavily leached and are likely to be poor in quality. Poorly drained soils are likely to be of uneven chemical texture, with several colors, such as gray, brown, and yellow, appearing in a spadeful of the material.

> Ask your County Extension Office for advice on how to take soil samples and where to send the samples for analysis. The cost of analysis is generally under $5 per sample. The analysis will tell you the amount of lime in the soil and the amount and type of fertilizer to apply.

Nutrients

At least 18 known elemental nutrients are required for normal green plant growth (Table 12–1). Three of these, **nonmineral nutrients,** are obtained from air and water. The mineral nutrients must be absorbed through the plant root system as solutes in water. The 15 known mineral nutrients fall into three groups: **primary nutrients, secondary nutrients,** and **micronutrients,** depending on the amounts necessary for healthy plant growth.

Table 12–1
Essential Plant Nutrients

Nonmineral	Primary	Secondary	Micronutrients
Carbon	Nitrogen	Calcium	Boron
Hydrogen	Phosphorus	Magnesium	Chlorine
Oxygen	Potassium	Sulfur	Copper
			Iron
			Manganese
			Molybdenum
			Sodium
			Vanadium
			Zinc

Nonmineral Nutrients

Carbon, hydrogen, and oxygen are available from the air and water. Carbon comes to plants as carbon dioxide, and hydrogen and oxygen come as water; in addition, plant roots absorb some free oxygen dissolved in water. During photosynthesis, green plants produce an excess of oxygen, which is released through the leaves and other green tissue.

Primary Nutrients

The primary nutrients are nitrogen, phosphorus, and potassium. Although bathed in an atmosphere of nitrogen, most plants are unable to use the air as a supply of this vital element. **Nitrogen fixation** is the process of changing atmospheric nitrogen into the compounds of this element that can be dissolved in water, absorbed through the plant roots, and assimilated by the plant (Fig. 12–2). Most plants thrive on soils rich in nitrates, but many plants that grow in swamps, where there is a lack of oxidized materials, can use reduced forms of nitrogen such as the ammonium ion. The nitrate ion is the most highly oxidized form of combined nitrogen, and the ammonium ion is the most reduced form of nitrogen.

Nature fixes nitrogen on a massive scale in two ways. Nitrogen is oxidized under highly energetic conditions, such as in the discharge of lightning or to a lesser extent, such as in a fire. The reaction is shown below:

$$N_2 + O_2 \longrightarrow 2\,NO$$

Nitric oxide, NO, is easily oxidized in air to nitrogen dioxide, which dissolves in water to form nitric acid, HNO_3, and nitrous acid, HNO_2:

$$2\,NO + O_2 \longrightarrow 2\,NO_2$$

$$H_2O + 2\,NO_2 \longrightarrow HNO_3 + HNO_2$$

Nitric acid is readily soluble in rain, clouds, or ground moisture and thus increases nitrate concentration in soil. A legume plant such as beans or clover lives in a symbiotic

Sir Humphrey Davies argued the Humus Theory, "Carbon for plants came from humus." A Swiss, de Saussure, showed the carbon came from carbon dioxide.

The air above each acre of Earth's surface contains 36,000 tons of nitrogen.

Chlorophyll requires nitrogen and magnesium from the soil.

Nitrogen is the distinguishing element in amino acids and proteins.

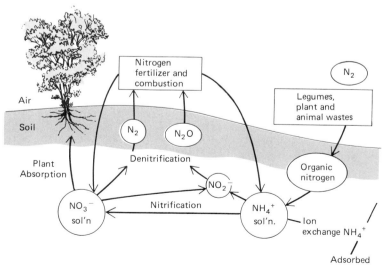

Figure 12–2
Nitrogen pathways through soil.

A German, Hellriegel, showed in 1886 that leguminous plants "fix" nitrogen.

relationship with bacteria, which live in nodules on the root system of the host plant. The bacteria are able to fix atmospheric nitrogen in far greater amounts than their own requirements, or even those of the host plant. As a result of legume fixation, more than 100 pounds of nitrogen can be added to an acre of soil in one growing season. In fact, if a legume is grown often enough on a particular plot of ground, there will be no need to supplement the soil with nitrogen fertilizer.

The protein content of corn and other grains is directly related to the amount of nitrogen in the soil.

Another major source of nitrogen replenishment in soil is dead organisms and animal wastes. Even in the absence of legumes, this can be an adequate source of nitrogen.

Phosphorus is concentrated in fast-growing tissue. A plant can translocate phosphorus from old to growing tissue.

Like nitrogen, phosphorus has to be in mineral or inorganic form before it can be used by plants. Unlike nitrogen, phosphorus has a natural species that comes totally from the mineral content of the soil. Orthophosphoric acid, H_3PO_4, loses hydrogen ions to form the dihydrogen and monohydrogen ions $H_2PO_4^-$ and HPO_4^{2-}, which are the dominant forms in soils of normal pH (Fig. 12-3). Because of the great concentration of electrical charge associated with the trivalent phosphate ion, phosphates tend to be held to positive centers in the soil structure and are not as easily leached by groundwater as are the nitrate compounds. The nitrate ion is a rather large ion with only one negative charge.

Potassium is absorbed as the free ion, $K^+(aq)$.

Potassium, if available, is constantly released by ion exchange reactions.

Potassium is a key element in the enzymatic control of the interchange of sugars, starches, and cellulose. Although potassium is the seventh most abundant element in the Earth's crust, soil used heavily in crop production can be depleted of this important metabolic element, especially if it is regularly fertilized with nitrate with no regard to potassium content. Some fungus plants in the soil produce chemicals that cause bound potassium to be released into a soluble form that can be taken in through the plant root system in excessive amounts or simply leached out by the flow of soil water.

Secondary Nutrients

Magnesium deficiencies, like nitrogen deficiencies, cause chlorosis, a low chlorophyll content.

Calcium and magnesium are available as Ca^{2+} and Mg^{2+} ions in small amounts as well as in complex ions and crystalline formations. These abundant elements are bound tightly enough so they are not readily leached yet loosely enough to be available to plants. When in the soil as sulfate (SO_4^{2-}), sulfur is readily available.

Micronutrients

Only very small amounts of micronutrients are required by plants; therefore, unless extensive cropping or other factors deplete the soil of the nutrients, sufficient quantities are usually available.

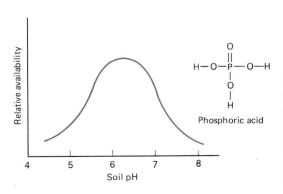

Figure 12-3
The availability of phosphate in the soil is a function of pH. The dominant species present for phosphoric acid at pH 5 to pH 8 are $H_2PO_4^-$ and HPO_4^{2-}. At very low pH values, the phosphorus is in the form of the acid H_3PO_4 (all three protons on the acid structure). At a very high pH, all three protons are removed, and the phosphorus is in the form of the PO_4^{3-} ion. Low soil temperatures in temperate regions significantly reduce phosphorus uptake by plants.

Iron is an essential component of the catalyst involved in the formation of chlorophyll, the green plant pigment. When the soil is iron deficient or when too much "lime," $Ca(OH)_2$, is present in the soil, iron availability decreases. This condition is usually indicated by plant leaves lightening in color or even turning yellow. Often a gardener or lawn worker will apply phosphate and lime to adjust soil acidity, only to see green plants turn yellow. What happens in such cases is that both phosphate and the hydroxide from the lime tie up the iron and make it unavailable to the plants:

$$Fe^{3+} + 2\ PO_4^{3-} \longrightarrow \underset{\text{Tightly Bound Complex}}{Fe(PO_4)_2^{3-}}$$
$$\underset{\text{Phosphate}}{}$$

$$Fe^{3+} + 3\ OH^- \longrightarrow \underset{\text{Insoluble Hydroxide}}{Fe(OH)_3}$$

Boron is absolutely necessary in trace amounts, but there is a relatively narrow concentration range above which boron is toxic to most plants.

Why a plant may turn yellow

Lime, as CaO or $Ca(OH)_2$, is the number-five commercial chemical. See inside the front cover.

The addition of lime, a basic substance, raises the pH of the soil.

SELF-TEST 12-A

1. Order the following types of soils from those that generally have the smallest soil particles to those that generally have the largest: silts, sandy soils, loams, clays.

2. Acidic soils are described as _____ because of the common taste of aqueous acids.
3. Carbon dioxide causes soils to be ( ) acidic or ( ) basic. Limestone, $CaCO_3$, causes soils to be ( ) acidic or ( ) basic.
4. The two factors that determine the percolation of a soil are _____ and _____.
5. Which is more acidic, a monovalent ion like Na^+ or a trivalent ion like Fe^{3+}? _____
6. A well-decomposed, dark-colored plant residue that is relatively resistant to further decomposition is known as _____.
7. The bulk of most soil horizons is composed of _____ materials.
8. The secondary elemental plant nutrients are _____, _____, and _____.
9. Nitrogen fixation involves the ( ) oxidation or ( ) reduction of nitrogen.
10. Some micronutrients can poison plants.   True ( ) or False ( )

Fertilizers Supplement Natural Soils

Primitive people raised crops on a cultivated plot until the land lost its fertility; then they moved to a virgin piece of ground. In many cases, the slash-burn-cultivate cycle was no more than a year in length, and few found a piece of ground anywhere that could support successful cropping for more than five years without fertilization. Farming villages,

Crop yield explosions: (1) U.S. corn—25 bushels per acre in 1800, 110 bushels per acre in the 1980s; (2) English wheat—below 10 bushels per acre from 800 A.D. to 1600, above 75 bushels per acre in the 1980s; (3) Rice in Japan, Korea, and Taiwan—fourfold increase in the last 40 years.

Chinese farmers added calcined bones to their soil 2000 years ago.

Mixed fertilizers were first produced in Baltimore in 1850.

Peruvian guano, a natural source of nitrates, was first imported into the United States in 1824.

Manure releases about half of its total nitrogen in the first growing season.

Manure as a fertilizer is graded less than 1-2-1.

developed in ancient times and prevalent throughout the Middle Ages, demanded innovation in fertilization, since they had to use the same land for many years. With the use of legumes in crop rotations, manures, dead fish, or almost any organic matter available, the land was kept in production. Primitive farming techniques continue in two thirds of the world today.

An estimated 4 billion acres are used worldwide in the cultivation of crops for food, less than 0.8 acre per person. This acreage would likely be sufficient if modern chemical fertilization were employed on all of it. If about $40 were spent on fertilizer for each cultivated acre, world crop production would increase by 50%, the equivalent of having 1.7 billion more acres under cultivation. However, the cost to produce this additional food would approach $160 trillion, a prohibitive cost.

Fertilizers that contain only one nutrient are called **straight** fertilizers; those containing a mixture of the three primary nutrients are called **complete,** or **mixed,** fertilizers. Urea for nitrogen and potassium chloride for potassium are examples of straight products. The macronutrients are absorbed by plant roots as simple inorganic ions: nitrogen in the form of nitrates (NO_3^-), phosphorus as phosphates ($H_2PO_4^-$ or HPO_4^{2-}), and potassium as the K^+ ion. Organic fertilizers can supply these ions, but only when used in large quantities over a long period of time. For example, a manure might be a 0.5-0.24-0.5 fertilizer, in contrast to a typical chemical fertilizer, which might carry the numbers 6-12-6. These numbers indicate the **grade** or **analysis,** in order, of the percentage of nitrogen, phosphorus as P_2O_5, and potassium as K_2O in the fertilizer (Fig. 12–4). In addition to having the desired ion, the chemical fertilizer places the ion in the soil in a form that can be absorbed directly by plants. The problem is that these inorganic ions are relatively easily leached from the soil and may pose pollution problems if not contained. The much slower organic fertilizer tends to stay put. **Quick-release** fertilizers are water-soluble, as opposed to **slow-release** products, which require days or weeks for the material to dissolve completely. Table 12–2 lists the necessary plant nutrients and suitable chemical sources of each.

Table 12–3 lists the quantities of nutrients known to be necessary to produce 150 bushels of corn.

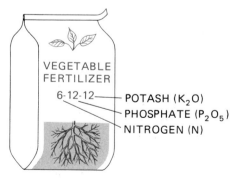

Figure 12–4
Fertilizer analysis numbers refer to the percentage by weight of N (nitrogen), P_2O_5 (phosphate), and K_2O (potash). Following the lead of Liebig, Samuel William Johnson, an American student of Liebig and the author of *How Crops Grow,* burned plants and analyzed their ashes. He expressed the nutrient concentrations in the oxide form present in the ashes as P_2O_5, K_2O, and so on, a practice that has continued to this day. The numbers for some of the chemicals and common commercial fertilizers are: ammonia, 82-0-0; urea, 46-0-0; ammonium nitrate, 33.5-0-0; ammonium sulfate 21-0-0; normal superphosphate, 0-18-0 to 0-20-0; phosphoric acid, 0-52-0 to 0-54-0; superphosphoric acid, 0-68-0 to 0-72-0; phosphorus pentoxide 0-100-0; potash, 0-0-100; and leading commercial fertilizers, 6-24-24, 10-10-10, and 13-13-13.

Table 12-2
Some Chemical Sources for Plant Nutrients

Element	Source Compound(s)
Nonmineral Nutrients	
C	CO_2 (carbon dioxide)
H	H_2O (water)
O	H_2O (water)
Primary Nutrients	
N	NH_3 (ammonia), NH_4NO_3 (ammonium nitrate), H_2NCONH_2 (urea)
P	$Ca(H_2PO_4)_2$ (calcium dihydrogen phosphate)
K	KCl (potassium chloride)
Secondary Nutrients	
Ca	$Ca(OH)_2$ (calcium hydroxide [slaked lime]), $CaCO_3$ (calcium carbonate [limestone]), $CaSO_4$ (calcium sulfate [gypsum])
Mg	$MgCO_3$ (magnesium carbonate), $MgSO_4$ (magnesium sulfate [epsom salts])
S	Elemental sulfur, metallic sulfates
Micronutrients	
B	$Na_2B_4O_7 \cdot 10\ H_2O$ (borax)
Cl	KCl (potassium chloride)
Cu	$CuSO_4 \cdot 5\ H_2O$ (copper sulfate pentahydrate)
Fe	$FeSO_4$ (iron (II) sulfate, iron chelates)
Mn	$MnSO_4$ (manganese (II) sulfate, manganese chelates)
Mo	$(NH_4)_2MoO_4$ (ammonium molybdate)
Na	$NaCl$ (sodium chloride)
V	V_2O_5, VO_2 (vanadium oxides)
Zn	$ZnSO_4$ (zinc sulfate, zinc chelates)

Table 12-3
Approximate Amounts of Nutrients Required to Produce 150 Bushels of Corn

Nutrient	Approximate Pound Per Acre	Source
Oxygen	10,200	Air
Carbon	7800	Air
Water	3225–4175 tons	29–36 inches of rain
Nitrogen	310	1200 pounds of high-grade fertilizer
Phosphorus	120 (as phosphate)	1200 pounds of high-grade fertilizer
Potassium	245 (as K_2O)	1200 pounds of high-grade fertilizer
Calcium	58	150 pounds of agricultural limestone
Magnesium	50	275 pounds of magnesium sulfate (epsom salt)
Sulfur	33	33 pounds of powdered sulfur
Iron	3	15 pounds of iron sulfate
Manganese	0.45	1.3 pounds of manganese sulfate
Boron	0.05	1 pound of borax
Zinc	Trace amounts	Small amount of zinc sulfate
Copper	Trace amounts	Small amount of copper sulfate
Molybdenum	Trace amounts	Trace of ammonium molybdate

Nitrogen

The first commercial U.S. ammonia plant began production in 1921.

Fixed nitrogen refers to nitrogen present in chemical compounds, that is, in combinations other than N_2.

Major energy problems in the fertilizer industry began in the 1970s.

The production of nitrogen through the distillation of air is described in Chapter 7. The reaction of steam with hydrocarbons is discussed in Chapter 10.

The fixed nitrogen in commercial fertilizers is obtained by the direct reaction of nitrogen with hydrogen to produce ammonia by the Haber process (Fig. 12-5).

$$N_2 + 3 H_2 \rightleftharpoons \underset{\text{Ammonia}}{2 NH_3}$$

Pure nitrogen is obtained for the process by distillation of oxygen and other gases from liquid air. Hydrogen is more difficult to obtain. At present, petroleum products such as propane ($CH_3CH_2CH_3$) react with steam in the presence of catalysts to produce hydrogen:

$$\underset{\text{Steam}}{CH_3CH_2CH_3 + 6 H_2O} \xrightarrow{\text{catalysts}} 3 CO_2 + 10 H_2$$

This is one of the principal reasons why ammonia fertilizer costs are so closely tied to petroleum prices. Hydrogen can also be prepared by the electrolysis of water:

$$2 H_2O \xrightarrow[\text{KOH}]{\text{electricity}} 2 H_2 + O_2$$

Several other methods are also available, but all of them require considerable energy per

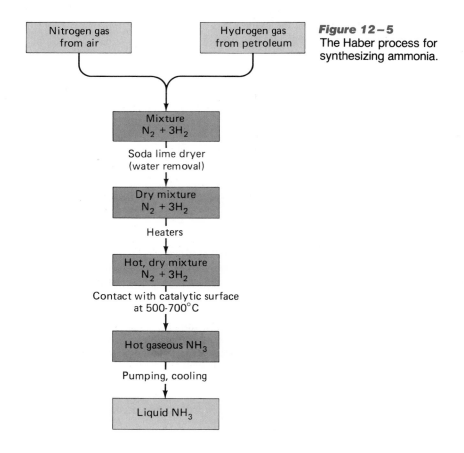

Figure 12-5
The Haber process for synthesizing ammonia.

pound of hydrogen produced. So as energy costs rise, the costs of fertilizers and therefore also the costs of food will necessarily rise.

At normal temperatures and pressures, ammonia is a gas. At higher pressures, ammonia is a liquid, known as **anhydrous ammonia.** Anhydrous ammonia can be injected directly into the soil (Fig. 12–6); it is retained there because polar ammonia has a very high affinity for moisture in the soil. Some danger is involved if liquid ammonia is released from a pressure tank into the atmosphere; there it will immediately evaporate and create a poisonous atmosphere.

Water solutions of ammonia, containing as much as 30% nitrogen by weight, can also be applied successfully to the soil. Again, special equipment and training are required to prevent the loss of ammonia through volatilization.

Anhydrous ammonia was first applied commercially to a field in 1944.

Research shows solid and liquid fertilizers to be equally effective.

Ammonia to Nitrates. Nitrogen in the form of ammonia is readily available for plant growth. Under usual soil conditions, the ammonia molecules pick up a hydrogen ion to form the ammonium ion:

$$NH_3 + H^+ \longrightarrow NH_4^+$$

The soil, rich in oxygen, is an oxidizing medium, so that the ammonium ion, through a

Figure 12–6
Liquid ammonia is injected directly into the ground to provide nitrogen for the growing plants.

280 *Chemistry of Food Production*

<small>Synthetic sodium nitrate was first produced in 1929.</small>

<small>Nitrogen fertilizers were not "cheap" until after World War II. In 1943 nitrate production was greater than war demands, and the nitrogen went into fertilizers.</small>

<small>Nitric acid is the number-13 commercial chemical. See inside the front cover.</small>

process called **nitrification,** can be oxidized to the nitrate ion, NO_3^-. The nitrate ion is then taken up and used by the plants.

Solid nitrate fertilizers are prepared from ammonia (Fig. 12–7). The ammonia from the Haber process is burned in oxygen over a platinum catalyst to obtain nitric oxide (NO):

$$4\ NH_3 + 5\ O_2 \longrightarrow 4\ NO + 6\ H_2O$$

The NO reacts readily with O_2 from the air to form NO_2:

$$2\ NO + O_2 \longrightarrow 2\ NO_2$$

NO_2 in turn reacts with water to yield nitric acid and nitrous acid. The nitrous acid, being unstable, is decomposed with heat, and the NO can be volatilized and recycled:

$$H_2O + 2\ NO_2 \longrightarrow HNO_3 + HNO_2$$

$$2\ HNO_2 \longrightarrow H_2O + 2\ NO$$

Additional ammonia then reacts with the nitric acid to produce ammonium nitrate, NH_4NO_3. Solid ammonium nitrate contains 35% nitrogen and should be handled with caution. It is an oxidizing agent and can explode when mixed with reducing materials.

A flooded soil quickly becomes a reducing medium as the air supply of oxygen is cut off. **Denitrification** occurs when conditions are such that nitrate is reduced to elemental nitrogen, which escapes into the atmosphere. It is estimated that soil fertilized with a soluble nitrate and then flooded for three to five days will lose 15% to 30% of its nitrogen as a result of denitrification.

<small>Solid urea was placed on the market in 1935.</small>

Urea. Urea (NH_2CONH_2) is one of the world's most important chemicals because of its wide use as a fertilizer and as a feed supplement for cattle. Ammonia and carbon dioxide

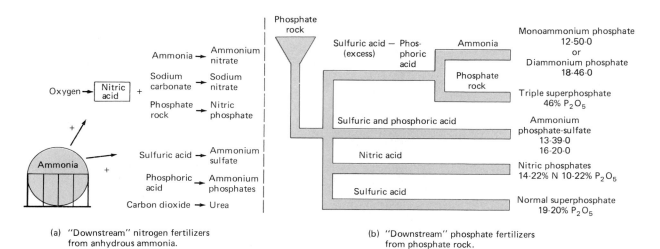

(a) "Downstream" nitrogen fertilizers from anhydrous ammonia.

(b) "Downstream" phosphate fertilizers from phosphate rock.

Figure 12–7
"Downstream" nitrogen fertilizers, produced from anhydrous ammonia, and "downstream" phosphate fertilizers, produced from phosphate rock. (From R.D. Young and F.J. Johnson, "Fertilizer Products," Chapter 3, of the *Fertilizer Handbook* published by the Fertilizer Institute, 1015 18th St, NW, Washington, D.C., 20036 in 1982.)

react under high pressure near 200°C to produce ammonium carbamate, which then decomposes into urea and water:

$$2\ NH_3 + CO_2 \longrightarrow H_2N-\underset{\underset{O^-NH_4^+}{|}}{\overset{\overset{O}{\parallel}}{C}} \longrightarrow H_2N-\overset{\overset{O}{\parallel}}{C}-NH_2 + H_2O$$

Ammonium Carbamate — Urea

Sulfur-coated urea (SCU) slows the release of nitrogen from urea.

A slurry of water, urea, and ammonium nitrate is often applied to crops under the name of "liquid nitrogen." Such a solution can contain up to 30% nitrogen and is easy to store and apply.

When applied to the surface of the ground around plants, urea is subject to considerable nitrogen loss unless it is washed into the soil by rain or irrigation. When urea hydrolyzes (is decomposed by water), ammonia is formed and is lost to the air unless surrounded by moist soil particles. As much as half of the nitrogen applied can be lost in this way.

Phosphorus and Potassium (Phosphate Rock and Potash)

Phosphate rock and potash are two minerals that can be mined, pulverized, and dusted directly onto deficient soil. Often they are specially treated to produce desirable mixing properties. Phosphorus, for example, is found scattered throughout the world in deposits of **phosphate rock.** The phosphate rock itself is not very useful to a growing plant because of its very low solubility. When treated with sulfuric acid, however, mined phosphate becomes more soluble and is called "superphosphate."

$$Ca_3(PO_4)_2 + 2\ H_2SO_4 \longrightarrow \underline{Ca(H_2PO_4)_2 + 2\ CaSO_4}$$

Phosphate Rock — "Superphosphate"

The world has limited deposits of phosphate rock, which is essential to the manufacture of fertilizers.

Within a few years, large new deposits of phosphate rock will be needed to meet the demand. A major problem is an increasing concentration of impurities in the available phosphate rock. See Figure 12–7 for products produced from phosphate rock.

Potassium in the form of **potash** (K_2CO_3) exists in enormous quantities throughout the world. A major problem involved in obtaining potash is depth of mining required. Another soluble form of potassium is its chloride (KCl), called muriate of potash. Because KCl is often found in combination with sodium chloride, which is toxic to plants, some process (such as recrystallization) must be used to separate the two compounds.

Sir John Lawes (England) set up a superphosphate fertilizer factory in 1842. By 1853 there were 14 superphosphate factories in England. By 1930 superphosphate supplied over 90% of U.S. phosphate.

Potash was first obtained from Searles Lake, California, in 1915.

Calcium, Magnesium, and Sulfur

Calcium is rarely in short supply as a plant nutrient. In addition to its widespread occurrence in clays and minerals, it is a major component of lime, which is used to make soil alkaline. Calcium is insoluble enough in neutral and alkaline media to prevent dangerously excessive uptake by plants.

Magnesium deficiencies occur in acid sandy soils and in other soils in which the magnesium is bound so strongly that an insufficient amount of the element is released in solution form. Magnesium sulfate or a double salt of magnesium and potassium sulfate is effective in supplying the element. Liming with dolomitic limestone has the advantage of adding magnesium from the dolomite while also adjusting the pH and adding calcium.

Sulfur deficiencies usually occur in sandy and well-drained soils and have been most often noted in the southeastern and northwestern parts of the United States. Positive

"Acidulated" bones (bones in dilute H_2SO_4) were sold commercially by Sir James Murray (Ireland) in 1808.

A *dolomite* is a soil mixture containing both calcium and magnesium carbonate.

crop responses to added sulfate have been documented in 29 of the states. Several sulfur chemicals are effective, including ammonium sulfate, ammonium thiosulfate, potassium sulfate, potassium magnesium sulfate, and ordinary superphosphate.

Micronutrients

The biggest difficulty in fertilizing with micronutrients is knowing what element is needed and in what amount. Precise analytical methods will tell the analyst the concentration of each element in the soil, but these tests will not indicate how much of the elements are available for plant absorption. Such analytical tests can be calibrated in terms of crop response in greenhouse and field studies, but this is a costly and laborious process, and the results are likely to vary from soil type to soil type and from crop to crop.

Micronutrients can be applied to the soil, to the plant, or to the seed before it is planted. Iron is often applied to the leaves of plants, and molybdenum compounds are dusted onto the seeds. Micronutrients are mixed with macronutrients in many applications, but in these cases one must allow for possible chemical reactions among the ingredients in the mixture. For example, manganese sulfate is effective as a source of manganese, but not in the presence of more soluble polyphosphate fertilizers. Another complication of micronutrient fertilization is the need for uniform application. In the case of boron, with which it is easy to get into the toxic range, nonuniform application of the correct amount per acre would result in both deficient and poisonously excessive amounts in the same plot.

Considerable chemical innovation has been required to address some of the micronutrient deficiencies that have surfaced in crop production. We shall illustrate with iron. In the 1950s severe iron deficiencies appeared on thousands of acres of both acid and alkaline soils in the Florida citrus groves. Applications of iron sulfate to the extent of 100 pounds per acre were ineffective in correcting the problem, since the iron was bound by soil chemicals and was therefore unavailable to the plants. It occurred to researchers to introduce into the soil another complexing agent that would hold the iron with just the right tenacity; these new complexes were meant to keep the iron from the "gripping" complexing agents already in the soil and at the same time to release enough iron to the roots. Ethylenediaminetetraacetic acid (EDTA) was tried, and it worked! The iron is encircled by EDTA, which chelates the iron ion from six different directions (Fig. 12-8). The word *chelate* comes from the Greek *chela*, meaning "claw." (The iron is bonded at six different bonding sites.) A number of micronutrients are now introduced into plants by this technique.

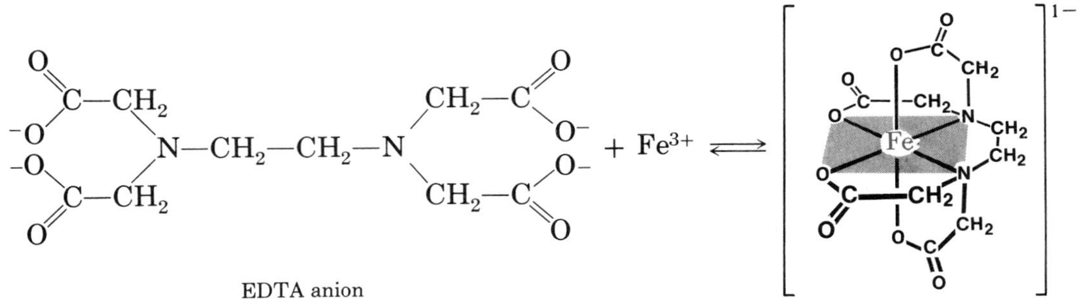

> It is a problem to keep micronutrients uniform when they are mixed with granulated fertilizers.
>
> Waxes and oils have been used to bind micronutrients to the surface of fertilizer particles.

EDTA anion + Fe^{3+} ⇌ [iron-EDTA complex]$^{1-}$

Figure 12-8
The EDTA anion formula and the structure of the iron-EDTA complex ion.

Super Growth in Plants

In addition to fertilizers, which help the natural biochemical growth processes of plants, so-called plant growth enhancers cause phenomenal growth. In 1932 it was noticed that ethylene or acetylene promotes flowering in pineapples. Two years later it was discovered that a group of compounds that came to be known as auxins induce elongation in shoot cells. Auxins are used to thin apples and pears, to increase yields in beans, sugar cane, and potatoes, to assist the rooting of cut plants, and to increase flower formation. Perhaps the most remarkable plant growth regulator is a group of compounds known as the gibberellins. Figure 12–9 presents the basic structure for gibberellic acid, and the following quotation from the *Farm Chemical Handbook,* published by the Meister Publishing Company of Willoughby, Ohio, illustrates its many uses:

> The gibberellins, the acid and its derivatives, are used to elongate cluster, increase berry size, and reduce bunch rot of grapes; to maintain color, delay yellowing, and reduce percentage of small tree-ripe fruit in lemons; reduce rind-staining water spots and tacky rind in Navel oranges; to overcome certain symptoms of cherry yellow virus diseases in sour cherries; to produce taller, thicker stalks of celery harvested in cooler seasons; to prevent head formation and induce production of the seed stalk in lettuce; to increase fruit set and yields of the Orlando tangelo; to accelerate maturity of artichokes and to shift the harvest to an earlier date; to stimulate uniform sprouting of seed potatoes that have not had a full rest period; to delay harvesting, to produce a brighter colored, firmer fruit, and to increase size of sweet cherries; for reduction of internal browning and watery pits of the Italian prune and to increase yields of marketable forced rhubarb and to break dormancy on plants receiving insufficient chilling; to increase yield and pickability of Fluggle hops; to improve fruit set in blueberries where natural honeybee pollination may be insufficient, and to overcome cold stress effects on certain grasses.

Super growth can kill plants. See the section on Herbicides.

Gibberlins and auxins are plant hormones, chemical messengers that coordinate the activities of different cells in multicellular organisms.

Protecting Plants in Order to Produce More Food

The natural enemies of plants include over 80,000 diseases due to viruses, bacteria, fungi, algae, and similar organisms, 30,000 species of weeds, 3000 species of nematodes, and about 10,000 species of plant-eating insects. One third of the food crops in the world are lost to pests each year, with the loss going above 40% in some developing countries. Crop losses to pests amount to $20 billion per year.

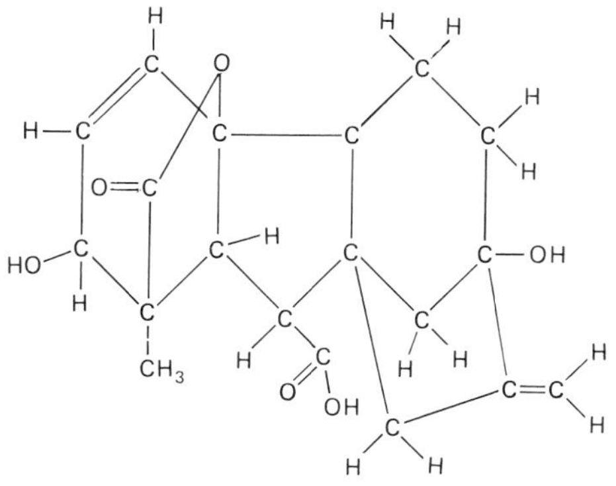

Figure 12–9
Gibberellic acid.

Organic farming is farming without chemical fertilizers and pesticides.

Because there are so many varieties and such large quantities of them are produced, (about 300,000 tons yearly in the United States), herbicides are considered a separate category from pesticides.

DDT

Half-life is the time required for half of the substance to disappear (see Chapter 4).

Pesticides are the chemical answer to pest control. The 18 classes of pesticides (Table 12–4) are fortified with more than 2600 active ingredients to fight the battle with pests. The agricultural market consumes about three fourths of the nearly $4.5 billion spent annually on pesticides in the United States. Without pesticides, crop production, on average, would be 20% lower than at present. In the following brief account, we discuss two of the major classes of pesticides, the insecticides and the herbicides.

Insecticides

Before World War II, the list of insecticides included only a few compounds of arsenic, petroleum oils, nicotine, pyrethrum, rotenone, sulfur, hydrogen cyanide gas, and cryolite. DDT, the first of the chlorinated organic insecticides, was originally prepared in 1873, but it was not until the beginning of World War II that it was recognized as an insecticide.

The use of synthetic insecticides increased enormously on a worldwide basis after World War II (Table 12–5). As a result, insecticides such as DDT have found their way into lakes and rivers. There is a great variety of pesticides, and their use frequently leads to severe damage to other forms of animal life, such as fish and birds. The toxic reactions and peculiar biological side effects of many of the pesticides were not thoroughly studied or understood prior to their widespread use.

A case in point is **DDT**. This insecticide, which has not been shown to be toxic to humans in doses as high as those received by factory workers involved in its manufacture (400 times the average exposure), does have peculiar biological consequences. The structure of DDT is such that it is *not* metabolized (broken down) very rapidly by animals; instead it is deposited and stored in the fatty tissues. The biological half-life of DDT is about eight years; that is, it takes about eight years for an animal to metabolize half of the amount it assimilates. If ingestion continues at a steady rate, DDT will build up

Table 12–4
Classes of Pesticides

Class	Function: Kills	Word Origin (Latin [L] or Greek [G])
Acaricide	Mites	(G) *akari*, mite or tick
Algicide	Algae	(L) *alga*, seaweed
Avicide	Birds	(L) *avis*, bird
Bactericide	Bacteria	(L) *bacterium*
Fungicide	Fungi	(L) *fungus*
Herbicide	Plants	(L) *herbum*, grass or plant
Insecticide	Insects	(L) *insectum*
Larvicide	Larvae	(L) *lar*, mask
Molluscicide	Snails, slugs, etc.	(L) *molluscus*, soft or thin shell
Nematicide	Round worms	(G) *nema*, thread
Ovicide	Eggs	(L) *ovum*, egg
Pediculicide	Lice	(L) *pedis*, louse
Piscicide	Fish	(L) *piscis*, fish
Predicide	Predators (coyotes, wolves, etc.)	(L) *praeda*, prey
Rodenticides	Rodents	(L) *rodere*, to gnaw
Silvicide	Trees and brush	(L) *silva*, forest
Slimicide	Slimes	English
Termiticide	Termites	(L) *termes*, wood-boring worm

within the animal over a period of time. For many animals this is not a problem, but for some predators, such as the eagle and osprey, that feed on other animals and fish, the consequences are disastrous. The DDT in the fish eaten by such a bird is concentrated in the bird's body, which attempts to metabolize the insecticide by altering its normal metabolic pattern. This alteration involves the use of compounds that normally regulate the calcium metabolism of the bird and are vital to its ability to lay eggs with thick shells. When these compounds are diverted to their new use, they are chemically modified and are no longer available for the egg-making process. As a consequence, the eggs the bird does lay are easily damaged, and the survival rate of the species decreases drastically. This process has led to the nearly complete extinction of eagles and ospreys in some parts of the United States.

The buildup of DDT in natural waters is not an irreversible process; the Environmental Protection Agency reported a 90% reduction of DDT in Lake Michigan fish by 1978 as a result of a ban on the use of the insecticide.

DDT and other insecticides such as **dieldrin** and **heptachlor** are referred to as **persistent pesticides.*** Other substances with biodegradable structures are now substituted as much as possible; the compound **chlordan** is an example of just such a substitution. It is interesting to note that the structural differences between heptachlor

Sidebar: Do we really have to decide between pests and eagles?

Sidebar: City water treatment generally does not remove DDT.

Dieldrin

and aldrin

were banned by the courts from their major uses (such as on corn) in 1974.

Heptachlor and chlordan were banned for most garden and home use in December 1975.

* The use of DDT was banned in the United States in 1973, although it is still in use in some other parts of the world. Over 1.8 billion kg of DDT have been produced and used.

Table 12-5
Insecticide Types and Examples

Type	Example(s)	Comment(s)
Organochlorines	DDT	Insecticide of greatest impact (banned in U.S. in 1973)
	Chlordane	
	Aldrin	Persistent, choice for termites (banned by EPA for agriculture, 1975–1980)
	Dieldrin	
	Endrin	
Polychloroterpenes	Toxaphene	Persistent, use peaked in 1976
Organophosphates	Malathion	Effective for plant and animal use
	Diazinon	Very versatile
Organosulfurs	Aramite	Choice for mite control
Carbamates	Carbaryl (Sevin)	Lawn and garden choice, low toxicity
Formamidines	Chlordimeform	Promising new group for resistant insects
Thiocyanates	Lethane	Limited use in aerosols
Dinitrophenols	Dinoseb	Choice for mildew fungi
Organotins	Cyhexatin	Most selective acaricide
Botanicals	Pyrethrum	Extracted from chrysanthemum
Synergists	Sesamex	Increases insecticide activity
Inorganics	Sulfur	Oldest insecticide
	Arsenicals	Stomach poisons
Fumigants	Methyl bromide	Kills insects in stored grain
Microbials	Heliothis virus	Specific for corn earworm and cotton ballworm
Insect growth regulators	Ecdysone	Environmentally sound
Insect repellents	Delphene	Superior to other repellents

(persistent) and chlordan (short-lived) are relatively slight (look at the chlorine atom on the lower five-membered ring):

Heptachlor   Chlordan

Many insecticides are much more toxic to humans than is DDT. These include inorganic materials based on arsenic compounds, as well as a wide variety of phosphorus derivatives based on structures of the type

$$R\!-\!\underset{R'}{\overset{Z}{\underset{\|}{P}}}\!-\!X$$

where Z is oxygen or sulfur, R and R' are alkyl, alkoxy, alkylthio, or amide groups, and X is a group that can be split easily from the phosphorus. Insecticides of this type include **parathion,** which is effective against a large number of insects but is also very poisonous to human beings. Called anticholinesterase poisons, these compounds are readily hydrolyzed to less toxic substances that are not residual poisons.

A much publicized case of a poorly handled insecticide is **kepone** ($C_{10}Cl_{10}O$, a complex molecule containing several fused rings), which was manufactured in a converted filling station. The resulting contamination of the James River in Virginia was extensive, as was the personal suffering of the workers. The dredging costs to clean the river are estimated by the EPA to be $1 billion, and it is expected to take one dredge 120 years to complete the job; total cleanup costs are estimated at $7.2 billion. Because of the expense and time involved, there are no immediate plans to clean up this chemical spill.

Aldicarb oxime is a systemic insecticide, acaricide, and nematicide. The pesticide is a solid material designed for soil use. Methyl isocyanate ($CH_3\!-\!N\!=\!C\!=\!O$) is used as an intermediate in the preparation of aldicarb oxime. It was methyl isocyanate gas that killed 2000 people in 1984 in Bhopal, India (see Chapters 1 and 15). Although considerably less dangerous than methyl isocyanate, aldicarb oxime, like all pesticides, must be treated with care. On August 11, 1985, aldicarb oxime was released in a cloud from a chemical plant in West Virginia. The release caused an area alert and sent 125 people to the hospital, although it apparently produced no lasting injuries.

Parathion

Malathion

The goal of the insecticide quest: a selectively toxic chemical that is quickly biodegradable.

The tank in the West Virginia spill contained 65% methylene chloride and 35% aldicarb oxime.

$$CH_3\!-\!S\!-\!\underset{CH_3}{\overset{CH_3}{\underset{|}{\overset{|}{C}}}}\!-\!CH\!=\!N\!-\!O\!-\!\overset{O}{\underset{\|}{C}}\!-\!\underset{H}{\overset{}{\underset{|}{N}}}\!-\!CH_3$$

Aldicarb Oxime (trade name, Temik)

The choice of solutions to the problems of pesticides is not an easy one. By using insecticides, we introduce them into our environment and our water supplies. If we refuse to use them, we must tolerate malaria, plague, sleeping sickness, and the consumption of a large part of our food supply by insects. Continuing research is needed on new methods and materials for the control of insect populations.

Herbicides

Herbicides kill plants. They may be selective and kill only a particular group of plants, such as the broad-leaved plants or the grasses, or they may be nonselective, making the ground barren of all plant life. Herbicides may come in a granular form, which is worked into the soil before planting, or in a liquid spray, which is applied at various stages after planting. The liquid spray may be applied either before or after the crop emerges; the choice depends on the particular chemical, weed, soil type, and crop involved. Even different species of plants in the same class may respond differently, with some requiring only one application of a herbicide and others requiring as many as three applications. The use of herbicides is thus a complicated matter.

Nonselective herbicides usually interfere with photosynthesis and thereby starve the plant to death. On application, the plant quickly loses its green color, withers, and dies. **Selective herbicides** act like a hormone, a very selective biochemical catalyst that controls a particular chemical change in a particular type of organism at a particular stage in its development. Most selective herbicides in use today are growth hormones; they cause cells to swell, such that leaves become so thick that chemicals cannot be transported through them and roots become so thick that they are unable to absorb needed water and nutrients. In theory, selective herbicides can become far more specific as the particular biochemistry of each plant species is understood, and it is expected that many new products will be developed as a result of intense research in this field.

The traditional method for the control of weeds in agriculture was **tillage.** Only in the early 1900s was it recognized that some fertilizers were also weed killers. For example, when calcium cyanamide (CaNCN) was used as a source of nitrogen, it was found to retard the growth of weeds. Arsenites, arsenates, sulfates, sulfuric acid, chlorates, and borates have also found use as weed killers. A typical product still in commercial use contains 40% sodium chlorate ($NaClO_3$), 50% sodium metaborate ($NaBO_2$), and 10% inert filler. These herbicides are nonselective and must be used with considerable care to protect the desired plants.

> Tillage is land prepared for agricultural use by plowing, disking, harrowing, etc.

Nitrophenol was used in 1935 as the first selective, organic herbicide. It was also in the 1930s that work began on the auxins, or hormone-type weed killers. The most widely used herbicide today came out of this work. It has the common name 2,4-D; the chemical name and formula are shown below:

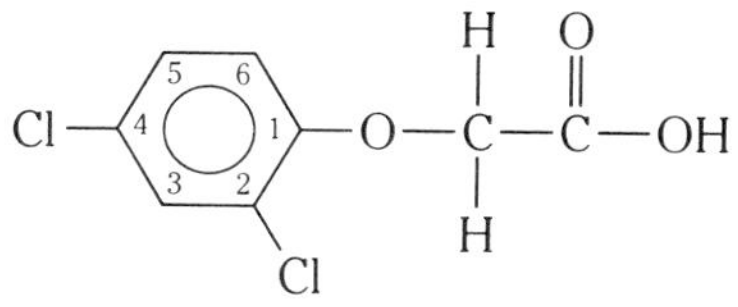

> Nitrophenols:
> Ortho
> Meta
> Para

2,4-D
2,4-Dichlorophenoxyacetic acid

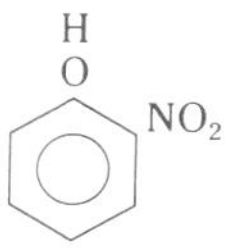

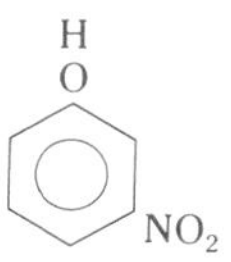

The corresponding trichloro- compound (common name: 2,4,5-T) has also been shown to be highly effective. The only difference between it and 2,4-D is the additional chlorine atom on the benzene ring in the fifth position. Agent Orange, widely used as a defoliant in the Vietnam War, is a mixture of these two compounds. The second compound, 2,4,5-T, has been removed by law from many markets because of a number of health problems associated with its use. It is probable that most of the problems were caused by the presence of an impurity, dioxin. Described in Chapter 10, dioxin is a severe poison with a toxicity equaled by few compounds. It will be interesting to see whether 2,4,5-T, which is now commercially produced free of dioxin, will be reestablished as a herbicide. Both of these compounds result in an abnormally high level of RNA in the cells of the affected plants, causing the plants to grow themselves to death.

Several different triazines have been effective as herbicides, the most famous one being atrazine. Atrazine is widely used in no-till corn production or for weed control in minimum tillage:

1,3,5-Triazine

Atrazine
2-Chloro-4-ethylamino-6-isopropyltriazine

No-till farming is the control of weeds by herbicides without cultivation.

Atrazine is a poison to any green plant if it is not quickly changed into another compound. Corn and certain other crops have the ability to render atrazine harmless, which weeds cannot do. Hence, the weeds die and the corn shows no ill effect.

A typical fact sheet by a state agricultural extension service lists six formulations using atrazine for no-till corn and nine mixtures for preplant and preemergence applications on tilled ground. For example, one formulation for no-till corn calls for a mixture of 2 to 3 pounds of atrazine, 0.25 to 0.50 pounds of paraquat, and a surfactant. The surfactant lowers the surface tension of the liquid spray and makes it easier for the liquid to wet and penetrate the plant surface.

Paraquat is also used as a contact herbicide. When applied directly to susceptible plants, it quickly causes a frostbitten appearance and death. Paraquat has received considerable attention because of its use by governmental aircraft on illegal poppy and marijuana fields in Mexico and elsewhere. Like atrazine, paraquat has a nitrogen atom in each aromatic ring of the two-ring system:

Paraquat
1,1'-Dimethyl-4,4'-Bipyridinium Dichloride

The amount of energy saved by herbicides used in no-till farming is enormous. The saving of topsoil is also considerable, since the cover from the previous crop holds the soil against wind and water runoff. However, agriculturists who use herbicides are highly dependent on agricultural research institutions for the selection of herbicides that will do the desired job without harmful side effects. Such selections depend on considerable research, much of which is carried out on a trial-and-error basis on test plots. A procedure that is recommended today may be outdated by the next growing season.

SELF-TEST 12-B

1. Most virgin soils can support crop production for a decade or more before fertilization is needed.   True ( ) or False ( )
2. What do the numbers 6-12-8 on a fertilizer mean? The 6 is the percentage of _____; the 12 is the percentage of _____; and the 8 is the percentage of _____ in the fertilizer.
3. Which would be properly termed a quick-release fertilizer, potassium nitrate for potassium or manure for nitrogen? _____
4. Would the nitrogen in ammonia be considered "fixed" nitrogen? _____
5. Pure ammonia under ordinary conditions is a ( ) solid, ( ) liquid, or ( ) gas. _____
6. If ammonia is oxidized to NO_2 and dissolved in water, two acids are produced: _____ and _____.
7. Flooded soils cause ( ) nitrification or ( ) denitrification. _____
8. The chemical formula for potash is _____.
9. Which is more likely to be in short supply as a plant nutrient in soils, calcium or magnesium? _____
10. Approximately what percentage of the food crops of the world is lost to pests each year? _____
11. The first chlorinated organic insecticide was _____.
12. Which of the following is not a persistent insecticide: DDT, dieldrin, heptachlor, or chlordan? _____
13. Which is more likely to be a hormone, a selective or a nonselective herbicide? _____
14. The most widely used herbicide today is _____.
15. Gibberellic acid is a _____.

MATCHING SET

___	1. Humus	a.	Solid nitrogen compound
___	2. Liebig	b.	Friable material
___	3. Loam	c.	Result of dissolution
___	4. Sweet soil	d.	Herbicide
___	5. Percolation	e.	Source of humus
___	6. Leaching	f.	Insecticide
___	7. Selenium	g.	Provides nitrogen to soil
___	8. Compost pile	h.	Central element in soil
___	9. Silicon	i.	Likely in sandy soils
___	10. Zinc	j.	Applied organic chemistry to agriculture
___	11. Lime	k.	Calcium compound that sweetens soil
___	12. Ammonia	l.	Alkaline
___	13. Haber process	m.	Micronutrient
___	14. Urea	n.	Fortunately leached
___	15. Sulfur deficiency	o.	Gas; fixed nitrogen
___	16. EDTA	p.	Passage of water
___	17. DDT	q.	Ammonia production
___	18. 2,4-D	r.	Complexing agent for micronutrients
		s.	Described seed selection

QUESTIONS

1. Explain how Liebig's Law of the Minimum was a consideration of macro and trace plant nutrients.
2. Describe the horizons that would be found in a typical soil.
3. Why is the air in soil of a different chemical composition than the air around us?
4. Which will contain more air per soil volume, a clay or a loam? Give a reason for your answer.
5. What causes a soil to be sour? Sweet?
6. If crushed limestone is spread on soil, will it raise or lower the pH of the soil? Explain.
7. How many pounds of water are typically required to produce 1 pound of food?
8. Give the approximate dates for the beginning of the world population explosion and the beginning of the application of scientific information to agriculture. In your opinion, is there a direct cause-and-effect relationship between these phenomena?
9. Guano was used as a fertilizer in colonial America. If you do not know the meaning of this word, look it up in the dictionary. What nutrients does guano add to the soil?
10. What are three principal elements in the soils of the Earth? Contrast the elements in this list to the elemental composition of the crust of the Earth.
11. Which groups of elements are first leached from soils, the alkali and alkaline earth metals or the Group III and transition metals? What is the effect of this selective leaching on soil pH?
12. What is the effect of selenium in the soil on plant and animal life? How is the concentration of selenium reduced through natural processes?
13. What are two important roles of humus in the soil? Do leaves turned into the soil to produce humus raise or lower the soil pH?

14. Relate the colors of soils to the plant nutrients that they likely contain.
15. The oxide of what element predominates in the soils of the Earth?
16. What three elements are obtained from the air and water as nonmineral plant nutrients?
17. What are the three primary mineral plant nutrients that are considered first in fertilizer formulations?
18. Explain the necessity of nitrogen fixation for plant growth. Give a physical, a biological, and a chemical method of nitrogen fixation.
19. Are nitrates or phosphates more easily leached from the soil? Explain.
20. Soil phosphates are in different ionic forms, depending on the pH. What are the predominant forms of ionic phosphate in sweet soils?
21. Which is more likely to be a problem in farming, a soil shortage of N, P, and K, or a shortage of Ca, Mg, and S? Give a reason for your answer.
22. Explain the numbers 6-12-6 as found on a fertilizer bag.
23. What is the danger in the use of anhydrous ammonia as a chemical fertilizer?
24. What is superphosphate? How is it made?
25. What two herbicides were formulated to produce Agent Orange? Which of these herbicides is presently banned in the United States for agricultural use?
26. How is it possible that your great-grandfather might have been happy with 25 bushels of corn per acre, whereas farmers today who cannot average over 100 bushels per acre cannot adequately compete in the corn market?
27. In the period after World War II, most farmers fertilized "enough to be sure." Farmers today are likely to have the soil analyzed and have a fertilizer formulated on prescription. What is the cause of this change? Can you see an effect that this change in farming practice might have on water pollution?
28. Investigate a no-till farming operation. What herbicides are used? How is energy saved and, at the same time, how is additional energy required? What is the effect of no-till farming on the conservation of topsoil?
29. Trace the rise and fall of the use of DDT in agriculture. Debate the question of whether it has been more good than bad for the human race.
30. Debate the proposition that herbicides such as paraquat should be sprayed from airplanes to destroy crops grown to produce illegal drugs.
31. Contact the U.S. Department of Agriculture through the Soil Conservation Service in your area. Find out if there is documentation of a micronutrient problem in the agriculture in your state. Define the problem if one is found, and outline a chemical solution.

CHAPTER 13

Nutrition: The Basis of Healthy Living

Nutrition is the science that deals with diet and health. The old saying "we are what we eat" is true in the sense that we are continually replacing parts of our bodies and that the material to make these replacements comes from our food. The skin that covers us now is not the same skin that covered us seven years ago. The fat beneath our skin is not the same fat that was there just a year ago. Our oldest red blood cells are 120 days old. The entire linings of our digestive tracts are renewed every 3 days. Many chemical reactions are required to replace these tissues, and all of these reactions are supplied ultimately by what we eat.

Nutrition, then, is concerned with the chemical requirements of the body — the nutrients. The six classes of nutrients are carbohydrates, fats, proteins, vitamins, minerals, and water. The preparation, molecular structures, and fundamental properties of these nutrients were discussed in Chapters 10 (fats) and 11. In this chapter we focus on the health effects of too much or too little of these nutrients, why we need these nutrients (their physiological functions), general ways to assess nutritional status, and the recommended intake of nutrients.

Chapter 11 and part of Chapter 10 (fats) are basic to material in this chapter.

In our society, a discussion of needed food would be incomplete without mention of food additives. Although the vitamins, minerals, and sugars added to natural foods often have nutritional value, chemicals added to preserve food, to make food taste and look better, or to give food a certain consistency often have no nutritional value. It is difficult to find food that has not been doctored with palate-pleasing food additives. We shall discuss how these additives act chemically and why they are added to food.

The Setting of Modern Nutrition

Concern about nutrition differs in different parts of the world. In the United States we no longer have diseases such as beriberi, scurvy, and rickets, which are due to the lack of a particular nutrient, but we do have cancer, heart and circulatory disorders, and so on, some forms of which are perhaps caused by an excess of certain nutrients.

By the beginning of the 20th century, the existence of carbohydrates, fats, and proteins was well recognized, and the heat-producing values of the various foodstuffs had been determined. The role of iron and several other minerals in human nutrition had also been recognized. At the turn of the century, an increasing amount of food was obtained from stores rather than from fresh farm products. In the store-bought food there were often appreciable amounts of dirt and other filth; furthermore, false or misleading labels frequently described the food. As a result of these problems, a federal food and drug law was passed in 1906. This law was replaced in 1938 by a more comprehensive act, known as the Pure Food and Drug law. Although these laws cleaned up food and made labeling more reliable, they also caused food to be too clean chemically. In the process of food purification, nutrients such as vitamins and minerals were removed or destroyed. The replacement of needed nutrients began the food additive process.

As early as 1915, E. V. McCollum had discovered the fat-soluble group of closely related organic compounds known as vitamin A and a water-soluble collection of organic compounds known as the B vitamins. Thereafter other investigators discovered other vitamins, essential minerals, essential fats, and essential amino acids.

The usual method of testing the effect of a particular nutrient on the human body is to give an excess of the nutrient and observe its effect or to withhold the nutrient and observe the effect of its absence. However, there are large individual variations with respect to the need for many of the essential nutrients. There are also anatomical and physiological differences among animals — such as rats, dogs, cats, guinea pigs, and monkeys — used for nutritional studies. For example, human beings require vitamin C in their diets, whereas rats and dogs, which can produce vitamin C in their bodies, do not. Although valuable information can be gained from animal studies, final conclusions can be reached only about the effects of the ingested food on the species involved and, to some extent, only about the effects on the individual involved.

General Nutritional Needs

Basically we need the elements that compose the tissues, organs, and systems of the body. The major elements in the human body, on a weight percentage basis, are oxygen (65%), carbon (18%), hydrogen (10%), and nitrogen (3%). Practically all of these elements are in the form of water or organic compounds. Other elements present (and required) are calcium (2%), phosphorus (1%), potassium (0.35%), sulfur (0.25%), sodium (0.15%), chlorine (0.15%), and magnesium (0.05%). The total thus far equals over 99.9% of the total body weight. The rest of the body is composed of trace amounts of other elements. Table 13-1 lists all of the elements found in the human body, the total amount of each element in a 70-kg man, and average daily intake. Either a lack of proper nutrients or an excess of improper nutrients can produce *malnutrition*.

General nutritional needs are determined by one's nutritional status. There are many ways to assess the nutritional status of a human being: (1) size and weight — a person should be neither too thin nor too fat; (2) effect of stress — if a person is well nourished, stress is more bearable; (3) intelligence — undernourished and malnourished individuals are dull and unresponsive; (4) ability to reproduce — undernourished individuals are sometimes unable to reproduce; and (5) biochemical and clinical analysis — analysis of urine, blood, appearance, weight change, posture, and other chemical and medical tests and observations. Items 1 and 5 are used most often to determine nutritional status because they are most reliable (do not involve many variables) and are determined quickly.

How is nutritional status assessed?

Elements in the body are combined into many chemical compounds in a precise fashion to build the various tissues and organs of the body. Knowledge of the composition and chemistry of these organs and tissues is needed to understand the nutritional requirements of humans. For example, a person's nutritional status can be determined by examination of that person's liver, fat tissue, and muscle tissue. If one consumes an excess of alcohol and has a poor diet, the liver grows and the cells become fatty. Fat tissue in humans can vary from a low of 13% to a high of 70% of body weight; the amount of fat tissue depends on the amount and type of food consumed, the age of the person, and certain inherited traits. Muscle tissue ranges from 25% to 45% by body weight. A

Table 13-1
*Chemical Elements in the Adult Human Body**

Element	Total Amount in Body (mg)	Daily Intake (mg)	Element	Total Amount in Body (mg)	Daily Intake (mg)
Aluminum	61	34	Manganese	12	3.7
Antimony	ca. 8	ca. 50	Mercury	15	0.015
Arsenic	18	1	Molybdenum	9	0.3
Barium	22	ca. 0.8	Nickel	10	0.4
Beryllium	0.036	0.012	Niobium	ca. 120	0.6
Bismuth	ca. 0.2	0.02	Nitrogen	1,800,000	16,000
Boron	20	1.3	Oxygen	43,000,000	3,500,000
Bromine	200	7.5	Phosphorus	780,000	1400
Cadmium	50	0.15	Potassium	140,000	3300
Calcium	1,000,000	1000	Radium	3×10^{-8}	$(1-7) \times 10^{-15}$
Carbon	16,000,000	300,000	Rubidium	680	2.2
Cesium	1.5	0.01	Selenium	15	0.15
Chlorine	95,000	5000	Silver	0.8	0.07
Chromium	ca. 6	0.15	Sodium	100,000	ca. 5000
Cobalt	ca. 1.5	0.30	Strontium	32	1.9
Copper	72	3.5	Sulfur	140,000	850
Fluorine	2600	1.8	Tellurium	9	0.6
Gold	ca. 9	—	Tin	ca. 16	4
Hydrogen	7,000,000	3,500,000	Titanium	9	0.9
Iodine	13	0.20	Uranium	0.09	0.002
Iron	4200	15	Vanadium	ca. 10	2
Lead	120	0.44	Zinc	2300	13
Lithium	80	2	Zirconium	ca. 450	4.2
Magnesium	19,000	340			

* The approximate values given here refer to a 70-kg man. (Adapted from W.S. Synder, M.J. Cook, E.S. Nasset, L.R. Karhausen, G.P. Howells, and I.H. Tipton: *Report of the Task Group on Reference Man*, ICRP Pub. 23. Oxford, England, Pergamon Press, 1975.)

weightlifter may possess a large percentage of muscle tissue. As one grows older, the amount of muscle tissue generally decreases. Despite common belief, the amount of muscle tissue is not determined by the amount of protein eaten but rather by the amount of exercise.

Nutrient Requirements: RDA and USRDA

Recommended Dietary Allowance (RDA) per day

A specific list of recommendations for nutrient intake is the Recommended Dietary Allowance (RDA), published by the Food and Nutrition Board of the National Academy of Sciences and the National Research Council. Sample data are shown in Table 13-2. RDAs are given according to age, sex, energy requirements, pregnancy and lactation (milk production). The RDA is the intake level of a nutrient that should ensure adequate nutrition among as large a percentage of the population as possible, with account being taken of the effects of some stress and biochemical differences. The listing was first published in 1968 and was revised in 1974 and 1980. For most nutrients, each revision produced slightly lower values.

United States recommended dietary allowances (USRDA) per day

The set of recommendations used for labeling products is the United States Recommended Dietary Allowances (USRDA), published by the Food and Drug Administration

Table 13-2
A Selection of Recommended Dietary Allowances (RDA), 1980

Group	Age (Years)	Weight (Pounds)	Height (Inches)	Protein (g)	Vitamins				Calcium (mg)
					D (μg)	C (mg)	E (mg)	B_6 (mg)	
Infants	0.5–1	20	28	18	10	4	35	0.6	540
Children	4–6	44	44	30	10	6	45	1.3	800
Males	19–22	154	70	56	7.5	10	60	2.2	800
Females	19–22	120	64	44	7.5	8	60	2.0	800
Pregnant women				+30	+5	+2	+20	+0.6	+400
Lactating women				+20	+5	+3	+40	+0.5	+400

(FDA). The USRDA is based on the 1968 version of the RDA but lists slightly higher values than the RDA and does not give ranges of recommendations (Table 13–3). On packages of products, such as cereal boxes, the datum for each nutrient is usually the percentage of the USRDA recommendation contained in one serving of the product. International Units (IUs) are commonly listed for vitamins A, D, and E on packaged, enriched foods. A remnant from the past, IUs were employed before chemical analyses

Table 13-3
United States Recommended Dietary Allowances (USRDA) per Day for Adults and Children over 4 Years Old

Nutrient	Amount
Protein	45 or 65 g*
Vitamin A	5000 IU
Vitamin C (ascorbic acid)	60 mg
Thiamine (vitamin B_1)	1.5 mg
Riboflavin (vitamin B_2)	1.7 mg
Niacin	20 mg
Calcium	1.0 mg
Iron	18 mg
Vitamin D	400 IU
Vitamin E	30 IU
Vitamin B_6	2.0 mg
Folic acid (folacin)	0.4 mg
Vitamin B_{12}	6 μg
Phosphorus	1.0 g
Iodine	150 μg
Magnesium	400 mg
Zinc	15 mg
Copper	2 mg
Biotin	0.3 mg
Pantothenic acid	10 mg

* 45 g if protein quality is equal to or greater than milk protein,
65 g if protein quality is less than milk protein.

Nutrition: The Basis of Healthy Living

$1 \mu g$ (microgram) $= 10^{-6}$ g
1 mg (milligram) $= 10^{-3}$ g

of the specific vitamins were possible. In modern units, 1 IU of vitamin A is 0.344 μg of crystalline vitamin A acetate; therefore, 5000 IU of vitamin A is the same as 1.72 mg of vitamin A acetate. Another unit sometimes used for vitamin A is the *retinal equivalent* (RE); 1 RE equals 5 IU. For vitamin D, 1 IU is 0.025 μg of cholecalciferol; the USRDA of 400 IU is the same as 0.01 mg. For vitamin E, 1 IU is 1 mg; the USRDA of 30 IU is the same as 30 mg of vitamin E. The more important information on a cereal box is the percentage of the USRDA provided by one serving of the cereal.

Dietary guidelines for Americans

A generalized summary of dietary recommendations was proposed in 1980 by the Department of Health and Human Services and the United States Department of Agriculture:

1. Eat a variety of foods.
2. Maintain ideal body weight.
3. Avoid too much saturated fat and cholesterol.
4. Eat foods with adequate starch and fiber.
5. Avoid excess sugar.
6. Avoid excess sodium.
7. Drink alcoholic beverages in moderation.

Caloric Needs

Heat is needed by humans to maintain body temperature at about 37°C (98.6°F under the tongue) and to energize endothermic chemical reactions. The principal source of this heat is the oxidation of fats and carbohydrates. The oxidation of proteins and various other exothermic reactions provides the rest of the heat for the body.

Since fat contains less oxygen per gram than do carbohydrates, the exothermic reaction of a fat with oxygen to form carbon dioxide and water produces more heat per gram of fat. Some specific oxidations representative of the three major sources of energy from food are the oxidation of glucose (a sugar),

A food Calorie is a kilocalorie (kcal).

$$C_6H_{12}O_6 + 6\ O_2 \longrightarrow 6\ CO_2 + 6\ H_2O + 670 \text{ kcal (3.7 kcal/g glucose)}$$
Glucose   Oxygen   Carbon   Water
                          Dioxide   (Liquid)

the oxidation of a fatty acid (representing a fat),

$$C_{16}H_{32}O_2 + 23\ O_2 \longrightarrow 16\ CO_2 + 16\ H_2O + 2385 \text{ kcal (9.3 kcal/g palmitic acid)}$$
Palmitic Acid

and the oxidation of an amino acid (representing a protein),

Calorie values for specific foods are listed in Table 13-6.

$$2\ C_3H_7O_2N + 6\ O_2 \longrightarrow CO(NH_2)_2 + 5\ CO_2 + 5\ H_2O + 416 \text{ kcal (2.3 kcal/g alanine)}$$
Alanine   Urea

For the purposes of comparison, the oxidation of ethanol is shown below:

$$C_2H_5OH + 3\ O_2 \longrightarrow 2\ CO_2 + 3\ H_2O + 327 \text{ kcal (7.1 kcal/g ethanol)}$$
Ethanol

The values from Table 13-4 can be used to calculate the calorie value of a food, if the composition of the food is known. For example, if a steak is 49% water, 15% protein, 0% carbohydrate, 36% fat, and 0.7% minerals, 3.5 ounces (about 100 g) would produce about 384 kcal, or 384 food Calories.

Table 13-4
Calorie Data for Fats, Carbohydrates, and Proteins

Foodstuff	kcal/g	RDA	Actually Consumed in U.S. Daily Diet	kcal Produced by Daily Intake	Percentage of Daily Calorie Output
Fat	9	—	100–150 g	900–1350	30–50
Carbohydrate	4	—	300–400 g	1200–1600	35–45
Protein	4	46–56 g (10 oz. meat)	80–120 g	320–480	10–15

Nutrient	Weight	kcal/g	Total
Water	49 g ×	0 kcal/g =	0 kcal
Protein	15 g ×	4 kcal/g =	60
Carbohydrate	0 g ×	4 kcal/g =	0
Fat	36 g ×	9 kcal/g =	324
Minerals	0.7 g ×	0 kcal/g =	0
		Total	384 kcal

Calorie values of most foods are calculated by this method, and these are the values that are listed in diet books.

Physical activity is one way to consume the foods that would be stored as fat (use them to produce heat and energy). Some average calorie values for various activities are listed in Table 13-5.

Energy spent for normal maintenance activities of the body is the **basal metabolic rate (BMR)**. These maintenance activities include the beating of the heart, breathing, maintenance of life in each cell, maintenance of body temperature, and the sending of

Basal metabolic rate is energy required to do nothing willfully.

Table 13-5
Approximate Energy Expenditure by a 150-Pound Person in Various Activities

Activity	Energy (kcal/hour)	Activity	Energy (kcal/hour)
Bicycling, 5.5 mph	210	Roller skating	350
13 mph	660	Running, 10 mph	900
Bowling	270	Skiing, 10 mph	600
Domestic work	180	Square dancing	350
Driving an automobile	120	Squash and handball	600
Eating	150	Standing	140
Football, touch	530	Swimming, 0.25 mph	300
tackle	720	Tennis	420
Gardening	220	Volleyball	350
Golf, walking	250	Walking, 2.5 mph	210
Lawn mowing (power mower)	250	3.75 mph	300
Lying down or sleeping	80	Wood chopping or sawing	400

nerve impulses from the brain to direct these automatic activities. Energy for these activities must be supplied before energy can be taken for digesting food, running, walking, talking, or other activities. BMR, usually expressed as kcal per hour, is defined as the energy spent by a body at rest after a 12-hour fast. To get a rough estimate of your BMR (kcal/day), multiply your weight (in pounds) by 10.

The BMR can be affected by many factors. An increased BMR can come from anxiety, stress, lack of sleep, low food intake, congestive heart failure, fever, increased heart activity, and the ingestion of drugs, including caffeine, amphetamine, and epinephrine. A decreased BMR can result from malnutrition, menopause, inactive tissue due to obesity, and low-functioning adrenal glands.

SELF-TEST 13-A

1. The six classes of nutrients are _____, _____, _____, _____, _____, and _____.
2. The most abundant element in the human body is _____, and the most abundant mineral is _____.
3. Which set of recommendations for nutrient intake is printed on packaged products? _____
4. The following statement is ( ) completely true, ( ) partially true, or ( ) completely false: All humans are sufficiently alike for nutrition studies to be extrapolated from one to the other.
5. The nutrient that produces the most heat per gram is _____.
6. A food Calorie is the same value as _____ kilocalorie(s), and it will raise the temperature of 1000 g of water _____ degree(s) centigrade.
7. The amount of heat required to operate the body at rest is the _____, which in kcal is about ten times your _____.

Individual Nutrients: Why We Need Them in a Balanced Amount

Proteins

Properties and structures of proteins are given in Chapter 11.

Histidine is required for wound healing.

See Table 11-1, and relate the essential amino acids to the letters TV Till PM HA.

Of the some 22 amino acids identified in human protein, 10 are considered essential in that the human body cannot synthesize them and therefore must obtain them from ingested food. Infants require arginine because they cannot make it fast enough to have a supply for both protein synthesis and urea synthesis. The lack of an essential amino acid in one meal is not supplied by an excess of the amino acid in another meal, since excess amino acids are not stored very long except in functioning proteins. If proteins are eaten at only one meal per day, the liver must store a full day's supply from that one meal.

Functions in the Human Body. Humans must have proteins, which provide the structural tissue for muscles and most organs. Proteins are part (the apoenzyme) of the some 80,000 known enzymes. Some hormones, transport molecules (such as hemoglobin and transferrin), antibodies, and fibrinoginin (for blood clotting) contain proteins.

Daily Needs. Proteins are nearly the only source of nitrogen in the diet. An adult male has about 10 kg of protein, about 300 g of which is replaced daily. Part of the 300 g is recycled, and part comes from intake. Various studies indicate that, on the average, 25 g to 38 g of high-quality protein (as in meat, chicken eggs, and cow's milk) or 32 g to 42 g of lower quality proteins (as in corn and wheat) are required in the daily diets of healthy adult humans in order to maintain nitrogen equilibrium in the body. The average daily intake has remained near 100 g of protein per person since 1910, although there was a small drop in protein intake during the Depression of the 1930s. Methionine is the essential amino acid required in the greatest amount (2 g of the total of 7.1 g of all of the essential amino acids). Protein is lost in urine (as urea, a byproduct of protein metabolism), fecal material, sweat, hair and nail cuttings, and sloughed skin.

Food Sources. Table 13-6 lists some foods that are relatively high in protein content. Generally speaking, persons who are reasonably well fed and eat meat, fish, eggs, or dairy products every day have no worry about their protein intake.

Protein-Related Problems. If the diet does not contain the proper balance of the essential amino acids, protein synthesis is curtailed. *Kwashiorkor* (pronounced: kwash-ee-OR-core) is a protein-deficiency disease. To Ghanaians, who named the disease, *kwashiorkor* originally meant "the evil spirit that infects the first child when the second child is born." Traditionally, a first Ghanaian child would nurse until a second child was born, at which time the first child would be weaned from the mother's protein-rich milk to a starchy, protein-poor sustenance of gruel. The first child would then begin to sicken and would often die within a few years. If kwashiorkor set in around the age of 2, by the time the child was four, growth would be stunted, hair would have lost its color, skin would be patchy and scaly with sores, the belly, limbs, and face would be swollen by the collection of fluid in intercellular spaces (edema), and the child would sicken easily (because of a lowered supply of antibodies) and would be weak, fretful, and apathetic. If a child with kwashiorkor is given nutritional therapy before the disease has progressed to its last stages, the chances of recovery are good.

If proteins occupy too large a proportion of dietary intake (and carbohydrates too low a proportion), some of the excess amino acids are consumed for energy or are converted into glucose and then glycogen in the liver. If the condition continues, **uremia** can occur. Uremia is marked by nausea, vomiting, headache, vertigo, dimness of vision, coma or convulsions, and a urinous odor of the breath and perspiration. Protein metabolized completely forms ammonia. The liver converts the ammonia to urea, some of which is used to make "nonessential" amino acids. Excess urea is excreted in the urine. If insufficient carbohydrates are available to be oxidized for energy, too much urea from the oxidation of proteins is sent to the kidneys, which may become overworked, and uremia sets in.

Athletes often believe that their diets should be extremely high in protein. Athletes do, in fact, use a little more protein than nonathletes, but only a little more — perhaps 10%. Most young people's diets already contain about twice as much protein as they can possibly use to build muscle; the excess is used for energy — a purpose any other energy nutrient could serve just as well, and less expensively. Cells do not respond to what is given to them, but rather they select from what is offered when they need nutrients in order to perform. So the way to make muscle cells grow is to put a demand on them, that is, to make them work.

Hormone (Greek *hormaein*, "to set in motion, spur on"): a chemical substance, produced by the body, that has a specific effect on the activity of a certain organ.

The USRDA requirement for protein is 46 g for young female adults and 56 g for adult males.

Table 13-6
The Approximate Percentages of Carbohydrates, Fats, Proteins, and Water in Some Whole Foods as Normally Eaten*

Food	Water	Protein	Fat	Carbohydrates	kcal/100 g	Food	Water	Protein	Fat	Carbohydrates	kcal/100 g
Vegetables						*Meats and Fish (cont'd)*					
Spinach, raw	90.7	3.2	0.3	4.3	26	Cod, raw	81.2	17.6	0.3	0	78
Collard greens, cooked	89.6	3.6	0.7	5.1	33	Salmon, broiled	63.4	27.0	7.4	0	182
Lettuce, Boston, raw	91.1	2.4	0.3	4.6	25	Freshwater perch, raw	79.2	19.5	0.9	0	91
Cabbage, cooked	93.9	1.1	0.2	4.3	20	Oysters, raw	84.6	8.4	1.8	3.4	66
Potatoes, cooked	75.1	2.6	0.1	21.1	93	*Grains and Grain Products*					
Turnips, cooked	93.6	0.8	0.2	4.9	23	Wheat grain, hard	13.0	14.0	2.2	69.1	330
Carrots, raw	88.2	1.1	0.2	19.7	42	Brown rice, dry	12.0	7.5	1.9	77.4	360
Squash, raw summer	94.0	1.1	0.1	4.2	19	Brown rice, cooked	70.3	2.5	0.6	25.5	119
Tomatoes, raw	93.5	1.1	0.2	4.7	22	Whole-wheat bread	36.4	10.5	3.0	47.7	243
Corn kernels, cooked on cob	74.1	3.3	1.0	21.0	91	White bread	35.8	8.7	3.2	50.4	269
Snap beans, cooked	92.4	1.6	0.2	5.4	25	Whole-wheat flour	12.0	14.1	2.5	78.0	361
Green peas, cooked	81.5	5.4	0.4	12.1	71	White cake flour	12.0	7.5	0.8	79.4	364
Lima beans, cooked	70.1	7.6	0.5	21.1	111	*Dairy Products and Eggs*					
Red kidney beans, cooked	69.0	7.8	0.5	21.4	118	Milk, whole	87.4	3.5	3.5	4.9	65
Soybeans, cooked	73.8	9.8	5.1	10.1	118	Yogurt, whole-milk	89.0	3.4	1.7	5.2	50
Meats and Fish						Ice cream	62.1	4.0	12.5	20.6	207
Lean beef, broiled	61.6	31.7	5.3	0	183	Cottage cheese	79.0	17.0	0.3	2.7	86
Beef fat, raw	14.4	5.5	79.9	0	744	Cheddar cheese	37.0	25.0	32.2	2.1	398
Lean lamb chops, broiled	61.3	28.0	8.6	0	197	Eggs	73.7	12.9	11.5	0.9	163
Lean pork chops, broiled	69.3	17.8	10.5	0	171	*Fruits, Berries, and Nuts*					
Lard, rendered	0	0	100.0	0	902	Apples, raw	84.4	0.2	0.6	14.5	58
Calf's liver, cooked	51.4	29.5	13.2	4.0	261	Pears, raw	83.2	0.7	0.4	15.3	61
Beef heart, cooked	61.3	31.3	5.7	0.7	188	Oranges, raw	86.0	1.0	0.2	12.2	49
Brains	78.9	10.4	8.6	0.8	125	Cherries, sweet	80.4	1.3	0.3	17.4	70
Chicken, whole, broiled	71.0	23.8	3.8	0	136	Bananas, raw	75.7	1.1	0.2	22.2	85
						Blueberries, raw	83.2	0.7	0.5	15.3	62
						Red raspberries, raw	84.2	1.2	0.5	13.6	57
						Strawberries, raw	89.9	0.7	0.5	8.4	37
						Almonds	4.7	18.6	54.2	19.5	598
						Pecans	3.4	9.2	71.2	14.6	689
						Walnuts	3.5	14.8	64.0	15.8	651

* The caloric value per 100 g of food is listed for each food.

Fats

> Properties and structures of fats and fatty acids are given in Chapter 10.
>
> *Triacylglycerol* is used by some sources for the older term *triglyceride*.
>
> Structures of glycerol and triglycerides are in Chapter 10, and the structure of cholesterol is in Chapter 16.

A **lipid** is an organic substance that has a greasy feel and is insoluble in water but soluble in organic solvents. Lipids include neutral fats and oils, waxes, steroids, phospholipids, and similar compounds. When we refer to fats, we are usually referring to triglycerides, composed of one glycerol molecule esterified by three fatty acid molecules. Ninety-five percent of the lipids in the diet are triglycerides. The other 5% are phospholipids (lecithin is an example) and steroids (cholesterol is the major one in food).

The only truly essential fatty acid is **linoleic acid;** it cannot be synthesized in the body and therefore must be eaten in the diet. Arachidonic and linolenic fatty acids were thought to be essential until it was discovered that they can be synthesized in the body from linoleic acid.

Functions in the Human Body. Fats are essential structural parts of cell membranes. They provide the highest energy per gram of the nutrients and serve as energy storage reservoirs in the body. They insulate thermally, pad the body, and are packing material for various organs. Fatty, or adipose, tissue is composed mainly of specialized cells, each featuring a large globule of triglycerides.

Fatty acids are precursors of prostaglandins; the oxidation of arachidonic acid yields several possible prostaglandins. The prostaglandins function as bioregulators to influence the action of certain hormones and nerve transmitters. They inhibit high blood pressure, ulcer formation, and inflammation.

Structures of prostaglandins are in Chapter 10.

Daily Needs. The daily consumption of fat in the United States has risen continuously from about 125 g per person in 1910 to about 155 g per person today. The fat in today's diet is about 40% saturated, 40% monounsaturated, and 20% polyunsaturated. There are no RDA or USRDA recommendations for fats. Dietary Guidelines for Americans recommend that we avoid too much fat, saturated fat, and cholesterol:

Dietary Guidelines for Americans were stated earlier in this chapter.

> Choose low-fat protein sources such as lean meats, fish, poultry, dry peas and beans; use eggs and organ meats in moderation; limit intake of fats on and in foods; trim fats from meats; broil, bake, or boil — don't fry; read labels for fat contents.

Food Sources. Table 13–6 lists the fat content of some foods. The fat in the edible portions of the food, as normally eaten, is difficult to ascertain, since people trim varying amounts of fat off their food before eating it. Animals fats (oils) are high in saturated and monounsaturated fatty acids. Vegetable fats (oils) have a high percentage of polyunsaturated fatty acids. Nearly all diets supply enough linoleic acid to meet the needs of the human body. Pork (lard) and chicken fat contain mostly linoleic acid. Even in a totally fat-free diet, 1 teaspoon (5 g) of corn oil supplies the daily need of linoleic acid. Two of the highest sources of cholesterol are brains (2.5 g/100 g) and egg yolk (1.15 g/100 g).

Problems Associated with Eating Fats. Too much fat in the diet can lead to obesity. After digestion of a fat, if the components glycerol and fatty acids are not used otherwise, they are resynthesized in the liver into fats and stored as such.

According to standards set by insurance companies, anyone over the "ideal" weight for his or her age group is considered obese. According to these standards, 10% to 25% of teenagers and 25% to 50% of adults in the United States are overweight. One test for obesity is the skinfold test, in which one measures the thickness of a big pinch of skin (skinfold) on the back of the upper arm, the back, or the waist; in this test fatness is defined as a skinfold thicker than 1 inch. The problem with both of these criteria and other such tests is that each individual has his or her own "set point," or ideal weight for optimum health, which may appear fat (or skinny) to others. The set-point weight depends on heredity, bone density, muscle conformations, occupation, number of fat cells, and other such factors and may well differ from other general recommendations.

The number of fat cells is fixed by adulthood. The more fat cells one has, the harder it is to lose weight.

When dieting, there is one immutable law: for each pound lost there must be an expenditure of 3500 kcal. There is no magic escape from this principle. The kilocalories can be used by the body for the BMR or for additional activities, but activity is required to use the energy. A person or a machine moving your muscles does not expend nearly as much of your energy as you do moving yourself; you must move your muscles to expend the energy. Before participating in a diet plan, it may be wise to consult an authority who can examine the plan to verify that it is based on sound, scientific studies. One such authority is the Committee on Nutritional Misinformation of the Food and Nutrition Board, National Academy of Sciences, National Research Council.

If there is too much fat in the diet and too little carbohydrate, **ketosis** can develop. Ketosis is the combination of high blood ketone levels **(ketonemia)** and ketones in the urine **(ketonuria)**. Ketones are formed when fats are broken down to form glucose when no glucose is readily available to the body. Glycerol derived from fat destruction forms pyruvate, then glucose. The fatty acids form ketones (the simplest being acetone [CH_3COCH_3]) and keto-acids (such as acetoacetic acid [CH_3COCH_2COOH] and the substance in largest amount, 3-hydroxybutanoic acid, [$CH_3CH(OH)CH_2COOH$]). One noticeable characteristic of ketosis is "acetone" breath. Although some cells in the body can use ketones for fuel, other cells must have glucose. There is a small amount of ketones in the blood normally. Excess ketones and keto-acids lead to **ketoacidosis,** a potentially fatal condition.

> Problems related to cholesterol, triglycerides, atherosclerotic plaque, and heart attack are discussed in Chapters 10 and 16.

If too little fats are eaten, especially the essential fatty acid, one can develop coarsened, sparse hair and eczema (a skin disease characterized by lesions, watery discharge, and crusts and scales).

Ingestion of hydrogenated polyunsaturated fats (oils) causes problems. Hydrogenation is used to convert oils into solid fats to make margarine, cooking fats, and similar products. However, hydrogenation of vegetable oils (liquid fats) decreases some of the double bonds, forms unnatural *trans*-fatty acids from natural *cis*-fatty acids, and moves double bonds around to form conjugated structures. The *trans*-fatty acids are not metabolized in the human system, but they can be stored for the life of the individual. The conjugated structures are too active chemically for the human system. The problem is solved by ingestion of fewer processed vegetable fats.

> Conjugated structures have alternating double and single bonds in the carbon chain or ring.

When saturated or unsaturated fats are used in cooking, they should not be heated to temperatures at which they smoke. Under these conditions fats produce toxic peroxides, and unsaturated fats can polymerize.

All fats and oily foods should be smelled for rancidity when the package or bottle is opened and should be returned to the store for credit if there is any evidence of rancidity.

Carbohydrates

Carbohydrates in foods include digestible simple sugars (glucose, fructose, galactose), disaccharides (sucrose, maltose, lactose), and polysaccharides (amylose, amylopectin, glycogen). Indigestible carbohydrates consumed include cellulose, insulin, hemicellulose, lignin, plant gums, sulfated polysaccharides, carrageenan, and cutin.

> Properties and structures of carbohydrates are given in Chapter 11.

Functions in the Human Body. The only beneficial function of digestible carbohydrates is to provide energy at the rate of approximately 4 kcal per gram of glucose oxidized. Excess digestible carbohydrates are stored first as glycogen, principally in the liver; further excesses are converted into fats and stored as such. The indigestible carbohydrates serve as roughage in the diet, along with bran and fruit pulp.

> Table 13–6 lists the carbohydrate contents of some foods.

Daily Needs. A daily caloric intake of 2000 kcal would require the ingestion of about 500 g of glucose (or its equivalent). Fats and proteins are also oxidized for energy, so less digestible carbohydrate is required. Daily consumption of digestible carbohydrate has declined from 500 g per person in 1910 to 380 g per person in the 1980s. The decline in total amount of carbohydrates is really a rise in the amounts of refined sugars: 150 g in 1910 to 200 g in the 1980s. There is no RDA or USRDA for carbohydrates. The 1980 Dietary Guidelines for Americans recommend that we decrease our ingestion of concentrated sweets (candy, soft drinks, cookies, etc.) and substitute with starches (complex carbohydrates), fresh fruits, and fiber.

Problems Associated with Carbohydrates. Problems can be encountered when the glucose level is too low in the blood and when there is too little roughage in the diet; the general medical term for too low a concentration of glucose in the blood is **hypoglycemia**. Persons with hypoglycemia need a regular intake of sugar to avoid lows, characterized by an inability to think clearly, emotional disturbances, and a feeling of general indisposition.

Diseases associated with lack of dietary fiber (roughage) are appendicitis, diverticular disease (herniation of the mucous membrane lining of a tubular organ), and benign or malignant tumors of the colon and rectum. In the gastrointestinal tract fiber absorbs water, swells, facilitates regular bowel movements, and prevents the stagnation of foods, particularly refined foods, in the intestines. Fiber also absorbs cholesterol and inhibits the use of glucose in the body. Bran is a good source of dietary fiber. Fruit pulp is another good source if the whole fruit is eaten. Some bakers incorporate wood or cotton cellulose into their high-fiber foods.

Clinical problems associated with a diet rich in carbohydrates, particularly refined sugar, are obesity, dyspepsia, atherosclerosis, and diabetes mellitus.

Diabetes mellitus is characterized by elevated blood glucose levels, multiple hormonal and metabolic disturbances in the secretion of insulin and growth hormone, thirst, hunger, weakness, low resistance to infection, slowness to heal, and, in later stages, blindness and coma. About 15% of people in high age groups have diabetes, but the disease is generally rare in parts of the world where the people eat no refined or processed food. Although several types of diabetes are due to a decrease in glucose metabolism, diabetes mellitus is caused by too little insulin due to defective *islets of Langerhans* in the pancreas. In this condition the pancreas does not produce sufficient insulin because of hereditary capacity, a disease or injury that destroyed part or all of the pancreas, interruption of the mechanism of insulin production by lack of proper reactants or the presence of toxicants, or too much demand for insulin invoked by too high glucose intake. If the pancreas produces even a small amount of insulin, diabetes can be held at bay by a diet low or absent in sugar. If the pancreas produces no insulin, daily injections of insulin are required.

The yearly sugar consumption in the United States in 1750 was 2 pounds per person; today it is 110 to 135 pounds per person. Sixty percent of the sugar comes from sugar cane; the other 40% comes from sugar beets. The sugar content of some commercially processed foods is given in Table 13–7.

Dyspepsia is the name applied to chronic indigestion caused by the consumption of large amounts of sugar.

Atherosclerosis is discussed in Chapters 10 and 16.

Insulin escorts glucose to the fatty cell membrane.

Insulin, a protein, is hydrolyzed (digested) in the gastrointestinal tract if taken orally.

Table 13–7
Refined Sugar Added to Some Commercially Processed Foods*

Food	Sugar (%)
Cherry Jello	82.6
Coffeemate	65.4
Shake'N Bake, Barbecue Style	50.9
Wishbone Russian Dressing	30.2
Heinz Ketchup	28.9
Sealtest Chocolate Ice Cream	21.4
Libby's Peaches (in Heavy Syrup)	17.9
Skippy Peanut Butter	9.2
Coca Cola	8.8

* According to *Consumer Reports* (1978); percents by weight.

Some problems with refined sugar are due to the removal of required nutrients during the refining process and the dumping of too much refined sugar into the bloodstream too quickly. The production of white sugar (almost pure sucrose) removes all other nutrients, such as B vitamins, manganese, and chromium, which generally coexist in natural foods with sucrose in the appropriate amounts for proper metabolism in the human body. Therefore, a large refined sugar intake means that the B vitamins and certain minerals must be obtained from another food source. Brown sugar supplies more minerals than white sugar because brown sugar is darkened with molasses, the residue from sugar cane that is rich in essential minerals.

Unrefined and unprocessed sugar is often contained in cellular structures, which are not easily digestible. Sugar such as from sugar cane or from apples goes to the bloodstream much more slowly than refined sugar. The slower transfer allows the body to metabolize the sugar for energy more efficiently and avoids both the buildup of glucose in the bloodstream and the consequent storage of excess glucose as fat. If we obtain our sugar from an unrefined, unprocessed source, we are likely to eat less sugar; for example, it takes four apples to supply the same amount of sugar present in one 12-ounce cola drink.

> Unrefined carbohydrates contain factors that destroy the bacteria that consume carbohydrates and produce acid and storage carbohydrates (tooth plaque).

Milled white flour constitutes about two thirds of all of the grains consumed by humans in the United States. The milling of flour removes some lysine and fat and reduces the fiber content to 10% of that found in wheat grains. Vitamins are reduced to between 10% and 50% of the original content, mostly by removal of wheat germ. Enriched flour has some of the removed vitamins and minerals added back. Some millers add white paper pulp to replace the roughage provided by the removed bran. Some people supplement white flour with wheat germ, though this practice is inadvisable, since wheat germ separated from wheat degrades nutritionally and becomes rancid easily.

Refined flour, like refined sugar, is digested more quickly and more completely than unrefined (whole-wheat) flour. Therefore, digested, refined flour and refined sugar have less bulk. This causes the stomach to empty more slowly, and the digested food stays in the intestines longer. As a result, wastes in the intestines have more time to effect any toxicity they might have. Studies have shown that it takes 40 to 140 hours for refined carbohydrates to pass through the human body and only 15 to 45 hours for a traditional diet (not supplemented by Western foods) to pass through the body.

The process of bleaching to make white flour also destroys vitamins. Commonly used bleaching agents are chlorine dioxide and benzoyl peroxide. This problem can be averted by the use of unbleached flour.

SELF-TEST 13-B

1. Proteins compose the _____ part of enzymes.
2. Kwashiorkor is a _____ deficient disease.
3. Muscles are built primarily by eating excess proteins.   True ( ) or False ( )
4. Uremia is caused by excess protein intake and the excretion of _____.
5. A common phospholipid in food is _____, and a common steroid in food is _____.
6. Most of the lipids in the diet are _____.
7. The one essential fatty acid is _____.

8. Natural sources rich in linoleic acid are _____ and _____.
9. Ketosis is caused by too much _____ and too little _____ in the diet.
10. Eczema is caused by too little _____ in the diet.
11. Heating fats until they smoke can produce _____.
12. All ingested carbohydrates are digestible.   True ( ) or False ( )
13. Refined _____ and refined _____ are absorbed more quickly than unrefined carbohydrates.

Minerals

As nutrients, minerals are substances that are needed for good health and that contain elements other than C, H, O, and N. On vitamin and mineral supplement labels and elsewhere, nutrient elements are called minerals, the two terms being used interchangeably in nutrition. Most of the elements needed for nutrition are obtained from soluble inorganic salts either in foods or in food supplements. Magnesium is an exception in that it is obtained primarily from organic chlorophyll.

Carbon, hydrogen, oxygen, and nitrogen are supplied by organic fats, carbohydrates, and proteins.

The required inorganic nutrients can be grouped into two classes. Calcium, phosphorus, and magnesium are required in amounts of 1 g or more per day. Trace elements such as chromium, chlorine, cobalt, copper, fluorine, iodine, iron, manganese, molybdenum, nickel, selenium, sulfur, vanadium, and zinc are needed in milligram or microgram quantities each day.

That the human body needs minerals is borne out both by the functions of the minerals and by the effects of mineral deficiencies. Table 13–8 lists the functions and deficiency effects of some minerals, as well as a few food sources for each.

The nutrient minerals have varied functions, including as components of enzymes, as structural components (calcium and phosphorus in bones and teeth), in electrolyte balance in body fluids, and as transport vehicles (iron in hemoglobin transports oxygen; iron and cobalt transport electrons in electron transport cycles). Not only does the human body need minerals for its functions, but the minerals must be maintained in balanced amounts, with no deficiencies and no excesses. Many of the body's minerals are excreted daily in the feces, urine, and sweat and must therefore be replenished. For most of the elements, the amount excreted each day is very nearly the amount ingested.

One way to ensure ingestion of an ample supply of each mineral nutrient, particularly the trace nutrients, is to eat a variety of whole foodstuffs grown in different places. Mineral supplements are also available.

Calcium slows down the heartbeat by increasing electrical resistance across nerve membranes. The movement of potassium and sodium ions across the membrane is constrained, and the nerve impulse rate is thus decreased. Calcium is metabolized in the body by a hormone synthesized from calciferol (vitamin D). The calciferol also brings about synthesis of a substance called calcium-binding protein (CBP), which carries calcium through the small intestine wall. Fat slows down the transfer, and lactose speeds up calcium absorption.

The role of calcium in regulating the heartbeat is discussed in Chapter 16.

We could not support ourselves nor eat without calcium. The hard part of bone is a calcium compound, and the structure of teeth is mostly hydroxyapatite, a calcium compound.

Excess calcium may lead to the formation of kidney stones, but the body has a protein, **calmodulin,** that collects excess calcium and then binds to a number of

The structure of apatite is given and the role of fluoride in tooth decay is discussed in Chapter 17.

Table 13-8
Need for and Sources of Some Nutrient Minerals

Element (Amount Ingested per Day)	USRDA*	Function	Deficiency Effects	Food Sources
Calcium (Ca) (1.0 g)	1 g	In bone apatite, collagen	Bone dissolution	Milk products
Chromium (Cr) (0.15 mg)	—	In collagen, glucose tolerance factor	Increase in cholesterol	Honey
Cobalt (Co) (0.30 mg)	—	In vitamin B_{12}	Wasting disease, pernicious anemia	Meat, eggs
Copper (Cu) (3.5 mg)	2 mg	Coenzymes (4)	Anemia, infertility	Seafood, spinach, molasses (copper water pipes)
Fluorine (F) (1.8 mg)	—	Fluorapatite	—	Toothpaste, drinking water
Iodine (I) (0.20 mg)	0.15 mg	In thyroxine hormone	Goiter, cretin children	Kelp, seafood, iodized salt
Iron (Fe) (15 mg)	18 mg	In hemoglobin, myoglobin	Anemia	Meat, eggs, raisins
Magnesium (Mg) (340 mg)	400 mg	In bone, dentine, coenzyme	Circulatory and mental problems, red nose of alcoholics	Green vegetables
Manganese (Mn) (3.7 mg)	—	In melanin (skin pigment) coenzymes	Growth, skeletal, reproduction abnormalities	Spinach, beans, grains
Molybdenum (Mo) (0.3 mg)	—	Coenzymes (2)	Cancer of esophagus, sexual impotency	Legumes, liver
Phosphorus (P) (1.4 g)	1 g	In bone, teeth, ATP	Weak bones, lack of energy	Beans, grain, meat, milk
Selenium (Se) (0.15 mg)	—	Coenzymes, growth stimulator	Degeneration of skeletal muscles	Wheat
Sulfur (S) (0.85 g)	—	In amino acids	Unhealthy hair, muscles	Eggs, meat, mustard
Zinc (Zn) (13 mg)	15 mg	In insulin, coenzymes; heals wounds	Dwarfism, stretch marks, painful joints, finickiness in appetite	Oysters, meat, nuts

* USRDA values for adults and children over 4 years old.

enzymes to mediate their activity. By the use of calmodulin, the body monitors the amount of calcium in the bloodstream. A possible benefit of excess calcium is that it will make a person taller.

A deficiency in calcium can occur in postmenopausal women, who produce less estrogen than premenopausal women. The estrogen suppresses bone dissolution. Further bone dissolution can be suppressed by long-term medication with estrogen, but this has produced some toxic side effects in some women and therefore seems unwise. Taking calcium supplements may help.

The principal function of **iodine** in the human body is the proper operation of the thyroid glands located at the base of the neck. Two of the thyroid hormones are T_3 and T_4:

3,5,3'-Triiodothyronine (T_3)

Thyroxine (T_4)

These hormones and other similar ones, collectively known as **thyroxine,** go into every cell and regulate the rate at which the cell uses oxygen. Thyroxine thus regulates the BMR and the Krebs cycle. Iodine is absolutely necessary to the production of thyroxine. If there is a deficiency of iodine, the thyroid glands sometimes swell to as large as a person's head; this swelling is called a goiter. In 1960 it was estimated that 7% of the world's population (200 million) had goiters. Treatment with iodized salt (0.1% KI) with the hormone thyroxine decreases the size of or even eliminates small goiters. Larger goiters may require surgery. Since an excess amount of iodine also causes goiters, balance is the key to health.

Anemia can be caused by a deficiency of **iron,** but it can also be due to heredity, an improper level of vitamin B_6, lack of folic acid (vitamin B_9), and lack of vitamin B_{12} (pernicious anemia). Iron-deficient anemia is not necessarily fatal; a person with only 20% of the normal amount of hemoglobin still has the energy and strength to walk.

> The lack of either vitamin B_9 or vitamin B_{12} can cause pernicious anemia.

In the case of some nutrient elements, good health depends on the element's being present in the proper amount and in the *proper ratio* to one or more other elements. An example of an important ratio is the potassium/sodium ratio (K/Na ratio).

Typical values of the **K/Na ratio** are greater than 1. Some K/Na ratios for specific tissues follow: muscle, 4; liver, 2.5; heart, 1.8; brain, 1.7; and kidney, 1.0. For individual cells, potassium ions (K^+) concentrate inside the cell while sodium ions (Na^+) concentrate outside in the fluid that bathes the cell. Natural, unprocessed food has high K/Na weight ratios. Fresh, leafy vegetables average a K/Na ratio of 35. Fresh, nonleafy vegetables and fruits average a ratio of 360, with extreme values of 3 for beets and 840 for bananas. K/Na ratios in meats range from 2 to 12. Thus, when such foods are eaten, the body has K/Na ratios greater than 1. However, problems occur with processed and cooked foods. Potassium and sodium compounds are quite soluble in water. During processing (and cooking, if foods are boiled), both potassium and sodium compounds are dissolved by water and discarded. The sodium is replenished by "salting" of the food (addition of sodium chloride). Potassium is usually not added to the food. One solution to the problem is to eat unprocessed, natural food, which "naturally" has the proper K/Na ratios. Another solution is to "salt" food with a commercial product that contains both potassium and sodium, such as Morton's Lite Salt. In summary, do not add much NaCl, if any, to food, and eat fresh vegetables and fruits high in potassium.

Normal daily urinary excretion of **sodium** is in the range of 1.4 g to 7.8 g for adults. If excess sodium is not eliminated, water is retained, which may lead to edema (swollen legs and ankles). Various clinical studies have shown that increased levels of sodium raise the blood pressure of some individuals but have no effect on the blood pressure of others. The high-salt diets of 70 g NaCl per day in certain areas of Japan have traditionally produced an unusually high frequency of heart attacks. Sodium levels in the bloodstream are regulated by **aldosterone,** which is secreted from the adrenal gland. Aldosterone works in the kidney to reabsorb sodium from the urine. The secretion of aldosterone is controlled by receptors that measure salt concentration in the blood. If the blood sodium concentration is too high, less aldosterone is excreted and less sodium is reabsorbed from the urine.

Sodium is also excreted in sweat as sodium chloride. Salt concentration in sweat depends on dietary sodium intake, environmental temperature, the amount of sweating, and the degree of acclimation to the environment. Abrupt overheating is a problem for the body, involving an increase in skin and rectal temperatures, rapid beating of the heart, and a greatly increased sweat rate. After about a week of working at high temperatures, the body adapts by lowering the pulse rate and body temperature to normal levels. A high rate of sweating continues, but the concentration of NaCl in the sweat is decreased. As a person becomes acclimated to the heat, he or she needs more water but no more salt than at normal temperatures. Salt tablets may be helpful during the acclimation period but are not advisable after the body becomes acclimated. It should be noted that some persons never adapt to heat and should avoid overheating.

> Salary derives from the Latin *sal*, for "salt." Roman soldiers were given an allowance for salt.

Vitamins

A vitamin is an organic constituent of food that is consumed in relatively small amounts (less than 0.1 g/kg of body weight per day) and is essential to the maintenance of life. Vitamins are not synthesized in the cells of human beings; they are synthesized by plants, our principal natural source of them. Of the some million organic compounds eaten in a normal diet, only about 100 are of proper size and stability to be absorbed from the digestive tract into the bloodstream without digestion or breakdown. Vitamins are included in this group of 100 or more compounds.

> USRDA amounts of vitamins are given in Table 13-3 and Table 13-9.

The structures of vitamins divide them into two classes: oil-soluble and water-soluble. The oil-soluble vitamins — A, D, E, F, and K — tend to be stored in the fatty tissues of the body (especially the liver). The structures of oil-soluble vitamins have nonpolar hydrocarbon chains and rings that are compatible with nonpolar oil and fat. For good health and nutrition, it is important to store enough oil-soluble vitamins, but not too much.

> Vitamin D is the most toxic of all of the vitamins; avoid excesses.

The water-soluble vitamins tend to pass through the body and are not stored readily. Water-soluble vitamins are the B group (called vitamin B complex) and C. The structures of these vitamins have polar hydroxy (—OH) and carboxyl (—COOH) groups, which are attracted to polar water. Fewer problems are caused by excessive intake of water-soluble than of oil-soluble vitamins.

Retinol
Oil-Soluble Vitamin A

Ascorbic Acid
Water-Soluble Vitamin C

Table 13–9 lists the vitamins, their USRDAs, some food sources, and their deficiency effects. Since vitamins are synthesized by plants, a good natural source is plants that are not overcooked.

Contrary to popular belief, carrots provide the **provitamin** β-carotene, not retinol (vitamin A). The body converts β-carotene into retinol during the transfer of the provitamin through the intestinal wall. Night blindness is prevented by regeneration of rhodopsin (visual purple) from retinol. **Vitamin A** aids in the prevention of infection by barring bacteria from entering and passing through cell membranes. The vitamin performs its sentinel duty by producing and maintaining mucus-secreting cells. Bacteria stick to the mucus and are thus trapped.

The function of **vitamin E** as an antioxidant has been well established. Vitamin E is particularly effective in preventing the oxidation of polyunsaturated fatty acids, which readily form peroxides. Perhaps this is why vitamin E is always found distributed among fats in nature. The fatty acid peroxides are particularly damaging because they can lead to runaway oxidation in the cells. Vitamin E protects the integrity of cell membranes,

Eat polar bear liver sparingly. Thirty grams contain 450,000 IU of retinol; continued ingestion causes peeling of the skin from head to foot.

Vitamin E is the only vitamin destroyed by the freezing of food.

Table 13–9
Vitamin Summary Chart

Name	USRDA*	Deficiency Effect	Sources
Water-Soluble			
Thiamine (B_1)	1.5 mg	Beriberi	Seeds, pork, whole-wheat bread
Riboflavin (B_2)	1.7 mg	Cheilosis (shark skin)	Organ meats, yeast, wheat germ
Niacin (B_3) (nicotinic acid)	2 mg	Pellagra	Meat, yeast, legumes
Pantothenic acid (B_5)	10 mg	Neuromotor disturbance	Yeast, liver, eggs
Vitamin B_6 (pyridoxine)	2 mg	Skin lesions, anemia	Liver, nuts, wheat germ
Biotin (B_7)	0.3 mg	Dermatitis	Liver, yeast, grains
Folic acid (B_9) (folacin)	0.4 mg	Anemia, gastrointestinal changes	Green leafy vegetables, liver
Vitamin B_{12} (cobalamin)	6 μg	Pernicious anemia	Intestinal bacteria, organ meats
Ascorbic acid (vitamin C)	60 mg	Scurvy	Fruits, vegetables
Oil-Soluble			
Vitamin A (retinol)	5000 IU	"Night blindness," (xerophthalmia)	Liver, fruits, vegetables
Vitamin D (calciferol)	400 IU	Rickets	Fish liver oil
Vitamin E (tocopherol)	30 IU	Lack of hemoglobin in blood cells (hemolysis)	Plant oils
Vitamin F (linoleic acid)	—	Lesions, scales (eczema)	Pork lard, fatty foods
Vitamin K (phylloquinone)	†	Blood loss	Green leafy vegetables

* USRDA values are for adults and children over 4 years old.
† There is no USRDA value for vitamin K, but an estimated need is 0.1 mg per day.

which contain considerable fat. In addition, vitamin E helps maintain the integrity of the circulatory and central nervous systems, and it is involved in the proper functioning of the kidneys, lungs, liver, and genital structures. Vitamin E also detoxifies poisonous materials absorbed into the body.

According to some theories that view aging as the cumulative effects of the action of free radicals running wild, vitamin E, with its antioxidant properties, is considered a good candidate as an agent to inhibit aging or at least to help avoid premature aging.

The B group of vitamins (the **B-complex vitamins**) work together, primarily as coenzymes in biochemical reactions leading to growth and to energy production. Their place of action is in the mitochondria of the cells. Being water-soluble, the B vitamins are easily eliminated during the processing and cooking of food. The effectiveness of vitamins B_3 and B_6 is diminished in the presence of light, especially if the food is hot.

Pyridoxine (B_6), considered the "master vitamin," is involved in 60 known enzymatic reactions, mostly in the metabolism and synthesis of proteins.

Vitamin C is involved in the destruction of invading bacteria, in the synthesis and activity of interferon, which prevents the entry of viruses into cells, and in decreasing the ill effects of toxic substances, including drugs and pollutants. The question of whether vitamin C will decrease the incidence of the common cold has been studied for many years. Results of the studies show an average decrease of about 30% in illness (particularly upper respiratory infection) as a result of ingestion of vitamin C supplements. Not as well publicized or studied, vitamin A in large doses also decreases colds and the effects of colds. In avoiding or breaking colds, some persons respond better to vitamin A than to vitamin C, others respond better to vitamin C than to vitamin A, and still others respond to neither. In any case, for either vitamin to be effective, it must be taken preferably before but no later than at the early onset of a cold. It is recommended that the vitamins not be taken in combination, since this seems to prolong the cold symptoms.

Is vitamin E the fountain of youth?

A free radical has one or more unpaired valence electrons. An example is the methyl group, $CH_3 \cdot$.

English sailors are called Limeys because the British admiralty ordered a daily ration of lime juice (vitamin C) to prevent scurvy.

SELF-TEST 13-C

1. Our mineral needs can be obtained by eating a _____ of _____ foodstuffs grown in _____.
2. Anemia can be caused by a deficiency of the mineral _____, which is used to make _____.
3. What mineral is required in melanin? _____ Melanin is involved in _____.
4. A deficiency of the mineral _____ causes stretch marks in the skin.
5. Finicky eaters may have a deficiency of the mineral _____.
6. Minerals involved in transmitting a nerve impulse are _____ and _____.
7. For proper balance, the K/Na weight ratio in the body should be slightly ( ) greater than or ( ) less than 1.
8. The mineral obtained from chlorophyll in green vegetables is _____.
9. The mineral with the largest weight in the body is _____, which is used mostly in the _____ and _____.

10. Calmodulin, a protein, collects excess calcium from the bloodstream and helps prevent _____ stones.
11. Goiter is caused by a deficiency of _____.
12. Vitamins are synthesized by cells of the body.   True ( ) or False ( )
13. Vitamins A, D, E, F, and K are _____-soluble, whereas vitamins B-complex and C are _____-soluble.
14. The relationship between β-carotene and vitamin A is that β-carotene is a _____ of vitamin A.
15. A good vegetable source of vitamin A is _____; a good animal source of the vitamin is _____.
16. Polar bear liver is exceptionally rich in vitamin _____.
17. Vitamin E is effective as a(n) _____, particularly in preventing the deterioration of polyunsaturated fatty acids.
18. What vitamin is destroyed in the freezing of foods? _____
19. Rickets is caused by a deficiency of vitamin _____.
20. Because it is involved in so many biochemical reactions, vitamin _____ is considered the master vitamin.
21. The most toxic vitamin is _____.
22. The nickname "Limeys" for English sailors came from the sailors' having been given limes to prevent _____.
23. B-complex vitamins function generally as _____ involved generally in the process of _____.
24. A deficiency of thiamin, vitamin _____, causes the disease _____.
25. Pellagra is caused by a deficiency of the vitamin called _____.

Food Additives

Many chemicals with little or no nutritive value are added to food for a variety of reasons (Fig. 13–1). The chemicals are added during the processing and preparation of food for the purpose of preserving the food from oxidation, microbes, and the effects of metals. Food additives add and enhance flavor. They color the food, control pH, prevent caking, stabilize, thicken, emulsify, sweeten, leaven, and tenderize among other effects.

The GRAS List

The Food and Drug Administration lists about 600 chemical substances **"generally recognized as safe" (GRAS)** for their intended use. A small portion of this list is given in Table 13–10. It must be emphasized that an additive on the GRAS list is safe *only if it is used in the amounts and in the foods specified*. The GRAS list was published in several installments in 1959 and 1960. It was compiled from the results of a questionnaire asking experts in nutrition, toxicology, and related fields to give their opinions about the safety of various materials used in foods. Since its publication, few substances have been added to the GRAS list, and some, such as the cyclamates, carbon black, safrole, and Red Dye No. 2, have been removed.

The GRAS list is a noble effort—but it is not foolproof.

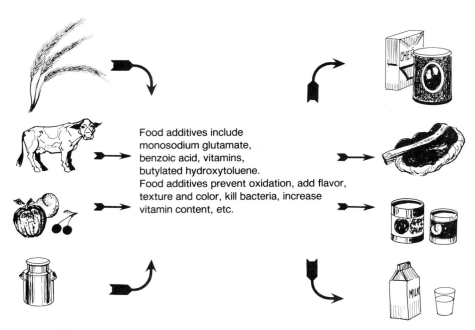

Figure 13-1
Between the harvested and the consumer-ready food, one often finds the addition of a large variety of food additives.

There are more than 2500 known food additives, and many more chemicals than those that appear on the GRAS list are approved (or at least, not banned) for use as food additives by the FDA. It is quite expensive to introduce a new food additive with the approval of the FDA. Allied Chemical Corporation began research in 1964 on a new synthetic food color, Allura Red AC, which was approved by the FDA and went on the market in 1972. The cost of introducing this product was $500,000, and about half of this amount was spent on safety testing.

Preservation of Foods

Foods generally lose their usefulness and appeal a short time after harvest. Bacterial decomposition and oxidation are the prime reasons steps must be taken to lengthen the time that a foodstuff remains edible. Any process that prevents the growth of microorganisms or retards oxidation is generally an effective preservation process. Perhaps the oldest technique is the drying of grains, fruits, fish, and meat. Water is necessary for the growth and metabolism of microorganisms, and it is also important in oxidation. Dryness thus thwarts both the oxidation of food and the microorganisms that feed on it.

Chemicals may also be added as preservatives. Salted meat, and fruit preserved in a concentrated sugar solution, are protected from microorganisms. The abundance of sodium chloride or sucrose in the immediate environment of the microorganisms forms a **hypertonic** condition in which water flows by **osmosis** from the microorganism to its environment. Salt and sucrose have the same effect on microorganisms as does dryness; both dehydrate them.

The canning process for preserving food, developed around 1810, involves first heating the food to kill all bacteria and then sealing it in bottles or cans to prevent access of other microorganisms and oxygen. Some canned meat has been successfully pre-

---

Dry foods tend to be stable.

A hypertonic solution is more concentrated than solutions in its immediate environment.

Osmosis is the flow of water from a more dilute solution through a membrane into a more concentrated solution.

Table 13-10
A Partial List of Food Additives Generally Recognized as Safe*

Anticaking Agents
 Calcium silicate
 Iron ammonium citrate

Acids, Alkalies, and Buffers
 Acetic acid
 Calcium lactate
 Citric acid
 Lactic acid
 Phosphates, Ca^{2+}, Na^+
 Potassium acid tartrate
 Sorbic acid
 Tartaric acid

Surface Active Agents (Emulsifying Agents)
 Glycerides: mono- and diglycerides of fatty acids
 Sorbitan monostearate

Polyhydric Alcohols
 Glycerol
 Mannitol
 Propylene glycol
 Sorbitol

Preservatives
 Benzoic acid
  Sodium benzoate
 Propionic acid
  Propionates, Ca^{2+}, Na^+
 Sorbic acid
  Sorbates, Ca^{2+}, K^+, Na^+
 Sulfites, Na^+, K^+

Antioxidants
 Ascorbic acid
  Ascorbates, Ca^{2+}, Na^+
 Butylated hydroxyanisole (BHA)
 Butylated hydroxytoluene (BHT)
 Lecithin
 Sulfur dioxide and sulfites

Flavor Enhancers
 Monosodium glutamate (MSG)
 5'-Nucleotides
 Maltol

Sweeteners
 Aspartame
 Mannitol
 Saccharin
 Sorbitol

Sequestrants
 Citric acid
 EDTA, Ca^{2+}, Na^+
 Pyrophosphate, Na^+
 Sorbitol
 Tartaric acid
  NaK (tartrate)

Stabilizers and Thickeners
 Agar-agar
 Algins
 Carrageenin

Flavorings (1700)
 Amyl butyrate (pearlike)
 Bornyl acetate (piney, camphor)
 Carvone (spearmint)
 Cinnamaldehyde (cinnamon)
 Citral (lemon)
 Ethyl cinnamate (spicy)
 Ethyl formate (rum)
 Ethyl vanillin (vanilla)
 Geranyl acetate (geranium)
 Ginger oil (ginger)
 Menthol (peppermint)
 Methyl anthranilate (grape)
 Methyl salicylate (wintergreen)
 Orange oil (orange)
 Peppermint oil (peppermint)
 Wintergreen oil (wintergreen) (methyl salicylate)

If past history is a guide, at least some of these compounds will be taken off the GRAS list in the future.

* For precise and authoritative information on levels of use permitted in specific applications, consult the regulations of the U.S. Food and Drug Administration and the Meat Inspection Division of the U.S. Department of Agriculture.

served for over a century. Newer techniques for the preservation of food include vacuum freezing, pasteurization, cold storage, irradiation, and chemical preservation.

Antimicrobial Preservatives. Food spoilage caused by microorganisms is a result of the excretion of toxins. A preservative is effective if it prevents multiplication of the microbes during the shelf life of the product. Sterilization by heat or radiation, or inactivation by freezing, is often undesirable, since it impairs the quality of the food. Chemical agents seldom achieve sterile conditions but can preserve foods for considerable lengths of time.

Antimicrobial preservatives are widely used in a large variety of foods. For example, in the United States sodium benzoate is permitted in nonalcoholic beverages and in some fruit juices, fountain syrups, margarines, pickles, relishes, olives, salads, pie fillings,

A preservative must interfere with microbes but be harmless to the human system—a delicate balance.

jams, jellies, and preserves. Sodium propionate is legal in bread, chocolate products, cheese, pie crust, and fillings. Depending on the food, the weight of the preservative permitted ranges up to a maximum of 0.1% for sodium benzoate and 0.3% for sodium propionate.

Sodium Benzoate

Sodium Propionate

Postulated mechanisms for the action of food preservatives may be grouped into three categories: (1) interference with the permeability of cell membranes of the microbes in foodstuffs, so the bacteria die of starvation; (2) interference with bacterial genetic mechanisms, so the reproduction processes are hindered; and (3) interference with intracellular enzyme activity, so that metabolic processes such as the Krebs cycle cease.

Atmospheric Oxidation. Microbial activity results in oxidative decay of food, but it is not the only means of oxidizing food. The direct action of oxygen in the air, **atmospheric oxidation,** is the chief cause of the destruction of fats and fatty portions of food. Foods kept wrapped, cold, and dry are relatively free of air oxidation. An antioxidant added to the food can also hinder oxidation. Antioxidants most commonly used in edible products contain various combinations of butylated hydroxyanisole (BHA) and butylated hydroxytoluene (BHT).

BHA   BHT

To prevent the oxidation of fats, the antioxidant can donate the hydrogen atom (H·) in the —OH group to reactive species; this effectively stops the reaction between fats and oxygen. If antioxidants are not present, the oxidation of fats leads to a complex mixture of volatile aldehydes, ketones, and acids, which cause a rancid odor and taste.

Sequestrants

Metals get into food from the soil and from machinery during harvesting and processing. Copper, iron, and nickel, as well as their ions, catalyze the oxidation of fats. However, molecules of citric acid bond with the metal ions, thereby rendering them ineffective as catalysts. With the competitor metal ions tied up, antioxidants such as BHA and BHT can accomplish their task much more effectively.

Citric acid belongs to a class of food additives known as **sequestrants.** For the most part sequestrants react with trace metals in foods, tying them up in complexes so the metals will not catalyze the decomposition or oxidation of food. Sequestrants such as sodium and calcium salts of EDTA (ethylenediaminetetraacetic acid) are permitted in beverages, cooked crab meat, salad dressing, shortening, lard, soup, cheese, vegetable oils, pudding mixes, vinegar, confectioneries, margarine, and other foods. The amounts

To sequester means "to withdraw from use." The sequestering ability of EDTA accounts for its use in treating heavy metal poisoning (Chapter 14).

range from 0.0025% to 0.15%. The structural formula of EDTA bonded to a metal ion is shown in Figure 13-2.

Flavor in Foods

Flavors result from a complex mixture of volatile chemicals. Since we have only four tastes — sweet, sour, salt, and bitter — much of the sensation of taste in food is smell. For example, the flavor of coffee is determined largely by its aroma, which in turn is due to a very complex mixture of over 100 compounds, mostly volatile oils.

Most flavor additives originally came from plants. The plants were crushed and the compound extracted with various solvents such as ethanol or carbon tetrachloride. Sometimes a single compound was extracted; more often the residue contained a mixture of several compounds. By repeated efforts, relatively pure oils were obtained. Oils of wintergreen, peppermint, orange, lemon, and ginger, among others, are still obtained in this way. These oils, alone or in combination, are then added to foods to produce the desired flavor. Gradually, analyses of the oils and flavor components of plants have revealed the active compounds responsible for the flavors. Today synthetic extracts of the same flavors actively compete with natural extracts.

The FDA has banned some of the naturally occurring flavoring agents that used to be used, including safrole, the primary root beer flavor, found in the root of the sassafras tree.

Flavor Enhancers

Flavor enhancers have little or no taste of their own but amplify the flavors of other substances. They exert synergistic and potentiation effects. Synergism is the cooperative action of discrete agents such that the total effect is greater than the sum of the effects of each used alone. Potentiators do not have a particular effect themselves but exaggerate the effects of other chemicals. The 5′-nucleotides, for example, have no taste but enhance the flavor of meat and the effectiveness of salt. Potentiators were first used in meat and fish but now are also used to intensify flavors or cover unwanted flavors in vegetables, bread, cakes, fruits, nuts, and beverages. Three commonly used flavor enhancers are *monosodium glutamate (MSG), 5′-nucleotides* (similar to inosinic acid; see Chapter 11), and *maltol.*

Some 1700 natural and synthetic substances are used to flavor foods, making flavors the largest category of food additives.

Safrole

In some people MSG causes the so-called Chinese restaurant syndrome, an unpleasant reaction characterized by headaches and sweating that usually occurs after an MSG-rich Chinese meal. Tomatoes and strawberries affect some individuals in the same way.

Maltol
(From Pine Needles)

Figure 13-2
The structural formula for the metal chelate of ethylenediaminetetraacetic acid (EDTA).

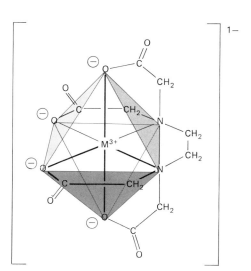

Another example of chelation by EDTA is the Fe-EDTA chelate in soil nutrient application.

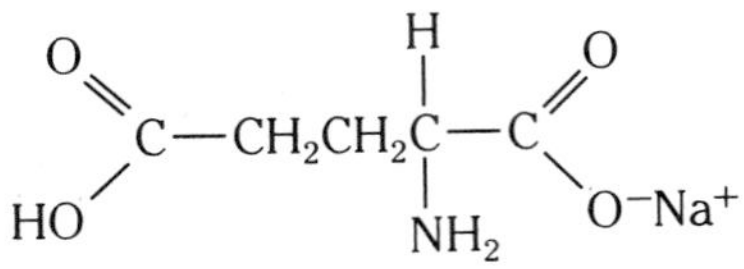

Monosodium Glutamate

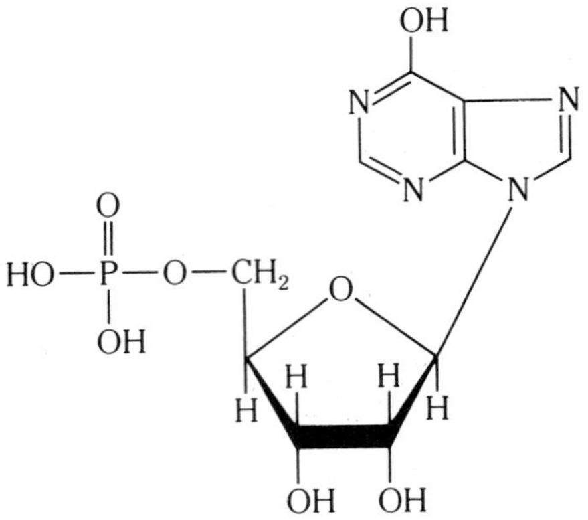

Inosinic Acid
(a 5'-nucleotide)

MSG is a natural constituent of many foods, such as tomatoes and mushrooms.

When MSG is injected in very high doses under the skin of 10-day-old mice, it causes brain damage. When these laboratory results were reported, considerable discussion ensued concerning the merits of MSG. National investigative councils have suggested that MSG be removed from baby foods, since infants do not seem to appreciate enhanced flavor. However, in the absence of hard evidence that MSG is harmful in the amounts used in regular food, no recommendations were made relative to its use.

Sweeteners

Sweetness is characteristic of a wide range of compounds, many of which are completely unrelated to sugars. Lead acetate ($Pb(CH_3COO)_2$) is sweet but poisonous. A number of **artificial sweeteners** are allowed in foods. These are primarily used for special diets such as those of diabetics. Artificial sweeteners have no known metabolic use in the body and do not need to be offset by insulin.

Insulin is a hormone that regulates glucose metabolism.

Saccharin. The most common artificial sweetener is saccharin:

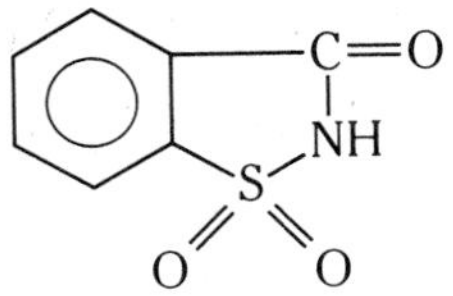

Saccharin

Saccharin is a synthetic chemical.

Saccharin is about 300 times sweeter than ordinary sugar (sucrose). When ingested, saccharin passes through the body unchanged; it therefore has no food value other than to render an otherwise bland mixture more tasty. Saccharin has a somewhat bitter aftertaste, which renders it unpleasant to some users. Glycine, the simplest amino acid, which is also sweet, is often added to counteract this bitter taste.

Laboratory studies have shown that high doses of saccharin cause cancer in mice. After months of consideration, the Institute of Medicine of the National Academy of Science joined the FDA in its 1978 statement that saccharin should be banned in U.S. foods. The U.S. Congress passed the ban and then suspended it pending the results of some ongoing studies in Canada. As of 1987, saccharin was still being sold, but with a warning label, "Use of this product may be hazardous to your health. This product contains saccharin which has been determined to cause cancer in laboratory animals."

Aspartame. A new entry in the sweetener market, aspartame is chemically an ester of a two-amino acid peptide with the name *N-L-α-aspartyl-L-phenylalanine methyl ester* and trade names of *NutraSweet, Equal,* and *Tri-Sweet.* The sweetener was approved by the FDA in 1974 and subsequently withdrawn by its maker, G. D. Searle Co., when toxicity questions were raised. When these questions were resolved in 1981, aspartame again received FDA approval.

Aspartame is about 180 times sweeter than table sugar (sucrose). The caloric value of aspartame is similar to that of proteins. The caloric intake of consumers using this product is reduced, since much smaller amounts of aspartame are needed to produce the same sweetening effect as sugar. Aspartame does not have the bitter aftertaste associated with other artificial sweeteners. It is normally metabolized in the body as a peptide.

Mannitol and sorbitol, polyhydric alcohols, are sweeteners used in such products as sugarless gum. These sweeteners have some caloric value in the body.

Food and Esthetic Appeal

Food Colors. There are about 30 chemical substances used to color food. All are under investigation by the FDA, and some may be prohibited as the investigations progress. About half of the food colors are laboratory synthesized, and half are extracted from natural materials. Most food colors are large organic molecules with several double bonds and aromatic rings. The electrons of these conjugated structures can absorb certain wavelengths of light and pass the rest; the wavelengths passed give the substances their characteristic colors. β-Carotene, an orange-red substance in a variety of plants that gives carrots their characteristic color, has a conjugated system of electrons and is used as a food color. β-Carotene is a precursor (provitamin) of vitamin A.

Colored organic substances often are *conjugated* molecules, having alternating double and single bonds in the carbon chain or ring.

Because one of the food colors, Yellow No. 5, causes allergic reactions (mainly rashes and sniffles) in an estimated 50,000 to 90,000 Americans, the FDA has required manufacturers to list Yellow No. 5 on the labels of any food products containing it.

pH Control in Foods. Weak organic acids are added to such foods as cheese, beverages, and dressings to give a mild acidic taste. They often mask undesirable aftertastes. Weak acids and acid salts, such as tartaric acid and potassium acid tartrate, react with bicarbonate to form CO_2 in the baking process.

Some acid additives control the pH of food during the various stages of processing as well as in the finished product. In addition to single substances, there are several combinations of substances that will adjust and then maintain a desired pH; these mixtures are called **buffers.** An example of one type of buffer is potassium acid tartrate, $KHC_4H_4O_6$.

Buffer solutions resist change in acidity and basicity; pH remains constant.

Adjustment of fruit juice pH is allowed by the FDA. If the pH of the fruit is too high, it is permissible to add acid (called an **acidulant**). Citric acid and lactic acid are the most

Small amounts of certain acids are allowed to be added to some foods.

common acidulants used, since they are believed to impart good flavor, but phosphoric, tartaric, and malic acids are also used. These acids are often added at the end of the cooking time to prevent extensive hydrolysis of the sugar. In the making of jelly they are sometimes mixed with the hot product immediately after pouring. To raise the pH of a fruit that is unusually acid, buffer salts such as sodium citrate or sodium potassium tartrate are used.

The versatile acidulants also function as preservatives to prevent the growth of microorganisms, as synergists and antioxidants to prevent rancidity and browning, as viscosity modifiers in dough, and as melting point modifiers in such food products as cheese spreads and hard candy.

Hygroscopic substances absorb moisture from the air.

Anticaking Agents. Anticaking agents are added to hygroscopic foods — in amounts of 1% or less — to prevent caking in humid weather. Table salt (sodium chloride) is particularly subject to caking unless an anticaking agent is present. The additive (magnesium silicate, for example) incorporates water into its structure as water of hydration and does not appear wet as sodium chloride does when it absorbs water physically on the surface of its crystals. As a result, the anticaking agent keeps the surface of sodium chloride crystals dry and prevents crystal surfaces from co-dissolving, and joining together.

Stabilizers and thickeners are types of emulsifying agents.

Stabilizers and Thickeners. Stabilizers and thickeners improve the texture and blends of foods. The action of carrageenan (a polymer from edible seaweed) is shown in Figure 13-3. Most of this group of food additives are polysaccharides (Chapter 11), which have numerous hydroxyl groups as a part of their structure. The hydroxyl groups form hydrogen bonds with water to prevent the segregation of water from the less polar fats in the food and to provide a more even blend of the water and oils throughout the food. Stabilizers and thickeners are particularly effective in icings, frozen desserts, salad dressings, whipped cream, confectioneries, and cheeses.

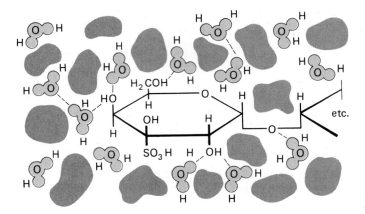

Figure 13-3
The action of carrageenan to stabilize an emulsion of water and oil in salad dressing. An active part of carrageenan is a polysaccharide, a portion of which is shown here. The carrogeenan hydrogen-bonds to the water, which keeps it dispersed. The oil, not being very cohesive, disperses throughout the structure of the polysaccharide. Gelatin (a protein) undergoes similar action in absorbing and distributing water to prevent the formation of ice crystals in ice cream.

Surface Active Agents. Surface active agents are similar to stabilizers, thickeners, and detergents in their chemical action. They cause two or more normally incompatible (nonpolar and polar) chemicals to disperse in each other. If the chemicals are liquids, the surface active agent is called an **emulsifier.** If the surface active agent has a sufficient supply of hydroxyl groups, as does cholic acid, the groups form hydrogen bonds to water. Cholic acid and its associated group of water molecules are distributed throughout dried egg yolk in a manner quite similar to that of carrageenan and water in salad dressing.

Some surface active agents have both hydroxyl groups and a relatively long nonpolar hydrocarbon end. Examples are diglycerides of fatty acids, polysorbate 80, and sorbitan monostearate. The hydroxyl groups on one end of the molecule are anchored by hydrogen bonds in the water, and the nonpolar end is held by the nonpolar oils or other substances in the food. This provides tiny islands of water held to oil. These islands are distributed evenly throughout the food.

Polyhydric Alcohols. Polyhydric alcohols are allowed in foods as humectants, sweetness controllers, dietary agents, and softening agents. Their chemical action is based on their multiplicity of hydroxyl groups that hydrogen-bond to water. They thus hold water in food, soften it, and keep it from drying out. Tobacco is also kept moist by the addition of polyhydric alcohols such as glycerol. An added feature of polyhydric alcohols is their sweetness. The two polyhydric alcohols mentioned earlier for their sweetness are mannitol and sorbitol. The structures of these alcohols are strikingly similar to the structure of glucose (Chapter 11), and all three have a sweet taste.

Kitchen Chemistry

Leavened Bread. Sometimes cooking causes a chemical reaction that releases carbon dioxide gas, and the trapped carbon dioxide causes breads and pastries to rise. Yeast has been used since ancient times to make bread rise, and remains of bread made with yeast have been found in Egyptian tombs and the ruins of Pompeii. The metabolic processes of the yeast furnish gaseous carbon dioxide, which creates bubbles in the bread and makes it rise:

$$C_6H_{12}O_6 \xrightarrow[\text{from yeast}]{\text{zymase}} 2\ CO_2 + 2\ C_2H_5OH$$
$$\text{Glucose} \qquad \text{(Gas)} \qquad \text{Ethanol}$$

When the bread is baked, the CO_2 expands even more to produce a light, airy loaf.

Carbon dioxide can be generated in cooking by other processes. For example, baking soda (which is simply sodium bicarbonate ($NaHCO_3$), a base) can react with acidic ingredients in a batter to produce CO_2:

$$NaHCO_3 + H^+ \longrightarrow Na^+ + H_2O + CO_2\ \text{(gas)}$$

Baking powders contain sodium bicarbonate and an added acid salt or a salt that hydrolyzes to produce an acid. Some of the compounds used for this purpose are potassium hydrogen tartrate ($KHC_4H_4O_6$), calcium dihydrogen phosphate monohydrate ($Ca[H_2PO_4]_2 \cdot H_2O$), and sodium acid pyrophosphate ($Na_2H_2P_2O_7$). The reactions of these white, powdery salts with sodium bicarbonate are similar, although the compounds all have somewhat different appearances. For example:

$$KHC_4H_4O_6 + NaHCO_3 \xrightarrow{\text{water}} KNaC_4H_4O_6 + H_2O + CO_2\ \text{(gas)}$$

Cholic Acid

Hydrogen bonding plays a major role in stabilizers, thickeners, surface active agents, and humectants.

D-Sorbitol   D-Mannitol

Leavened bread is as old as recorded history.

About 350 million pounds of phosphates are added to foods in the United States each year: This is 20% to 25% of our phosphorus intake. Some phosphates are used as leavening agents. Sodium phosphate thickens puddings, retains juices, makes hams tender, and prevents canned milk from thickening on standing.

Cooking and Precooking: "Preliminary Digestion." The cooking process involves the partial breakdown of proteins or carbohydrates by means of heat and hydrolysis. The polymers that must be degraded if cooking is to be effective are the carbohydrate cellular wall materials in vegetables and the collagen, or connective tissue, in meats. Both types of polymers are subject to hydrolysis in hot water or moist heat. In either case, only partial depolymerization is required.

In recent years several precooking additives have become popular; the **meat tenderizers** are a good example. These are simply enzymes that catalyze the breaking of peptide bonds in proteins via hydrolysis at room temperature. As a consequence, the same degree of "cooking" can be obtained in a much shorter heating time. Meat tenderizers are usually plant products such as papain, a proteolytic (protein-splitting) enzyme from the unripe fruit of the papaw tree. Papain has a considerable effect on connective tissue, mainly collagen and elastin, and shows some action on muscle fiber proteins. On the other hand, microbial protease enzymes (from bacteria, fungi, or both) have considerable action on muscle fibers. A typical formulation for the surface treatment of cuts of beef consists of 2% commercial papain or 5% fungal protease, 15% dextrose, 2% monosodium glutamate (MSG), and salt.

> Cooking starts the digestive process, although it is rarely needed for this purpose. Since it may destroy nutrients, it is done mostly for esthetic reasons.

SELF-TEST 13-D

1. Flavor enhancers exert a(n) _____ or _____ effect on the flavors of foods.
2. Name two of the oldest means for preserving food. _____ and _____
3. Antimicrobial preservatives make foods sterile.   True ( ) or False ( )
4. Flavors result from ( ) volatile or ( ) nonvolatile compounds.
5. Antioxidants are ( ) more or ( ) less easily oxidized than the food into which they are placed.
6. Citric acid is an example of a(n) _____. Such compounds tie up metals in stable complexes.
7. A flavor in a food can usually be traced to a single compound.   True ( ) or False ( )
8. Monosodium glutamate is a(n) _____.
9. Salt is effective in preserving foods because it kills microorganisms by _____ them.
10. GRAS is an acronym for _____.
11. Which has the sweetest taste when an equal weight of each is tasted: table sugar, aspartame, or saccharin? _____
12. What molecular characteristic do most food colors have? _____
13. What acids are added to foods to lower the pH? _____ and _____
14. The gas released by leavening agents is _____.
15. Cooking _____ some chemical bonds.

16. Hydrogen bonding generally plays a very important role in the action of surface _____ agents.

MATCHING SET I

Match each nutrient with the disease(s) caused by a *deficiency* in the nutrient. Use all of the diseases.

	Nutrient		Disease
_____	1. Cobalamin		a. Xerophthalmia
_____	2. Iodine		b. Kwashiorkor
_____	3. Vitamin A		c. Goiter
_____	4. Vitamin D		d. Scurvy
_____	5. Vitamin C		e. Pernicious anemia
_____	6. Thiamine		f. Rickets
_____	7. Riboflavin		g. Anemia
_____	8. Niacin		h. Finickiness in appetite, stretch marks
_____	9. Protein		i. Pellagra
_____	10. Fat		j. Hypoglycemia
_____	11. Sugar		k. Eczema
_____	12. Iron		l. Night blindness
_____	13. Zinc		m. Beriberi
			n. Shark skin

MATCHING SET II

_____	1. β-Carotene		a. Nutrient supplement in food
_____	2. Monosodium glutamate		b. Food color
_____	3. Copper, nickel, and iron		c. Flavor enhancer
_____	4. Sodium benzoate		d. Catalyze oxidation of fats
_____	5. Potentiator		e. Antimicrobial preservative
_____	6. Mineral		f. Exaggerates some chemical effects
_____	7. Mannitol		g. Sequestering agent
			h. Sweetener
			i. pH adjuster

QUESTIONS

1. What two factors influenced the United States government to pass laws on food? What problem was caused by the government regulations?

2. If all of our blood cells are renewed each 120 days, must new nutrients be ingested to make the new cells? Explain.

3. Distinguish between RDA and USRDA recommendations on the basis of the following:
   a. Which has the higher and which has the lower recommendations?
   b. Which has the greater breakdown with respect to age, sex, and so on?
   c. Which is used on package labels?
4. What are two sources of dietary fiber, and why is fiber important in the diet?
5. What problem is caused
   a. By hydrogenation of polyunsaturated fats (oils)?
   b. By heating, especially of unsaturated fats, to temperatures at which they smoke?
6. What is the name of the essential fatty acid? What vitamin is designated as this fatty acid?
7. If the USRDA is ingested each day, is good health assured? Explain.
8. Give two examples of dietary components in which balance is especially important.
9. a. What activities go on during a determination of the basal metabolic rate (BMR)?
   b. What three factors affect BMR in addition to weight, height, and age?
10. What are the functions of fat, protein, and carbohydrates in the body?
11. Based on their solubilities, what are the two classes of vitamins? An excess of which class of vitamins causes fewer problems?
12. Why is vitamin B_6 called the "master vitamin"?
13. Why should one eat the whole fruit rather than, for example, sucking the juice out and throwing the rest away?
14. What are some good food sources of complex carbohydrates?
15. Name the ten essential amino acids.
16. What is the cause of kwashiorkor?
17. What foods are good sources of thiamine, vitamin B_{12}, niacin, ascorbic acid, vitamin A, vitamin D, and vitamin E?
18. What diseases or symptoms are caused by a deficiency of niacin, thiamine, proteins, calciferol, and ascorbic acid?
19. What group of vitamins is most easily destroyed or removed during food processing and cooking?
20. a. Distinguish between refined sugar and complex carbohydrates with respect to how they are assimilated by the body.
    b. What problems may arise from consuming too much
       (1) refined sugar?
       (2) refined grains (white flour)?
21. Use Table 13-4 to calculate the Calories in some fast-food hamburgers, French fries, and milkshakes.

	Burgers	Shakes	Fries
Chain A			
Protein (g)	11	10	2
Carbohydrates (g)	29	72	20
Fat (g)	9	9	19
Chain B			
Protein (g)	12	10	3
Carbohydrates (g)	30	66	26
Fat (g)	10	9	12
Chain C			
Protein (g)	13	11	2
Carbohydrates (g)	29	55	25
Fat (g)	11	7	10

Which chain of fast-food restaurants offers the *lowest* Calorie total for a hamburger, milkshake, and an order of French fries? What is the total number of kcal?

22. Should salt tablets be taken after the body is acclimated to the heat? Why should or why should they not be taken?
23. a. What is the typical K/Na weight ratio in the body?
    b. What is a problem in maintaining this ratio?
    c. What can be done to maintain the proper K/Na ratio?
24. Why may a calcium deficiency occur in some women after menopause?
25. Why do we need each specific mineral and each specific vitamin? List one function or one deficiency effect.
26. Which vitamins are produced by bacteria in the intestine?
27. Does vitamin C decrease the symptoms of the common cold? Does vitamin A?
28. How does salt preserve food?
29. a. Why does it take less time to cook food in a pressure cooker than in an open pot of boiling water?
    b. Why does cooking aid digestion?
30. A label on a brand of breakfast pastries lists the following additives: dextrose, glycerin, citric acid, potassium sorbate, vitamin C, sodium iron pyrophosphate, and BHA. What is the purpose of each substance?
31. What is a common flavor enhancer? How do flavor enhancers work?
32. Choose a label from a food item, and try to identify the purpose of each additive.
33. Describe some of the chemical changes that occur during the cooking of

a. A carbohydrate   c. A fat
   b. A protein
34. What causes bread to rise?
35. What do the letters *FDA* represent, and what does this government agency do?
36. What are the pros and cons of eating "natural" foods, as opposed to foods containing chemical additives?
37. Many consumer products are almost identical in chemical composition but are sold at widely different prices under different trade names. Do you think the products should be identified by their chemical names or their trade names? Why?
38. What is the GRAS list?
39. What foods have you eaten during the past week that did not have chemicals added or applied to them?

CHAPTER 14

Toxic Chemicals

Without a doubt, some chemical substances are hazardous. Some chemicals are **explosive;** others, such as natural gas and gasoline, are **flammable.** Acids and bases are **corrosive.** Perhaps the most troublesome property of many chemical substances is their **toxicity** — their ability to react with the chemistry of living organisms in such a way as to impair or threaten life. Table 14-1 gives some information on chemical hazards.

Many different government agencies regulate chemicals; that is, they give users of dangerous chemicals guidelines concerning the safe use of those chemicals. Table 14-2 shows the hazards associated with several commonplace chemicals and how these chemicals are regulated.

Since most of the hazards of harmful chemicals in our environment are related to their toxic properties, this chapter examines toxicity and the chemical mechanisms by which toxic substances act.

Toxicity

Toxic substances are materials that upset the incredibly complex system of chemical reactions that occur in the human body. Sometimes toxic substances cause mere discomfort; sometimes they cause illness, disability, or even death. Toxic symptoms can be produced by very small amounts of extremely toxic materials (such as sodium cyanide) or larger amounts of less toxic substances. The term *toxic substances* usually is limited to materials that are dangerous in small amounts. However, as most of us know, ill effects can be caused by the excessive intake of substances that are normally considered harmless (too much candy, for example). Fortunately, in most cases the human body is capable of recognizing "foreign" chemicals and ridding itself of them.

A large enough dose of any compound can result in poisoning.

Table 14-1
Hazards of Some Chemicals

Substance	Reaction Type	Hazard
Benzoyl peroxide	Decomposition at elevated temperatures	Explosiveness
Methane	Oxidation by oxygen in air	Flammability
Sulfuric acid	Reaction with protein (skin)	Corrosiveness
Carbon monoxide	Reaction with hemoglobin	Toxicity

Table 14-2
Common Chemicals, Their Hazards, and How They Are Regulated

Substance (Use)	Major Hazard	Secondary Hazard(s)	How Regulated*
Gasoline (fuel)	Flammability	Contains benzene, a carcinogen	By U.S. Dept of Transportation (DOT)
Nitric acid (used for metal treatment)	Corrosiveness	Causes fire on contact with certain combustible materials	By DOT as a corrosive
Chloroform (solvent)	Central nervous system depressant	Causes certain types of cancer	By U.S. Environmental Protection Agency (EPA) in water, wastes and waste water
Asbestos (used in brake linings, insulation)	Causes lung cancer	Causes lung diseases	By EPA as an environmental carcinogen
Dynamite (explosive)	Explosiveness	Causes dilation of blood vessels	By DOT and the Alcohol, Tobacco and Firearm branch of the Justice Department
Used motor oil (used in recycling)	Pollution of surface and groundwater	May contain toxic solvents and heavy metals	By EPA as a hazardous waste

* Only certain regulatory aspects are mentioned here for each chemical. Often several federal agencies regulate a single chemical; state and local regulations may also apply.

Dose

Lethal doses of toxic substances are customarily expressed in milligrams (mg) of substance per kilogram (kg) of body weight of the subject. For example, the cyanide ion (CN^-) is generally fatal to human beings in a dose of 1 mg of CN^- per kilogram of body weight. For a 200-pound (90.7-kg) person, about $\frac{1}{10}$ g of cyanide is a lethal dose. Examples of somewhat less toxic substances and their ranges of lethal doses for human beings are shown below:

Morphine	1–50 mg/kg
Aspirin	50–500 mg/kg
Methyl alcohol	500–5000 mg/kg
Ethyl alcohol	5000–15,000 mg/kg

One can measure toxicity by introducing various dosages of substances to be tested into laboratory animals (such as rats). The dosage that is found to be lethal in 50% of a large number of the animals is called the LD_{50} (lethal dosage, 50%) and is reported in

Dosis sola facit venenum —"the dose makes the poison."

Metabolism (from the Greek *metaballein*, "to change or alter") is the sum of all the physical and chemical changes by which living organisms are produced and maintained.

milligrams of poison per kilogram of body weight. Thus, if a statistical analysis of data on a large population of rats showed that a dosage of 1 mg/kg was lethal to 50% of the population tested, the LD_{50} for this poison would be 1 mg/kg. Obviously, metabolic variations and other differences among species will produce different LD_{50} values for a given poison in different kinds of animals. For this reason such data cannot be extrapolated to human beings with any assurance, but it is safe to assume that a substance with a low LD_{50} value for several animal species will also be quite toxic to humans (Table 14–3).

Toxic substances can be classified according to the way in which they disrupt the chemistry of the body — toxic substances can be **corrosive, metabolic, neurotoxic, mutagenic, teratogenic,** or **carcinogenic.**

Corrosive Poisons

Toxic substances that destroy tissues are corrosive poisons. Examples include strong acids and alkalies, as well as many oxidants, such as those found in laundry products. Sulfuric acid (found in auto batteries) and hydrochloric acid (also called muriatic acid; it is used for cleaning purposes) are very dangerous corrosive poisons. So is sodium hydroxide, which is used to clear clogged drains. Ingestion of 1 ounce of concentrated (98%) sulfuric acid causes death, and much smaller amounts can cause extensive damage and severe pain.

Concentrated mineral acids such as sulfuric acid act by first dehydrating cellular structures. Cells die because their protein structures are destroyed by the acid-catalyzed hydrolysis of the peptide bonds.

$$R-\underset{\text{Peptide Link (in Protein)}}{\underbrace{C(=O)-N(H)-}}R' + H_2O \xrightarrow[\text{from acid}]{H^+} \underset{\text{Carboxyl End of Smaller Peptide or Amino Acid}}{R-C(=O)-OH} + \underset{\text{Amine End of Smaller Peptide or Amino Acid}}{H-N(H)-R'}$$

Strong acids and bases destroy cell protoplasm.

In the early stages of this process, a large proportion of large fragments are present. Subsequently, as more bonds are broken, smaller and smaller fragments result, and the tissue ultimately disintegrates completely.

Table 14–3
Approximate Comparison of LD_{50} Values with Lethal Doses for Human Adults

Oral LD_{50} for Any Animal (mg/kg)	Probable Lethal Oral Dose for Human Adult
Less than 5	A few drops
5 to 50	"A pinch" to 1 teaspoonful
50 to 500	1 teaspoonful to 2 tablespoonfuls
500 to 5000	1 ounce to 1 pint (1 pound)
5000 to 15,000	1 pint to 1 quart (2 pounds)

Some poisons act by undergoing chemical reactions in the body to produce corrosive poisons. Phosgene, the deadly gas used during World War I, is an example. When inhaled, it is hydrolyzed in the lungs to hydrochloric acid, which has a hydrolyzing effect on tissues and thereby causes pulmonary edema (a collection of fluid in the lungs). The victim dies of suffocation because the flooded and damaged tissues cannot effectively absorb oxygen.

Chemical "warfare gases," such as phosgene, were outlawed by an international conference in 1925.

$$\underset{\text{Phosgene}}{Cl_2C=O} + H_2O \longrightarrow \underset{\substack{\text{Hydrochloric} \\ \text{Acid}}}{2HCl} + \underset{\substack{\text{Carbon} \\ \text{Dioxide}}}{CO_2}$$

Sodium hydroxide (NaOH; caustic soda that is a component of drain cleaners) is a very strongly alkaline, or basic, substance that can be just as corrosive to tissue as are strong acids. The hydroxide ion also catalyzes the splitting of peptide linkages:

$$R-\underset{\substack{\| \\ H}}{C(=O)}-N-R' + H_2O \xrightarrow[\text{base}]{OH^-} R-C(=O)-OH + H-\underset{H}{N}-R'$$

Both acids and bases, as well as other types of corrosive poisons, continue their action until they are consumed in chemical reactions.

Some corrosive poisons destroy tissue by oxidizing it. This is characteristic of substances such as ozone, nitrogen dioxide, and possibly iodine, which destroy enzymes by oxidizing their functional groups. Specific groups, such as the —SH and —S—S— groups in the enzyme, are believed to be converted by oxidation to nonfunctioning groups; alternatively, the oxidizing agents may break chemical bonds in the enzyme, leading to its inactivation.

A summary of some common corrosive poisons is presented in Table 14–4.

Metabolic Poisons

Metabolic poisons are more subtle than the tissue-destroying corrosive poisons. In fact, many of them do their work without indicating their presence until it is too late. Metabolic poisons can cause illness or death by interfering with vital biochemical mechanisms to such an extent that they cease to function.

Carbon Monoxide. The interference of carbon monoxide with extracellular oxygen transport is one of the best understood processes of metabolic poisoning. As early as 1895, it was noted that carbon monoxide deprives body cells of oxygen (asphyxiation), but it was much later before it was known that carbon monoxide, like oxygen, combines with hemoglobin:

$$O_2 + \text{hemoglobin} \longrightarrow \text{oxyhemoglobin}$$

$$CO + \text{hemoglobin} \longrightarrow \text{carboxyhemoglobin}$$

Laboratory tests show that carbon monoxide reacts with hemoglobin to give a compound (carboxyhemoglobin) that is 140 times more stable than the compound of hemoglobin and oxygen (oxyhemoglobin) (Fig. 14–1). Since hemoglobin is so effectively tied up by carbon monoxide, it cannot perform its vital function of transporting oxygen.

Table 14-4
Some Corrosive Poisons

Substance	Formula	Toxic Action	Use/Exposure
Hydrochloric acid	HCl	Acid hydrolysis	As tile and concrete floor cleaner; concentrated acid used to adjust acidity of swimming pools
Sulfuric acid	H_2SO_4	Acid hydrolysis, oxidization	In auto batteries
Phosgene	ClCOCl	Acid hydrolysis	In combustion of chlorine-containing plastics (PVC or Saran)
Sodium hydroxide	NaOH	Base hydrolysis	As caustic soda, drain cleaner
Trisodium phosphate	Na_3PO_4	Base hydrolysis	As detergent, household cleaner
Sodium perborate	$NaBO_3 \cdot 4H_2O$	Base hydrolysis, oxidation	As laundry detergent, denture cleaner
Ozone	O_3	Oxidation	In air, electric motors
Nitrogen dioxide	NO_2	Oxidation	In polluted air, automobile exhaust
Iodine	I_2	Oxidation	As antiseptic
Hypochlorite ion	OCl^-	Oxidation	As bleach
Peroxide ion	O_2^{2-}	Oxidation	As bleach, antiseptic
Oxalic acid	$H_2C_2O_4$	Reduction, precipitation of Ca^{2+}	As bleach, ink eradicator, leather tanner; in rhubarb, spinach, tea
Sulfite ion	SO_3^{2-}	Reduction	As bleach
Chloramine	NH_2Cl	Oxidation	In mixture of household ammonia and chlorinated bleach
Nitrosyl chloride	NOCl	Oxidation	In mixture of household ammonia and bleach
Chlorine	Cl_2	Oxidation	As water purifier in swimming pools and water supplies

An organic material that undergoes incomplete combustion always liberates carbon monoxide. Sources include auto exhaust, smoldering leaves, lighted cigars or cigarettes, and charcoal burners. In the United States alone, combustion sources of all types dump about 200 million tons of carbon monoxide per year into the atmosphere.

The maximum global background level of carbon monoxide is estimated to be in the order of 0.1 ppm, but the background concentration in cities is higher. In heavy traffic sustained levels of 100 ppm or more are common; for offstreet sites in large cities an average of about 7 ppm is typical. A concentration of 30 ppm for 8 hours is sufficient to cause headache and nausea. Breathing an atmosphere that is 0.1% (1000 ppm) carbon monoxide for 4 hours converts approximately 60% of the hemoglobin of an average adult to carboxyhemoglobin (Table 14-5) and is likely to cause death (Fig. 14-2).

Since both carbon monoxide and oxygen reactions with hemoglobin are easily reversed, the concentrations, as well as the relative strengths of the bonds, affect the direction of the reactions. In air that contains 0.1% CO, oxygen molecules outnumber CO molecules 200 to 1. The larger concentration of oxygen helps counteract the greater

ppm—parts per million—is a measure expressing concentration. *50 ppm CO* means 50 mL of CO for every 1 million mL of air.

To convert ppm to percent, divide by 10,000.

Air is 21% O_2 by volume; in 1 million "air molecules" there would be 210,000 O_2 molecules.

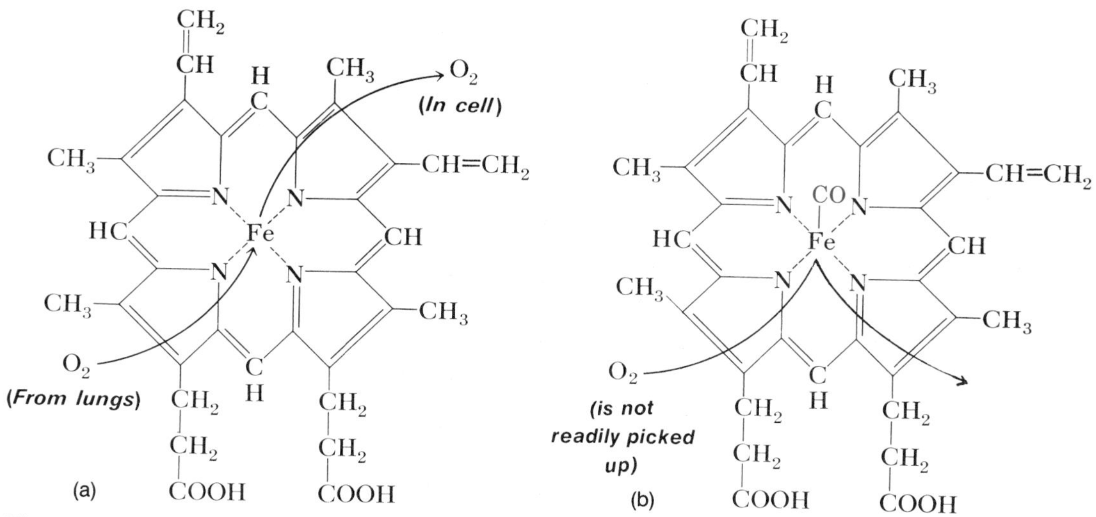

Toxicity

Figure 14-1
Structure of the heme portion of hemoglobin. (a) Normal acceptance and release of oxygen. (b) Oxygen blocked by carbon monoxide.

Table 14-5
*Concentration of CO in Atmosphere Versus Percentage of Hemoglobin (Hb) Saturated**

CO concentration in air	0.01% (100 ppm)	0.02% (200 ppm)	0.10% (1000 ppm)	1.0% (10,000 ppm)
Percentage of hemoglobin molecules saturated with CO†	17	20	60	90

* A few hours of breathing time is assumed.
† Normal human blood contains up to 5% of the hemoglobin as carboxyhemoglobin (HbCO).

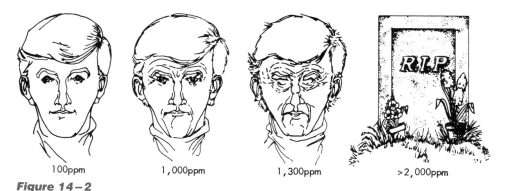

Figure 14-2
A healthy adult can tolerate 100 ppm carbon monoxide in air without suffering ill effect. A 1-hour exposure to 1000 ppm causes a mild headache and a reddish coloration of the skin develops. A 1-hour exposure to 1300 ppm causes the skin to turn cherry red and a throbbing headache to develop. A 1-hour exposure to concentrations greater than 2000 ppm will likely cause death.

330   *Toxic Chemicals*

The equilibrium concept was discussed in Chapter 6.

combining power of CO with hemoglobin by shifting the reaction equilibrium to the right. Consequently, if a carbon monoxide victim is exposed to fresh air or, still better, pure oxygen (provided he or she is still breathing), the carboxyhemoglobin (HbCO) is gradually decomposed because of the greater concentration of oxygen:

$$HbCO + O_2 \rightleftharpoons HbO_2 + CO$$

Equilibrium Shifted to Right Because of Greater Concentration of Oxygen

Although carbon monoxide is not a cumulative poison, permanent damage can occur if certain vital cells (e.g., brain cells) are deprived of oxygen for more than a few minutes.

Individuals differ in their tolerance of carbon monoxide, but generally those with anemia or an otherwise low reserve of hemoglobin (e.g., children) are most susceptible. No one is helped by carbon monoxide, and smokers suffer chronically from its effects. It is a subtle poison, since it is odorless and tasteless.

Cyanide. The cyanide ion, CN^-, is the toxic agent in cyanide salts such as sodium cyanide (used in electroplating). Since cyanide is a relatively strong base, it reacts easily with many acids (weak and strong) to form volatile hydrogen cyanide (HCN):

$$CH_3COOH + Na^+CN^- \longrightarrow HCN + Na^+CH_3COO^-$$

Acetic Acid   Sodium Cyanide   Hydrogen Cyanide   Sodium Acetate

HCN boils at a relatively low temperature (26°C) and is therefore a gas at temperatures slightly above room temperature. It is often used as a fumigant in storage bins and holds of ships because it is toxic to most forms of life and in gaseous form can penetrate into tiny openings, even into insect eggs.

The cyanide ion is one of the most rapidly working poisons. Lethal doses taken orally act in minutes. Cyanide poisons by asphyxiation, as does carbon monoxide, but the mechanism of cyanide poisoning is different (Fig. 14-3). Instead of preventing cells from getting oxygen, cyanide interferes with oxidative enzymes such as cytochrome oxidase. Oxidases are enzymes containing a metal, usually iron or copper; they catalyze the oxidation of substances such as glucose:

A metabolite *is any substance produced in a life process.*

$$Fe^{2+} \longrightarrow Fe^{3+} + e^-$$

Oxidation

$$\text{Metabolite } (H)_2 + \tfrac{1}{2}O_2 \xrightarrow{\text{oxidase}} \text{Oxidized metabolite} + H_2O + \text{energy}$$

The iron atom in cytochrome oxidase is oxidized from Fe^{2+} to Fe^{3+} to provide electrons for the reduction of O_2. The iron regains electrons from other steps in the process. The cyanide ion forms stable cyanide complexes with the metal ion of the oxidase and renders the enzyme incapable of reducing oxygen or oxidizing the metabolite.

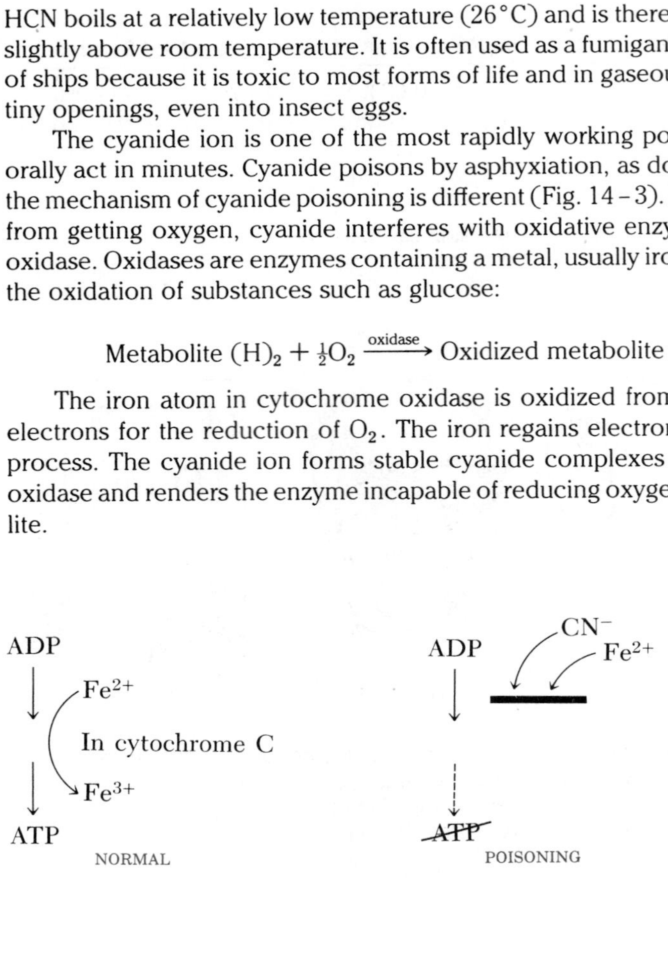

Figure 14-3 The mechanism of cyanide (CN^-) poisoning. Cyanide binds tightly to the enzyme cytochrome C, an iron compound, thus blocking the vital ADP-ATP reaction in cells.

$$\text{Cytochrome oxidase (Fe)} + CN^- \longrightarrow \underbrace{\text{Cytochrome oxidase (Fe)} \cdots CN^-}_{\text{complex}}$$

In essence the electrons of the iron ion are "frozen"—they cannot participate in the oxidation-reduction processes. Plenty of oxygen gets to the cells, but the mechanism by which the oxygen is used in the support of life is stopped. Hence the cells die, and if this occurs fast enough in the vital centers, the victim dies.

The body can rid itself of a large number of toxic substances if the dose is small enough and there is sufficient time.

The body has a mechanism for ridding itself slowly of cyanide ions. The cyanide-oxidative enzyme reaction is reversible, and other enzymes such as rhodanase, found in almost all cells, can convert cyanide ions to relatively harmless thiocyanate ions. For example,

$$CN^- + \underset{\text{Thiosulfate}}{S_2O_3^{2-}} \xrightarrow{\text{rhodanase}} \underset{\text{Thiocyanate}}{SCN^-} + SO_3^{2-}$$

However, this mechanism is not as effective in protecting a cyanide-poisoning victim as it might appear, since there is only a limited amount of thiosulfate available in the body at a given time.

Fluoroacetic Acid. The synthesis of fluoroacetic acid is part of a defense mechanism of certain plants. The South African *gilfbaar* plant contains lethal quantities of fluoroacetic acid; cattle that eat leaves of this plant usually sicken and die.

$$F-\underset{H}{\overset{H}{C}}-C\underset{O-H}{\overset{\displaystyle O}{\Big\Vert}}$$

Fluoroacetic Acid

Sodium fluoroacetate, the sodium salt of this acid, is a potent rodenticide (rat poison). Because it is odorless and tasteless it is especially dangerous, and its sale in this country is strictly regulated by law. Fluoroacetate enters the Krebs cycle and is converted to the fluorocitrate ion, which mimics the citrate ion and blocks the Krebs cycle reaction.

Heavy Metals. Heavy metals are perhaps the most common of all the metabolic poisons. These include such common elements as lead and mercury, as well as many less common ones such as cadmium, chromium, and thallium. In this group we should also include the infamous poison, arsenic, which is not a metal but has many metal-like properties, including its toxic action.

Most heavy metals are cumulative poisons.

The effects of cumulative poisons add up.

Arsenic, a classic homicidal poison, occurs naturally in small amounts in many foods. Shrimp, for example, contain about 19 ppm arsenic, and corn may contain 0.4 ppm arsenic. Because some agricultural insecticides contain arsenic (Table 14–6), arsenic is observed in very small amounts on some fruits and vegetables. The Federal Food and Drug Administration (FDA) has set a limit of 0.15 mg arsenic per pound of food, an amount that apparently causes no harm. Several drugs, such as arsphenamine, which has found use in the treatment of syphilis, contain covalently bonded arsenic. In its ionic forms, arsenic is much more toxic.

Arsenic and heavy metals owe their toxicity primarily to their ability to react with and inhibit sulfhydryl (—SH) enzyme systems, such as those involved in the production of cellular energy. For example, glutathione (a tripeptide of glutamic acid, cysteine, and

Toxic Chemicals

Table 14-6
Some Arsenic-Containing Insecticides

Name	Formula
Lead arsenate	$Pb_3(AsO_4)_2$
Monosodium methanearsenate	$CH_3-\underset{\underset{OH}{\vert}}{\overset{\overset{O}{\Vert}}{As}}-O^-Na^+$
Paris green (copper acetoarsenite)	$3CuO \cdot 3As_2O_3 \cdot Cu(C_2H_3O_2)_2$

glycine) occurs in most tissues, where it is involved in maintaining healthy red blood cells, and its behavior with metals illustrates the interaction of metals with sulfhydryl groups. The metal replaces the hydrogen on two sulfhydryl groups on adjacent molecules (Fig. 14-4), and the strong bond that results effectively eliminates the two glutathione molecules from further reaction.

The typical forms of toxic arsenic compounds are inorganic ions such as arsenate (AsO_4^{3-}) and arsenite (AsO_3^{3-}). The reaction of an arsenite ion with sulfhydryl groups results in a complex in which the arsenic unites with two sulfhydryl groups, which may be on two different molecules of protein or on the same molecule:

$$^-O-As\begin{smallmatrix}O^-\\O^-\end{smallmatrix} + \begin{smallmatrix}H-S-C-\\H-S-C-\end{smallmatrix} \longrightarrow {}^-O-As\begin{smallmatrix}S-C-\\S-C-\end{smallmatrix} + 2OH^-$$

Arsenite   Sulfhydryl Groups   Arsenic Complex

The problem of developing a compound to counteract *Lewisite*, an arsenic-containing poison gas used in World War I, led to an understanding of how arsenic acts as a poison and subsequently to the development of an antidote. Once it was understood that Lewisite poisoned people by the reaction of arsenic with protein sulfhydryl groups, British scientists set out to find a suitable compound that contained highly reactive sulfhydryl groups that could compete with sulfhydryl groups in the natural substrate for the arsenic, and thus render the poison ineffective. Out of this research came a compound now known as British Anti-Lewisite (BAL).

The BAL, which bonds to the metal at several sites, is called a **chelating agent** (Greek *chela*, "claw"), a term applied to a reacting agent that envelops a species such as a metal ion. BAL is one of a large number of compounds that can act as chelating agents for metals (Fig. 14-5).

With the arsenic or heavy metal ion tied up, the sulfhydryl groups in vital enzymes are freed and can resume their normal functions. BAL is a standard therapeutic item in a hospital's poison emergency center and is used routinely to treat heavy metal poisoning.

Mercury deserves some special attention because it has a rather peculiar fascination for some people, especially children, who love to touch it. It is poisonous and accumu-

$\begin{matrix}CH_2-CH-CH_2\\|||\\OHSHSH\end{matrix}$
BAL
British Anti-Lewisite

A chelating agent encases an atom or ion similar to the way a crab or an octopus surrounds a bit of food.

$$2 \text{ Glutathione} + \text{Metal ion (M}^{2+}) \longrightarrow \text{M(Glutathione)}_2 + 2\text{H}^+$$

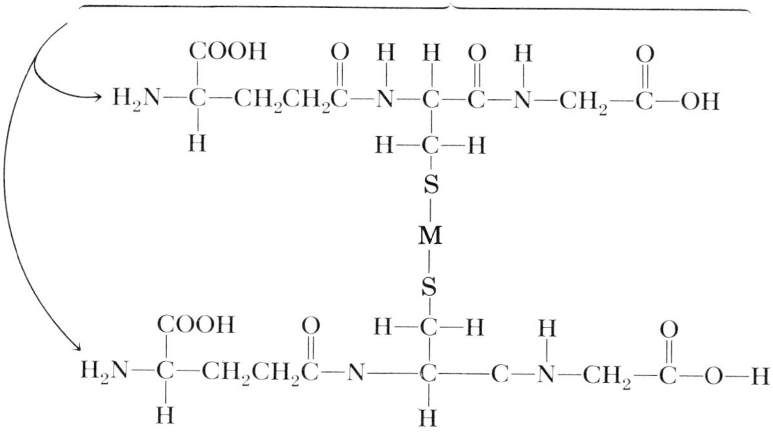

Glutathione-Metal Complex

Figure 14–4
Glutathione reaction with a metal (M).

lates in the body, as do its salts. This means the body has no quick means of ridding itself of mercury, so there tends to be a buildup of the toxic effects, leading to **chronic** poisoning.

Although mercury is rather unreactive compared with other metals, it is quite volatile and easily absorbed through the skin. In the body the metal atoms are oxidized to Hg_2^{2+} [mercury (I) ion] and Hg^{2+} [mercury (II) ion]. Compounds of both Hg_2^{2+} and Hg^{2+} are known to be toxic.

Today mercury poisoning is a potential hazard to dentists (who use mercury to make amalgams for fillings), various medical and scientific laboratory personnel (who routinely use mercury compounds or mercury pressure gauges), and some agricultural workers (who employ mercury salts as fungicides).

Mercury can also be a hazard when it is present in food. It is generally believed that mercury enters the food chain through small organisms that feed at the bottom of bodies of water that contain mercury from industrial waste or mercury minerals in the sediment. These in turn are food for bottom-feeding fish. Game fish that eat these bottom-feeding fish accumulate the largest concentration of mercury, the accumulation of poison building up as the food chain progresses.

Lead is another widely encountered heavy metal poison. The body's method of handling lead provides an interesting example of a "metal equilibrium" (Fig. 14–6).

A vivid characterization of someone affected by mercury poisoning can be found in the Mad Hatter, in Lewis Carroll's *Alice in Wonderland*. The fur-felt industry once used mercury (II) nitrate, $Hg(NO_3)_2$, to stiffen felt. In addition to causing odd behavior, chronic mercury poisoning gave the workers in hat factories symptoms known as "hatter's shakes."

An *amalgam* is any mixture or alloy of metals of which mercury is a constituent.

Figure 14–5
BAL chelation of arsenic or a heavy metal ion such as lead.

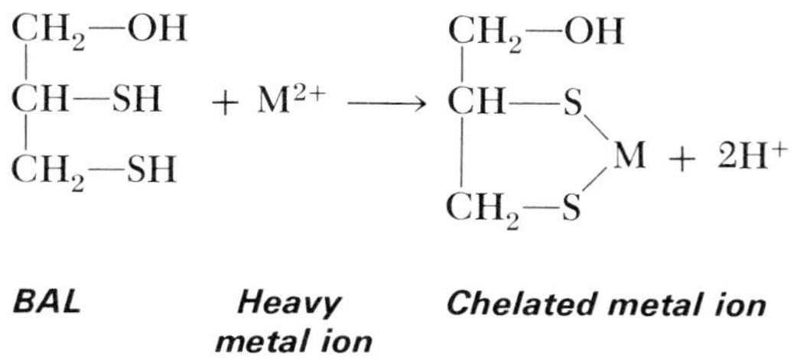

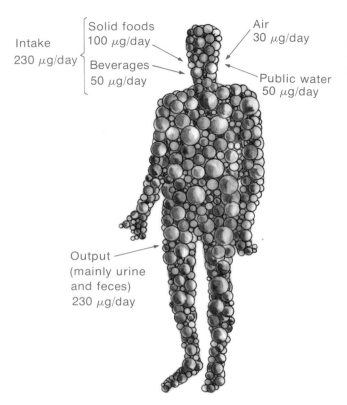

Figure 14–6
Lead equilibrium in humans. Figures chosen for intake are probable upper limits.

$1 \, \mu g$ (microgram) $= 10^{-6}$ g

$1 \, mg$ (milligram) $= 10^{-3}$ g $= 1{,}000 \, \mu g$

Lead is often present in foods (100–300 μg/kg), beverages (20–30 μg/liter), public water supplies (100 μg/liter, from lead-sealed pipes), and even air (2.5 μg/m³, from lead compounds in auto exhausts). With this many sources and contacts per day, it is obvious that the body must be able to rid itself of this poison; otherwise everyone would have died long ago of lead poisoning! The average person can excrete about 2 mg of lead a day through the kidneys and intestinal tract; daily intake is normally less than this. However, if intake exceeds this amount, accumulation and storage result. In the body lead resides in soft tissues, where it behaves like other heavy metal poisons such as mercury and arsenic, as well as in bone, where it acts on the bone marrow. Like mercury and arsenic, lead can also affect the central nervous system.

Lead salts, unless they are very insoluble, are always toxic, and their toxicity is directly related to the salt's solubility. One common covalent lead compound, tetraethyllead; ($Pb[C_2H_5]_4$), until recently a component of most gasolines, is different from most other metal compounds in that it is readily absorbed through the skin. Even metallic lead can be absorbed through the skin; cases of lead poisoning have resulted from repeated handling of lead foil, bullets, and other lead objects.

Even though lead-pigmented paints have not been used in this country for interior painting in the past 30 years, children are still poisoned by lead from old paint. Health experts estimate that up to 225,000 children become ill from lead poisoning each year, with many experiencing mental retardation or other neurological problems. The reason for this is twofold. First, lead-based paints still cover the walls of many old dwellings.

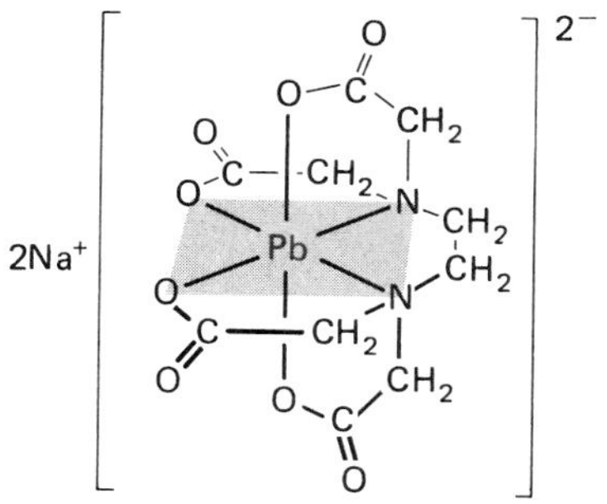

Figure 14-7
The structure of the chelate formed when the anion of EDTA envelops a lead (II) ion.

Second, many children in poverty-stricken areas, who are ill fed and anemic, develop a peculiar appetite trait called **pica,** and among the items that satisfy their cravings are pieces of flaking paint, which may contain lead. Lead salts also have a sweet taste, which many contribute to the consumption of lead-based paint. In 1969 about 200 children in the United States alone died of lead poisoning, and thousands have suffered permanent neurological damage.

Toxicologists have discovered an effective chelating agent to remove lead from the human body: ethylenediaminetetraacetic acid, also called EDTA (Fig. 14-7).

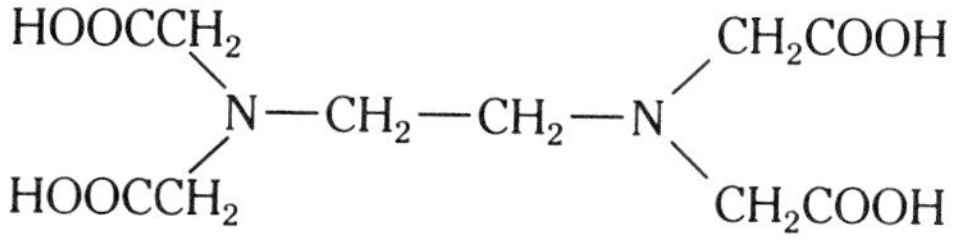

$$\begin{array}{c} HOOCCH_2 \\ \diagdown \\ N-CH_2-CH_2-N \\ \diagup \\ HOOCCH_2 \end{array} \begin{array}{c} CH_2COOH \\ \diagup \\ \diagdown \\ CH_2COOH \end{array}$$

EDTA
(Ethylenediaminetetraacetic Acid)

The calcium disodium salt of EDTA is used in the treatment of lead poisoning because EDTA by itself would remove too much of the blood serum's calcium. In solution, EDTA has a greater tendency to complex with lead (Pb^{2+}) than with calcium (Ca^{2+}). As a result, calcium is released and lead is tied up in the complex:

$$[CaEDTA]^{2-} + Pb^{2+} \longrightarrow [PbEDTA]^{2-} + Ca^{2+}$$

The lead chelate is then excreted in the urine.

Halogenated Solvents. Solvents like methylene chloride (CH_2Cl_2), chloroform ($CHCl_3$), and carbon tetrachloride (CCl_4) are metabolic poisons. These and other chlorinated solvents that destroy kidney tissue are **nephrotoxins.** In the kidney, chlorinated solvents are converted into substances that bind to proteins and cause a decrease in kidney function.

One sign that halogenated solvents are having toxic effects on the kidney is **polyuria,** increased production of urine. In severe cases of toxicity, **anuria**—no urine production—and death can occur.

Chlorinated solvents are useful for degreasing metal parts such as those found in machine shops and garages, for certain stain removal techniques, and for dry cleaning. These solvents should be used only with adequate ventilation and should never be allowed to touch the skin.

SELF-TEST 14-A

1. Corrosive poisons such as sulfuric acid destroy tissue by _____ followed by _____ of proteins.
2. Corrosive poisons, such as ozone, nitrogen dioxide, and iodine, destroy tissue by _____ it.
3. Carbon monoxide poisons by forming a strong bond with iron in _____ and thus preventing the transport of _____ from the lungs to the cells throughout the body.
4. CO is a cumulative poison.   True ( ) or False ( )
5. The cyanide ion has the formula _____. It poisons by complexing with iron in the enzyme _____, thus preventing the use of _____ in the oxidative processes in the cells.
6. Give an example of a metabolic poison that is toxic because its structure is so similar to a useful substance that it can mimic the useful substance. _____
7. BAL is an antidote for _____. BAL is effective because its sulfhydryl (—SH) groups _____ arsenic and heavy metals and render them ineffective toward enzymes.
8. Mercury is a cumulative poison.   True ( ) or False ( )
9. What organ is destroyed by chlorinated solvents? _____

Other Toxins

Neurotoxins

Investigations of the actions of neurotoxins have provided insight into how the nervous system works.

Some metabolic poisons are known to limit their action to the nervous system. These include strychnine and curare (a South American Indian dart poison), as well as the dreaded nerve gases developed for chemical warfare. The exact modes of action of most neurotoxins are not known for certain, but investigations have discovered the actions of a few.

A nerve impulse, or stimulus, is transmitted along a nerve fiber by electrical impulses. The nerve fiber connects with either another nerve fiber or with some other cell (such as a gland, or cardiac, smooth, or skeletal muscle) capable of being stimulated by the nerve impulse (Fig. 14–8). Neurotoxins often act at the point at which two nerve fibers come together, called a **synapse.** When the impulse reaches the end of certain nerves, a small quantity of **acetylcholine** is liberated. This activates a receptor on an adjacent nerve or organ. The acetylcholine is thought to activate a nerve ending by changing the permeability of the nerve cell membrane. How it does this is not clear, but it may be related to an ability to dissociate fat-protein complexes or to penetrate the surface films of fats. Such effects, which can be brought about by as little as 10^{-6} mole of acetylcholine, alter the permeability of a cell such that ions can cross the cell membrane more freely.

Permeability is the ability of a membrane to let chemicals pass through it.

10^{-6} of a mole of acetylcholine is 6×10^{17} molecules.

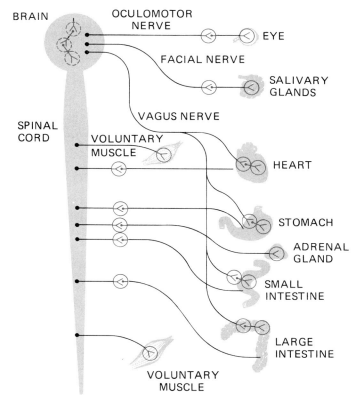

Figure 14-8 "Cholinergic" nerves, which transmit impulses by means of acetylcholine, include nerves controlling both voluntary and involuntary activities. Exceptions are parts of the "sympathetic" nervous system that utilize norepinephrine instead of acetylcholine. Sites of acetylcholine secretion are circled in color; poisons that disrupt the acetylcholine cycle can interrupt the body's communications at any of these points. The role of acetylcholine in the brain is uncertain, as is indicated by the broken circles.

To enable the receptor to receive further electrical impulses, the enzyme **cholinesterase** breaks down acetylcholine into acetic acid and choline (Fig. 14-9):

$$CH_3COCH_2CH_2\overset{CH_3}{\underset{CH_3}{N^+}}-CH_3, OH^- + H_2O \xrightarrow{cholinesterase} CH_3COH + HOCH_2CH_2\overset{CH_3}{\underset{CH_3}{N^+}}-CH_3, OH^-$$

Acetylcholine   Water   Acetic Acid   Choline

In the presence of potassium and magnesium ions, other enzymes such as acetylase resynthesize new acetylcholine from the acetic acid and the choline within the incoming nerve ending:

$$\text{Acetic acid} + \text{Choline} \xrightarrow{acetylase} \text{Acetylcholine} + H_2O$$

The new acetylcholine is available for transmitting another impulse across the gap.

Neurotoxins can affect the transmission of nerve impulses at nerve endings in a variety of ways. The **anticholinesterase poisons** prevent the breakdown of acetylcholine by deactivating cholinesterase. These poisons are usually structurally analogous to acetylcholine, so they bond to the enzyme cholinesterase and deactivate it. The cholinesterase molecules bound by the poison are held so effectively that the restoration of proper nerve function must await the manufacture of new cholinesterase. In the meantime, the excess acetylcholine overstimulates nerves, glands, and muscles, producing irregular heart rhythms, convulsions, and death. Many of the organic phosphates

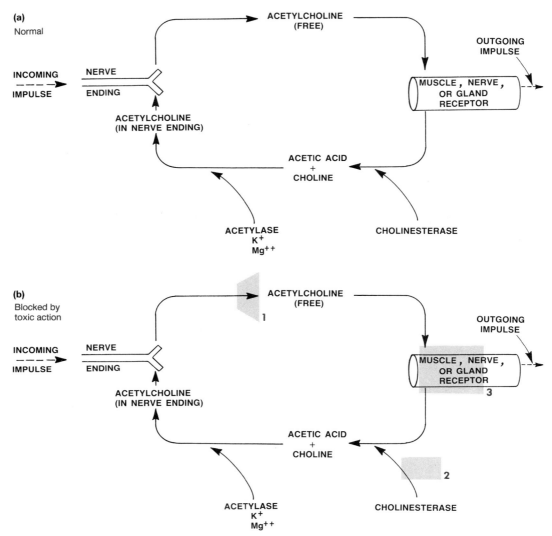

Figure 14-9
The acetylcholine cycle, a fundamental mechanism in nerve impulse transmission, is affected by many poisons. An impulse reaching a nerve ending in the normal cycle *(a)* liberates acetylcholine, which then stimulates a receptor. To enable the receptor to receive further impulses, the enzyme cholinesterase breaks down acetylcholine into acetic acid and choline; other enzymes resynthesize these into more acetylcholine. *(b)* Botulinus and dinoflagellate toxins inhibit the synthesis, or the release, of acetylcholine (1). The "anticholinesterase" poisons inactivate cholinesterase and therefore prevent the breakdown of acetylcholine (2). Curare and atropine desensitize the receptor to the chemical stimulus (3).

Curare, used by South American Indians in poison darts, was brought to Europe by Sir Walter Raleigh in 1595. It was purified in 1865, and its structure was determined in 1935.

that are widely used as insecticides are metabolized in the body to produce anticholinesterase poisons. For this reason they should be treated with extreme care. Some poisonous mushrooms also contain an anticholinesterase poison. Figure 14-10 shows the structures of some anticholinesterase poisons.

Neurotoxins such as **atropine** and **curare** are able to occupy the receptor sites on nerve endings of organs that are normally occupied by the impulse-carrying acetylcho-

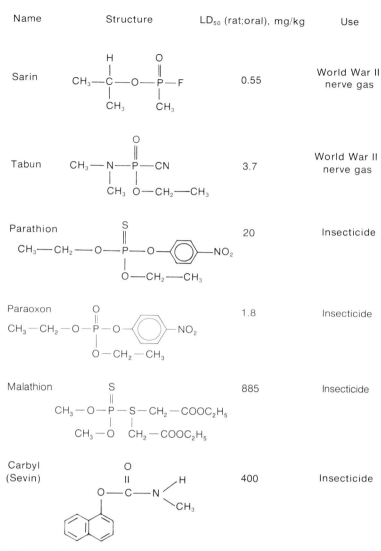

Figure 14–10
Some anticholinesterase poisons. In animals parathion is converted into paraoxon in the liver. Carbyl and malathion do not bind to cholinesterase as strongly. Malathion was the insecticide used in California in July 1981 to eradicate medflies.

line. When atropine or curare occupies the receptor site, no stimulus is transmitted to the organ. Acetylcholine in excess causes a slow heartbeat, a decrease in blood pressure, and excessive production of saliva; atropine and curare produce excessive thirst and dryness of the mouth and throat, a rapid heartbeat, and an increase in blood pressure. These neurotoxins cause the normal responses to acetylcholine activation to be absent and, if they are present in sufficient quantity, the opposite responses to occur.

Neurotoxins of this kind can be extremely useful in medicine. For example, atropine is used to dilate the pupil of the eye to facilitate examination of its interior. Applied to the skin, atropine sulfate and other atropine salts relieve pain by deactivating sensory nerve

Toxic Chemicals

endings on the skin. Atropine is also used as an antidote for anticholinesterase poisons. Curare has long been used as a muscle relaxant.

A well-known, natural organic compound that blocks receptor sites in a manner similar to that of curare and atropine is **nicotine.** This powerful poison causes stimulation and then depression of the central nervous system. The probable lethal dose for a 70-kg person is less than 0.3 g. It is interesting to note that pure nicotine was first extracted from tobacco and its toxic action observed only after tobacco use had been established as a habit.

Mutagens

> A mutagen is a chemical that can change the hereditary pattern of a cell.

Mutagens are chemicals capable of altering genes and chromosomes sufficiently to cause abnormalities in offspring. Chemically, mutagens alter the structures of DNA and RNA, which compose the genes (and in turn the chromosomes) that transmit the traits of parents to offspring. Mature sex, or germinal, cells of humans normally have 23 chromosomes; body, or somatic, cells have 23 *pairs* of chromosomes.

Although many chemicals are under suspicion because of their mutagenic effects on laboratory animals, it should be emphasized that no one has yet shown conclusively that any chemical induces mutations in human germinal cells. Part of the difficulty of determining the effects of mutagenic chemicals in humans is the extreme rarity of mutations. A specific genetic disorder may occur as infrequently as only once in 10,000 to 100,000 births; for meaningful statistical data to be obtained, a carefully controlled study of the entire population of the United States would be required. In addition, the very long time between generations presents great difficulties, and there is also the

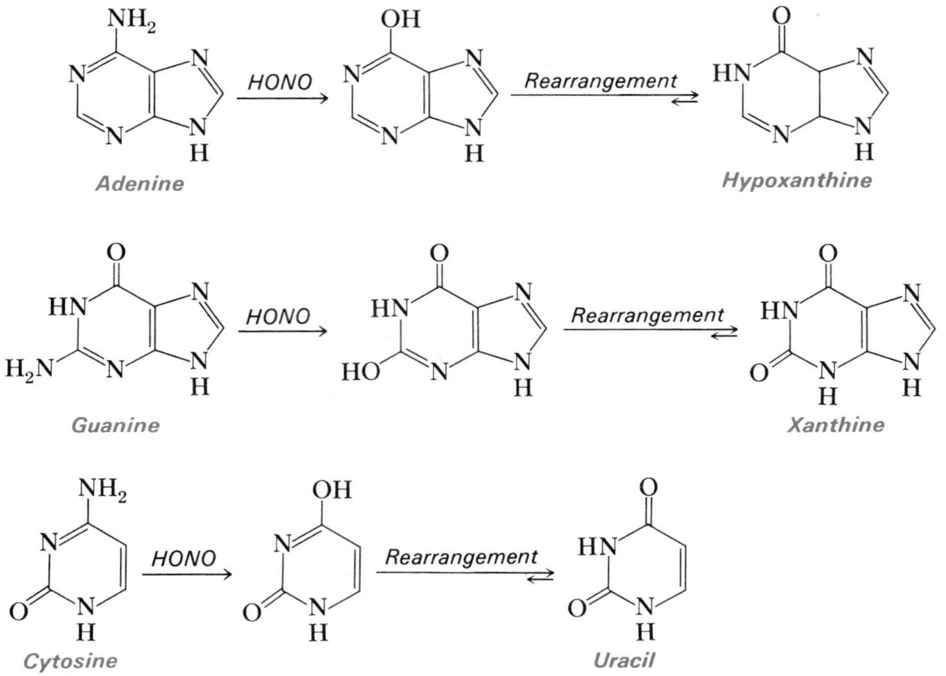

Figure 14–11
Reaction of nitrous acid (HONO) with nitrogenous bases of DNA. Nitrogen and water are also products of each reaction.

Table 14-7
Mutagenic Substances As Indicated by Experimental Studies on Plants and Animals

Substance	Experimental Results
Aflatoxin (from mold, *Aspergillus flavus*)	Mutations in bacteria, viruses, fungi, parasitic wasps, human cell cultures, mice
Benzo(α)pyrene (from cigarette and coal smoke)	Mutations in mice
Caffeine	Chromosomal changes in bacteria, fungi, onion root tips, fruit flies, human tissue cultures
Captan (fungicide)	Mutations in bacteria and molds; chromosome breaks in rats and human tissue cultures
Chloroprene	Mutations in male sex cells; spontaneous abortions in rats
Dimethyl sulfate (used extensively in chemical industry to methylate amines, phenols, and other compounds)	Methylation of DNA base guanine; mutations in bacteria, viruses, fungi, higher plants, fruit flies
LSD (lysergic acid diethylamide)	Chromosome breaks in somatic cells of rats, mice, hamsters, and in white blood cells of humans and monkeys
Maleic hydrazide (plant growth inhibitor; trade names, Slo-Gro, MH-30)	Chromosome breaks in many plants and in cultured mouse cells
Mustard gas (dichlorodiethyl sulfide)	Mutations in fruit flies
Nitrous acid (HNO_2)	Mutations in bacteria, viruses, fungi
Ozone (O_3)	Chromosome breaks in root cells of broadleaf plants
Solvents in glue (toluene, acetone, hexane, cyclohexane, ethyl acetate)	Breaks and abnormalities in human white blood cells
TEM (triethylenemelamine; anticancer drug, insect chemosterilant)	Mutations in fruit flies, mice

problem of tracing a medical disorder to a single specific chemical out of the tens of thousands of chemicals with which we come in contact.

Although there is no direct evidence for specific mutagenic effects in human beings, there is a great deal of interest in the subject. The possibility of a deranged, deformed human race is frightening, there is hope for the chance of an improved human body, and there is clear evidence of chemical mutations in plants and lower animals. A wide variety of chemicals are known to alter chromosomes and to produce mutations in rats, worms, bacteria, fruit flies, and other plants and animals. Some of these are listed in Table 14-7.

Experimental work on the chemical basis of the mutagenic effects of nitrous acid (HNO_2) has been very revealing. Repeated studies have shown that nitrous acid is a potent mutagen in bacteria, viruses, molds, and other organisms. In 1953 Dr. Stephen Zamenhof of Columbia University demonstrated experimentally that nitrous acid attacks DNA. Specifically, nitrous acid reacts with the adenine, guanine, and cytosine bases of DNA by removing the amino group of each of these compounds. The eliminated group is replaced by an oxygen atom (Fig. 14-11). The changed bases may garble a part of DNA's genetic message, and in the next replication of DNA the new base may not form a base pair with the proper nucleotide base.

Sodium nitrite produces nitrous acid in the stomach.

For example, adenine (A) typically forms a base pair with thymine (T) (Fig. 14-12). However, when adenine is changed to hypoxanthine, the new compound forms a base pair with cytosine (C). In the second replication, the cytosine forms its usual base pair with guanine (G). Thus, where an adenine-thymine (A-T) base pair existed originally, a guanine-cytosine (G-C) pair now exists. The result is an alteration in the DNA's genetic coding, so that a different protein is formed later.

Do all of these findings mean that nitrous acid is mutagenic in humans? Not necessarily. We do know that **sodium nitrite** has been widely used as a preservative, color enhancer, and color fixative in meat and fish products for at least the past 30 years. It is currently used in such foods as frankfurters, bacon, smoked ham, deviled ham, bologna, Vienna sausage, smoked salmon, and smoked shad. In the human stomach the sodium nitrite is converted to nitrous acid by hydrochloric acid:

$$NaNO_2 + HCl \longrightarrow HNO_2 + NaCl$$

The Food and Drug Administration (FDA) now considers the mutagenic effects of nitrous acid in lower organisms sufficiently ominous to suggest strongly that the use of sodium nitrite in foods be severely curtailed, and a complete ban of this use of sodium nitrite is being considered. A number of European countries already restrict the use of sodium nitrite in foods. The concern is that this compound, after being converted in the body to nitrous acid, may cause mutations in somatic cells (and possibly in germinal cells) and thus could possibly produce cancer in the human stomach. However, some scientists doubt that nitrous acid is present in germinal cells and therefore seriously question whether this compound could be a cause of genetically produced birth defects in humans. The uncertainty of extrapolating results obtained in animal studies to human beings hovers over the mutagenic substances.

Thus far research has concentrated on the mutagenic properties of chemicals in bacterial viruses, molds, fruit flies, mice, rats, human white blood cells, and so on. Perhaps in the next 10 to 20 years it will be demonstrated that these chemicals can produce transmissible alterations of chromosomes in human germinal cells. Meanwhile, many scientists are pressing for more vigorous research to expand our knowledge of chemically induced mutations and of their potentially harmful effects. One intriguing

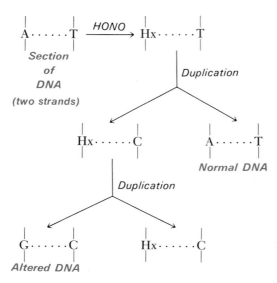

Figure 14-12
Alteration of DNA genetic code by base-pairing of nitrous acid-converted nitrogenous bases. The bases are adenine (A), cytosine (C), guanine (G), hypoxanthine (Hx), and thymine (T).

theory is that some compounds cause cancer because they are first and foremost mutagenic; the supporting evidence at present is still extremely inconclusive.

Teratogens

The effects of chemicals on human reproduction are a frightening aspect of toxicity. The study of birth defects produced by chemical agents is the discipline of **teratology.** The word root *terat* comes from the Greek for "monster." There are three known classes of teratogens: radiation, viral agents, and chemical substances.

Birth defects occur in 2% to 3% of all births. About 25% of these are due to genetic causes (perhaps including contact with mutagens), and 5% to 10% are due to teratogens. The remaining 60% or so result from unknown causes.

In the development of the newborn, there are three basic periods during which the fetus is at risk. For a period of about 17 days between conception and implantation in the uterine wall, a chemical "insult" may result in cell death. The rapidly multiplying cells often recover, but if a lethal dose of a toxin is administered, the embryo dies and is then aborted or reabsorbed. During the critical embryonic stage (18 to 55 days) of organogenesis, or development of the organs, the fetus is extremely sensitive to teratogens. During the fetal period (56 days to term) the fetus is less sensitive. Contact with teratogens results in a reduction in cell size and number. This is manifested in growth retardation and the failure of vital organs to reach maturity.

The horrible thalidomide disaster in 1961 focused worldwide attention on chemically induced birth defects. Thalidomide, prescribed to pregnant women to overcome morning sickness, caused gross deformities (flipperlike arms, shortened arms, no arms or legs, and other defects) in children whose mothers used the drug during the first two months of pregnancy. The use of this drug resulted in more than 4000 surviving malformed babies in West Germany, more than 1000 in Great Britain, and about 20 in the United States. (Lack of Food and Drug Administration approval of thalidomide kept the United States figure from being higher; see Chapter 16.) With shattering impact, this incident demonstrated that a compound can appear on the basis of animal studies to be remarkably safe and yet cause catastrophic effects in humans.

Since thalidomide has an asymmetric carbon atom, two optical isomers are possible. After the thalidomide disaster, it was discovered that D-thalidomide is safe and that L-thalidomide is a teratogen.

Thalidomide
*Indicates the asymmetric carbon atom

Although the thalidomide tragedy focused attention on chemical mutagens, the drug presumably does not cause genetic damage in the germinal cells and is really not mutagenic. Rather, when taken by a woman during early pregnancy, it causes direct injury to the developing embryo.

Any chemical substance that can cross the placenta is a potential teratogen, and any activity resulting in the uptake into the mother's blood of such a chemical may prove dangerous to the health and well-being of the fetus. Smoking a cigarette results in higher-than-normal blood levels of such substances as carbon monoxide, hydrogen cyanide, cadmium, nicotine, and benzo(α)pyrene. Of course, many of these substances are present in polluted air as well. Table 14-8 lists a number of chemical substances known to be teratogenic in humans and laboratory animals.

Carcinogens

Carcinoma is cancer of the epithelial tissue.

Sarcoma is cancer of the connective tissue.

Carcinogens are chemicals that cause **cancer,** an abnormal growth condition in an organism that manifests itself in at least three ways. First, the rate of cell growth (that is, the rate of cellular multiplication) in cancerous tissue differs from that in normal tissue; cancerous cells may divide more rapidly or more slowly than normal cells. Second, cancerous cells spread to other tissues; they know no bounds. Normal liver cells divide and remain a part of the liver; cancerous liver cells may leave the liver and later be found, for example, in the lung. Third, most cancer cells show partial or complete loss of specialized functions. Although located in the liver, for example, cancer cells no longer perform the functions of the liver.

Attempts to determine the cause of cancer have evolved from early studies in which the disease was linked to a person's occupation. It was first noticed in 1775 that persons employed as chimney sweeps in England had a higher rate of skin cancer than the general population. It was not until 1933 that **benzo(α)pyrene** ($C_{20}H_{12}$, a 5-ringed aromatic hydrocarbon) was isolated from coal dust and shown to be metabolized in the body to produce one or more carcinogens. In 1895 the German physician Rehn noted three cases of bladder cancer among employees of a factory that manufactured dye intermediates in the Rhine Valley. Rehn attributed these cancers to his patients' occupation. These and other cases confirmed that at times as many as 30 years passed between

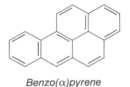

Benzo(α)pyrene

Table 14-8
Teratogenic Substances

Substances	Species	Effects on Fetus
Metals		
Arsenic	Mice, hamsters	Increase in males born with eye defects, renal damage
Cadmium	Mice, rats	Abortions
Cobalt	Chickens	Eye, lower limb defects
Gallium	Hamsters	Spinal defects
Lead	Humans, rats, chickens	Low birth weights, brain damage, stillbirth, early and late deaths
Lithium	Primates	Heart defects
Mercury	Humans	Minamata disease (Japan)
	Mice	Fetal death, cleft palate
	Rats	Brain damage
Thallium	Chickens	Growth retardation, abortions
Zinc	Hamsters	Abortions
Organic Compounds		
DES (diethylstilbestrol)	Humans	Uterine anomalies
Caffeine (15 cups per day equivalent)	Rats	Skeletal defects, growth retardation
PCBs (polychlorinated biphenyls)	Chickens	Central nervous system and eye defects
	Humans	Growth retardation, stillbirths

the time of the initial employment and the occurrence of bladder cancer. The principal product of these factories was aniline. Although aniline was first thought to be the carcinogenic agent, it was later shown to be noncarcinogenic. It was not until 1937 that continuous long-term treatment with up to 5 g a day of **2-naphthylamine,** one of the suspected dye intermediates, was found to produce bladder cancer in dogs. Since then other dye intermediates have also been shown to be carcinogenic.

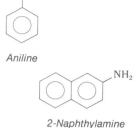

Aniline

2-Naphthylamine

A vast amount of research effort has verified the carcinogenic behavior of a large number of diverse chemicals (Table 14-9). Carcinogenic effects on lower animals are commonly extrapolated to humans. The mouse has come to be the classic animal for studies of carcinogenicity. Strains of inbred mice and rats have been developed that are genetically uniform and show a standard response.

Some carcinogens are relatively nontoxic in a single large dose but are quite toxic, often increasingly so, when administered continually. Thus, much patience, time, and money must be expended in carcinogenic studies. The development of a sarcoma in humans, from the activation of the first cell to the clinical manifestation of the cancer, takes from 20 to 30 years. With life expectancy of an average person in the United States now set at about 70 years, it is not surprising that the number of deaths due to cancer is increasing.

Cancer does not occur with the same frequency in all parts of the world. Breast cancer occurs with a lower frequency in Japan than in the United States or Europe. Cancer of the stomach, especially in males, is more common in Japan than in the United States. Cancer of the liver is not widespread in the Western Hemisphere but accounts for a high proportion of the cancers among the Bantu in Africa and in certain populations in the Far East. The widely publicized incidence of lung cancer is higher in the industrialized world and is increasing at an appreciable rate.

Some compounds cause cancer at the point of contact. Other compounds cause cancer in an area remote from the point of contact. The liver, the site at which most toxic chemicals are removed from the blood, is particularly susceptible to the latter compounds. Since such a compound does not cause cancer on contact, some other compound made from it must be the cause of cancer. For example, it appears that the substitution of an ⊃NOH group for an ⊃NH group in an aromatic amine derivative produces at least one of the active intermediates for carcinogenic amines. If R denotes a two- or three-ring aromatic system, then the process can be represented as follows:

Cancer spreads from one tissue to another via *metastases*.

An abnormal growth is classified as cancerous or malignant when examination shows it is invading neighboring tissue. A growth is benign if it is localized at its original site.

$$\underset{\substack{\text{Inactive} \\ \text{on Contact}}}{\text{RNCOCH}_3^{\text{H}}} \longrightarrow \underset{\substack{\text{Active on} \\ \text{Contact}}}{\text{RNCOCH}_3^{\text{OH}}} \longrightarrow \underset{\substack{\text{Other Unknown} \\ \text{Intermediates}}}{\text{RX?}} \longrightarrow \text{RY?} \xrightarrow{\text{tissue}} \text{Tumor cell}$$

As indicated by the variety of chemicals in Table 14-9, a variety of molecular structures produce cancer. Some structures closely related to carcinogenic structures are noncarcinogenic. For example, the 2-naphthylamine mentioned earlier is carcinogenic, but repeated testing gives negative results for 1-naphthylamine.

1-Naphthylamine
(Noncarcinogenic)

2-Naphthylamine
(Carcinogenic)

Table 14-9
Chemicals Carcinogenic for Humans

Compound	Formula	Use or Source	Site Affected	Confirming Animal Tests*
Inorganic Compounds				
Arsenic (and compounds)	As	Insecticides, alloys	Skin, lung, liver	—
Asbestos	$Mg_6(Si_4O_{11})(OH)_6$	Brake linings, insulation	Respiratory tract	+
Beryllium	Be	Alloy with copper	Bone, lung	+
Cadmium	Cd	Metal plating	Kidney, lung	+
Chromium	Cr	Metal plating	Lung	+
Organic Compounds				
Benzene	(benzene ring)	Solvent, chemical intermediate in syntheses	Blood (leukemia)	+
Acrylonitrile	$CH_2{=}CH(CN)$	Monomer	Colon, lung	+
Aflatoxins B_1 (shown)	(structure)	Mold or peanuts	Liver	+
Carbon tetrachloride	CCl_4	Solvent	Liver	+
Diethylstilbestrol	$HO{-}\text{(ring)}{-}C(C_2H_5){=}C(C_2H_5){-}\text{(ring)}{-}OH$	Hormone	Female genital tract	+
Benzo(α)pyrene	(structure)	Cigarette and other smoke	Skin, lung	+
Benzidine	$H_2N{-}\text{(ring)}{-}\text{(ring)}{-}NH_2$	Dye manufacture, rubber compounding	Bladder	+
Ethylene oxide	$CH_2{-}CH_2$ with O bridge	Chemical intermediate used to make ethylene glycol, surfactants	Gastrointestinal tract	+
Soots, tar, and mineral oils		Roofing tar, chimney soot, oils of hydrocarbon nature	Skin, lung, bladder	+
Vinyl chloride	$CH_2{=}CHCl$	Monomer for making PVC	Liver, brain, lung, lymphatic system	+

* +, positive supporting data; —, lack of supporting data; ±, conflicting data.

For some types of cancer there are distinct stages that ultimately result in cancer. These may be identified as the *initiation period*, the *development* or *promotion period*, and the *progression period*. A single minute dose of a carcinogenic polynuclear aromatic hydrocarbon, such as benzo(α)pyrene, applied to the skin of a mouse produces the permanent change of a normal cell to a tumor cell. This is the initiation step. No noticeable reaction occurs unless further treatment is made. If the area is painted repeatedly with noncarcinogenic croton oil, even up to one year later, carcinomas appear (Fig. 14–13); this is the development period. Additional fundamental alterations in the nature of the cells occur during the progression period. If there is no initiator, there are no tumors. If there is initiator but no promoter, there are no tumors. If the initiator is followed by repeated doses of promoter, tumors appear. This seems to indicate that cancer cannot be contracted from chemicals unless repeated doses are administered.

Just how do these toxic substances work? One theory is that cancer is caused when a carcinogen combines with growth control proteins, rendering them inactive. During the normal growth process, the cells divide and the organism grows to a point and stops. Cancer is abnormal in that cells continue to divide and portions of the organism continue to grow. One or more proteins are thought to be present in each cell with the specific duty of preventing replication of DNA and cell division. Virtually all of the carcinogens bind firmly to proteins, but so do some similar compounds that are noncarcinogenic. The specific growth proteins involved are not yet known for any of the carcinogens, despite considerable efforts to find them.

Another theory suggests that carcinogens react with and alter nucleic acids so the proteins ultimately formed on the messenger RNA are sufficiently different to alter the cells' functions and growth rates. The carcinogen may be included in the DNA or RNA strands by covalent bonding or it may be entangled in the helix and held by weak intermolecular bonding. The carcinogenic compounds nitrosodimethylamine and mustard gas have been shown to react with nucleic acids.

While researchers collect data in their laboratories and speculate on the theoretical structural causes of cancer, we can studiously avoid compounds known to cause cancer

Smoking is thought to both initiate and promote cancer.

Almost all chemical carcinogens have an induction period.

Smoking is associated with over 20% of all cancers; asbestos with between 3% and 18%.

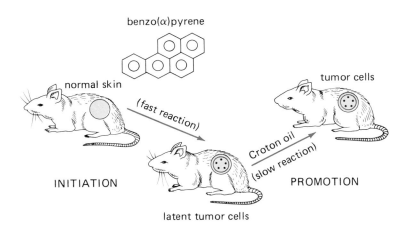

Figure 14–13
A second chemical can promote tumor growth in mice after an initiation period. Treatment of mice with croton oil produces no tumor, and neither does treatment with small quantities of benzo(α)pyrene alone. Croton oil is an irritant similar to castor oil; both are derived from plants.

in humans. It has been proposed that as much as 80% of all human cancer has its origin in carcinogenic chemicals.

The Liver: The Body's Detoxification Factory

The normal functions of the liver include carbohydrate metabolism and storage, synthesis of blood proteins, urea formation, bile formation, and the metabolism of fats, hormones, and foreign chemicals. Because of its position in the circulatory system, the liver receives most of the blood flow returning to the heart (Fig. 14-14).

The liver is the primary organ for **biotransformation,** metabolic processes that alter foreign compounds into inactive biological forms. In this process foreign chemicals usually become more polar and hence more water-soluble and are eventually excreted from the body. If the liver did not detoxify benzene, the time required to remove a single small dose of it from the body would be about 100 years.

Two types of biotransformation reactions take place in the liver and, to a lesser extent, in the lungs and the kidneys. The first type of reaction involves a group of

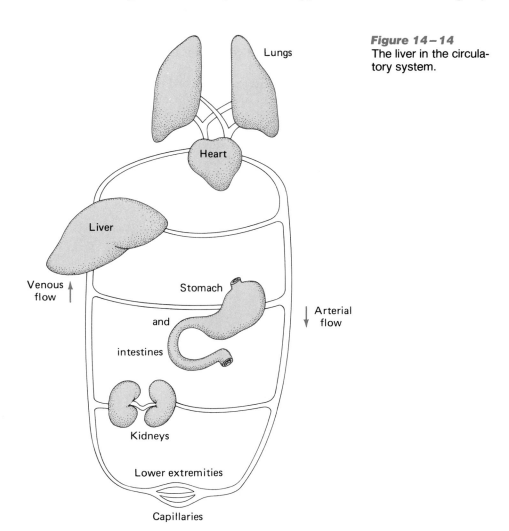

Figure 14-14
The liver in the circulatory system.

enzymes called cytochrome P-450 enzymes. These enzymes can catalyze the alteration of a foreign chemical by attaching some functional group to the molecule. An example of this type of adding-on reaction is the conversion of bromobenzene to the corresponding hydroxy compounds, which are more polar, more readily soluble in polar water, and more easily excreted in the urine.

Bromobenzene $\xrightarrow[\text{liver}]{\text{cytochrome P-450}}$ hydroxy products

A second type of biotransformation reaction involves the bonding of a high-molecular-weight molecule to the foreign molecule. Presumably, this immobilizes the harmful substance. An example of such a biotransformation is the conversion of acetaminophen (the active ingredient in Tylenol) to a glucuronide by the enzyme glucuronyl transferase:

Acetaminophen + Glucuronic acid $\xrightarrow[\text{liver}]{\text{glucuronyl transferase}}$ glucuronide

The liver's biotransformations do not always lead to the detoxification of chemicals. It is even possible for harmful reactions to take place; for example, the liver **bioactivates** some chemicals to make them more toxic.

The cholinesterase poison parathion is both detoxified and bioactivated in the liver. In the reaction shown in Figure 14–15, the detoxification step of ester hydrolysis produces two chemicals that are less toxic, whereas the bioactivation step produces an organophosphate cholinesterase poison, which on the basis of LD_{50} values for mice appears to be paraoxon, about 10 times more toxic than parathion.

Conclusions

This chapter has not covered all the toxic properties of chemical substances, but it does suggest the risks associated with exposure to toxic chemicals. The problem is that you do not always know which chemicals are toxic and which are not — a good argument for **avoiding exposure** to all chemicals unless you know they are nontoxic.

If you have a low body weight, are pregnant, or have some health problems, exposure to chemicals can put you at even greater risk than is the population as a whole. Some chemicals cause immediate (acute) symptoms upon exposure, some cause only

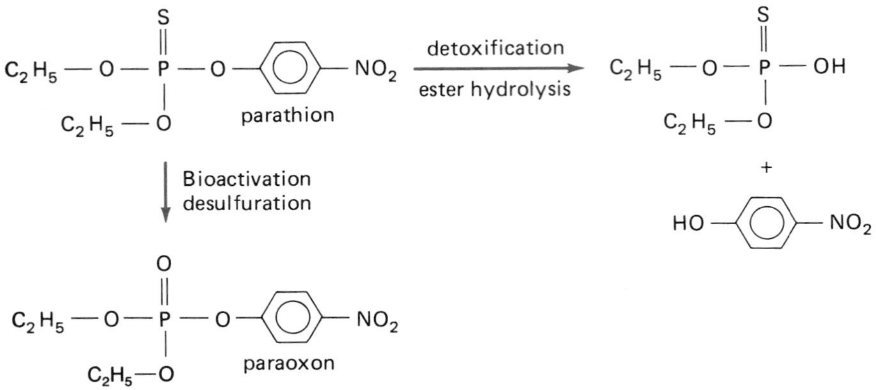

Figure 14–15
Detoxification and bioactivation of parathion in the liver.

chronic symptoms, and some cause delayed symptoms, such as the appearance of cancer many years after chronic exposure.

We do not need to be afraid of toxic chemicals, but we need to treat them with respect. We also need to control them and limit their presence in the air we breathe, the water we drink, and the food we eat.

SELF-TEST 14-B

1. Substrates that poison the nervous system are called _____.
2. Most neurotoxins affect chemical reactions that occur in the region between two nerve cells. These regions are called _____.
3. The electrical impulse is carried across a synapse by the chemical _____.
4. Mutagens alter the structures of _____ or _____.
5. If a substance is mutagenic in test animals, particularly dogs, it must necessarily be mutagenic in human beings. True ( ) or False ( )
6. The first occupation definitely linked to cancer was _____.
7. The organ where most biotransformations take place is the _____.
8. The process whereby a chemical is converted into a more toxic product is called _____.
9. Two dangers associated with smoking are _____ and _____.
10. A chemical that can cross the placenta and harm the fetus is called a _____.

MATCHING SET

_____ 1. Metabolic poison
_____ 2. Metabolism
_____ 3. Corrosive poison
_____ 4. Neurotoxin
_____ 5. Mutagen
_____ 6. Melanoma
_____ 7. Carcinogen
_____ 8. Metastases
_____ 9. Sarcoma
_____ 10. Chelating agent
_____ 11. Biotransformation
_____ 12. Lead
_____ 13. Nephrotoxin

a. Process carried out by the liver
b. Cyanide ion
c. Cancer in connective tissue
d. Benzo(α)pyrene metabolite
e. Cancerous growths of lung tissue located in liver
f. Use of chemicals in the body
g. Kidney toxin
h. Sodium hydroxide (caustic soda)
i. Atropine
j. Alters DNA
k. EDTA
l. Acetylcholine
m. Ingestion associated with pica
n. Cancer of the skin

Questions

1. Give an example of a toxic substance that is toxic as a result of the following:
   a. Binding to an oxygen-carrying molecule
   b. Disguising itself as another compound
   c. Attack on an enzyme
   d. Hydrolysis
2. True/False. Explain each answer concisely.
   a. Lead is a corrosive poison.
   b. Carbon monoxide and the cyanide ion poison in the same way.
   c. There are no known chemical compounds that cannot be toxic under some circumstances.
3. The application of a single minute dose of a fused-ring hydrocarbon such as benzo(α)pyrene fails to produce a tumor in mice. Does this mean this compound is definitely noncarcinogenic? Give a reason for your answer.
4. Describe the chemical mechanism by which the following substances show their toxic effects:
   a. Fluoroacetic acid
   b. Phosgene
   c. Curare
5. Should any laws and regulations be placed on the use of any of the following? Justify your answers.
   a. Lead in paint   c. Chlorinated solvents
   b. Mercury   d. Thalidomide
6. Discuss some of the pros and cons of testing toxic substances on animals.
7. Give chemical reactions in words for the following:
   a. Action of NaOH on tissue
   b. Action of carbon monoxide in blood
   c. Reaction of EDTA and the lead ion
8. What questions do you think need to be answered before the effects of smoking and cancer are understood?
9. Assume a normal diet has the quantity of lead in a given quantity of food stated in the text. What would be a person's total food intake of lead per day?
10. What is the meaning of the symbol LD_{50}?
11. Write chemical equations for the following:
    a. The hydrolysis of acetylcholine
    b. Acid hydrolysis of a protein having a glycine-glycine primary structure
12. Describe how the corrosive poisons lye (NaOH), NO_2, and the hypochlorite ion (OCl^-) destroy tissue.
13. What are some common sources of carbon monoxide?
14. If a relatively small amount of carbon monoxide is inhaled, are the chemical reactions reversible, or is carbon monoxide a cumulative poison?
15. What poisons can be rendered ineffective by wrapping a large molecule around them?

16. What is the cause of pica?
17. What organ detoxifies most toxic chemicals?
18. Give two examples each of a corrosive poison, a metabolic poison, a neurotoxic poison, a mutagenic poison, and a carcinogenic poison.
19. Phosgene hydrolyzes in the lungs to produce what acid?
20. What concentration of carbon monoxide in the air is likely to cause death in 1 hour?
21. Is it possible that one molecule of a mutagen could pose a threat to human life?
22. Two new chemicals are prepared in the lab. One is a relatively simple acid with corrosive properties; the other has carcinogenic properties. Which chemical's toxic property is more likely to be discovered?
23. An old laboratory chemical is discovered to have a new property: it reacts with amine groups to produce —OH groups. Could this chemical be a mutagen?
24. If a poisoned victim has pinpoint pupils and is salivating excessively, what type of poison did he or she probably ingest?
25. If your drinking water contains numerous toxic substances, why are you not normally harmed by drinking it?
26. What is bioactivation? Explain.

CHAPTER 15

Water and Air Pollution

Two of the necessary ingredients of all forms of life, water and air, are in danger of being made unfit for life by the presence of toxic chemicals.

Accidental chemical spills are devastating to the aquatic environment. The November, 1986, spill into the Rhine River involved mercury and a wide variety of toxic organic compounds such as pesticides, and it killed virtually all marine life from Basel, Switzerland, to the coast of the Netherlands. In recent years other spills by large, ocean-going oil tankers sinking or running aground have also caused damage to marine life. In 1983, 11,000 polluting incidents discharged about 30 million gallons of pollutants into the waters of the United States.

In addition to chemical spills, there are toxic chemicals that migrate into surface and groundwaters from hazardous waste dump sites and underground storage tanks (Fig. 15-1). We face the risk of being harmed by toxic pollutants in water before we succeed in cleaning them up.

Our atmosphere is also polluted. This is nothing new. In *Hamlet* (Act II, Scene 2), Shakespeare wrote about air pollution in the 17th century:

> this most excellent canopy, the air,
> look you, this excellent o'erhanging
> firmament, this magestical roof fretted
> with golden fire, why, it appears
> no other thing to me but a foul
> and pestilent congregation of vapors.

Nature pollutes the air on a massive scale with volcanic ash, mercury vapor, and hydrogen sulfide from volcanoes and reactive and odorous organic compounds from coniferous plants such as pine trees. But automobiles, power plants, smelting plants, other metallurgical plants, and petroleum refineries add significant quantities of toxic chemicals to the atmosphere, especially in heavily populated areas. Atmospheric pollutants cause burning eyes, coughing, acid rain, smog, the destruction of ancient monuments, and even the destruction of the atmosphere itself.

In this chapter we look at how chemicals pollute the water and the atmosphere. We also look at some of the ways in which these chemicals get into the environment and how they can be removed. Preventing these chemicals from entering the atmosphere is not always easy, and getting rid of them is often seemingly impossible. Air and water pollutants know no international boundaries, so prevention and cleanup require cooperation among nations. Even if we stopped polluting today, the enormous cleanup tasks would probably take the rest of our lives.

Pollution of water is a major problem. Who causes these problems? Who pays for the cleanup?

Pollutants are unwanted substances found in our water and air.

Figure 15-1
Use of polymer-lining material is one way to help ensure that a leaking underground storage tank will not contaminate groundwater. Many states require monitoring of underground tanks as well as secondary containment such as a liner. (Photo courtesy of the Engineering Polymers Division of DuPont.)

Water: A Special Compound

There would be no life on Earth without water and its unique properties. What are the unique properties of water, and what is their effect on life as we know it?

1. The density of solid water (ice) is less than that of liquid water; that is, water expands when it freezes. If ice were a normal solid, it would be more dense than liquid water, and lakes would freeze from the bottom up. This would have disastrous consequences for marine life, which could not survive in areas with winter seasons.
2. Water is a liquid at room temperature. However, the hydrogen compounds of all nonmetals around oxygen in the periodic table are toxic, corrosive gases such as NH_3, H_2S, and HF.
3. Water has a high heat capacity per unit of weight. This means it can absorb relatively large quantities of heat without undergoing large changes in temperature. For comparison, the heat capacity of water is about ten times that of copper or iron for equal weights. This property accounts for the moderating influence of lakes and oceans on the climate. Huge bodies of water absorb heat from the Sun and release it at night or in cooler seasons. The Earth would have extreme temperature variations if it weren't for this unique property of water.

Heat capacity is defined as the amount of heat required to raise the temperature of a sample of matter 1°C (Celsius).

4. Water has the highest heat of vaporization of all known substances. The heat needed to vaporize 1 g of water at 100°C is 540 cal. A consequence of this is the cooling effect that occurs with perspiration because of the heat absorbed from the skin when water evaporates.
5. Water has a large surface tension. This property and the ability of water to wet surfaces are the bases for capillary action, which carries water to leaves in plants and trees.
6. Water is an excellent solvent, often referred to as the universal solvent. As a result, natural water is not pure water but a solution of substances dissolved by contact with water.

Surface molecules of a liquid are pulled inward by the intermolecular interactions with molecules below the surface. Surface tension is a measure of this force.

The causes of these unique properties of water are hydrogen bonding between water molecules and the polarity of the water molecule. The extensive hydrogen bonding that occurs between water molecules in liquid water and in ice was described in Chapter 5.

The high boiling point, high heat of vaporization, and large heat capacity of water are a result of the energy needed to break the hydrogen bonds in liquid water as it is heated or vaporized.

Review the discussion of hydrogen bonding in Chapter 5.

Water: The Most Abundant Compound

Water is the most abundant substance on the Earth's surface. Oceans cover about 72% of the Earth and have an average depth of 2.5 miles. The oceans are the source of 97.2% of all water; glaciers provide 2.16%; fresh water in lakes and rivers, 0.0197%; groundwater, 0.61%; brine wells and brackish waters, 0.01%; atmospheric water, 0.001%.

Water is the major component of all living things. The water content of a human adult is 70%, the same proportion as the Earth's surface (Table 15–1).

It is estimated that an average of 4350 billion gallons of rain and snow fall on the contiguous United States each day. Of this amount, 3100 billion gallons return to the atmosphere by evaporation and transpiration. Discharge to the sea and to underground reserves amounts to 800 billion gallons daily, leaving 450 billion gallons of surface water

Brackish water contains dissolved salts but at a lower level than sea water.

Table 15–1
Water Content (%)

Marine invertebrates	97
Human fetus (1 month)	93
Adult human	70
Body fluids	95
Nerve tissue	84
Muscle	77
Skin	71
Connective tissue	60
Vegetables	89
Milk	88
Fish	82
Fruit	80
Lean meat	76
Potatoes	75
Cheese	35

each day for domestic and commercial use. The 48 contiguous states withdrew from natural sources 40 billion gallons per day in 1900 and 430 billion gallons per day in 1980, and it is estimated that the demand will be at least 900 billion gallons per day by the year 2000. The demand for water by our growing population is already greater than the resupply by natural resources in many parts of the country.

Public Use of Water

In the United States average daily use of water is 60-gallons per person. Table 15-2 gives the average amounts for various personal uses. One factor that contributes to the water crisis is that only **one half gallon** of water per person per day must be of drinkable quality. Although urban water delivery systems are not set up to deliver two types of water — drinkable water and water for other purposes — dual water supply systems are feasible. Only a small fraction of municipal water would have to be of drinking water quality. The largest portion of a nonpotable water supply would be disinfected and bacteriologically safe but would not meet drinking water regulations; this second source would be suitable for the irrigation of lawns, parks, and golf courses, as well as for air conditioning, industrial cooling, and toilet flushing.

Where do we get the water we drink? Groundwater and surface water sources each provide half of the more than 35 billion gallons of drinking water per day used in the United States. The water listed for public supplies in Table 15-2 is potable water. How pure is this water? The U.S. Public Health Service has set standards for potable water (Table 15-3). Waste water from sewage treatment plants, industries, and agricultural runoff goes into lakes and rivers, so water purification is necessary.

The Clean Water Act of 1977 shifted the burden of producing water suitable for reuse from the user to the waste water discharger. This action was a crucial step in improving the quality of our rivers and lakes, since it is easier to clean waste water before it is dumped than to clean river water after untreated waste has been discharged into it. In addition, the quality of the waste water effluent is often high enough to be used as a source of water for other purposes, such as for irrigation or in cooling towers.

The Scope of Water Pollutants

There was a time when polluted water could be thought of in terms of dissolved minerals, natural silt, and contaminants associated with the natural wastes of animals and humans. As our use of water has increased, water pollution has become more

Table 15-2
Average Water Usage (by Gallons) per Person per Day

Flushing toilets	24
Bathing	18
Laundering	9
Dishwashing	3.6
Drinking and cooking	3.0
Miscellaneous	2.4
Total	60.0

Table 15-3
U.S. Public Health Service Standards for Potable Water

Contaminating Ion(s)	Maximum Concentration (mg/L)
Arsenic	0.05
Barium	1.00
Cadmium	0.01
Chloride	250.00
Chromium	0.05
Copper	1.00
Cyanide	0.20
Fluoride	2.00
Iron	0.30
Lead	0.05
Manganese	0.05
Nitrate	45.00
Organics	0.20
Selenium	0.01
Silver	0.05
Sulfate	250.00
Zinc	5.00
Total dissolved solids	500.00

diversified. The U.S. Public Health Service now classifies water pollutants into the eight broad categories listed in Table 15-4.

Impact of Industrial Wastes on Water Quality

Industrial wastes can be an especially difficult problem; often they either are not removed or are removed only very slowly by naturally occurring purification processes, and they are generally not removed at all by typical municipal water treatment plants. Table 15-5 lists some of the industrial pollutants that result from the manufacture of products that are important to us.

Table 15-4
Classes of Water Pollutants, with Some Examples

Oxygen-demanding wastes	Plant and animal material
Infectious agents	Bacteria, viruses
Plant nutrients	Fertilizers, such as nitrates and phosphates
Organic chemicals	Pesticides such as DDT, detergent molecules
Other minerals and chemicals	Acids from coal mine drainage, inorganic chemicals such as iron from steel plants
Sediment from land erosion	Clay silt on stream bed, which may reduce or even destroy life forms living at the solid-liquid interface
Radioactive substances	Waste products from mining and processing of radioactive material, used radioactive isotopes
Heat from industry	Cooling water used in steam generation of electricity

Table 15–5
Important Industrial Products and Consequent Hazardous Wastes

The Products We Use	The Potentially Hazardous Waste They Generate
Plastics	Organic chlorine compounds
Pesticides	Organic chlorine compounds, organic phosphate compounds
Medicines	Organic solvents and residues, heavy metals like mercury and zinc
Paints	Heavy metals, pigments, solvents, organic residues
Oil, gasoline, and other petroleum products	Oils, phenols, and other organic compounds, lead, salts, acids, alkalies
Metals	Heavy metals, fluorides, cyanides, acids, and alkaline cleaners, solvents, pigments, abrasives, plating salts, oils, phenols
Leather	Chromium, zinc
Textiles	Heavy metals, dyes, organic chlorine compounds, organic solvents

Industrial wastes that contain certain toxic, reactive, or flammable chemicals are classified by the U.S. Environmental Protection Agency (EPA) as **hazardous wastes.**

Landfills have been the principal method of disposal for industries, agriculture, and municipalities for decades. In the 1970s, after incidents such as the Love Canal disaster, which drew attention to the serious contamination of groundwater by hazardous wastes, action began on local, state, and federal levels to solve problems caused by past disposal and to develop workable methods for the future disposal of hazardous wastes. In 1980 Congress established the "Superfund," a $1.6 billion program designed to clean up hazardous waste sites that were threatening to contaminate the nation's underground water supplies. By October 1985 the EPA had placed 850 hazardous waste sites on its National Priorities List for cleanup under the Superfund law (Fig. 15–2). In 1985 the Office of Technology Assessment estimated that the number of hazardous waste sites requiring cleanup will increase, perhaps to as high as 10,000, and that the cost of cleanup may reach $100 billion.

Although only 1% of aquifers have been polluted by hazardous wastes, many of these aquifers are near large population centers, so the problem is a serious one. The basic problem with land disposal of hazardous wastes is the contamination of groundwater as it moves through the disposal area (Fig. 15–3). Water pollution from these sites generally occurs as seepage into an underlying aquifer.

An *aquifer* is a water-bearing stratum of permeable rock, sand, or gravel, as illustrated in Figure 15–3.

The EPA has designated 34 industrial categories that produce polluted waste water and 65 classes of pollutants (Table 15–6). Under the law it is illegal for these pollutants to be discharged into U.S. waterways without a permit, and where permits are granted, strict monitoring of the amounts of pollutants discharged is required.

In 1976 the federal government passed the Resource Conservation and Recovery Act (RCRA). This law is designed to give "cradle-to-grave" (origin-to-disposal) responsibility to generators of hazardous wastes. The RCRA regulations cover the generation, transportation, storage treatment, and disposal of hazardous wastes.

Considerable attention has been given to the safe disposal of hazardous wastes, the monitoring of groundwater near hazardous waste sites, and the reduction in quantity of hazardous wastes by recycling chemicals. The technology for safe disposal exists, but the costs are high. Data reported by the EPA in 1980 (Table 15–7) indicated that 90% of

Impact of Industrial Wastes on Water Quality 359

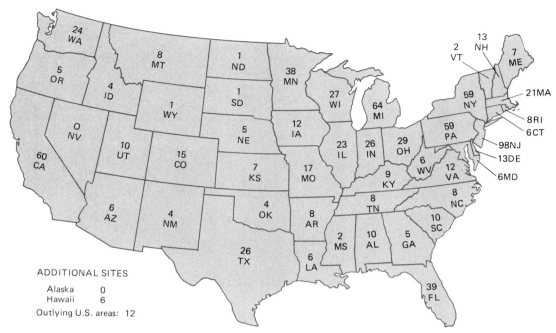

Figure 15–2
Distribution of 850 Superfund priority sites as of October 1985.

hazardous wastes were being disposed of by environmentally unsound methods. By 1985 this figure had fallen to 10%. Acceptable disposal methods, such as incineration, secured landfills, and recycling, have increased in importance. New government regulations and the increased public awareness are making current disposal methods safer, but the cleanup of Superfund sites and other landfills that are contaminating groundwater will take time.

The following are examples of groundwater contamination by seepage from hazardous waste sites:

> In 1978 the area around an old chemical dump site in Love Canal, a community in southeastern Niagara Falls, New York, was declared a disaster area by President Carter. Record rainfall caused leaching of chemicals from corroding waste disposal barrels buried in the old chemical dump site. Over 230 families were relocated, and the area was fenced off. In 1980 new boundaries that affected an additional 800 families were established. The emergency declarations by President Carter in 1978 and 1980 provided federal funds to assist the state in relocating families. This was the first use of federal emergency funds for something other than a "natural" disaster.
>
> Groundwater supplies in Toone and Teague, Tennessee, were contaminated by organic wastes from a nearby landfill in 1978. The landfill, closed six years earlier, held 350,000 drums, and pesticide wastes were leaking from many of them. The towns must now pump water from other locations.
>
> Groundwater in a 30-mile-square area near Denver was contaminated by pesticide waste disposed of between 1943 and 1957 in unlined disposal ponds.

Water and Air Pollution

Figure 15–3
(a) Groundwater in the water cycle. Groundwater—water that saturates soil and rock formations below the water table—plays an integral role in the hydrologic cycle and supplies the drinking water for half of this country's population. Bodies of groundwater stored in underground geologic formations, known as aquifers, may range in thickness from a few feet to several hundred feet and extend in area for many square miles. Shallow aquifers close to the land surface may be recharged by rainfall and surface runoff that percolates through pores and cracks in overlying soil and rock. Deep aquifers may contain large quantities of water, though they typically receive no surface recharge. All groundwater tends to move toward an area of discharge in lowlands where the water table intersects the land surface in streams, lakes, or wetlands. But rates of movement vary greatly, depending on the local geology. Groundwater moves more quickly in and through sandy aquifers but is slowed by clay soils or lightly fractured limestone. Threats to groundwater quality can come from the full spectrum of human activities on or near the surface, including septic tanks, agricultural fertilizers and pesticides, waste landfills and ponds, underground storage tanks, and other sources of soluble material. (b) Flow within groundwater systems. The time required for groundwater (and contaminants that may be in the groundwater) to move through aquifers and reach surface streams or drinking-water wells can range from days and years to decades and centuries. Many factors affect the travel time, including distance, hydraulic gradient, the nature of local geological media, and the chemical nature of the contaminants themselves. Groundwater in sandy or coarse rock formations, for example, may move as much as several feet a day, while less permeable geology may limit the rate of movement to a few inches a year. Water in some very deep aquifers has remained virtually in situ for thousands of years. (Courtesy of Electric Power Research Institute; reprinted from *EPRI Journal,* October 1985, p. 6. Original source: Wisconsin Bureau of Water Resources Management.)

Figure 15–4
"Valley of the Drums" site in Kentucky. (Courtesy of the Environmental Protection Agency.)

Table 15–6
EPA Categories of Industrial Materials Producing Pollutants and Related Toxic Pollutant Classes

34 Industrial Categories

Adhesives	Plastics processing	Organic chemicals
Leather tanning and finishing	Porcelain enamel	Pesticides
Soaps and detergents	Gum and wood chemicals	Pharmaceuticals
Aluminum forming	Paint and ink	Plastic and synthetic materials
Battery manufacturing	Printing and publishing	
Coil coating	Pulp and paper	Rubber
Copper forming	Textile mills	Auto and other laundries
Electroplating	Timber	Mechanical products
Foundries	Coal mining	Electric and electronic components
Iron and steel	Ore mining	
Nonferrous metals	Petroleum refining	Explosives manufacturing
Photographic supplies	Steam electric plants	Inorganic chemicals

65 Toxic Pollutant Classes

Acenapthene	DDT and metabolites	Nitrobenzene
Acrolein	Dichlorobenzenes	Nitrophenols
Acrylonitrile	Dichlorobenzidine	Nitrosamines
Aldrin/dieldrin	Dichloroethylenes	Pentachlorophenol
Antimony and compounds	2,4-Dimethylphenol	Phenol
Arsenic and compounds	Dinitrotoluene	Phthalate esters
Asbestos	Diphenylhydrazine	Polychlorinated biphenyls (PCBs)
Benzene	Endosulfan and metabolites	
Benzidine	Endrin and metabolites	Polynuclear aromatic hydrocarbons
Beryllium and compounds	Ethylbenzene	
Cadmium and compounds	Fluoranthene	Selenium and compounds
Carbon tetrachloride	Haloethers	Silver compounds
Chlordane	Halomethanes	2,3,7,8-Tetrachlorodibenzo-p-dioxin (TCDD)
Chlorinated benzenes	Heptachlor and metabolites	
Chlorinated ethanes	Hexachlorobutadiene	Tetrachloroethylene
Chloralkyl ethers	Hexachlorocyclopentadiene	Thallium and compounds
Chlorinated phenols	Hexachlorocyclohexane	Toluene
Chloroform	Isophorone	Toxaphene
2-Chlorophenol	Lead and compounds	Trichloroethylene
Chromium and compounds	Mercury and compounds	Vinyl chloride
Copper and compounds	Napthalene	Zinc and compounds
Cyanides	Nickel and compounds	

At least 1500 drums containing wastes from a metal-finishing operation were buried near Bryon, Illinois, until 1972. Surface water, soil, and groundwater were contaminated with cyanide, heavy metals, and organic toxic compounds. About 17,000 waste drums littered a 7-acre site in Kentucky that became known as the "Valley of the Drums" (Fig. 15–4). Many drums leaked their contents onto the ground. In 1979 an EPA survey identified about 200 toxic organic chemicals and 30 heavy metals in soil and in water samples near the dump.

Table 15–7
Hazardous Waste Disposal Methods in 1980 and 1985

Method	Percentage of Total	
	1980	1985
Unacceptable		
Unlined surface impoundment	48	<1
Land disposal	30	2
Uncontrolled incineration	10	5
Other	2	2
Acceptable		
Lined surface impoundment	<2	22
Secure landfills	2	33
Recycling	2	5
Controlled incineration	6	30

Chemical contamination has forced the closing of more than 600 groundwater wells in the New York City area since 1980.

The Love Canal situation illustrated what many fear will be a recurring event in other communities across our nation. The canal into which the waste chemicals were placed in the 1940s and 1950s was never intended for any use other than as a dump site. As time passed, however, the land was sold to developers, who built homes over the drums of chemical waste. When the drums began to leak, rising groundwater carried the chemicals near the surface, where their vapors went into the homes.

The costs of this mistake will never be paid fully. In addition to causing the uprooting of families and ruining the health of individuals, the Love Canal situation necessitated relocation of an entire community. Technically, Love Canal was contained, but the persons who lived there will always be afraid of the future consequences of their exposure to those chemicals.

What About the Future of Our Water Supplies?

An EPA report on water quality in 1985 indicates a large improvement over water quality in 1972. However, additional improvements are necessary. Thirty-seven states reported elevated levels of toxic pollutants in some of their waters, as evidenced by elevated levels of toxic substances in fish tissue. Metals were the most frequently reported, followed by pesticides and other organic chemicals. Thirty-five states reported some problems with groundwater contamination from industrial and municipal landfills, underground storage tanks, pesticide applications, septic tanks, and chemical and oil spills. The contaminants included chlorinated solvents, pesticides, gasoline, salts, and radionuclides.

This chapter's discussion of water has focused on water quality in the United States. Contaminated water is a much more serious problem for the 75% of the world's population living in developing nations. It has been estimated that 80% of the sickness in the world is caused by contaminated water. For years many countries and international organizations have provided financial and technical aid to help improve water quality in developing nations; however, much work remains to be done.

SELF-TEST 15-A

1. Nature is responsible for some air pollution. True ( ) or False ( )
2. What is the most abundant compound on Earth? _____
3. Approximately how many gallons of water are used per person per day in the United States? _____
4. Name three kinds of water pollution. _____, _____, and _____
5. What is the name of the U.S. law dealing with hazardous waste sites? _____
6. The major concern in hazardous waste management involves control of the pollution of _____.

What is the Atmosphere?

The Earth is enveloped by a few vertical miles of chemicals, which compose the gaseous medium in which we exist — the atmosphere. Close to the Earth's surface and near sea level, the atmosphere is mostly nitrogen (80%) and life-sustaining oxygen (20%). It is the few little fractions of a percentage point of other chemicals that make a difference in the quality of life in various spots on Earth. Extra water in the atmosphere can mean a rain

Figure 15-5
Coppertown Basin (Ducktown), Tennessee, as photographed in 1943. Copper ore (principally copper sulfide, Cu_2S) had been mined and smelted in this area since 1847. In the early years large quantities of sulfur dioxide, a byproduct, were discharged directly into the atmosphere and killed all vegetation for miles around the smelter. Today the sulfur is reclaimed in the exhaust stacks to make sulfuric acid, but the denuded soil remains a monument to the misuse of the atmosphere.

forest, a little less water produces a balanced rainfall, and practically no water results in a desert.

Before 1960 there was little concern about air pollution. Most smoke, carbon monoxide, sulfur dioxide, nitrogen oxides, and organic vapors were emitted into the air with little apparent thought of their harmful nature, as long as they were scattered into the atmosphere and away from human smell and sight (Fig. 15–5). Humans acted as though the atmosphere were infinite, which it is not (Fig. 15–6). Ninety-nine percent of the estimated 5500 trillion tons of gases that compose the atmosphere is below an altitude of 19 miles. Sufficient oxygen to sustain life extends upward to only about 4 miles above sea level, and most of our weather takes place within an average altitude of

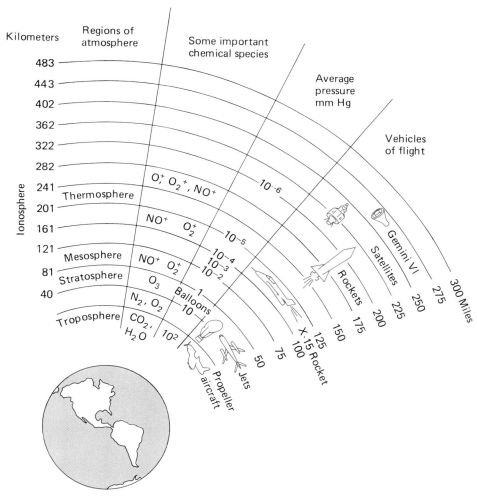

Figure 15–6
Some facts about our limited atmosphere. The troposphere was named by British meteorologist Sir Napier Shaw from the Greek word *tropos*, meaning "turning." The stratosphere was discovered by the French meteorologist Leon Philippe Teisserenc de Bort, who believed that this region consisted of an orderly arrangement of layers with no turbulence or mixing. The word *stratosphere* comes from the Latin word *stratum*, meaning "layer."

just 7 miles. Pollutants collect and react in this limited region; they do not escape Earth and venture into outer space.

What is Air Pollution?

Increased pollution of the air was brought on by increased urbanization, industrialization, and transportation via automobile and airplane. More public and governmental concern about air pollution led to the federal Clean Air Act of 1970 and additional state regulations. As a result, controls were placed on emissions from automobiles, industries, and power companies. The effects of the controls can be seen in Figure 15–7.

Although pollutants constitute relatively small proportions of the atmosphere compared with oxygen, nitrogen, and carbon dioxide, on a *clear* day a resident of Los Angeles still inhales about 200 quadrillion *pollutant* molecules per breath. An average breath of half a liter contains:

A quadrillion is 10^{15}.

Ozone is a pollutant in the lower troposphere, but ozone in the stratosphere protects us from harmful ultraviolet radiation.

Carbon monoxide	175 quadrillion molecules
Hydrocarbons	10 quadrillion molecules
Peroxides	5 quadrillion molecules
Nitrogen oxides	4 quadrillion molecules
Lower aldehydes	3.5 quadrillion molecules
Ozone	3 quadrillion molecules
Sulfur dioxide	2.5 quadrillion molecules

Air pollutants tend to attack the site where they first enter the body, that is, the lungs.

along with 1.0×10^{22} nitrogen molecules and 2.6×10^{21} oxygen molecules.

On a smoggy day the pollutants increase by a factor of five or more. The air you are now inhaling could contain pollutant molecules in comparable amounts, give or take a few quadrillion.

Pollutants may exist and react as single, isolated molecules, ions, or atoms. More often, because of the polar nature of pollutants such as SO_2 and NO_2, pollutants are

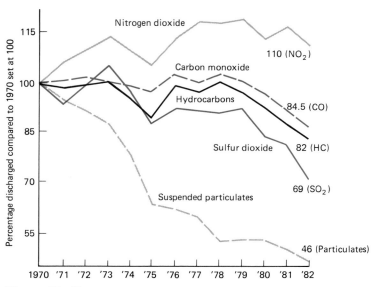

Figure 15–7
Changes in discharges of air pollutants since 1970, the birth date of the Clean Air Act. (Courtesy of *Fortune* magazine, May 4, 1981; updated 1985.)

attracted into water droplets, where they form **aerosols,** or onto larger particles called **particulates.**

Aerosols range upward in diameter from 1 nanometer to about 10,000 nanometers and may contain as many as 1 trillion atoms, ions, or small molecules per particle. They are small enough to remain suspended in the atmosphere for long periods of time. Smoke, dust, clouds, fog, mist, and sprays are typical aerosols. Since they are small, many aerosol particles can exist in a small volume of gas. Because of their vast combined surface area, aerosol particulates have an enormous capacity to *adsorb* and concentrate gases on the surfaces of the particles. At other times liquid aerosols *absorb* air pollutants, thereby concentrating them and providing a water medium in which reactions can occur readily. Because of these concentration and reaction effects, aerosols can be more devastating than isolated air pollutant molecules.

Particulates, which are generally large enough to be seen as individual particles, range in size from 1 micron to 10 microns in diameter. Millions of tons of soot, dust, and smoke particulates are deposited into the atmosphere of the United States each year. Average suspended particulate concentrations in the United States range from 0.00001 g/m^3 of air in remote rural areas to about 6 times that value in urban locations; in heavily polluted areas, concentrations up to 0.002 g/m^3, or 200 times the usual value, have been measured.

Particulates can cause damage in several ways. As small, solid particles, they may cause damage by abrasive action, by fouling and shorting electrical contacts and switches, and by affecting breathing passages.

Some particulates are intrinsically toxic to human beings and animals. The toxic effects of lead and arsenic were described in Chapter 14. Lead components are emitted from automobiles that use leaded gasoline. Arsenic compounds are used as insecticides to dust growing plants. Particulates containing fluorides, commonly emitted from aluminum-producing and fertilizer factories, have caused weakening of bones and loss of mobility in animals that have eaten plants covered with the dust. Asbestos particulates have been shown to be carcinogenic.

Like aerosols, particulates can adsorb and concentrate air pollutants. Sulfur dioxide, nitrogen oxides, hydrocarbons, and carbon monoxide do their greatest damage when concentrated on the surface of particulates or aerosols.

Particulates in the atmosphere can cool the Earth by partially shielding the Earth from the Sun. Large volcanic eruptions such as that from Mt. St. Helens in 1980 have a cooling effect on the Earth.

Particulates and aerosols are removed naturally from the atmosphere by gravitational settling and by rain and snow. They can be prevented from entering the atmosphere by the treatment of industrial emissions by one or more of a variety of physical methods, such as filtration, centrifugal separation, and spraying. Electrostatic precipitation, a common method, is better than 98% effective in removing aerosols and dust particulates even smaller than 1 micron from exhaust gases of industrial plants. A diagram of a Cottrell electrostatic precipitator is shown in Figure 15-8. The central wire is connected to a source of direct current at high voltage (about 50,000 volts). As dust or aerosols pass through the strong electrical field, the particles attract ions that have been formed in the field, become strongly charged, and are attracted to the electrodes. The solid grows larger and heavier and falls to the bottom, where it is collected.

Aerosol particles are intermediate in size between small molecules and easily visible small particles.

1 nanometer = 10^{-9} meter.

Adsorption is the attachment of particles to a surface.

Absorption is the pulling of particles inside.

1 micron = 10^{-6} meter.

Particulate effects depend heavily on the chemical nature of the particle.

Contributors to the present amount of atmospheric particulates are volcanic eruptions by: Hekla, Iceland, 1947; Mt. Spurr, Alaska, 1953; Bezymyannaya, U.S.S.R., 1956; Mt. St. Helens, Washington, 1980.

Smog: Infamous Air Pollution

The poisonous mixture of smoke, fog, air, and other chemicals was first called **smog** in 1911 by Dr. Harold de Voeux in his report on a London air pollution disaster that caused

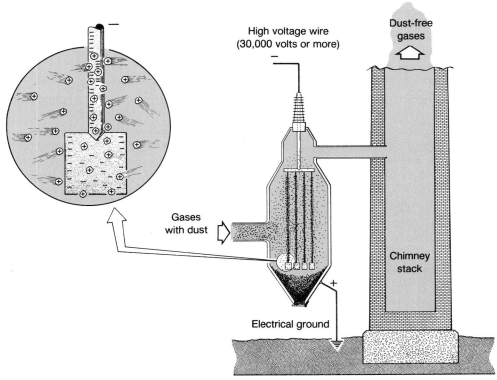

Figure 15-8
The Cottrell electrostatic precipitator.

Industrial or London-type smog: fog + SO_2

Photochemical smog: fog + NO_x + hydrocarbons.

Olefin is another name for an unsaturated hydrocarbon.

Organic peroxides, which contain the R—O—O—R′ structure, are produced by the reaction of ozone with organic molecules. Hydrogen peroxide is H—O—O—H.

Thermal inversion is a mass of warm air over a mass of cool air.

the deaths of 1150 people. Through the years, smog has been a technological plague in many communities and industrial regions (Fig. 15-9).

Two general kinds of smog have been identified. One is the chemically reducing type, which is derived largely from the combustion of coal and oil and contains sulfur dioxide mixed with soot, fly ash, smoke, and partially oxidized organic compounds. This is the **London type,** which is diminishing in intensity and frequency as less coal is burned and more controls are installed. A second type of smog is the chemically oxidizing type, typical of Los Angeles and other cities where exhausts from internal combustion engines are highly concentrated in the atmosphere. This type is called **photochemical smog** because light — in this instance sunlight — is important in initiating the photochemical process. This smog is practically free of sulfur dioxide but contains substantial amounts of nitrogen oxides, ozone, ozonated olefins, and organic peroxide compounds, together with hydrocarbons of varying complexity.

What general conditions are necessary to produce smog? Although the chemical ingredients of smogs often vary, depending on the unique sources of the pollutants, certain geographical and meteorological conditions exist in nearly every instance of smog.

There must be a period of windlessness so that pollutants can collect without being dispersed vertically or horizontally. Such a lack of movement in ground air can occur when a layer of warm air rests on top of a layer of cool air. This sets the conditions for a **thermal inversion,** an abnormal temperature arrangement for air masses (Fig. 15-10). Normally the warmer air is on the bottom nearer the warm Earth, and this warmer,

Figure 15-9
Smog over New York City. A heavy haze hangs over Manhattan Island, viewed from the roof of the RCA Building. The Empire State Building is barely visible in the background. (Wide World Photos, Inc.)

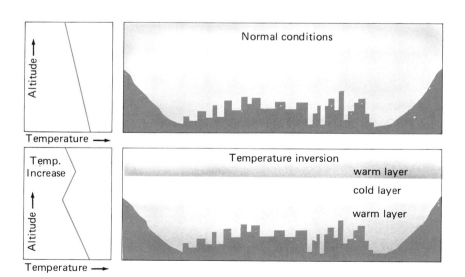

Figure 15-10
A diagram of a temperature inversion over a city. Warm air over a polluted air mass effectively acts as a lid, holding the polluted air over the city until the atmospheric conditions change. The line on the left of the diagram indicates the relative air temperature.

less dense air rises and transports most of the pollutants to the upper troposphere, where they are dispersed. In a thermal inversion the warmer air is on top, and the cooler, more dense air retains its position nearer the Earth. The air thus becomes stagnated. If the land is bowl shaped (surrounded by mountains, cliffs, or the like), this stagnant air mass can remain in place for some time.

Humans supply the pollutants through combustion and evaporation in automobiles, electrical power plants, space heating, and industrial plants. The chief pollutants are sulfur dioxide (from the burning of coal and some oils), nitrogen oxides, carbon monoxide, and hydrocarbons (chiefly from the automobile). Add to these ingredients radiation from the Sun, and a massive smog is in the offing.

A city's atmosphere is an enormous mixing bowl of frenzied chemical reactions. Learning the exact chemical reactions that produce photochemical smog has been a tedious job, but in 1951 insight into the formation process was gained when smog was first duplicated in the laboratory. Detailed studies have subsequently revealed that the chemical reactions involved in the smog-making process are photochemical and that aerosols serve to keep the reactants together long enough to form secondary pollutants. Ultraviolet radiation from the Sun is the energy source for the formation of this photochemical smog.

Primary pollutants are pollutants emitted into the air.

Secondary pollutants are pollutants formed in the air by chemical reactions.

The exact reaction scheme by which primary pollutants are converted into the secondary pollutants found in smog is still not completely understood, but the reactions shown in Figure 15–11 account for the major secondary pollutants. The process is thought to begin with the absorption of a quantum of light by nitrogen dioxide; this light causes the breakdown of nitrogen dioxide into nitrogen oxide and atomic oxygen, a chemical radical. The very reactive atomic oxygen reacts with molecular oxygen to form ozone (O_3), which is then consumed by a reaction with nitrogen oxide to form the original reactants, nitrogen dioxide and molecular oxygen. Atomic oxygen, however, also reacts with reactive hydrocarbons — olefins and aromatics — to form other chemi-

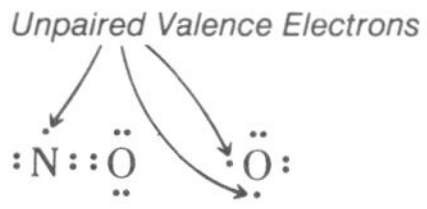

Unpaired Valence Electrons

:N::O :O:

A *chemical radical* is a species with an unpaired valence electron. Chemical radicals are usually very reactive and short-lived.

$$NO_2 + \text{light} \longrightarrow NO + O\cdot$$
$$O\cdot + O_2 + M \longrightarrow O_3 + M$$
$$O_3 + NO \longrightarrow NO_2 + O_2$$
$$O\cdot + Hc \longrightarrow HcO\cdot$$
$$HcO\cdot + O_2 \longrightarrow HcO_3\cdot$$

$$HcO_3\cdot + Hc \longrightarrow RCHO \text{ or } R\overset{O}{\overset{\|}{C}}R$$
$$HcO_3\cdot + NO \longrightarrow HcO_2\cdot + NO_2$$
$$HcO_3\cdot + O_2 \longrightarrow O_3 + HcO_2\cdot$$

$$HcO_3\cdot + NO_2 \longrightarrow R-\overset{O}{\overset{\|}{C}}-O-O-N\overset{O}{\underset{O}{\diagdown}} + \text{other products}$$
$$(PAN)$$

Figure 15–11
Simplified reaction scheme for photochemical smog. Ultraviolet light initiates the process to produce oxygen atoms. Hc is a hydrocarbon (unsaturated or aromatic); M is a third body to absorb the energy released from forming the ozone. Among many possibilities, M could be an N_2 molecule, an O_2 molecule, or a solid particle. A species with a dot, such as $HcO\cdot$, is a chemical radical. R is a saturated hydrocarbon group; PAN is peroxyacyl nitrate, a very reactive secondary pollutant that causes eyes to water and lungs to hurt.

cal radicals. These radicals in turn react to form other radicals and secondary pollutants such as aldehydes (e.g., formaldehyde). About 0.2 ppm of nitrogen oxides and 1 ppm of reactive hydrocarbons are sufficient to initiate these reactions. The hydrocarbons involved come mostly from unburned petroleum products like gasoline.

Major Air Pollutants: Sulfur Dioxide and Nitrogen Oxides

Sulfur dioxide is produced by the burning of sulfur or sulfur-containing substances in air:

$$S + O_2 \longrightarrow SO_2 \text{ (gas)}$$

Most atmospheric SO_2 comes from electrical power plants, smelting plants (which treat sulfide ores), sulfuric acid plants, and burning of coal or oil for home heating. Almost 50% of the nation's total SO_2 output is isolated in the seven industrialized states of New York, Pennsylvania, Michigan, Illinois, Indiana, Ohio, and Kentucky.

When coal or petroleum is burned, as in electrical power plants, the sources of sulfur are elemental sulfur (S), iron pyrite (FeS_2), and sulfur covalently bonded to carbon and hydrogen atoms in the coal structure. The average sulfur content of all coal mined in the United States is about 2%. Much of the petroleum used in the eastern United States is Caribbean residual fuel oil, which has an average sulfur content of 2.6%. A 1000-megawatt electrical power plant fueled by coal and petroleum containing up to 5% sulfur produces about 600 tons of SO_2 each day. More than 23 million tons of SO_2 have been released into the air annually since 1980.

What happens to the primary pollutant, SO_2, once it is in the air? It can be oxidized to a secondary pollutant, SO_3, in the presence of oxygen, sunlight, and water vapor:

$$2\ SO_2 + O_2 \longrightarrow 2\ SO_3$$

When SO_3 dissolves in water, sulfuric acid is formed:

$$SO_3 + H_2O \longrightarrow H_2SO_4$$

You will recall that H_2SO_4 is a strong acid:

$$H_2SO_4 \longrightarrow H^+ + HSO_4^-$$
Strong Acid

Sulfurous acid, a weak acid, can be formed in the presence of SO_2 and H_2O:

$$H_2O + SO_2 \rightleftharpoons H_2SO_3 \rightleftharpoons H^+ + HSO_3^-$$
Weak Acid

When dissolved in rivers, lakes, and streams, SO_2 causes the acidity of the water to increase considerably. If the pH varies much, aquatic life suffers. Salmon, for example, cannot survive if the pH is as low as 5.5. The lower limit of tolerance for most organisms is a pH of 4.0.

Sulfur dioxide and its attendant forms can damage vegetation, affect breathing, corrode metals, and decay building stones, in particular marble and limestone. Both marble and limestone are forms of calcium carbonate ($CaCO_3$), which reacts readily with acid (H^+) and with SO_2 and H_2O:

$$CaCO_3 + 2\ H^+ \longrightarrow Ca^{2+} + H_2O + CO_2$$

$$CaCO_3 + SO_2 + 2\ H_2O \longrightarrow CaSO_3 \cdot 2\ H_2O + CO_2$$
(Soluble)

A large modern power station (e.g., a station of 2000 megawatts capacity) annually produces about the same amount of SO_2 as an industrial city of 1 million inhabitants.

Sulfurous acid, H_2SO_3, is a weak acid; sulfuric acid, H_2SO_4, is a strong acid. Both can make the water in streams and lakes too acidic for fish.

An alarming example is the disintegration of marble statues and buildings on the Acropolis in Athens, Greece. All coatings have failed to protect the marble adequately, and the only known solution is to bring the prized objects into air-conditioned museums protected from SO_2 and other corroding chemicals.

Several efficient methods are available to trap SO_2. In one method, limestone is heated to produce lime. The lime reacts with SO_2 to form calcium sulfite, a solid particulate, which can be removed from an exhaust stack by an electrostatic precipitator.

$$CaCO_3 \xrightarrow{heat} CaO + CO_2$$
Limestone   Lime

$$CaO + SO_2 \longrightarrow CaSO_3 \text{ (solid)}$$
Calcium Sulfite

Another trapping method involves the passage of SO_2 through molten sodium carbonate. Solid sodium sulfite is formed by this method.

$$SO_2 + Na_2CO_3 \xrightarrow{800°C} Na_2SO_3 + CO_2$$
Sodium Carbonate   Sodium Sulfite

The least desirable method of dissipating SO_2 is by tall stacks. Although tall stacks emit SO_2 into the upper atmosphere away from the immediate vicinity and give SO_2 a chance to dilute itself on the way down, it still comes down, and the longer it stays up the greater chance it has to become sulfuric acid. A nickel smelter in Sudbury, Ontario, Canada, has a 1250-foot smokestack that emits about 2500 tons of SO_2 a day (Fig. 15-12). A ten-year study in Great Britain showed that although SO_2 emissions from power plants increased by 35%, the construction of tall stacks decreased the ground level concentrations of SO_2 by as much as 30%. The question is, who got the SO_2? In this case, Britain's solution was others' pollution. In the United States the EPA may have unwittingly added to a pollution problem with rules in 1970 that caused plants to increase the height of smokestacks and caused pollutants to be carried longer distances by winds. There are about 179 stacks in the United States that are 500 feet or higher and 20 stacks that are 1000 or more feet tall.

There are eight known oxides of nitrogen, two of which are recognized as important components of the atmosphere: dinitrogen oxide (N_2O) and nitrogen dioxide (NO_2).

Most of the nitrogen oxides emitted are in the form of NO, a colorless reactive gas. In a combustion process involving air, some of the atmospheric nitrogen reacts with oxygen to produce NO:

$$N_2 + O_2 + heat \longrightarrow 2\,NO$$

In the atmosphere NO reacts rapidly with atmospheric oxygen to produce NO_2, a brown gas:

$$2\,NO + O_2 \longrightarrow 2\,NO_2$$
Nitrogen Dioxide

Normally the atmospheric concentration of NO_2 is only a few parts per billion.

Fixed nitrogen (nitrogen oxides are one type) is necessary to perpetuate nature's cycle (Fig. 15-13). In this respect, nitrogen oxides are useful. However, too large a quantity of nitrogen oxides in the air can lead to photochemical smog and bronchial

---

Nitrogen oxide is formed in this manner during electrical storms. Since the formation of nitrogen oxide requires heat, it follows that a higher combustion temperature would produce relatively more NO.

About 97% of the nitrogen oxides in the atmosphere are produced naturally; only 3% result from human activity.

Fixed nitrogen is nitrogen chemically bonded to another element.

Figure 15-12
The world's largest chimney (as of 1972), standing 1250 feet high (as tall as the Empire State Building), was built at a cost of $5.5 million for the Copper Cliff smelter in the Sudbury District of Ontario, Canada.

problems for those who breathe this air. In these respects and many others, nitrogen oxides are harmful.

If NO_2 does not react photochemically, it can react with water vapor in the air to form nitric and nitrous acids:

$$2\ NO_2 + H_2O \longrightarrow \underset{\text{Nitric Acid}}{HNO_3} + \underset{\text{Nitrous Acid}}{HNO_2}$$

In addition, nitrogen dioxide and oxygen yield nitric acid:

$$4\ NO_2 + 2\ H_2O + O_2 \longrightarrow 4\ HNO_3$$

These acids in turn can react with ammonia or metallic particles in the atmosphere to produce nitrate or nitrite salts:

$$\underset{\text{Ammonia}}{NH_3} + HNO_3 \longrightarrow \underset{\substack{\text{Ammonium Nitrate}\\\text{(a Salt)}}}{NH_4NO_3}$$

Nitrates are important components of fertilizers.

The acids or the salts, or both, ultimately form aerosols, which eventually settle from the

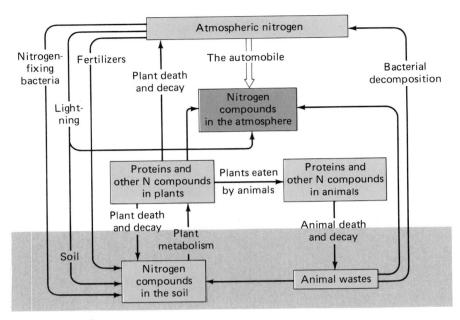

Figure 15-13
The nitrogen cycle.

air or dissolve in raindrops. Nitrogen dioxide, then, is a primary cause of haze in urban or industrial atmospheres because of its participation in the process of aerosol formation. Normally nitrogen dioxide has a lifetime in the atmosphere of about 3 days.

At present the emission of nitrogen oxides by human activities is significant in some urban areas but is minor globally compared with emissions by natural processes such as lightning fixation of nitrogen and oxygen and the natural decay of plants and animals (Fig. 15-13).

In laboratory studies, nitrogen dioxide in concentrations of 25 ppm to 250 ppm inhibits plant growth and causes defoliation. The growth of tomato and bean seedlings is inhibited by 0.3 ppm to 0.5 ppm NO_2 applied continuously for 10 to 20 days.

In a concentration of 3 ppm for 1 hour, nitrogen dioxide causes bronchioconstriction in humans, and short exposures at high levels (150-220 ppm) produce changes in the lungs that have fatal results. A seemingly harmless exposure one day can cause death a few days later.

Acid Rain

Most particulates removed from combustion gases are basic oxides; the acidic gaseous oxides, which are harder to remove, are the cause of acid rain.

Acids formed from sulfur oxides (sulfuric [H_2SO_4] and sulfurous [H_2SO_3]) and acids formed from nitrogen oxides (nitric [HNO_3] and nitrous [HNO_2]) can be leached from the air by rain or snow and produce precipitation with a pH below 5.6 (Fig. 15-14). The average annual pH of precipitation in much of northeastern Europe and large areas of the northeastern United States and southeastern Canada is between 4 and 4.5. Specific storms have dumped pH 2.7 precipitation on Kane, Pennsylvania, and pH 1.5 precipitation on Wheeling, West Virginia. In February, 1979, the average pH of precipitation in Canada was 4. "Clean" rain is slightly acidic because of dissolved carbon dioxide, which forms carbonic acid in water and produces a pH of 5.6. Any precipitation with a pH

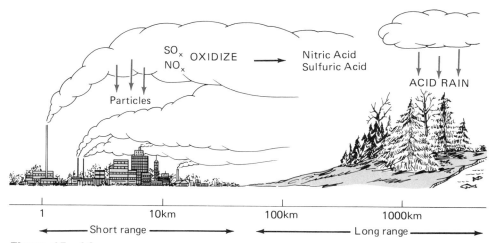

Figure 15–14
Major sources and components of acid rain. (From J. Turk and A. Turk: *Environmental Science,* 3rd ed. Philadelphia, Saunders College Publishing, 1984.)

below 5.6 is considered **acid rain.** For a review of the pH scale and a view of where acid rain fits into the scale, see Figure 6–8.

The extent of the acid rain problem can be seen in dead (fishless) ponds and lakes, dying or dead forests, and crumbling buildings. Because of wind patterns, Norway and Sweden have received the brunt of Western Europe's emissions of sulfur oxides and nitrogen oxides as acid rain. As a result, of the 100,000 lakes in Sweden, 4000 have become fishless and 14,000 others have been acidified to some degree. In the United States 6% of all ponds and lakes in the Adirondack Mountains of New York are now fishless, and 200 lakes in Michigan are dead. For the most part, these "dead" lakes are still picturesque, but no fish can live in the acidified water. Lake trout and yellow perch die at a pH below 5.0, smallmouth bass die at a pH below 6.0, and mussels die at a pH below 6.5

Trees have been affected by acid rain along the eastern coast of the United States, in southeastern Canada, and especially in West Germany's Black Forest. As the second highest recipient of SO_2 in Europe, West Germany receives about 2560 tons of sulfur dioxide per month in its precipitation. (Czechoslovakia is highest, with 2870 tons of SO_2 per month.) Fifty percent of the trees in the Black Forest show damage, such as yellowing of needle tips on conifers (fir, pine, spruce), premature defoliation on deciduous trees, and slowing of the growth rate and dieback on all types of trees. Trees at high elevations are most vulnerable. At the rate going in 1985, the fir and spruce in the Black Forest will all be dead before the end of this century.

Acid rain damages trees in several ways. It disturbs the stomata (openings) in tree leaves and causes increased transpiration and a water deficit in trees. Acid rain can acidify the soil, damaging fine root hairs, and thus diminish nutrient and water uptake. The acid can leach out needed minerals in the soil; these minerals are then carried off in groundwater. The surface structures of the bark and leaves can be destroyed by acid rain.

The effects of acid rain and other pollution on stone and metal structures are more subtle, but they are especially devastating because of their irreversibility. By damaging stone buildings in Europe, acid rain is slowly but surely dissolving the continent's historic heritage. The bas-reliefs on the Cologne (West Germany) cathedral are barely recognizable. The London Tower, St. Paul's Cathedral, and the Lincoln Cathedral in London have suffered the same fate. Other beautifully carved statues and bas-reliefs on buildings throughout Europe and the eastern part of the United States and Canada are slowly passing into oblivion by the action of pollutants, in particular acid rain.

What can be done about acid rain? Some stopgap measures are being taken, such as the spraying of lime ($Ca[OH]_2$) into acidified lakes to neutralize at least some of the acid and raise the pH toward 7.

$$Ca(OH)_2 + 2\ H^+ \longrightarrow Ca^{2+} + 2\ H_2O$$

Sweden is spending $40 million per year to neutralize the acid in some of its lakes. Some lakes in the problem areas have their own safeguard against acid rain: they have limestone-lined bottoms, which supply calcium carbonate ($CaCO_3$) to neutralize the acid (just as an antacid tablet relieves indigestion). Statues and bas-reliefs have been coated with a variety of plastics and other materials. However, none of these materials appear to be long-range protectors.

Ultimately, governments will have to act to diminish acid rain. Acid rain is presently a major issue in diplomatic relations between the United States and Canada. For several years the Canadian government has been urging the United States to reduce SO_2 emissions. Several meetings between leaders of the two countries in 1986–1987 led to endorsement of a five-year program that calls for the U.S. government and American industry to spend $5 billion to develop ways to burn coal cleanly. Bills have been introduced in the U.S. Congress, though none has yet been passed, to reduce the SO_2 emissions by 10 to 12 million tons per year. Even if no more sulfur dioxide and nitrogen oxides were emitted, lakes not fed and drained by streams would require 30 or more years to recover naturally from their present dead state. Some ornate, historic buildings and statues have deteriorated too much for repair.

Organic Compounds as Air Pollutants

As we saw in Chapter 9, hydrocarbons come in all shapes and sizes, beginning with methane (CH_4) and continuing to molecules containing hundreds of carbon atoms. Some have all single bonds, some have double bonds, and a few have triple bonds; some are aromatic. Hundreds of these hydrocarbons and their oxygen, sulfur, nitrogen, and halogen derivatives find their way into the atmosphere.

Trees and other plants silently release turpentine, pine oil, and thousands of other hydrocarbons into the air. We can smell some of them. Bacterial decomposition of organic matter emits very large amounts of marsh gas, principally methane. We contribute our share of 15% (of the total global emissions; a greater quantity in urban areas) through incomplete incineration, leakage of industrial solvents, unburned fuel from automobiles, evaporation of gasoline from tanks, incomplete combustion of coal and wood, and petroleum processing, transfer, and use. An estimated 163,000 pounds of polynuclear aromatic hydrocarbons pass from the air into Lake Superior each year.

In Chapter 14 we saw that polynuclear aromatic hydrocarbons such as benzo(α)pyrene (BaP) are capable of causing cancer in mice and humans. In the late 1950s the U.S. Public Health Service, Division of Air Pollution, surveyed 103 urban and 28 nonurban areas of the United States and found that the air in all of the 103 urban areas contained BaP. Concentrations ranged from 0.11 μg to 61 μg per 1000 m^3 of air, with the average concentration being 6.6 μg. In the 1980s the estimated annual emission of BaP in the United States was 422 tons from the burning of coal, oil, and gas, 20 tons from the

A commercial synthetic fuel plant, if not controlled, could emit as much as 10,000 kg of polynuclear aromatic hydrocarbons per day.

burning of refuse, 19 tons from industries (petroleum catalytic cracking, asphalt road mix, and the like), and 21 tons from motor vehicles. British researchers report that lung cancer in nonsmokers closely parallels the ten-times-greater amount of BaP in city air than in rural air; there is nine times more lung cancer in cities than in rural areas. A resident of a large town may inhale 0.20 g BaP a year. If he or she is a heavy smoker (two packs a day without filters), another 0.15 g can be added, for a total of 0.35 g. This is about 40,000 times the amount of BaP necessary to produce cancer in a mouse. Coal smoke contains about 300 ppm BaP. Every 1 million tons of coal burned in England in 1958 produced smoke laden with 750 tons of BaP; many authorities attribute England's high lung cancer rate today to this enormous production of BaP.

Other polynuclear aromatics have also shown evidence of being carcinogenic. Particulates from the atmosphere around Los Angeles, London, Newcastle, Liverpool, and eight other urban sites were extracted with organic solvents, and the extracts were found to produce cancer in mice.

Ozone: A Secondary Air Pollutant and Sunscreen

Ozone is a pungent-smelling gas that can be detected by the human nose at concentrations as low as 0.02 ppm. Sparking electrical appliances, lightning, and even silent electrical discharges convert oxygen into ozone, which is a more reactive form of oxygen.

$$\text{Energy} + 3\ O_2 \longrightarrow 2\ O_3$$
$$\text{Ozone}$$

Pure oxygen can be breathed for weeks by humans and animals without apparent injurious effects. Several studies have shown that concentrations of 0.3 ppm to 1.0 ppm ozone, well within the recorded range of photochemical oxidant levels, after 15 minutes to 2 hours cause marked respiratory irritation accompanied by choking, coughing, and severe fatigue. For these reasons, outdoor recreation classes in Los Angeles public schools are cancelled on days when the ozone level reaches 0.35 ppm. Ozone at this level for 1 hour depresses body temperature, perhaps by impairing the brain center that regulates body temperature or by opening the pores of the skin. Levels of 0.2 ppm to 0.5 ppm cause a considerable decrease in night vision, in addition to having other effects on vision.

Ozone attacks mercury and silver, which are not affected by molecular oxygen at room temperature. A typical reaction might be

$$6\ Ag + O_3 \longrightarrow 3\ Ag_2O$$

but

$$Ag + O_2 \longrightarrow \textit{No reaction}$$

Even with all the electrical sparks from lightning, electrical motors, and such, very little ozone is emitted into the air. It decomposes into molecular oxygen or reacts with other molecules too quickly to leave its source.

Ozone is found in the lower troposphere only as a secondary pollutant; that is, it is formed from other substances, as in photochemical smog. When sunlight impinges on automobile exhaust fumes, a considerable amount of ozone is produced. The stratosphere contains about 10 ppm of ozone in a layer that has the important function of filtering out some of the Sun's ultraviolet light, thus providing an effective shield against radiation damage to living things.

In the atmosphere, the *troposphere* is sea level to about 7 miles up (N_2, O_2, H_2O, CO_2), and the *stratosphere* is 7 to about 50 miles up (N_2, O_2, ozone).

Halogenated Hydrocarbons and the Ozone Layer

Most pollutants are adsorbed on surfaces or react with other pollutants in the troposphere and eventually wash out in rain. The halogenated hydrocarbons, reluctantly reactive pollutants, do not react quickly enough to be consumed in the troposphere and so may eventually mix with air in the stratosphere. The common halogenated hydrocarbon pollutants are listed in Table 15–8.

Chlorofluorohydrocarbons (CFCs) were described in Chapter 10. Recall that CFC-11 and CFC-12 are still the major chemicals used in refrigeration of buildings and vehicles. Measurements taken since 1974 indicate that the atmospheric concentration of CFCs is increasing. Eventually, CFC molecules reach the stratosphere, where the ultraviolet radiation is most intense. Bonds in CFC molecules are broken by the ultraviolet radiation, and species with unpaired electrons are formed. The most common reactive species produced is the chlorine atom ($Cl\cdot$), as illustrated by the breakdown of CFC-11.

Why is $Cl\cdot$ reactive?

$$\underset{\underset{Cl}{|}}{\overset{\overset{Cl}{|}}{F-C-Cl}} + light \longrightarrow \underset{\underset{Cl}{|}}{\overset{\overset{Cl}{|}}{F-C\cdot}} + Cl\cdot$$

The reactive chlorine atom then combines with ozone (O_3) in the stratosphere:

$$O_3 + Cl\cdot \longrightarrow ClO\cdot + O_2$$

Since many oxygen atoms are available in the upper atmosphere as participants in the production of ozone,

$$O_2 + light \longrightarrow \dot{O}\cdot + \dot{O}\cdot$$

$$\dot{O}\cdot + O_2 \longrightarrow O_3$$

the $ClO\cdot$ species can react with an oxygen atom to release the $Cl\cdot$ atom

$$ClO\cdot + \dot{O}\cdot \longrightarrow O_2 + Cl\cdot$$

to react with another ozone molecule.

The net effect of these reactions is the destruction of ozone in the ozone layer. The total number of ozone molecules destroyed by a single chlorine atom can run into the thousands. Eventually the chlorine atom reacts with a water molecule to form HCl, which mixes into the troposphere and washes out in rain.

Table 15–8
The Major Halogenated Hydrocarbons

Name	Formula	Uses
CFC-11	CCl_3F	As aerosol propellant
CFC-12	CCl_2F_2	As propellant, refrigerant
Carbon tetrachloride	CCl_4	In making fluorocarbons
Methyl chloroform	CH_3CCl_3	In metal cleaning
Perchloroethylene	C_2Cl_4	In dry cleaning

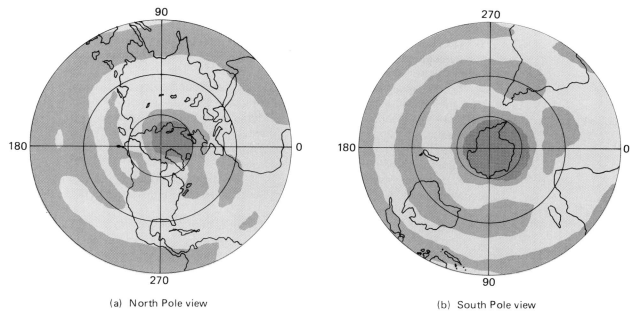

(a) North Pole view   (b) South Pole view

Figure 15-15
Computer-generated maps showing ozone concentrations between the end of 1978 and 1984. Areas of no change are unmarked.

In 1974 M. J. Molina and F. S. Roland of the University of California predicted that the increasing use of halogenated hydrocarbons could seriously deplete the stratospheric ozone, with a corresponding increase in ultraviolet rays reaching the Earth's surface. Since living things are sensitive to ultraviolet rays (for instance, ultraviolet light increases the risk of skin cancer in humans), depletion of stratospheric ozone would pose a serious threat to human safety and health. Further research has shown that this theory is probably correct.

Satellite and ground-based measurements since 1978 indicate the global ozone concentrations in the stratosphere have been decreasing. Near the North and South Poles ozone loss has been between 1% and 2.5% per year (Fig. 15-15). These losses are termed *holes* in the ozone layer. Scientists are not certain whether chlorofluorocarbons are responsible for the holes, but the evidence strongly suggests that they are.

Indoor Pollution

At Home

A five-year study by the EPA ending in 1985 concluded that pollution is greater indoors than outdoors for a number of toxic organic chemicals. According to the study, indoor pollution levels are about the same for houses in industrialized areas as for houses in rural areas.

What are the sources of indoor pollution? Tobacco smoke, if present, is an obvious source. Benzene, a known carcinogen, was 30% to 50% higher in the air of the homes of smokers in the study than in the homes of nonsmokers. Building materials and other consumer products are also sources of indoor pollution; as buildings have become more tight and energy-efficient, more emissions from materials in the building have been trapped inside.

The National Aeronautics and Space Administration has tested 10,000 materials for gaseous emissions because astronauts were experiencing "sick-building syndrome." Xylenes were found in 800 of the materials tested.

According to the EPA study, toxic chemicals found to be prevalent indoors are 1,1,1-trichloroethane (source: dry-cleaning solvent), tetrachloroethylene (solvent), benzene (paint, tobacco, gasoline), o-xylene (paint, gasoline, marking pens), m,p-xylene (paint, gasoline), ethylbenzene (paint, gasoline), carbon tetrachloride (solvent), trichloroethylene (solvent), chloroform (tap water), styrene (insulation, plastics), and p-dichlorobenzene (moth crystals, deodorants).

Radon gas, mentioned in Chapter 4, is also an indoor pollutant in many homes. In some homes the level of this radioactive gas exceeds 100 times the accepted level of 4 picocuries per liter of air.

Air conditioners, dehumidifiers, smoke removers, vacuum cleaners, and disposal of dish and bath cleaning water down the drain help remove some of the indoor pollution.

EPA estimates that as many as 12% of American homes may have more than the accepted level of radon (4 picocuries of radon per liter of air).

At Work

The workplace atmosphere is often very contaminated with pollutants such as solvents, welding fumes, corrosive vapors, aerosols and fumes, and particulates like asbestos. The U.S. Occupational Safety and Health Administration (OSHA) oversees regulations designed to limit worker exposure to harmful chemicals. OSHA has established concentrations called Permissible Exposure Limits (PELs) for approximately 400 chemicals commonly found in the workplace. PEL values are concentrations a worker may be exposed to during an 8-hour work day with no ill effects. Some PELs relate to skin contact by solids and liquids, but most relate to concentrations in the workplace air.

Table 15–9
OSHA's Permissible Exposure Limits (PEL) for Some Common Industrial Chemicals

Chemical	PEL (ppm)
Acetaldehyde	200
Allyl chloride	1
Ammonia	50
Carbon monoxide	50
Diborane	0.1
Cyclohexane	300
Formaldehyde	3
Hydrogen chloride	5
Isopropyl alcohol	400
Methyl alcohol	200
Naphthalene	10
Ozone	0.1
Propane	1000
1,1,1-Trichloroethane	350
Xylene	100

Those chosen represent a wide range of PEL values. All PEL values are defined by OSHA as concentrations a worker may be exposed to for an 8-hour day with no ill effects.

The PEL values of some workplace chemicals are given in Table 15–9. Those chemicals with low PEL values are very toxic or cause great discomfort upon exposure or both. (As you read the table, recall from Chapter 14 that the body can detoxify many chemicals.)

Could a Bhopal Tragedy Occur in the United States?

In December of 1984, over 2000 persons died and tens of thousands more were injured within hours of being exposed to methyl isocyanate vapors that had been released from a storage tank at a pesticide manufacturing plant in Bhopal, India (see Chapters 1 and 12). Methyl isocyanate is a liquid at room temperature and is toxic by a number of mechanisms.

$$H-\underset{\underset{H}{|}}{\overset{\overset{H}{|}}{C}}-N=C=O$$

Methyl Isocyanate
Boiling Point, 39.1°C
Reacts Strongly with Water

Toxicity

Denatures proteins by reacting with hydroxyl, amino, sulfhydryl and carboxyl groups

Causes hemorrhaging of kidneys, liver, and brain tissue

Causes severe lachrymation (tearing) and inflammation of the cornea

Has other toxic effects on the blood, heart, and central nervous system

The circumstances surrounding the release of methyl isocyanate at Bhopal were not highly unusual. A large quantity of a very toxic chemical was being stored in a container inside the fenced-in area of a chemical plant, but when the release occurred, the vapors, like any other pollutant, knew no boundaries and were carried by wind into nearby residential areas.

As a result of the Bhopal disaster, the EPA developed a list of 402 highly toxic chemicals that have the potential of causing a similar accident in the United States. Beginning in 1987, chemical users who have more of certain chemicals than the quantities on this EPA list must notify appropriate state agencies. These state agencies have developed plans for emergencies involving the sudden release of toxic chemicals. The emergency plans may include community awareness, so residents will know what the chemicals look or smell like if they are released. In addition, various alarms will be established in those communities at high risk of chemical exposure.

Some of the Bhopal List Chemicals

Name	Use	Emergency Planning Quantity (Pounds)
Ammonia	As fertilizer, refrigerant	500
Chlorine	As water disinfectant	100
Ethylene oxide	As sterilant in hospitals	1000
Methyl isocyanate	As chemical intermediate	500
Potassium cyanide	In metal plating operations	100

Carbon Dioxide: An Air Pollutant . . . Or Is It?

CO_2 is a necessary ingredient for photosynthesis.

How can carbon dioxide be considered a pollutant when it is a natural product of respiration and a required reactant for photosynthesis? CO_2 is not a pollutant per se; its increasing amount is. Between 1900 and 1970, the global concentration of CO_2 increased from 296 ppm to 318 ppm, an increase of 7.4%. The level in 1985 was 340 ppm, and if the present trend continues, by the year 2020 the concentration will be 640 ppm.

The increased amount of CO_2 comes primarily from electrical power plants, internal combustion engines, and cement manufacturing plants. However, there are numerous other sources: home heating, trash burning, forest fires, and bacterial oxidation of soil humus, to mention a few. CO_2 has been entering the atmosphere faster than oceans and growing plants can remove it.

A substantial increase of CO_2 in the air can have two detrimental effects: to increase the temperature of the atmosphere and to increase the acidity of the oceans.

Carbon dioxide increases the temperature of the atmosphere by the **greenhouse effect.** Like the glass or plastic of a greenhouse, CO_2 lets the shorter wavelengths of light through but absorbs the longer wavelengths as they are emitted from the surface of the Earth. In turn, the CO_2 (and the glass) emits the absorbed energy as heat radiation, some of which returns to Earth and some of which energizes other molecules. The net effect is an increase in the temperature of the lower atmosphere.

Particulates and aerosols counteract the warming effect of CO_2. They decrease the amount of solar energy reaching the Earth by scattering incoming sunlight of all wavelengths. The net effect is a cooling of the lower atmosphere. Calculations indicate that a 25% increase in aerosols would counteract a 100% increase in CO_2.

Carbon dioxide can make surface waters acidic by reacting with water to form carbonic acid, H_2CO_3, a weak acid:

$$CO_2 + H_2O \rightleftharpoons H_2CO_3 \rightleftharpoons H^+ + HCO_3^-$$

It would take an enormous amount of CO_2 in an ocean to affect the ocean's acidity. A major consumer of CO_2 (through photosynthesis) is the huge amount of phytoplankton (small plants) in the oceans.

Carbon dioxide is not the only atmospheric pollutant whose concentration is known to be rising. Concentrations of several other pollutants are also slowly increasing. Table 15–10 lists the atmospheric concentrations of several trace air pollutants as measured in Barbados; similar results have been obtained in Tasmania, Oregon, and Ireland.

What will be the effect of these increases in air pollutants? All of the chemicals listed in Table 15–10 are known to be adding to the greenhouse effect, perhaps cumulatively even doubling the effect of carbon dioxide. These chemicals are also involved in ozone depletion reactions. Since they are all industrial pollutants, the answer to the pollution problem they cause will probably be tighter controls on their manufacture and emission.

What Does the Future Hold for Clean Air?

Air pollution will undoubtedly decrease in the future; the sheer pressures of population increase will demand it. But life will undoubtedly also have to be different. Perhaps the first major change will be the disappearance of the automobile from the city, followed by a gradual modification of the automobile engine until it is relatively nonpolluting. One interesting effect of the lowering of the legal speed limit on our nation's highways to 55 mph was the reduction of NO_x emissions due to the lower operating temperatures of auto engines. However, NO_x emissions may increase again since the speed limit was raised in 1987 to 65 mph on rural interstates.

Table 15–10
Atmospheric Concentrations of Several Trace Air Pollutants Measured in Barbados

Pollutant	Concentration, ppt (Volume)		Percentage Increase
	1978	1985	
Chlorofluorocarbon-11	160	220	37.5
Chlorofluorocarbon-12	260	375	44.2
1,1,1-Trichloroethane	100	135	35
Carbon tetrachloride	115	130	13
Nitrous oxide	300	308	2.7

ppt = parts per trillion (10^{12}). Source: *Chemical and Engineering News*, November 24, 1986. L.R. Ember, P.L. Layman, W. Lepkowski, P.S. Zurer, "Tending the Global Commons," pp. 14–64.

Most pollution exists because we have demanded the benefits of technology while for the most part giving little consideration to the long-range effects of its products. When industrial plants, automobile manufacturers, or power plants add equipment to stop noxious waste products from getting into the air, the costs of this equipment are added to manufacturing expenses without adding 1 cent to the market value of the product being made. Thus, the cost to the consumer goes up; that is the price of clean air.

It is unclear whether knowledge of the harmful long-range effects of air pollution will cause people to willingly give up the immediate activities that give rise to the pollution. Not many people seem willing to use less energy or to give up the automobile.

Existing studies and regulations are not enough; society still needs to address other problems in air pollution, such as indoor air pollution, hazardous chemicals, transient fine particles, and SO_2 in acid rain.

There will be increased litigation to bring industry into conformance with the Clean Air Act. In 1981, of the estimated 6500 major sources of industrial pollution, 2400 had never complied or agreed to comply with the Clean Air Act.

In the final analysis, we all pollute the atmosphere. Much of the pollution is due to the misapplication of chemical techniques, yet the eradication of most forms of air pollution is within the capabilities of chemical technology. At present there is some awareness of the problems, but the tradeoffs among pollution, energy costs, and inflation are far from being settled.

In 1981, the estimated cost to bring 147 plants in the Chicago area under the level of 250 μg of NO_2 per liter was $130 million.

The cost of air pollution control in the United States is about $19.3 billion annually.

SELF-TEST 15–B

1. The stratum (layer) of the atmosphere that is closest to the Earth's surface is the _____.

2. The stratum of the atmosphere containing the ozone layer, which blocks the Sun's harmful ultraviolet rays, is the _____.

3. Since passage of the Clean Air Act, concentrations of air pollutants have been ( ) increasing or ( ) decreasing.

4. When particles draw chemicals onto their surfaces, the chemicals are said to be _____.

5. Sulfur dioxide is associated with _____ smog.
6. Nitrogen dioxide is associated with _____ smog.
7. Acid rain is caused by _____ and _____.
8. One can work without harm in an atmosphere containing the PEL concentration of a chemical.   True ( ) or False ( )

MATCHING SET

_____ 1. Cooling effect
_____ 2. Cause of ice floating
_____ 3. Percent water in human body
_____ 4. Amount of drinking water one person needs daily
_____ 5. Industrial wastes containing toxic substances
_____ 6. Major component of air
_____ 7. Suspended particles in air about 1 nanometer to 10,000 nanometers in size
_____ 8. Mixture of air, fog, and chemicals
_____ 9. Prime component of London-type smog
_____ 10. Acceptable method of disposal of hazardous wastes
_____ 11. Layer of warm air over layer of cooler air
_____ 12. One of the chemicals causing acid rain
_____ 13. Useful in removing SO_2
_____ 14. Formed by lightning
_____ 15. Found in smoke; causes cancer in mice
_____ 16. Cause a depletion of the ozone layer in the troposphere
_____ 17. Causes the greenhouse effect

a. Smog
b. Hazardous wastes
c. Nitrogen
d. Incineration
e. Thermal inversion
f. Benzo(α)pyrene
g. Heat of vaporization
h. $\frac{1}{2}$ gallon
i. NO
j. Chlorofluorocarbons
k. Hydrogen bonding
l. CO_2
m. 70
n. SO_2
o. Aerosols
p. CaO
q. Oxygen

QUESTIONS

1. Using several examples, explain what air pollution is.
2. Using several examples, explain what water pollution is.
3. Name two properties of water, and describe how these properties either hinder water from becoming polluted or make it more easily polluted.
4. Why are there allowed limits of pollution in drinking water? Why shouldn't regulations require that drinking water be 100% pure?
5. Describe how hazardous wastes can cause water pollution.
6. Describe the Love Canal pollution situation. You may wish to do some outside reading to become more familiar with the history of Love Canal.

7. What is an aerosol? How do aerosols play a role in air pollution?
8. What factors contribute to London-type smog?
9. What factors contribute to photochemical-type smog?
10. Where does SO_2 in smokestack gases come from?
11. Name two sources of oxides of nitrogen in the atmosphere.
12. What are the two causes of acid rain, and how is acid rain harmful to the environment?
13. Name two ways in which acid rain can be controlled.
14. What is benzo(α)pyrene, how is it produced, and how is it hazardous to our health?
15. How is ozone formed? Where in the atmosphere is it in greatest concentration?
16. Why is ozone considered a pollutant in the troposphere and beneficial in the stratosphere?
17. What chemicals cause a destruction of ozone in the stratosphere? How does this happen?
18. Where have ozone layer "holes" been found in the stratosphere?
19. List several indoor air pollutants and their sources.
20. What is a PEL value for a chemical? What is the significance of a high PEL?
21. What is the greenhouse effect? What chemicals contribute to this effect?

CHAPTER 16
Medicines, Drugs, and Drug Abuse

The average life expectancy for men in the United States has risen from 53.6 years in 1920 to 71.1 years in 1984, a rise of 32.6%. During this same period, the life expectancy for women has risen from 54.6 years to 78.3 years, a rise of 43.4% (Table 16–1).

The reasons for these differences between the sexes are not altogether clear, but it is known that the widespread use of a large assortment of new medicinal compounds has contributed to the longer life span of both sexes.

Medicines rise and fall in popularity.

The contents of the medicine cabinet have changed drastically in the past few decades. A survey of physicians shortly before World War I revealed the ten most essential drugs (or drug groups) to be ether, opium and its derivatives, digitalis, diphtheria antitoxin, smallpox vaccine, mercury, alcohol, iodine, quinine, and iron. When another survey was made at the end of World War II, at the top of the list were sulfonamides, aspirin, antibiotics, blood plasma and its substitutes, anesthetics and opium derivatives, digitalis, antitoxins and vaccines, hormones, vitamins, and liver extract. Today there is an even wider array of medicinal chemicals, but drugs for reducing fever, relieving pain, and fighting infection still head the list in all areas of medical practice (Table 16–2).

Many new drugs are tested each year, but few ever reach the marketplace.

Americans spent $25.8 billion on medicines in 1984. This amounts to about $105 per person. Today, the top-ten prescription drugs, based on the number of prescriptions written (Table 16–3), show an interesting cross section of medicinal uses.

Table 16–1
Life Expectancy in the United States for Men and Women (1920–1984)*

Year	Men	Women
1920	53.6	54.6
1930	58.1	61.6
1940	60.8	65.2
1950	65.6	71.1
1960	66.6	73.1
1970	67.1	74.8
1980	70.0	77.5
1982	70.8	78.2
1983	71.0	78.3
1984	71.1	78.3

* Data from U.S. Bureau of the Census, *Statistical Abstract of the United States:* 1986 (106th ed.) Washington, D.C.

Table 16–2
Widely Prescribed Drugs

Generic Name*	Medical Use
Tetracycline HCl	Antibiotic
Ampicillin	Antimicrobial (kills bacteria but is not derived from a plant, as is an antibiotic)
Phenobarbital	Sedative, hypnotic, anticonvulsant
Thyroid	Increases rates of metabolism
Prednisone	Anti-inflammatory, antiallergic agent similar to cortisone
Digoxin	Decreases the rate of the heartbeat, but increases the force of the heartbeat, similar to digitalis
Meprobamate	Tranquilizer
Erythromycin	Antimicrobial
Penicillin G potassium	Antibiotic
Nitroglycerin	Dilates the blood vessels of the heart
Penicillin VK	Antibiotic
Quinidine sulfate	Slows the heartbeat; also used for malaria and hiccups
Cephalexin	Antibiotic
Reserpine	Tranquilizer
Nicotinic acid (niacin)	Dilates blood vessels; also, essential B vitamin with antipellagra activity

* The generic name for a drug, not a specific brand name, is the drug's generally accepted chemical name. For example, the generic drug tetracycline HCl is sold under brand names such as *Acromycin V, Ambracyn, Artomycin, Diacycline*, and *Quatrex*. Medical doctors can prescribe either the generic name or a brand name. If the generic name is used, the prescription is often cheaper, particularly if the drug is not protected by patents and can be manufactured and marketed competitively by several companies.

The generic name for a drug is its widely accepted chemical name.

Table 16–4 summarizes the different classes of drugs. This chapter describes representatives of each of these classes.

Antacids

The walls of a human stomach contain thousands of cells that secrete hydrochloric acid, the main purposes of which are to suppress the growth of bacteria and to aid in the

The contents of the stomach are highly acidic.

Table 16–3
Ten Most Prescribed Drugs in 1986*

Trade Name	Generic Name	Use
1. Dyazide	Triamterene and hydrochlorothiazide	Diuretic/antihypertensive
2. Amoxil	Amoxicillin	Antibiotic
3. Inderal	Propranolol	Heart disease
4. Lanoxin	Digitalis glycoside	Heart disease
5. Tylenol with codeine	Acetaminophen and codeine	Pain relief
6. Tagamet	Cimetidine	Stomach ulcer control
7. Ortho-Novum	Norethindrone	Birth control
8. Valium	Diazepam	Tranquilizer
9. Lasix	Furosemide	Diuretic
10. Tenormin	Atenolol	Heart disease

* Data from *American Druggist*, vol. 195, no. 2, February 1987, p. 19.

Table 16–4
Classification of Drugs

Designation	Description	Examples
Over-the-counter (OTC)	Available to anyone	Antacids, aspirin, cough medicines
Prescription drugs	Available only by prescription	Antibiotics
Unregulated nonmedical drugs	Available in beverages, foods, or tobacco	Ethanol, caffeine, nicotine
Controlled substances*		
Schedule 1	Abused drugs with no medical use	Heroin, ecstasy, LSD, mescaline
Schedule 2	Abused drugs that also have medical uses	Morphine, amphetamines
Schedule 3	Prescription drugs that are often abused	Valium, phenobarbitol

* Drugs the sale, distribution, and possession of which are controlled by the Drug Enforcement Administration of the U.S. Department of Justice.

> Review the pH chart of common substances (Figure 6–8).
>
> If the reduction of acidity is too great, the stomach responds by secreting an excess of acid. This is "acid rebound."
>
> Analgesics relieve pain, but they are harmful in large doses.

hydrolysis (digestion) of certain foodstuffs. Normally the stomach's inner lining is not harmed by the presence of hydrochloric acid, since the mucosa, the inner lining of the stomach, is replaced at the rate of about a half million cells per minute. However, when too much food is eaten, the stomach often responds with an outpouring of acid, which lowers the pH to a point at which discomfort is felt.

Antacids are basic compounds used to decrease the amount of hydrochloric acid in the stomach. The normal pH of the stomach ranges from 0.9 to 1.5. Some alkaline compounds used for antacid purposes and their modes of action are given in Table 16–5.

Analgesics

Analgesics are painkillers. Most people need these compounds at one time or another. When we have a headache we take an over-the-counter painkiller, such as aspirin,

Table 16–5
The Chemistry of Some Antacids

Compound	Reaction in Stomach	Examples of Commercial Products
Milk of magnesia: $Mg(OH)_2$ in water	$Mg(OH)_2 + 2\,H^+ \longrightarrow Mg^{2+} + 2\,H_2O$	Phillips Milk of Magnesia
Calcium carbonate: $CaCO_3$	$CaCO_3 + 2\,H^+ \longrightarrow Ca^{2+} + H_2O + CO_2$	Tums, Di-Gel
Sodium bicarbonate: $NaHCO_3$	$NaHCO_3 + H^+ \longrightarrow Na^+ + H_2O + CO_2$	Baking soda, Alka-Seltzer
Aluminum hydroxide: $Al(OH)_3$	$Al(OH)_3 + 3\,H^+ \longrightarrow Al^{3+} + 3\,H_2O$	Amphojel
Dihydroxyaluminum sodium carbonate: $NaAl(OH)_2CO_3$	$NaAl(OH)_2CO_3 + 4\,H^+ \longrightarrow Na^+ + Al^{3+} + 3\,H_2O + CO_2$	Rolaids

acetaminophen (Tylenol), or ibuprofen (Advil). When we have a tooth filled or extracted, the dentist gives us Novocain. Intense suffering requires a strong painkiller, such as codeine or morphine. Although these compounds are immensely useful, they are nevertheless dangerous if taken or used improperly. They can even kill if taken in overdose.

Early societies used opium. Although not all opium derivatives have therapeutic value, most of them are efficient painkillers. Their chief disadvantage lies in their addictive properties.

Opium is obtained from the opium poppy by scratching of the seed pod with a sharp instrument. From this scratch flows a sticky mass that contains about 20 different compounds called **alkaloids** (organic nitrogenous bases containing basic nitrogen atoms). About 10% of this mass is the alkaloid **morphine,** which is primarily responsible for opium's effects.

Morphine and its derivatives are addictive drugs.

Analgesics may be habit forming.

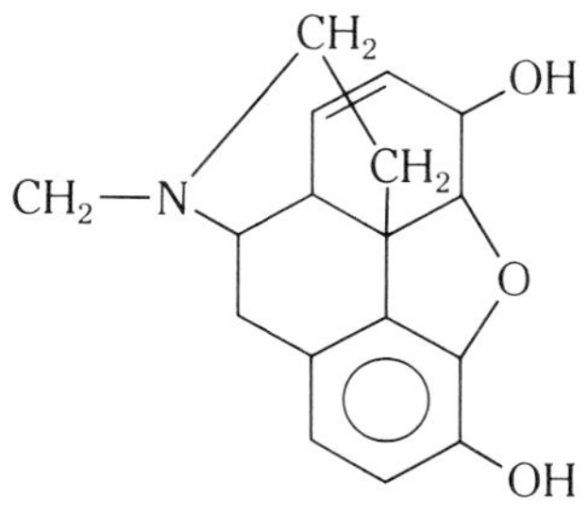

Morphine

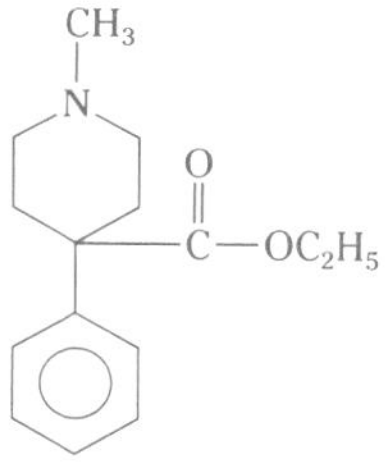

Two derivatives of morphine are of interest. One of these is **codeine,** a methyl ether of morphine, which is less addictive than morphine and is almost as powerful an analgesic. The other compound is **heroin,** the diacetate ester of morphine. Heroin is much more addictive than morphine and for that reason has no medical use in the United States.

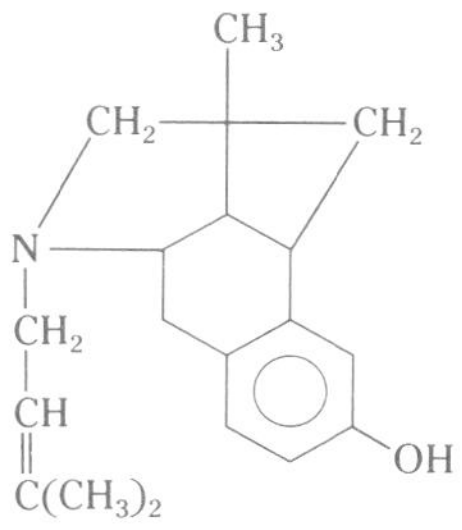

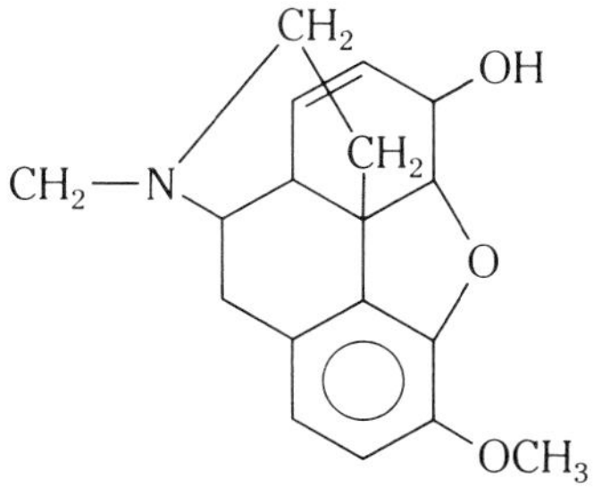

Codeine (Methyl Ether)

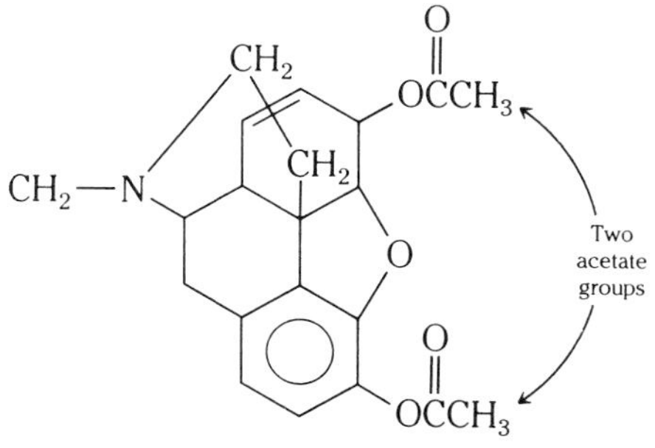

Heroin (Acetate Ester)

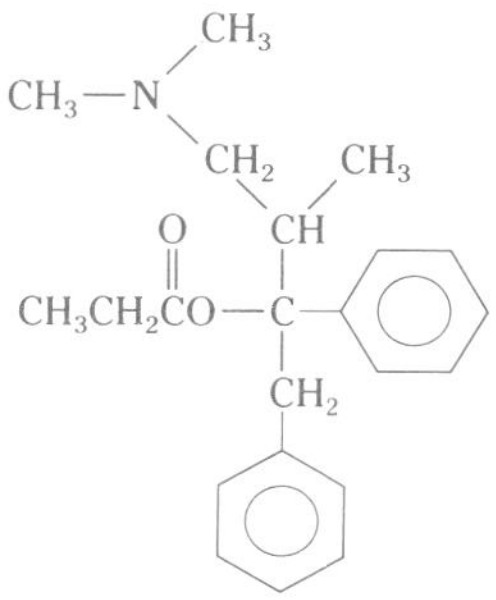

Meperidine (Demerol)

Pentazocine (Talwin)

Propoxyphene (Darvon)

One of the most effective substitutes for morphine is **meperidine,** first reported in 1931 and now sold as Demerol. It is less addictive than morphine. Two other relatively strong pain relievers used today are **pentazocine** (Talwin) and **propoxyphene** (Darvon). Talwin is slightly addictive, whereas Darvon has not been shown to be. However, Darvon has been much abused; there were approximately 600 Darvon-related deaths in 1977. Critics argue that Darvon is more dangerous and less effective in killing pain than other available opiates.

Considerable progress has been made in understanding the drug action of the opiates. Solomon Snyder and co-workers discovered in 1973 that the brain and spinal cord contain specific bonding sites into which the opiate molecules fit as a key fits into a lock. John Hughes and Hans Kosterlitz followed in 1975 with the discovery that vertebrates produce their own opiates, which they named **enkephalins** (from the Greek "en" "kephale," meaning "within the head"). Individuals with a high tolerance for pain produce more enkephalins and consequently tie up more receptor sites than normal; hence, they feel less pain. A dose of heroin temporarily bonds to a high percentage (or all) of the sites, resulting in little or no pain. Continued use of heroin causes the body to reduce or cease its production of enkephalins. If the use of the narcotic is stopped, the receptor sites become empty and withdrawal symptoms appear.

Many of the **local analgesics,** or local anesthetics, are nitrogen compounds like the alkaloids (Table 16–6). Local analgesics include the naturally occurring **cocaine,** derived from the leaves of the coca plant of South America, and the familiar **Novocain** (procaine). All of these drugs act by some blockage of the nerves that transmit pain. Acetylcholine appears to be the "opener of the gate" for sodium (Na^+) and potassium (K^+) ions to flow into a nerve cell (Fig. 16–1).

When milder general analgesics are required, few compounds work as well for as many people as **aspirin.** Not only is aspirin an analgesic, but it is also an antipyretic, or fever reducer. Each year about 40 million pounds of aspirin are manufactured in the United States. Aspirin is thought to inhibit cyclooxygenase, the enzyme that catalyzes the

Table 16–6
Some Local Analgesics

Cocaine	*(structure)*	Probably the first local analgesic used
Procaine (Novocain)	*(structure)*	Often used in dental work
Lidocaine (Xylocaine)	*(structure)*	More potent than procaine; can be applied to the skin

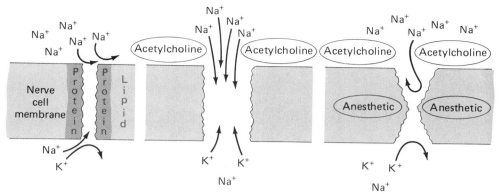

Figure 16-1
Action of acetylcholine and anesthetics in depolarizing the membrane of a nerve cell. Acetylcholine makes it possible for sodium and potassium ions to neutralize the negative charge associated with a nerve impulse so another impulse can be transmitted. Anesthetics block the action of acetylcholine and do not allow repetitive impulses to travel along the nerve.

reaction of oxygen with polyunsaturated fatty acids to produce prostaglandins. Excessive prostaglandin production causes fever, pain, and inflammation — just the symptoms aspirin relieves.

Review the discussion of prostaglandins in Chapter 10.

A danger presented by aspirin is stomach bleeding, caused when an undissolved aspirin tablet lies on the stomach wall. As the aspirin molecules pass through the fatty layer of the mucosa, they appear to injure the cells, causing small hemorrhages. The blood loss for most individuals taking two 5-grain tablets is between 0.5 mL and 2 mL. However, some persons are more susceptible than others. Early aspirin tablets were not particularly fast dissolving, which aggravated this problem greatly. Today's aspirin tablets are formulated to disintegrate quickly, and the process can be sped up by crushing the tablet in a little water before ingestion.

Many aspirin tablets contain starch to hasten their disintegration in the stomach.

A greater potential danger of aspirin is its possible link to **Reye's syndrome,** a brain disease that also causes fatty degeneration in organs such as the liver. Reye's syndrome can occur in children recovering from the flu or chicken pox. Vomiting, lethargy, confusion, and irritability are the symptoms of the disease. Studies have shown a strong correlation between aspirin ingestion and the onset of Reye's syndrome. About one quarter of the 200 to 600 cases per year have proved fatal. Beginning in 1982, aspirin products were required to contain a warning about the possible link between aspirin and Reye's syndrome. As of now, there is no explanation for the relationship between aspirin and this disease.

Aspirin Substitutes

Several over-the-counter alternatives are now available for pain sufferers who have trouble with aspirin. The two principal ones are acetaminophen (Tylenol) and ibuprofen (Advil). Like aspirin, acetaminophen and ibuprofen are both analgesics and antipyretics. Of course, no drug should be taken without proper caution.

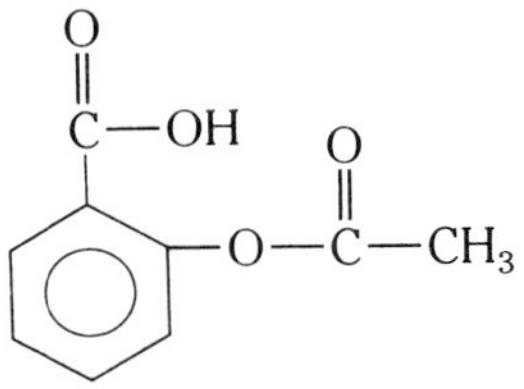

Acetylsalicylic Acid
(Aspirin)

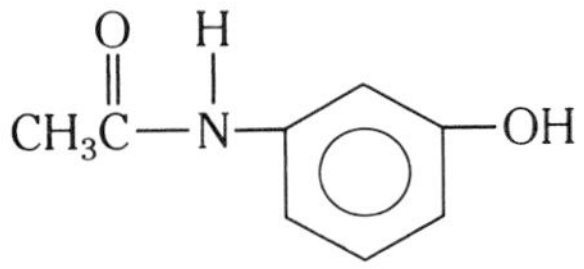

Acetaminophen
(Tylenol)

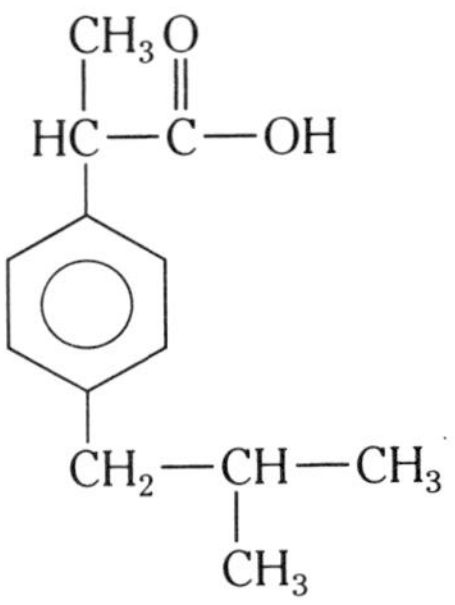

Ibuprofen
(Advil)

Anti-Ulcer Drugs

The principal anti-ulcer drug is Tagamet (number six in total prescriptions in 1986). Its generic name is **cimetidine.** This compound is used for the treatment of both duodenal and gastric ulcers; it acts by inhibiting gastric acid secretion.

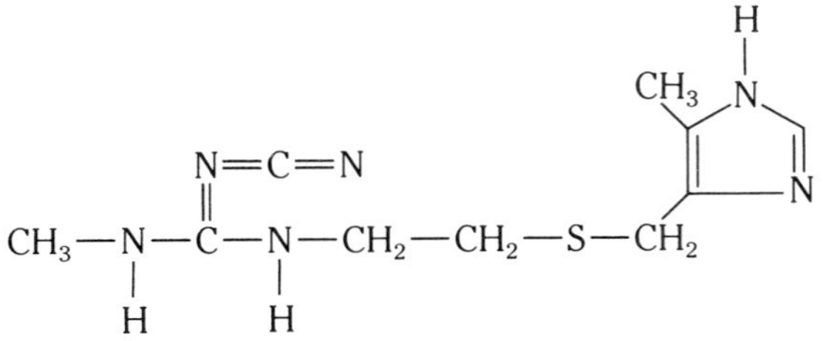

Cimetidine
(Tagamet)

Allergens and Antihistamines

> An *allergy* is a physiological response — such as sneezing, runny nose, coughing, or dermatitis — to a foreign substance. This foreign substance is called an *allergen*.
>
> Most allergens are high-molecular-weight substances.
>
> 10^{-12} is 0.000000000001
> A picogram is 1×10^{-12} g.

A person may have an unpleasant physiological response to poison ivy, pollen, mold, food, cosmetics, penicillin, aspirin, and even cold, heat, and ultraviolet light. In the United States about 5000 people die yearly from bronchial asthma, at least 30 from the stings of bees, wasps, hornets, and other insects, and about 300 from ordinary doses of penicillin. The reason is **allergies.** About one person in ten suffers from some form of allergy; more than 16 million Americans suffer from hay fever.

An allergy is an adverse response to a foreign substance or to a physical condition that produces no obvious ill effects in most other organisms, including humans. An **allergen** (the substance that initiates the allergic reaction) is in many cases a highly complex substance — usually a protein. Some allergens are polysaccharides or compounds formed by the combination of a protein and polysaccharide. Usually allergens have a molecular weight of 10,000 or more.

Scientists have succeeded in isolating the principal allergen of ragweed pollen, a major allergy producer, named ragweed antigen E. It is a protein with a molecular weight of about 38,000; it represents only about 0.5% of the solids in ragweed pollen but contributes about 90% of the pollen's allergenic activity. A mere 1×10^{-12} g (1 picogram) of antigen E injected into an allergic person is enough to induce a response.

The allergens come in contact with special cells in the nose and breathing passages to which is attached the IgE antibody, which has a molecular weight of about 196,000. Allergic persons have 6 to 14 times more IgE in their blood serum than nonallergic persons. The IgE is formed in the nose, bronchial tubes, and gastrointestinal tract and binds firmly to specific cells, called **mast cells,** in these regions.

Antigen E from ragweed reacts with the IgE antibody attached to the mast cells, forming antigen-antibody complexes. The formation of these antigen-antibody complexes leads to the release of so-called allergy mediators from special granules in the mast cells (Fig. 16–2). The most potent of these mediators found so far is **histamine.** Although it is widely distributed in the body, histamine is especially concentrated in the 250 to 300 granules of the mast cells. Histamine accounts for many, if not most, of the symptoms of hay fever, bronchial asthma, and other allergies.

$$H_2NCH_2CH_2-\text{[imidazole ring]}$$

Histamine

Antibodies are high-molecular-weight proteins, called immunoglobulins, that attack foreign proteins such as allergens.

Histamine causes runny noses, red eyes, and other hay fever symptoms.

Chemical mediators such as histamine must be released from the cell to cause the symptoms of allergy. The release mechanism is an energy-requiring process in which the granules may move to the outer edge of the living cell and, without leaving the cell, discharge their histamine contents through a temporary gap in the cell membrane. This sends the histamine on its way to produce the toxic effects of hay fever.

Treatment consists of three procedures: avoidance (Fig. 16–3), desensitization, and drug therapy. Desensitization therapy is costly and inconvenient, since 20 or more injections are required to achieve what is usually only a partial cure. One possible form of chemical desensitization is injection of a blocking antibody that preferentially reacts with the allergen so that it cannot react with the IgE allergy-sensitizing antibody. This breaks the chain of events leading to the release of histamine or other allergy-producing mediators. Many small injections, spaced over time, are required to build up a sufficient level of the blocking antibody.

Epinephrine (adrenalin), steroids, and antihistamines are effective drugs in treating allergies. The first two are particularly effective in treating bronchial asthma; the **antihistamines,** introduced commercially in the United States in 1945, are the most widely used drugs for treating allergies. More than 50 antihistamines are offered commercially

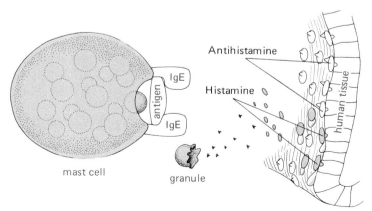

Figure 16–2
A postulated mechanism for the cause of and relief from hay fever. The details are described in the text.

Figure 16-3
In this Abbott Laboratories map, the size of each circle represents the amount of all late-summer and fall pollens found in the air in each city. Dark portions show the amount of ragweed pollen. Shaded areas are regions of low pollen count.

in the United States. Many of these contain, as does histamine, an ethylamine group ($-CH_2CH_2N\subset$):

Pyribenzamine
(An Important Antihistamine)

These drugs act competitively by occupying the receptor sites on cells that are normally occupied by histamine, effectively blocking the action of histamine.

Antimicrobial Drugs

In 1900 infectious diseases were the principal cause of death (Fig. 16-4). However, the successful development of a variety of **antibiotics** has reduced mortality due to infectious disease from 668 per 100,000 population to 21 per 100,000 population. In the original sense, an antibiotic is a substance such as penicillin, produced by a microorganism, that inhibits the growth of another organism. It has become common practice to include synthetic chemicals such as the sulfa drugs in a discussion of antibiotics.

Since the antibiotics are so efficient, they were the first of what came to be called miracle drugs. Their job generally is to aid the white blood cells by stopping bacteria from multiplying. When a person falls victim to or is killed by a disease, it means that the invading bacteria have multiplied faster than the white blood cells could devour them and that the bacterial toxins increased more rapidly than the **antibodies** could neutralize them. The action of the white blood cells and antibodies plus an antibiotic is generally enough to repulse an attack of disease germs.

An antibody is a specific protein produced to protect the organism from harmful invading molecules.

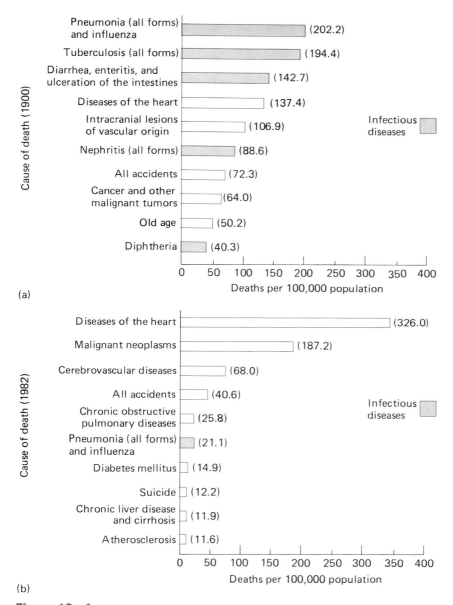

Figure 16-4
(a) In 1900 five of the ten leading causes of death in the United States were infectious diseases. *(b)* In 1982 only one infectious disease was among the top ten. (Data from the U.S. Bureau of the Census, 1986 *Statistical Abstract of the United States,* 106th ed., Washington, D.C.)

The Sulfa Drugs

Sulfa drugs represent a group of compounds discovered in a conscious search for antibiotics. In 1904 the German chemist Paul Ehrlich (1854–1915; 1908 Nobel prize recipient) realized that infectious diseases could be conquered if toxic chemicals could be found that attacked parasitic organisms within the body to a greater extent than they did host cells. Ehrlich achieved some success toward his goal; he found that certain dyes

Large doses of sulfa drugs are required compared with the doses of "true" antibiotics.

that were used to stain bacteria for microscopic examination could also kill the bacteria. This led to the use of dyes against organisms causing African sleeping sickness and arsenic compounds against those causing syphilis.

After experimenting with several drugs, Gerhard Domagk, a pathologist in the I. G. Farbenindustrie Laboratories in Germany, found in 1935 that Prontosil, a dye, was somewhat effective against bacterial infection in mice. Prontosil can be changed to **sulfanilamide,** which is very effective:

$$H_2N-\underset{}{\underset{}{\bigcirc}}-N=N-\underset{NH_2}{\underset{}{\bigcirc}}-SO_2NH_2 \xrightarrow{H_2O} H_2N-\underset{}{\underset{}{\bigcirc}}-SO_2-NH_2$$

Prontosil   Sulfanilamide

This discovery led to the synthesis and testing of many related compounds in the search for drugs that are more effective or less toxic to the infected experimental animal. By 1964 more than 5000 sulfa drugs had been prepared and tested.

Sulfa drugs inhibit bacteria by preventing the synthesis of folic acid, a vitamin essential to their growth. The drugs' ability to do this apparently lies in their structural similarity to a key ingredient in the folic acid synthesis, *para*-aminobenzoic acid.

A sulfa drug mimics an essential compound.

$$H_2N-\underset{}{\underset{}{\bigcirc}}-\underset{O}{\overset{O}{\underset{\|}{\overset{\|}{S}}}}-\underset{H}{\overset{H}{N}}-H \qquad H_2N-\underset{}{\underset{}{\bigcirc}}-C\underset{OH}{\overset{O}{\diagup}}$$

Sulfanilamide   p-Aminobenzoic Acid
(A Typical Sulfa Drug)

The close structural similarity of sulfanilamide and *p*-aminobenzoic acid permits sulfanilamide instead of *p*-aminobenzoic acid to be incorporated into the enzymatic reaction sequence. By bonding tightly sulfanilamide shuts off the production of the essential folic acid, and the bacteria die of vitamin deficiency. In humans and other higher animals in which *p*-aminobenzoic acid is not necessary for folic acid synthesis, sulfa drugs have no effect.

The Penicillins

Penicillin was discovered in 1928 by Alexander Fleming (Fig. 16–5), a bacteriologist at the University of London, who was working with cultures of *Staphylococcus aureus*, a germ that causes boils and some other types of infections. In order to examine cultures with a microscope, Fleming had to remove the covers of the culture plates for a while. One day as he started work he noticed that one culture was contaminated by a blue-green mold. For some distance around the mold growth, the bacterial colonies were being destroyed. Upon further investigation, Fleming found that the broth in which this mold had grown also had an inhibitory or lethal effect on many pathogenic (disease-causing) bacteria. The mold was later identified as *Penicillium notatum* (the spores sprout and branch out in pencil shapes; hence the name).

Penicillin, the name given to the antibacterial substance produced by the mold, apparently has no toxic effect on animal cells, and its activity is selective. The structure of penicillin (see margin) has now been determined. Many different penicillins exist, differing in the structure of the R group. Penicillin G is the most widely used in medicine. Several antibiotics, such as penicillin and bacitracin, are known to prevent cell wall synthesis in bacteria, as shown in Figure 16–6.

Penicillin G

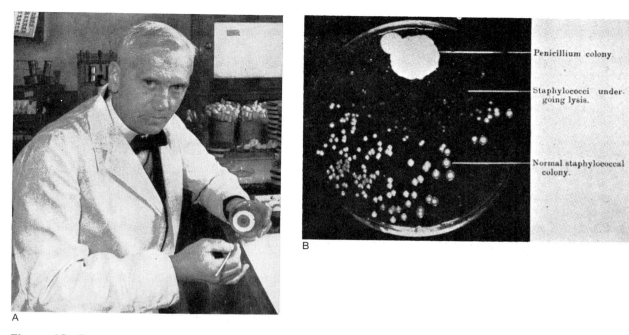

Figure 16-5
(a) Sir Alexander Fleming. *(b)* A culture plate showing the dissolution of staphylococcal colonies in the neighborhood of a colony of penicillium.

Streptomycin and Tetracyclines

In 1937, following collaboration with René Dubos, Selman Waksman isolated a compound from a soil organism, *Streptomyces griseus;* this compound, which came to be known as **streptomycin,** was released to physicians in 1947. It was quite successful in controlling certain types of bacteria but was later withdrawn because of adverse side effects.

Streptomycin has many undesirable side effects.

Figure 16-6
Penicillin kills bacteria by interfering with the formation of cross links in the cell wall.

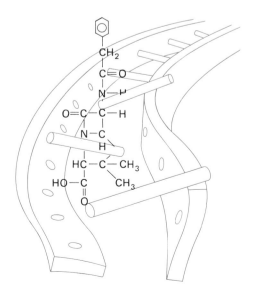

In 1945 B. M. Duggan discovered that a gold-colored fungus, *Streptomyces aureofaciens*, produced a new type of antibiotic, **Aureomycin,** the first of the **tetracyclines.** Research then stepped up to a fever pitch. Pfizer Laboratories tested 116,000 different soil samples before they discovered the next antibiotic, **Terramycin.**

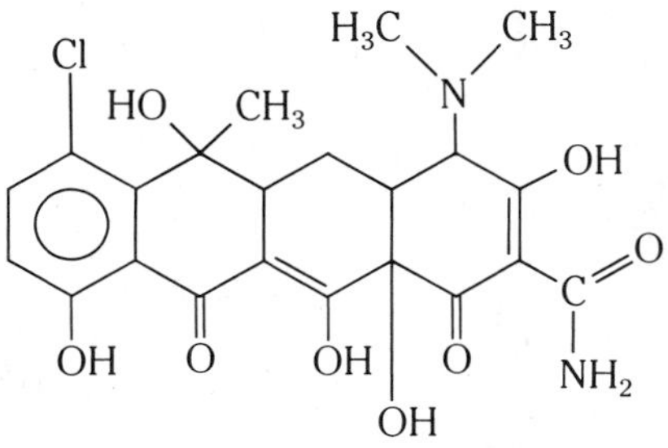

Chlortetracycline
(Aureomycin)

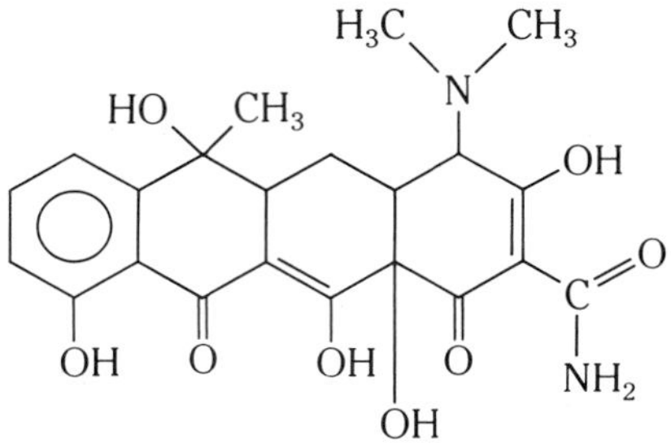

Oxytetracycline
(Terramycin)

Tetracyclines get their names from their four-ring structures.

Compounds of the tetracycline family are so named because of their four-ring structure. One side effect of these drugs is diarrhea, caused by the killing of the patient's intestinal flora (the bacteria normally residing in the intestines).

Cephalosporins

Cephalosporins, another class of antibiotics, were discovered in the 1950s and have become more widely used then penicillins. The most widely used cephalosporin is cephalexin (Keflex), which ranked 14th in total prescriptions in the United States during 1986.

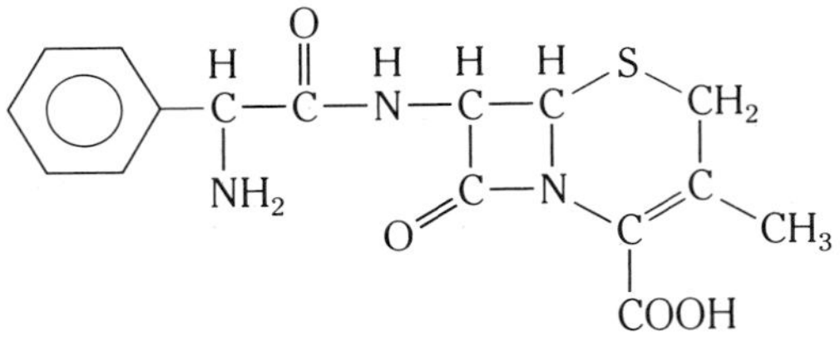

Cephalexin

Antiseptics and Disinfectants

Pathogenic bacteria cause many illnesses.

An antiseptic is a compound that prevents the growth of microorganisms. It now is legally considered a **germicide,** or a compound that kills microorganisms. A disinfectant is a compound that destroys pathogenic bacteria or microorganisms, but usually not bacterial spores. Disinfectants are generally poisonous and therefore suitable only for external use, as on the skin or a wound.

Common germicides, such as the halogens, sodium hypochlorite, and hydrogen peroxide, are effective because they can oxidize any kind of cell, including human cells. For this reason they are used mostly as disinfectants on nonliving objects. The quaternary ammonium compounds are surface active agents, and their bactericidal effect seems to be related to their ability to weaken the cell wall so the cell contents cannot be contained.

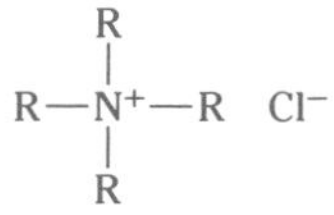

A Quaternary Ammonium Chloride

Iodophors, commonly used as disinfectants for restaurant glassware, are a complex of iodine with polyvinylpyrrolidone.

Because antiseptics and disinfectants are generally toxic, it is necessary to use dilute

An Iodophor

solutions that are applied only to the skin. Although they help prevent the spread of disease, germicides are practically useless in the treatment of disease because they act nonspecifically against all cells. However, there are a number of antibiotics that are used as germicides because they act more selectively against infecting bacteria. The most common one is **Neosporin,** which is applied topically to treat a number of skin conditions. Neosporin is a combination of three antibiotics: polymyxin B, bacitracin, and neomycin.

SELF-TEST 16-A

1. The accepted chemical name for a drug is called its _____ name.
2. The most widely prescribed drug in the United States is used for _____.
3. Which of the following antacids will not produce a gaseous byproduct? (a) magnesium hydroxide, Mg(OH)$_2$ (b) calcium carbonate, CaCO$_3$ (c) sodium bicarbonate, NaHCO$_3$
4. Another name for a pain-killing drug is a(n) _____.
5. Morphine comes from the _____ plant.
6. Codeine is less addictive than morphine.   True ( ) or False ( )
7. Another name for a fever reducer is a(n) _____.
8. The IgE antibody plays a role in what disorder? _____
9. A compound that can kill microorganisms can be called a(n) _____.
10. Drugs that inhibit the growth of microorganisms are called _____.
11. The drug Prontosil is converted to sulfanilamide, which resembles _____, which is used for folic acid synthesis.
12. Penicillin is produced from cultures of what kind of organism? _____
13. The tetracycline drugs are produced from cultures of what kind of organism? _____

Heart Disease Drugs

Heart disease is the number-one killer of Americans, claiming approximately 750,000 lives each year, or 38% of total yearly deaths. More than 100 years ago nitroglycerin was found to be effective in relieving pain associated with an insufficient flow of blood to the heart. Now many new drugs are being used to improve the quality of life of persons suffering from heart disease.

Medicines, Drugs, and Drug Abuse

Heart disease is an assortment of diseases, but the basic cause of many of them is **atherosclerosis**, which is the buildup of fatty deposits called **plaque** on the inner walls of arteries. Cholesterol, a lipid, is a major component of atherosclerotic plaque. Many scientists believe that a high level of cholesterol in the blood, along with high blood levels of triglycerides, which are also lipids, contributes to the buildup of this plaque. The plaque buildup reduces the flow of blood to the heart. If a coronary artery is blocked by plaque, a heart attack occurs as a result of the reduced blood flow carrying oxygen to the heart. If prolonged, such an attack can cause part of the heart muscle to be destroyed. Of about 1.5 million persons who experience heart attacks each year, half a million, or one-third, die. About 98% of all heart attack victims have atherosclerosis. The drugs used to treat heart disease ease the flow of blood to the heart. This lowers the force the heart must exert to pump the blood through the circulatory system. Some heart disease drugs lower the buildup of plaque by regulating the amount of lipids in the blood.

> One prominent theory contends that injury to a coronary artery and/or heredity causes the buildup of atherosclerotic plaque.

In 1948 Raymond P. Ahlquist of the Medical College of Georgia discovered that heart muscle contains receptors; he called these **beta receptors.** Stimulation of these receptors by epinephrine and norepinephrine results in an increase in the number of heart beats. In 1967 Alonzo M. Lands, a pharmacologist in Rensselaer, New York, discovered two different beta receptors, beta$_1$ and beta$_2$. Beta$_1$ sites are located primarily in the heart but also in the kidneys. Beta$_2$ receptors are involved in the relaxation of the peripheral blood vessels and the bronchial tube.

With the knowledge about beta receptors gained by Ahlquist, chemists began to explore the action of chemicals that would compete with epinephrine and norepinephrine at the beta receptor sites. If these sites could be blocked, the heart rate would decrease. For a heart already overworked from the buildup of plaque in the arteries supplying heart tissue with blood (and oxygen), this might just produce enough relaxation to allow recovery from an impending attack. In addition, these drugs might be able to relieve high blood pressure **(hypertension)** and **migraine** headaches.

The first drugs of this type, called **beta blockers** because of their action of blocking beta receptor sites, came into use in the late 1950s and early 1960s but were later withdrawn because of undesirable side effects. In 1967 the beta blocker propranolol (trade name *Inderal*; Table 16–7) was first prescribed. Its first use was for cardiac

Table 16–7
Beta-Blocking Drugs Used in Treating Heart Disease

Trade Name	Generic Name	Structure
Inderal	Propranolol	naphthyl—O—CH$_2$—CH(OH)—CH$_2$NHCH(CH$_3$)$_2$
Lopressor	Metoprolol	CH$_3$—O—CH$_2$—CH$_2$—C$_6$H$_4$—O—CH$_2$—CH(OH)—CH$_2$—NHCH(CH$_3$)$_2$
Corgard	Nadolol	(HO)(HO)-tetrahydronaphthyl—O—CH$_2$—CH(OH)—CH$_2$—NHC(CH$_3$)$_3$

arrhythmias, but now it has been approved by the FDA for the treatment of angina, hypertension, and migraine headache. Propranolol has high lipid solubility, so it passes through the blood-brain barrier and builds up the central nervous system, where it is more slowly metabolized. This buildup of the chemical in the central nervous system causes the side effects of fatigue, lethargy, depression, and confusion. In spite of these side effects, propranolol is one of the most widely prescribed drugs of any type in the United States (see Table 16–3). Sales of this blocking agent increased from almost $12 million in 1972 to over $225 million in the late 1980s.

A second beta blocker, metoprolol (trade name *Lopressor*) was introduced in the United States in 1978 for the treatment of hypertension. Because metoprolol is selective in its beta-blocking effects, blocking only $beta_1$ sites and not the $beta_2$ sites of the peripheral blood vessels or the bronchial tube, it is safe for asthma sufferers and for patients with severe blood vessel disorders. Metoprolol is not as soluble in lipids as propranolol and does not accumulate in the central nervous system. It is also more slowly metabolized by the liver, which allows more widely spaced doses, usually twice a day. A third beta blocker, nadolol (trade name *Corgard*), was introduced in the United States in 1979. This chemical is not metabolized to a great extent by the liver and passes mostly unchanged through the kidneys to the urine. Because it is largely unmetabolized, nadolol may be taken only once a day. Nadolol is also less soluble in lipids than either propranolol or metoprolol and thus produces very few side effects associated with accumulation in the central nervous system. The three agents have some structural similarities, as may be seen in Table 16–7.

Other heart disease drugs are the **calcium channel blockers.** Research has found that calcium ions move into the heart muscle by means of holes or channels in the phospholipid membrane surrounding the muscle. In the muscle cells the calcium ions cause an interaction between the parallel protein filaments myosin and actin, and this interaction causes the cell to contract (Fig. 16–7). In addition, the double positive charge on the calcium ion neutralizes some of the negative charge of the muscle cell, also causing the muscle to contract. Movement of the calcium ions out of the cell restores the negative charge, and the cell relaxes. Blocking the flow of calcium ions into the cell causes the muscles to relax. When the smooth muscles in the walls of the coronary arteries are relaxed, these arteries expand and increase the supply of blood to the heart. Some calcium blockers also decrease the force of contraction and thus decrease the oxygen requirements of the heart.

Calcium blockers, unlike beta blockers, can prevent spasms of the coronary arteries. These spasms cause a blockage of blood flow to the heart and the intense pain of the angina attack. The causes of these spasms are poorly understood, but the calcium blockers do dilate the arteries and lessen the possibility of an angina attack.

Calcium blockers are not structurally similar (Table 16–8). That such structurally dissimilar chemicals act on the same process may be due to different types of calcium channels in muscle cell walls; alternatively, the differently shaped molecules of the different calcium blockers may stop up the calcium ion channels at different places (see Fig. 16–7).

After 19 years of use in Europe, verapamil (trade names *Isoptin* and *Calan*) became available in the United States and was first used to treat angina in 1981. In the same year a second calcium channel blocker, nifedipine (trade name *Procardia*), was introduced. Nifedipine is a powerful dilator of coronary arteries that has an immediate effect on patients suffering from angina. The calcium blockers are not without their problems, however. Most produce the side effects of headaches, dizziness, flushing of the skin, and light-headedness.

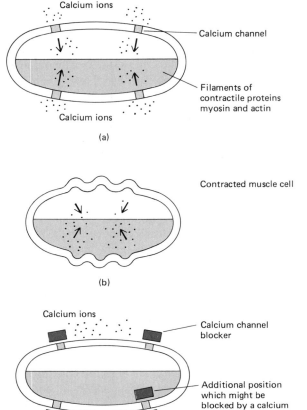

Figure 16–7
(a) Calcium ions flowing into a muscle cell. (b) Contracted muscle cell caused by the presence of calcium ions in contractile protein fiber bundles. (c) Relaxed muscle cell with calcium channels blocked by drug molecules.

The ultimate method of prevention of most forms of heart disease would be prevention of the buildup of atherosclerotic plaque in the arteries. In 1954 John W. Gofman, of the University of California at Berkeley, discovered four separate density categories of lipoproteins in the blood. About 65% of the cholesterol in the blood is carried by low-density lipoproteins, whereas only 25% is carried by high-density lipoproteins. In 1968 John A. Glomset, of the University of Washington, showed that high-density lipoproteins are effective in removing cholesterol from arterial walls and transporting it to the liver, where it is metabolized. This discovery opened up the possibility that already formed atherosclerotic plaque might be dissolved. For patients whose blood levels of lipids cannot be controlled by proper diet, scientists have been attempting to develop drugs that either raise the level of high-density lipoproteins or lower the levels of low-density lipoproteins.

Niacin (nicotinic acid) lowers blood lipid concentrations by interfering with the synthesis of cholesterol in the liver. Several drugs have been introduced that cause the liver to convert more cholesterol into bile acids by lowering the concentration of bile acids in the intestines. One of these, cholestyramine resin (Table 16–9), binds bile acids.

Table 16-8
Two Calcium Channel-Blocking Drugs Used in Treating Heart Disease

Generic Name	Trade Name	Structure
Verapamil	Isoptin, Calan	
Nifedipine	Procardia	

Anticancer Drugs

Cancer is not one but perhaps 100 different diseases, caused by a number of factors. A cancer begins when a cell in the body starts to multiply without restraint and produces descendants that invade tissues in the vicinity. It seems reasonable, then, that some drugs might exist that would be able either to stop this undesirable spreading of cancer cells or to prevent cancer from happening at all.

Cancers are treated by (1) surgical removal of whole areas affected by them, as well as the cancerous growths themselves; (2) irradiation to kill the cancer cells; and (3) chemicals to kill the cancer cells **(chemotherapy).** These treatment methods have resulted in some dramatic improvements in the rates of survival of patients with certain cancers. A group of cancer patients can be considered cured if, after their treatment, they die at about the same rate as the general population. Another way of judging success in cancer therapy is by the number of patients who survive for five years after the treatment.

Table 16-9
Two Drugs Used to Lower Cholesterol in the Blood

Generic Name	Trade Name	Structure
Niacin (Nicotinic acid) (vitamin B_3)	Nicolar	
Cholestyramine resin	Questran	

Medicines, Drugs, and Drug Abuse

Table 16-10
Five-Year Survival Rates Associated with Various Cancers*

Type	1960-1963 (%)	1977-1982 (%)
Breast	63	74
Bladder	53	76
Hodgkin's disease	40	73
Colon	43	53
Leukemia	14	33
Lung	8	13
Pancreas	1	2

* Source: National Cancer Institute, 1985.

As Table 16-10 shows, some cancers have shown marked increases in survival rates over the past two decades. However, the survival rate is still very low for lung cancer (13%), stomach cancer (16%), liver cancer (3%), pancreatic cancer (2%), and cancer of the esophagus (6%).

In World War I the toxic effects of a class of the chemical warfare gases called mustard gases were recognized. These gases were found to cause damage to the bone marrow and to be mutagenic. In these ways they were acting like X rays, which are also toxic to cells and cause mutations.

$$Cl-CH_2CH_2-S-CH_2CH_2-Cl$$
Mustard Gas

Beginning around 1935, other mustards of the nitrogen family were synthesized. They, too, caused mutations in some laboratory animals. In addition, they caused cancers in some animals.

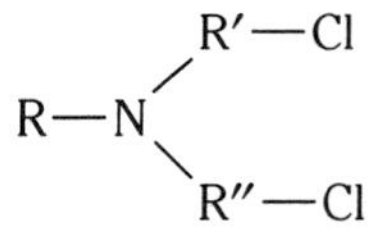

Nitrogen Mustard General Formula

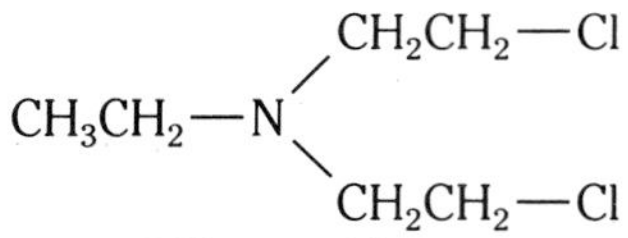

A Nitrogen Mustard

After World War II the secrecy surrounding the mutagenic nature of these chemicals was lifted, and it occurred to cancer researchers that cancers might be treated with chemicals that selectively destroy unwanted cells.

One of the most widely used anticancer drugs is cyclophosphamide, a compound that contains the nitrogen mustard group (shown in color).

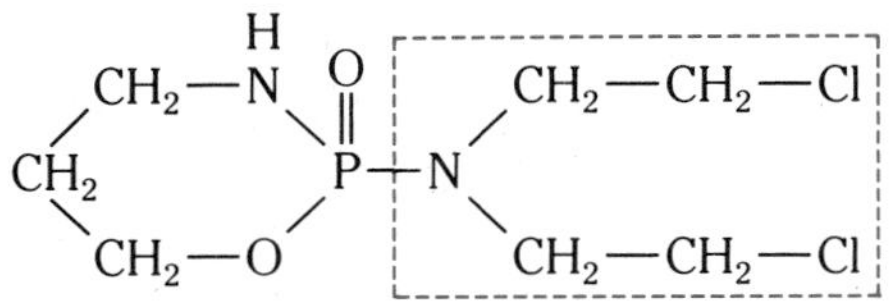

Cyclophosphamide

Compounds such as cyclophosphamide belong to the **alkylating class** of anticancer drugs. Alkylating agents are reactive organic compounds that transfer alkylating groups in chemical reactions. Their effectiveness as anticancer agents is due to the transfer of

Figure 16–8
The structural arrangement of atoms when a cis[Pt(NH$_3$)$_2$] fragment (shown in color) binds to two nitrogen atoms in the guanine rings of a dinucleotide.

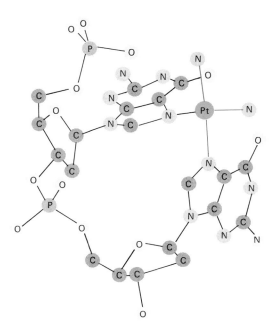

alkyl groups to the nitrogen bases in DNA, particularly guanine. The presence of the alkyl group in the guanine molecule blocks base pairing and prevents DNA replication, which stops cell division. Although alkylating agents attack both normal cells and cancer cells, the effect is greater for rapidly dividing cancer cells.

Another widely used anticancer drug, **cisplatin,** blocks DNA replication by a similar mechanism.

Cisplatin

Inside the cell, the Cl$^-$ ions are displaced, and the Pt(NH$_3$)$_2$ unit binds to the nitrogen sites on the guanine bases in DNA (Fig. 16–8). As with alkylating agents, the blocking action prevents base pairing and DNA replication.

Another group of chemotherapeutic agents, called **antimetabolites,** interferes with DNA synthesis. One of these chemicals, 5-fluorouracil, gets involved in the synthesis of a nucleotide, which inhibits the formation of a thymine-containing nucleotide necessary for DNA synthesis. Lack of proper DNA synthesis slows cell division. 5-Fluorouracil has proved useful in the treatment of cancers of the breast.

Uracil   *5-Fluorouracil*

Another antimetabolite is methotrexate, which has a structure similar to that of folic acid.

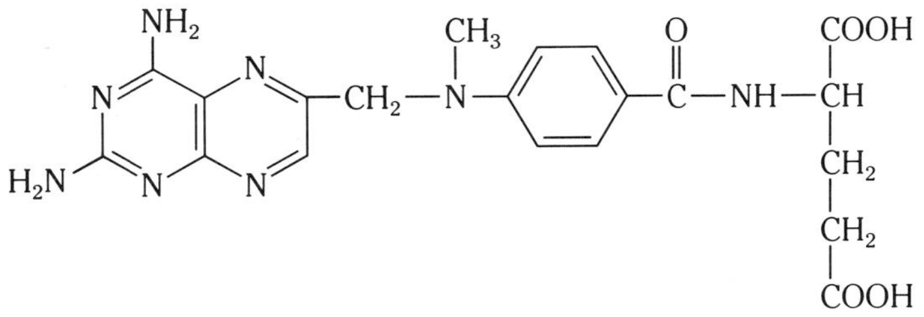

Methotrexate

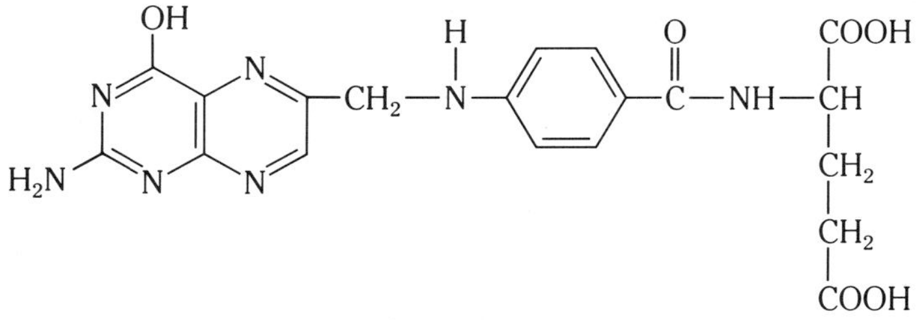

Folic Acid

In a manner similar to that in which sulfanilamide inhibits bacterial cell wall synthesis (see page 396), methotrexate takes the place of folic acid on an enzyme that promotes cell growth. This results in a slowdown of cell growth. Leukemia is treated with methotrexate.

All cancer chemotherapy is tedious and has its risks. In addition to being highly toxic, most of the useful chemotherapy agents are themselves carcinogenic (cancer-causing). Often very high doses are necessary to effect treatment. As a result, single-agent chemotherapy has largely given way to combination chemotherapy because of the success of additive, or even **synergistic,** effects when two or more anticancer drugs are used. For example, a combination that is used in the treatment of several cancers is cisplatin and cyclophosphamide. Because of their synergistic action, lower doses of each compound can be used when they are given together than if they were each given alone, and this reduces the harmful side effects of chemotherapy.

Synergism is the working together of two things to produce an effect greater than the sum of the individual effects.

Childhood cancers respond most favorably to chemotherapy. Most children with leukemia who are treated with chemotherapy drugs enter a period of relapse-free survival. In terms of the definition given earlier, they are cured. During the early 1950s about 1900 children under the age of five years died of cancer per year in the United States. Today the number is fewer than 700 per year. A few other cancers, such as Hodgkin's disease, have shown similar increases in survival rates due to chemotherapy. Not all cancers have been so treatable, however, because cancer is so many different diseases.

The Steroid Drugs

A large and important class of naturally occurring compounds is derived from the following tetracyclic structure:

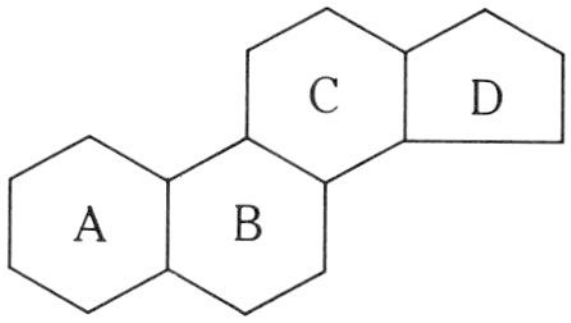

The Steroid Drugs

These compounds, known as **steroids,** occur in all plants and animals. The most abundant animal steroid is cholesterol, $C_{27}H_{46}O$. The human body synthesizes cholesterol and readily absorbs dietary cholesterol through the intestinal wall. It is associated with gallstones and atherosclerosis.

Biochemical alteration or degradation of cholesterol leads to many steroids of great importance in human biochemistry. **Cortisone,** one of the adrenal cortex hormones, acts as an anti-inflammatory agent when applied topically or injected into a diseased joint, and it is used in treating acute cases of arthritis.

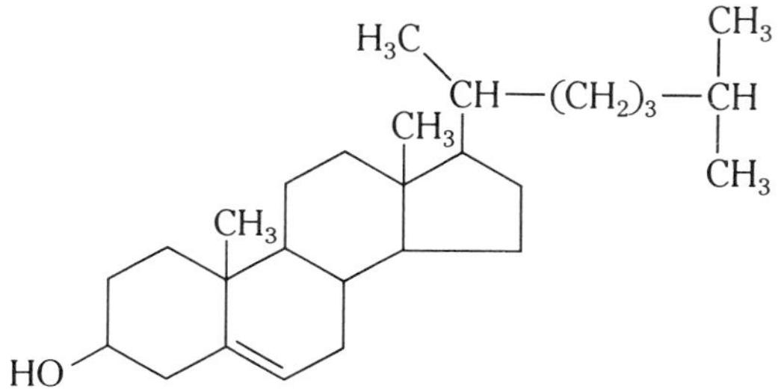

Cholesterol

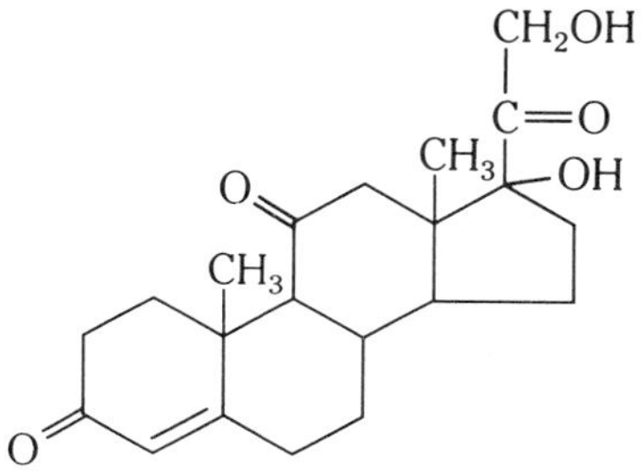

Cortisone

Structurally related to cholesterol and cortisone are the sex hormones. One female sex hormone, **progesterone,** differs only slightly in structure from an important male hormone, **testosterone.**

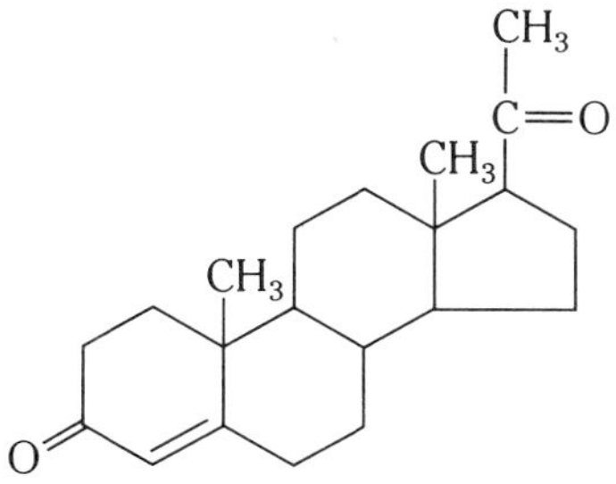

Progesterone

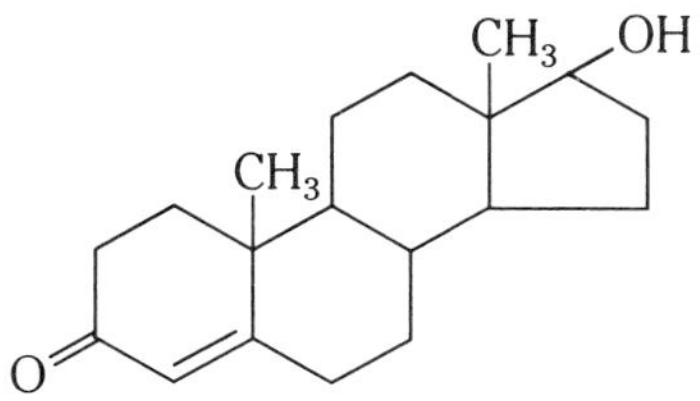

Testosterone

Other female hormones are estradiol and estrone, together called **estrogens.** The estrogens differ from the other steroids discussed earlier in that they contain an aromatic ring.

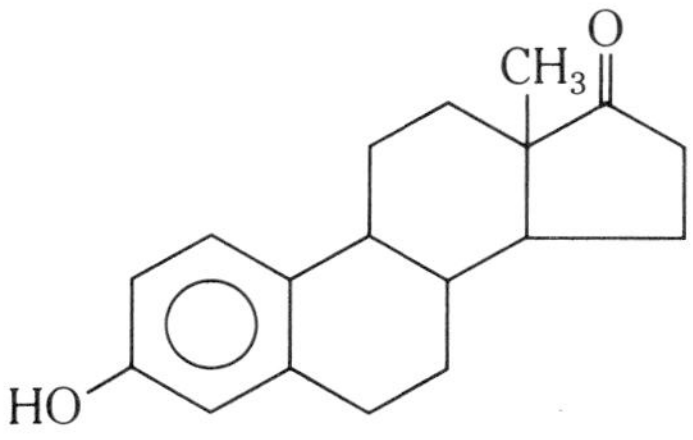

Estrone

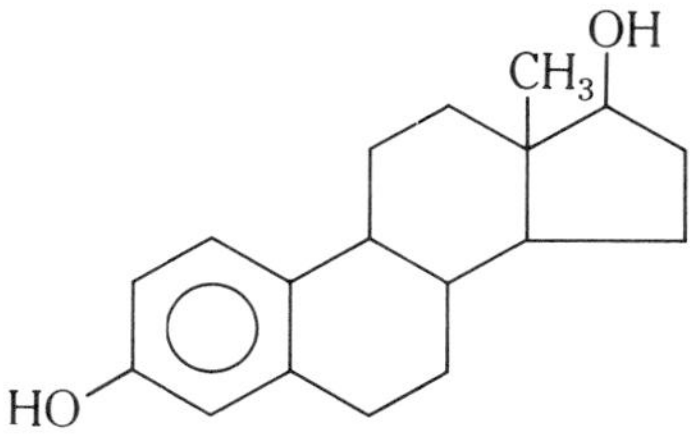

Estradiol

Medicines, Drugs, and Drug Abuse

The estrogens and progesterone are produced by the ovaries. Estrogens are important to the development of the egg in the ovary, whereas progesterone causes changes in the wall of the uterus and after pregnancy prevents release of a new egg from the ovary (ovulation). Birth control drugs use derivatives of estrogens and progesterone to simulate the hormonal processes resulting from pregnancy and thereby prevent ovulation.

Birth Control Pills

One of the most revolutionary medical developments of the 1950s was the worldwide introduction and use of "the pill." More than 10 million women currently use birth control pills. Now there are two types of oral contraceptives, "the pill" and "the mini-pill."

The pill contains small amounts of synthetic analogs of estrogens and progesterone. The common ones are mestranol, a synthetic estrogen derivative, and norethindrone, a synthetic progesterone derivative. The estrogen derivative regulates the menstrual cycle, while the progesterone derivative establishes a state of false pregnancy resulting in the prevention of ovulation.

Mestranol
(synthetic estrogen)

Norethindrone
(synthetic progesterone)

$1 \text{ mg} = 1 \times 10^{-3} \text{ g}$

Early versions of the pill contained larger doses of the estrogen and progesterone derivatives. However, since the 1960s researchers have succeeded in reducing the steroid content of the pill from about 200 mg to as little as 2.6 mg. In addition, "biphasic" and "triphasic" products were developed to reflect more accurately a woman's natural hormonal changes. These products are formulated to provide the lowest effective dosages of both progesterone and estrogen on different days of the menstrual cycle. An example of a biphasic oral contraceptive is Ortho-Novum 10/11, which provides 0.5 mg of norethindrone for 10 days, then 1.0 mg of norethindrone for 11 days; 35 μg of ethynyl estradiol is administered consistently for 21 days. Triphasic oral contraceptives change the amount of norethindrone three times instead of twice.

$1 \text{ }\mu\text{g} = 1 \times 10^{-6} \text{ g or}$
$1 \times 10^{-3} \text{ mg}$

Ethynyl estradiol

All oral contraceptives are prescription drugs and should be taken only after a checkup by a doctor. There are risks associated with the use of the pill. The risk of blood clots, the most common of the serious side effects, increases with age and with heavy smoking (more than 15 cigarettes per day). The chance of a fatal heart attack is about 1 in 10,000 in women between the ages of 30 and 39 who use oral contraceptives and smoke, compared with about 1 in 50,000 in users who do not smoke and 1 in 100,000 in nonusers who do not smoke.

Women are warned not to use oral contraceptives if they have or have had a heart attack or stroke, blood clots in the legs or lungs, angina pectoris, known or suspected cancer of the breast or sex organs, or unusual vaginal bleeding. In addition, they should not use the pill if they are pregnant or suspect they are pregnant.

The "mini-pill", developed in the 1970s, contains much smaller amounts of the synthetic progesterones (0.1–0.2 mg) and no synthetic estrogen. The mini-pill was introduced in the hope that its users would experience fewer side effects than users of the pill, since the estrogen component of the pill is regarded as the cause of many of the serious side effects. However, there is not sufficient information available to support this concept. Although the mechanism of mini-pills is not fully understood, these contraceptives are thought to stop conception by preventing release of the egg, by keeping the sperm from reaching the egg, and by making the uterus unreceptive to any fertilized egg that reaches it.

Steroid Drugs in Sports

The steroid testosterone is responsible for the muscle building that boys experience at puberty, in addition to the development of adult male sexual characteristics. Synthetic steroids have been developed in part to separate the masculinizing (androgenic) effects and muscle-building (anabolic) effects of testosterone. These steroids have been prescribed by physicians to correct hormonal imbalances or to prevent the withering of muscle in persons who are recovering from surgery or starvation.

Anabolic means muscle-building effect.

Healthy athletes discovered that synthetic steroids appeared to have an anabolic effect on them as well. Initially these anabolic steroids were used by weight lifters and by athletes in track-and-field events like the shot-put and hammer throw. Later, some inconclusive evidence surfaced suggesting that anabolic steroids increased endurance, and this caused runners, swimmers, and cyclists to begin using them.

Such sports organizations as the National Collegiate Athletic Association and the International Olympic Committee have banned the use of anabolic steroids and other drugs by athletes. Although few human studies have been carried out on the use of anabolic steroids by healthy individuals, a number of harmful side effects have been identified.

The side effects of anabolic steroid use include acne, baldness, and changes in sexual desire. Some men experience enlargement of the breasts. Accompanying these noticeable changes are testicular atrophy and decreased sperm production. This is caused by an imbalance among the testes, pituitary, and hypothalamus due to the increased concentration of these male sex hormones in the bloodstream. High levels of male sex hormones cause the hypothalamus to signal the pituitary gland to lower production of two other hormones, luteinizing hormone and follicle-stimulating hormone, which stimulate sperm production in the testes. Although these changes appear to be reversible, additional testing is needed. In women the use of anabolic steroids produces facial hair, male-pattern baldness, deepening of the voice, and changes in the menstrual cycle. Most of these changes are not reversible.

In addition to these problems, oral-dose anabolic steroids are toxic to the liver. Testosterone taken orally is not very effective, since most of it is rapidly metabolized by the liver before it reaches the bloodstream. However, several of the common anabolic steroids are active when taken orally, in part because of an alkyl group in addition to the hydroxyl group at the carbon-17 position of the steroid nucleus (Fig. 16–9). This alkyl structure slows metabolism in the liver and thus allows more of the dose to reach the bloodstream, but it also increases liver toxicity. Some liver cancer has been reported in anabolic steroid users.

SELF-TEST 16–B

1. What are two ingredients in atherosclerotic plaque? _____ and _____

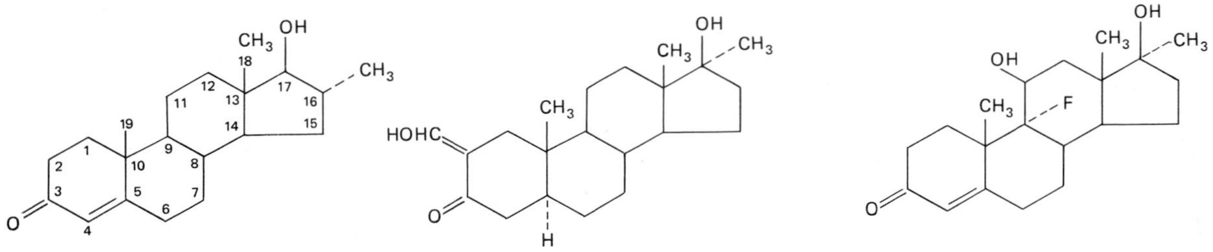

Figure 16-9
Structure of some anabolic steroids. The carbon-17 position in some oral-dose steroids is occupied by an alkyl group.

2. When beta sites in the heart are stimulated by epinephrine, the heart beats ( ) faster or ( ) slower.
3. When propranolol passes through the blood-brain barrier, it causes the side effects of _____ and _____.
4. A drug that blocked calcium ions from flowing into heart muscles would have the effect of ( ) exciting or ( ) relaxing the heart.
5. Cholesterol is carried in the blood by both high- and low-density lipoproteins. Which carries the greater percentage of cholesterol? _____
6. A drug that removed cholesterol from the bloodstream would be useful in treating some forms of heart disease.   True ( ) or False ( )
7. Most anticancer drugs can also cause cancer.   True ( ) or False ( )
8. The nitrogen mustards act on cancer cells by blocking _____ replication.
9. Chemicals that interfere with DNA synthesis are called _____. Which cancer has shown the greatest response to chemotherapeutic drugs? _____
10. The male sex hormone is called _____.
11. The female sex hormone is called _____.
12. Anabolic as used in the term *anabolic steroid* means _____.
13. Name three undesirable side effects in male athletes who use anabolic steroids.
_____
_____

Drug Abuse

Problems with drug abuse and drug addiction are as old as civilization itself. However, a new dimension has been added in the 1980s as a result of two developments, "crack" and "designer drugs."

Crack

Crack is a purified form of cocaine obtained by heating a mixture of cocaine with sodium bicarbonate for 15 minutes. More than 20 million Americans have used cocaine, and about 4 million are using or abusing it now. The appearance of crack is likely to cause a drastic increase in the number of cocaine addicts, since crack is regarded as the most addictive drug known. The reason for its addictiveness is that crack is more potent than cocaine and is smoked rather than sniffed, giving the user a much quicker, more intense high. Because the high lasts less than 10 minutes, users have a tendency to use crack repeatedly over a short period of time, and most users thus become addicted after only one try. The problem is enhanced by the ready availability and low cost of crack ($10–$15 per fix).

Designer Drugs

Designer drugs are chemical substances that are structurally similar to legal drugs. Because of their action, they are potential drugs of abuse. All the designer drugs that have been discovered so far are either narcotics or hallucinogens. For example, fentanyl (Fig. 16–10) is a powerful narcotic marketed under the trade name *Sublimaze*. Fentanyl is about 150 times more potent than morphine and just as addictive, but very short-acting. Fentanyl is used in up to 70% of all surgical procedures in the United States. The derivatives of the fentanyl molecule are also potent narcotics. These drugs were called designer drugs when they first appeared on the streets because they had obviously been designed by some unscrupulous chemists for consumption by drug addicts. These fentanyl derivatives were every bit as potent as heroin, but because they were not listed on the U.S. Drug Enforcement Administration (DEA) list of controlled substances, they could be sold legally. Until a compound is recognized as being abused and is classified as a dangerous drug—a process known as *scheduling*—no laws apply to it.

In the past few years several fentanyl derivatives (Fig. 16–10) have appeared in California. Samples ranged from pure white powder, sold as China White, to a brown material. First came α-methyl fentanyl, then p-fluoro fentanyl, then α-methyl acetyl fentanyl, and in early 1984 3-methyl fentanyl, a compound 3000 times more potent than morphine. Because of this potency and because heroin addicts can use the fentanyl

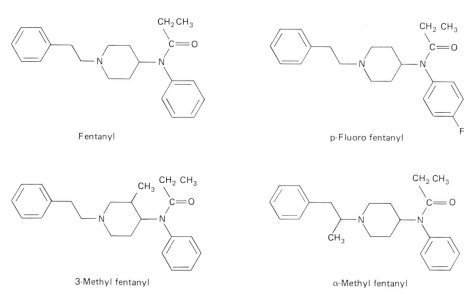

Figure 16–10
Fentanyl and several of its derivatives.

Doonesbury

BY GARRY TRUDEAU

DOONESBURY, by Garry Trudeau. Copyright, 1985, G. B. Trudeau. Reprinted with permission of Universal Press Syndicate. All rights reserved.

derivatives interchangeably with heroin, fentanyl derivatives, mostly 3-methyl fentanyl, have been responsible for over 100 overdose deaths in California.

Another group of designer drugs is derived from meperidine (Demerol). One of these is MPPP, which is short for 1-methyl-4-phenyl-4-propionoxy-piperidine. This compound was first synthesized in 1947, never used commercially, and never scheduled as a controlled substance. MPPP is about 3 times more potent than morphine and 25 times more potent than meperidine. It is structurally so close to meperidine that one has to look closely at the structures to see the difference (hint: look at the ester linkage). If the synthesis of MPPP is carried out at too high a temperature or at too low a pH, the product is MPTP, 1-methyl-4-phenyl-1,2,3,6-tetrahydropyridine.

In 1982 a batch of MPTP-tainted MPPP, sold in San Jose, California, as "synthetic heroin," produced terrible side effects. It seems that MPTP causes the symptoms of Parkinson's disease, which include stiffness, impaired speech, rigidity, and tremors. Users of this batch of synthetic heroin became victims of advanced Parkinson's disease, in which cells in the area of the brain called the substantia nigra no longer produce dopamine, which is necessary for normal muscle control. A substance that had been used to treat Parkinson's disease, L-dopa, also proved useful in treating the victims of MPTP toxicity. L-dopa could not be used to effect complete recovery, however, since it also causes hallucinations and exaggerated movements.

Ecstasy, also called XTC, Adam, or MDMA, is a hallucinogenic designer drug.

3,4-methylenedioxymethamphetamine
MDMA

MDA

Ecstasy is used mostly by college students and young professionals, who pay $20 a sample. Before its appearance on the streets as a designer drug, ecstasy was used by some psychiatrists who found it useful in the treatment of schizophrenia, depression, and anxiety. However, there is evidence for long-term, irreversible effects on the brain. Rats given MDMA have shown low levels of serotonin, the neurotransmitter that aids in the regulation of sleep, aggression, and mood. An analog of MDMA, MDA, has caused irreversible damage to nerve fibers in the brains of rats.

The control of designer drugs is easier since passage of the Comprehensive Crime Control Act of 1984, which gave the DEA emergency scheduling authority. Now any drug can be designated a controlled substance within 30 days. This scheduling lasts for one year while additional data are gathered to determine final scheduling authority. All the designer drugs described here have been placed in Schedule 1, which precludes their use for any legal purpose.

DEA stands for Drug Enforcement Administration.

Drugs in Combinations

Like some food additives, drugs can have enhanced effects when placed in certain chemical environments; these effects are sometimes harmful, sometimes helpful. Take the case of an aging business executive who took an antidepressant and then ate a meal that included aged cheese and wine. The antidepressant was an inhibitor of monoamine oxidase, an enzyme that helps control blood pressure. Both the aged cheese and the wine the man consumed contained pressor amines, which raise blood pressure. Without the controlling effect of the monoamine oxidase, the pressor amines skyrocketed the man's blood pressure and caused a stroke. Neither the amines nor the antidepressant alone would have been likely to cause the stroke, but the combination did.

Pressor amines tend to increase blood pressure.

Persons who take digitalis for heart trouble and for reducing blood sodium levels should take aspirin only under medical supervision, since aspirin can cause a 50% reduction in salt excretion for 3 or 4 hours after it is taken.

Alcohol increases the action of many antihistamines, tranquilizers, and drugs such as reserpine (for lowering blood pressure) and scopolamine (contained in many over-the-counter nerve and sleeping preparations), making such combinations extremely dangerous. Staying away from dangerous alcohol-drug combinations is not as easy as it may seem. Many people fail to realize that a large number of over-the-counter preparations, such as liquid cough syrup and tonics, contain appreciable amounts of alcohol.

Not all drug combinations are bad. The synergistic effect observed in the treatment of cancer by drugs in combination has already been mentioned. Doctors have been highly successful in prolonging the lives of leukemia and other cancer victims with combinations of drugs that individually could not do the job. Resistant kidney disease has also responded to drug combinations in cases in which single drugs were ineffective.

Perhaps the best advice is to take medicine only when you are seriously ill, making sure that a physician knows what you are taking.

The Role of the FDA

Since 1940 more than 1200 new drugs have been introduced into the U.S. market. Included in this number are almost all of the drugs you probably recognize on your medicine shelf at home. Drug companies must petition the U.S. Food and Drug Administration (FDA) with data showing that a new drug is safe and effective for its intended use. Usually this involves extensive animal tests. If the FDA gives its approval, the drug undergoes limited controlled testing on healthy human test subjects. A second phase involves testing on research subjects who have the disease that the drug is intended to treat. Further tests are then carried out on larger groups to gauge the drug's effectiveness and safety. After all of this testing, only about one in ten drugs passes. The FDA has in the past been accused of being overly cautious, but the thalidomide incident (see Chapter 14) showed that caution can often prevent tremendous human anguish and needless suffering.

Dr. Frances O. Kelsey, a new drug investigator at the FDA in Washington, refused to allow thalidomide to be listed as safe and effective in light of some evidence she read in the data supplied concerning the drug and its effects on laboratory animals. Only after her refusal to certify thalidomide did the teratogenic effects of the drug become known. Thousands of young people born in the early 1960s in the United States owe Dr. Kelsey gratitude for their good health.

The FDA often gets embroiled in controversy regarding drugs that have claimed effectiveness in other countries. However, if a drug is not tested properly, the FDA has no recourse but to withhold approval. One drug that falls into this category is laetrile, a cyanide-containing compound found in the seeds of apples and peaches. This compound has been reputed to cure certain cancers. Yet the FDA refuses to approve its use as a chemotherapy agent because of a lack of sufficient evidence. Laetrile clinics are run in a number of countries, where U.S. citizens travel and receive treatments without FDA approval.

Another chemical involved in controversy is cyclosporin A, a cyclic peptide consisting of 11 amino acid residues with a molecular weight of 1202. Cyclosporin was discovered in 1970 by J. F. Borel of Switzerland as a metabolite in a culture broth of the fungus *Tolypocladium inflatum Gams*. It was soon discovered that this peptide had powerful immunosuppressive effects, and by 1977 the chemical was being used in organ transplants in animals. A year later cyclosporin A was being used to suppress rejection of human liver transplants and marrow transplants in patients with leukemia. Because cyclosporin A so effectively involves itself with the immune system, it is not surprising to see it tried for the treatment of diseases in which the immune system is out of control (after all, the most effective anticancer drugs are themselves carcinogens). Cyclosporin A has recently been tested in the treatment of acquired immune deficiency syndrome (AIDS), although not in the United States, where it has not been approved for testing for that purpose. Whether cyclosporin A will ever be used in the treatment of AIDS in the United States depends on the data available to the FDA regarding its intended use.

If the FDA continues to do its job, we can be reasonably certain the drugs we take have been thoroughly tested before being approved for widespread human use.

Experimental Drugs with Promise

The search for new drugs that are effective against diseases is a continuing process. The three drugs discussed in this section, though still in clinical trials, have attracted considerable interest because of their potential in treating AIDS, Alzheimer's disease, and the early aging associated with diabetes.

AZT

Azidothymidine (AZT) has shown promise in clinical trials in the treatment of patients with AIDS.

AIDS is an abbreviation for acquired immunodeficiency syndrome.

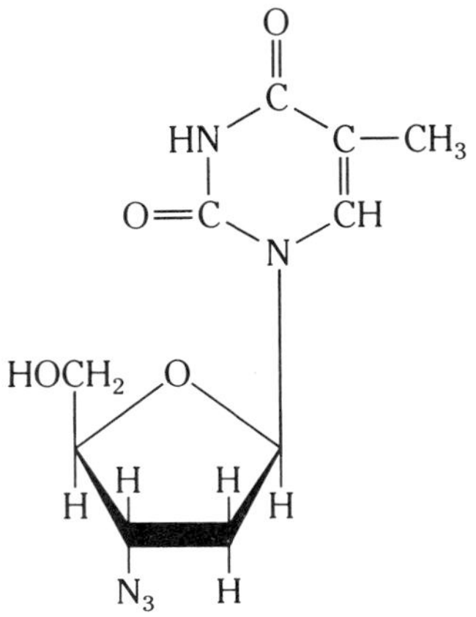

Azidothymidine (AZT)

AZT is a derivative of thymidine, a nucleoside. Nucleosides contain a base and a ribose unit, whereas nucleotides (described in Chapter 11) contain a base, a ribose unit, and a phosphate group. The structures of thymine, thymidine, and thymidine phosphate are shown below:

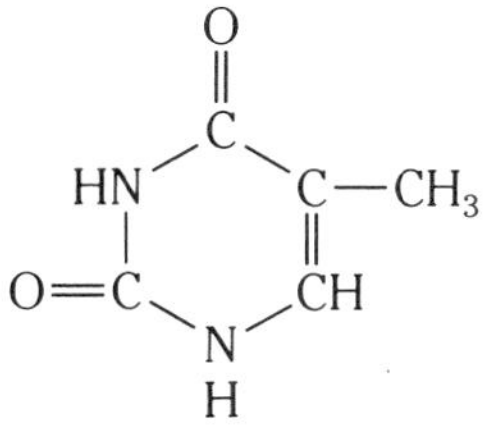

Thymine 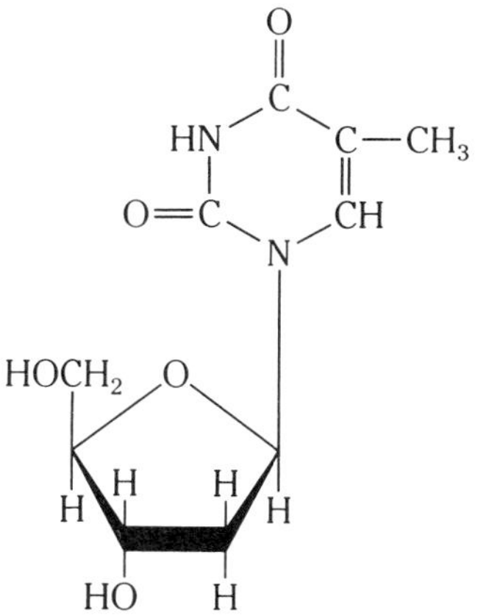   Deoxythymidine (Thymidine) 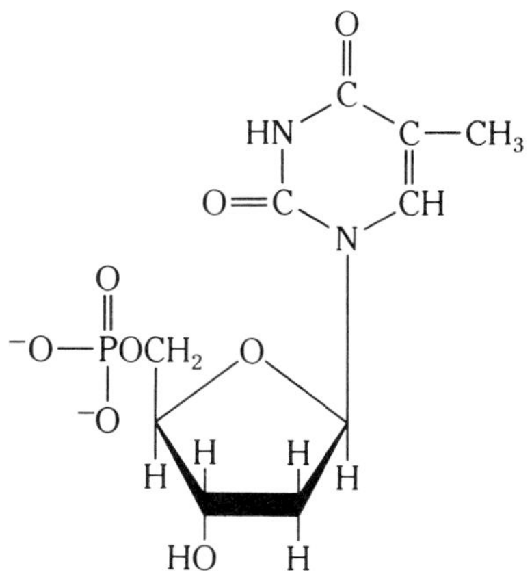   Thymidine 5'-phosphate

AIDS is caused by a **retrovirus,** a virus with an outer double layer of lipid material that acts as an envelope for several types of proteins, an enzyme called reverse transcriptase, and RNA. The term *retrovirus* is used because the virus enzyme carries out RNA-directed synthesis of DNA rather than the usual DNA-directed synthesis of RNA (see Fig. 11-23).

The AIDS retrovirus penetrates the T cell, a key cell in the immune system of the body. Once the retrovirus is inside the T cell, the reverse transcriptase of the AIDS virus

T cells are white blood cells that play a crucial role in controlling the immune response of the body.

translates the RNA code of the virus into the T cell's double-stranded DNA, directing the T cell to synthesize more AIDS viruses. Eventually the T cell swells and dies, releasing more AIDS viruses to attack other T cells.

Although scientists are still searching for a cure or vaccine for AIDS, AZT is one of several compounds showing some promise for use in the treatment of AIDS patients. In limited clinical trials AZT has been found to be effective in the treatment of AIDS patients who have recently had *Pneumocystis carinii* pneumonia. AZT apparently works by being accepted by reverse transcriptase in place of thymidine. After AZT has become a part of the DNA chain, its structure prevents additional nucleosides from being added onto the DNA chain.

Azidothymidine was first synthesized in the 1960s by Jerome Horwitz, who was looking for a compound that would stop cancer cells from multiplying. He reasoned that the incorporation of a "fake nucleoside" into a DNA chain would prevent additional nucleosides from being added and thus cause cell division to stop. The idea didn't work because tumor cells recognized that AZT was not thymidine and didn't incorporate AZT into DNA chains. However, it appears the enzyme in the AIDS virus is fooled by the fake nucleoside.

In 1987 the FDA approved the use of AZT for the treatment of AIDS.

THA

Tetrahydroaminoacridine (THA), an experimental drug that blocks the destruction of acetylcholine in the brain, significantly improved memory in 16 of 17 elderly people with Alzheimer's disease in clinical trials at two hospitals in Pasadena, California.

1,2,3,4-tetrahydro-9-aminoacridine

THA is also not a new drug, but its potential in the treatment of Alzheimer's disease is new. Discovered in 1909, the compound at one time was used to counteract overdoses of muscle relaxants, barbiturates, and anesthetics. It is not currently marketed in the United States.

Aminoguanidine

Animal tests indicate that aminoguanidine may prevent the vascular disease that leads to blindness, kidney problems, and other major side effects of diabetes.

$$H_2N-\underset{H}{N}-\underset{NH}{C}-NH_2$$

Aminoguanidine

Aminoguanidine appears to prevent the formation of cross links between proteins that are associated with the development of atherosclerosis, cataracts, and tissue stiffening. Although these maladies are all associated with aging in normal persons, they tend to develop prematurely in diabetics because of elevated glucose levels in the blood, which in turn elevate the rate of cross-linking proteins, especially collagen, the connective tissue protein.

SELF-TEST 16-C

1. Fentanyl derivatives that have been sold on the streets tend to act like what other drug of abuse? _____
2. "Crack" is a purified form of _____.
3. AZT is used to treat _____.
4. The difference between nucleotides and nucleosides is a _____ group.
5. The AIDS retrovirus penetrates _____, which are white blood cells that play a crucial role in controlling the immune response of the body.
6. An illegal drug, which may have pronounced psychological or addictive effects and differs little in a molecular sense from a legal drug, is often called a _____.

MATCHING SET

_____ 1. Histamine
_____ 2. Cortisone
_____ 3. Tetracycline
_____ 4. Penicillin
_____ 5. Sulfa drug
_____ 6. Hallucinogenic designer drug
_____ 7. Propranolol
_____ 8. Procaine (Novocain)
_____ 9. Opium poppy
_____ 10. Dihydroxyaluminum sodium carbonate
_____ 11. 3-Methyl fentanyl
_____ 12. Anabolic
_____ 13. Alkylating drug
_____ 14. Antimetabolite drug
_____ 15. Immune suppressant
_____ 16. Calcium ion
_____ 17. Nicotinic acid

a. 3000 times more powerful than heroin
b. Source of morphine
c. Interferes with cell wall synthesis
d. Analgesic used in dentistry
e. Related to Reye's syndrome
f. Lowers blood cholesterol
g. Causes symptoms of hay fever
h. Causes heart muscles to contract
i. Sulfanilamide
j. Muscle producing
k. Cyclosporin A
l. Methotrexate
m. Female sex hormone
n. Cyclophosphamide
o. Heart muscle relaxant
p. Antibiotic containing a four-ring structure
q. Antacid
r. Ecstasy
s. Steroid

QUESTIONS

1. Discuss why women on the average live longer than men.
2. Why does the stomach not dissolve itself?
3. Name two alkaloid narcotics that are derived from morphine.
4. What is Reye's syndrome? What common drug is associated with Reye's syndrome?
5. Name three ions that neutralize negative charges associated with nerve and muscle cells.
6. Nitrogen mustards are alkylating agents. These interfere with DNA replication. Explain.
7. Explain how the antimetabolite methotrexate works to kill cells.
8. If the chemotherapeutic agents kill living cells, why do they have a preferential effect on cancer cells?
9. What is one of the dangers of chemotherapy using an alkylating agent or an antimetabolite?
10. Describe the action of antihistamines.
11. Describe how the sulfa drug sulfanilamide works.
12. How was penicillin discovered?
13. What happens when beta receptor sites in heart muscles are stimulated?
14. What can happen to a patient when a drug passes through the blood-brain barrier?
15. What is atherosclerotic plaque? What are its two major ingredients?
16. Explain how a beta blocker can lower blood pressure.
17. (a) What effect does the calcium ion have on the heart? (b) How can different calcium channel blockers be structurally so dissimilar and yet affect the flow of calcium ions into muscle cells in such a similar way?
18. How do the estrogens differ structurally from the other female hormone progesterone?
19. Name an anti-inflammatory steroid produced in the body.
20. Anabolic steroids that can be taken orally are unlike testosterone with respect to what structural feature?
21. What designer drug is toxic to the part of the brain called the substantia nigra? What disease does this drug cause?
22. The Drug Enforcement Administration has three different classifications of controlled substances. Explain the differences among these and give an example of each.
23. Look up the drugs scheduled by the Drug Enforcement Administration as Schedule 1 drugs. How many are there?
24. What is a retrovirus? How does the AIDS retrovirus attack the immune system of the body?

CHAPTER 17

Chemical Formulations in Our Weekly Budget

An appreciable portion of our income is spent on chemical formulations applied to the skin, hair, clothes, home, automobile, and other useful things. A knowledge of the basic chemistry of cosmetics, cleansing agents, and surface coatings can be of considerable value in the selection of these products. An understanding of the chemical effect of a product will help us decide if a risk associated with it is worth the satisfaction gained from it. Allergenic reactions and the safe use of a product may vary from person to person. Furthermore, one brand of a product (in a pretty package with a lot of slick advertising) may cost several times more than another brand when both contain the same active ingredient. It is apparent, then, that to select chemical products that are "right" for us and to avoid the expensive and sometimes dangerous practice of trial and error, some chemical considerations are required.

We hope this chapter will lead you to consult product labels, a chemical formulary from the library, and a basic chemistry text to investigate products that are of interest to you.

Skin, Hair and Nails: A Chemical View

The skin, hair, and nails are protein structures. Skin (Figs. 17–1 and 17–2), like other organs of the body, is not composed of uniform tissue and has several functions made possible by its structure; they are protection, sensation, excretion, and body temperature control. The exterior of the epidermis, called the **stratum, corneum,** or **corneal layer,** is where most cosmetic preparations for the skin act. The corneal layer is composed principally of dead cells and has a moisture content of about 10%. The principal protein of the corneal layer is **keratin,** which is composed of about 22 different amino acids. Its structure renders it insoluble in, but slightly permeable to, water. Dry skin is uncomfortable, and excessively moist skin is a good host for fungus organisms. An oily secretion, **sebum,** is secreted by the sebaceous glands as protection against excessive moisture loss, which results in dry skin. To control the moisture content of the corneal layer so that it does not dry out and slough off too quickly, one may add moisturizers to the skin. Normal skin is slightly acidic, with a pH of about 4.

Hair is composed principally of keratin. An important difference between hair keratin and other proteins is hair keratin's high content of the amino acid **cystine.** About 16% to 18% of hair protein is cystine, whereas only 2.3% to 3.8% of the keratin in corneal cells is cystine. This amino acid plays an important role in the structure of hair.

The toughness of both skin and hair is due to the bridges between different protein chains, such as hydrogen bonds and —S—S— linking bonds, called **disulfide bonds:**

$$\begin{array}{c} NH_2 \\ | \\ HOOC-CH-CH_2-S \\ | \\ HOOC-CH-CH_2-S \\ | \\ NH_2 \end{array}$$

Cystine

420 Chemical Formulations in Our Weekly Budget

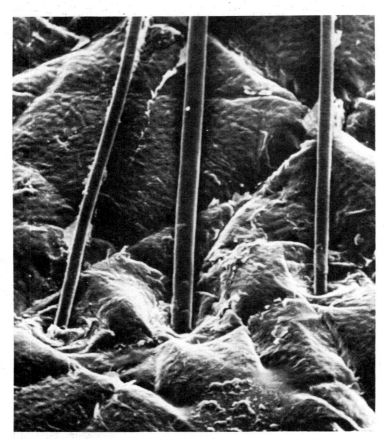

Figure 17–1
Replica of the surface of human forearm skin, showing three hairs emerging from the skin (×225). (Courtesy of E. Bernstein and C. B. Jones: "Skin Replication Procedure for the Scanning Electron Microscope" *Science,* Vol. 166, pp. 252–253, 1969. Copyright © 1969 by the American Association for the Advancement of Science.)

$$\begin{array}{c} | \\ O=C \\ | \\ H-CCH_2-S-S-CH_2C-H \\ | \quad\quad\quad\quad\quad\uparrow\quad\quad\quad | \\ H-N \quad\quad\quad\quad\quad\quad C=O \\ | \quad\quad\quad\quad\quad\quad\quad\quad\quad | \end{array}$$

$$\begin{array}{c} | \\ N-H \\ | \end{array}$$

Disulfide Bond (Cross Link)

The structures of protein tissues are due in part to disulfide cross links and to ionic bonds between "molecules."

Another type of bridge between two protein chains, important in keratin as well as in all proteins, is the **ionic bond.** Consider the interaction between an amine group $-NH_2$ of lysine and a carboxylic group $-COOH$ of glutamic acid on a neighboring protein chain. At pH 4.1, protons are added to the $-NH_2$ groups and removed from the $-COOH$ groups, resulting in $-NH_3^+$ and $-COO^-$ groups on adjacent chains. If the two charged groups approach closely, an ionic bond is formed:

$$\underset{\text{Lysine}}{\overset{|}{\underset{|}{\text{HCCH}_2\text{CH}_2\text{CH}_2\text{CH}_2\text{NH}_2}}} + \underset{\text{Glutamic Acid}}{\overset{|}{\underset{|}{\text{HOOCCH}_2\text{CH}_2\text{CH}}}} \xrightarrow{\text{at pH 4.1}}$$

$$\underset{\text{Ionic Bond}}{\overset{|}{\underset{|}{\text{HCCH}_2\text{CH}_2\text{CH}_2\text{CH}_2\text{NH}_3^+}}\ \overset{|}{\underset{|}{^-\text{OOCCH}_2\text{CH}_2\text{CH}}}}$$

As the pH rises above 4, these cross links are broken, and keratin swells and becomes soft. This is an important aspect of hair chemistry.

Fingernails and toenails are composed of **hard keratin,** a very dense type of keratin. These epidermal cells grow from epithelial cells lying under the white crescent at the growing end of the nail. Like hair, the nail tissue beyond the growing cells is dead.

Curling, Coloring, Growing, and Removing Hair

The curl, color, and presence of human hair are matters of personal choice that vary considerably from person to person. Chemical means are available to alter these conditions, except the growing of hair on skin where none is present or the prevention of natural baldness.

Changing the Shape of Hair

Hair that is wet can be stretched to one and a half times its dry length because water (pH 7) weakens some of the ionic bonds and causes swelling of the keratin. Imagine the disulfide cross links remaining between two protein chains in hair, as shown in Figure 17–3a. Winding the hair on rollers causes tension to develop at the cross links (Fig. 17–3b). In "cold" waving these cross links are broken by a reducing agent (Fig. 17–3c),

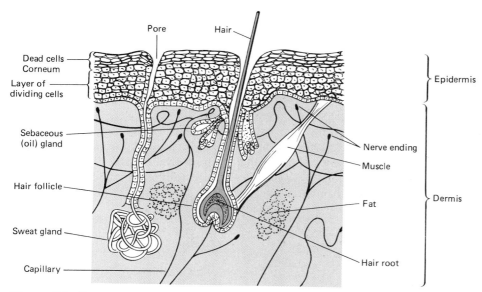

Figure 17–2
Cross section of the skin.

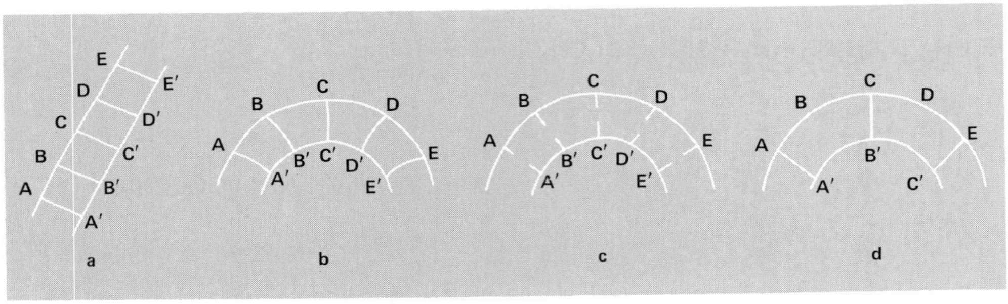

Figure 17-3
A schematic diagram of a permanent wave.

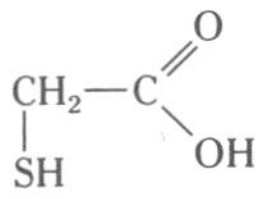

Thioglycolic Acid

Hair can be straightened by the same solutions. It is simply "neutralized" (or oxidized) while straight (no rolling up).

relaxing the tension. Then an oxidizing agent regenerates the cross links (Fig. 17-3d), causing the hair to hold the shape of the roller. The chemical reactions are shown in simplified form in Figure 17-4.

The most commonly used reducing agent is thioglycolic acid. The common oxidizing agents include hydrogen peroxide, perborates ($NaBO_2 \cdot H_2O_2 \cdot 3\,H_2O$), and sodium or potassium bromate ($KBrO_3$). A typical neutralizer solution contains one or more of the oxidizing agents dissolved in water. The presence of water and a strong base in the oxidizing solution also helps break and re-form hydrogen bonds between adjacent protein molecules. However, too frequent use of a strong base causes hair to become brittle and lifeless.

Various additives are present in both the oxidizing and the reducing solutions to control pH, odor, and color, as well as for general ease of application. A typical waving lotion contains 5.7% thioglycolic acid, 2.0% ammonia, and 92.3% water.

Coloring and Bleaching Hair

Melanin—black
Iron pigment—red

Hair contains two pigments: brown-black melanin and an iron-containing red pigment. The relative amounts of each determine the color of the hair. In deep black hair melanin predominates, and in light-blond hair the iron pigment predominates. The depth of the color depends on the size of the pigment granules.

Formulations for dyeing hair vary from temporary coloring (removable by shampoo), which is usually achieved by means of a water-soluble dye that acts on the surface of the hair, to semipermanent dyes, which penetrate the hair fibers to a great extent (Fig. 17-5). Semipermanent hair dyes often consist of cobalt or chromium complexes of dyes dissolved in an organic solvent. Permanent dyes are generally "oxidation" dyes. They penetrate the hair and then are oxidized to produce a colored product that is permanently attached to the hair by chemical bonds or that is much less soluble than the reactant molecule. Permanent hair dyes generally are derivatives of phenylenediamine, which dyes hair black. A blond dye can be formulated with *p*-aminodiphenylaminesulfonic acid or *p*-phenylenediaminesulfonic acid.

Some hair dyes are suspected of being carcinogenic.

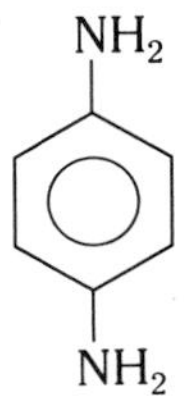

p-Phenylenediamine

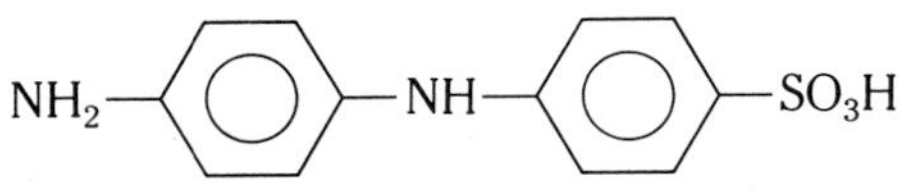

Diphenylaminesulfonic Acid

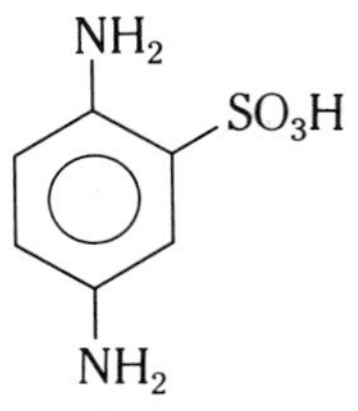

p-Phenylenediamine-sulfonic Acid

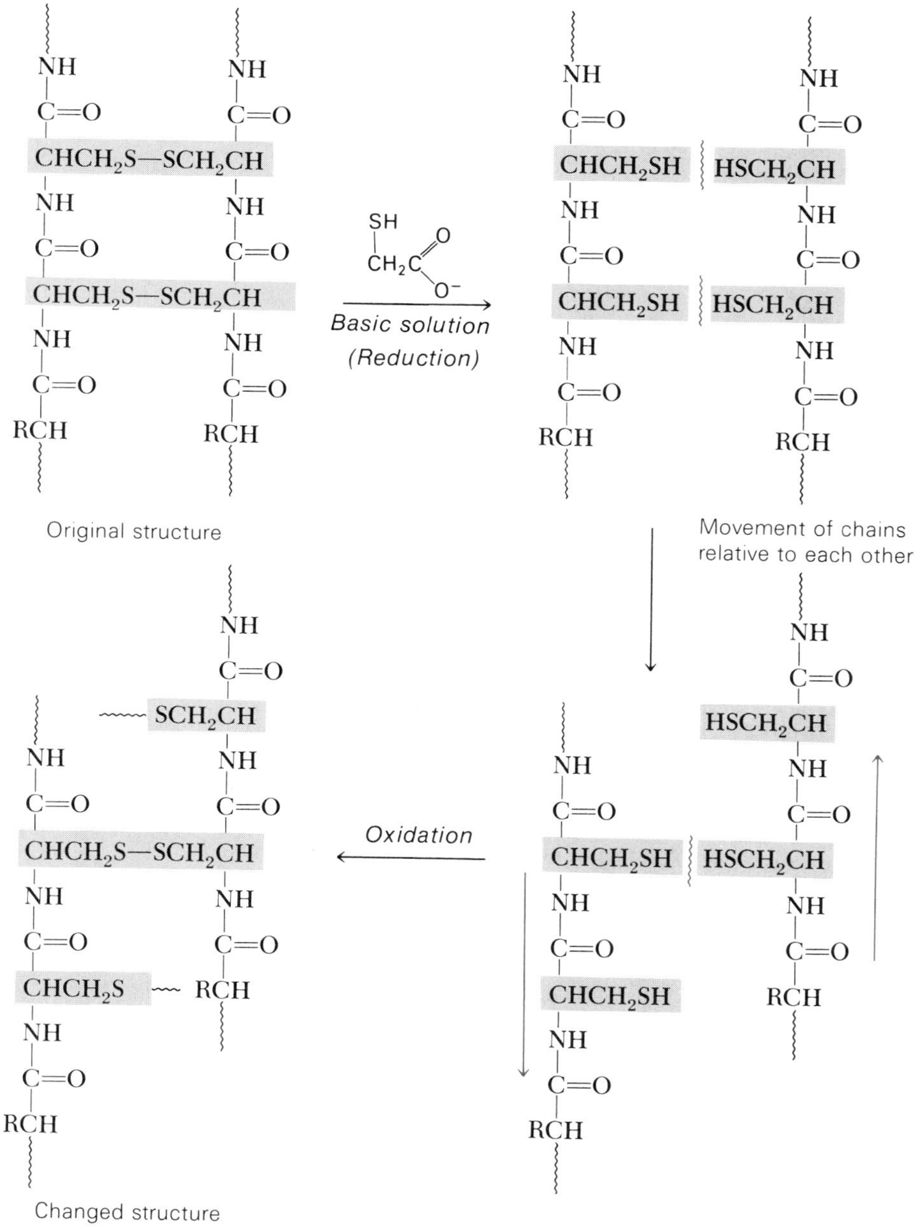

Figure 17-4
Structural changes that occur in hair during a permanent wave.

The active compounds are applied in an aqueous soap or detergent solution containing ammonia to make the solution basic. The dye material is then oxidized by hydrogen peroxide to produce the desired color. The amines are oxidized to nitro compounds.

$$-NH_2 + 3\ H_2O_2 \xrightarrow{\text{oxidation}} -NO_2 + 4\ H_2O$$
Amine → *Nitro Compound*

Just about any shade of hair color can be prepared by variations in the modifying groups on certain basic dye structures.

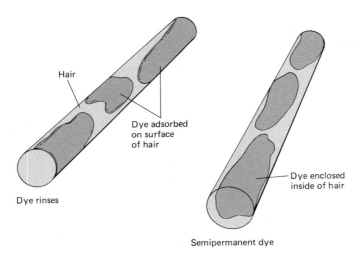

Figure 17-5
Methods of dyeing hair.

Hair can be bleached by more concentrated solutions of hydrogen peroxide, which destroy the hair pigments by oxidation. Such solutions are made basic with ammonia to enhance the oxidizing power of the peroxide. Parts of the chemical process are given in Figure 17-6. This drastic treatment of hair does more than just change the color; it may destroy sufficient structure to render the hair brittle and coarse.

Hair Sprays

Hair spray coats the hair with a plastic film.

Hair sprays are essentially solutions of a resin in a volatile solvent, the purpose of which, when sprayed on hair, is to furnish a film with sufficient strength to hold the hair in place after the solvent has evaporated (Fig. 17-7). A common resin in hair sprays is the addition polymer polyvinylpyrrolidone (PVP).

Polyvinylpyrrolidone (PVP)

The resin is blended in hair spray formulations with a plasticizer, a water repellent, and a solvent-propellant mixture. The plasticizer makes the plastic more pliable. The resin concentration of hair sprays is of the order of 4%. Other additives, such as silicone oils, are often put into hair sprays to give hair a sheen.

Since PVP tends to pick up moisture, other less hygroscopic polymers are beginning to replace PVP in hair sprays. For example, significantly better moisture control is obtained with a copolymer made from a 60:40 ratio of vinylpyrrolidone and vinyl acetate.

Breathing the vapor of hair sprays may cause carcinogens to act on delicate lung tissue or may lead to asphyxiation by the plastic coating lining the lungs, among other dangers.

Growing Hair

Minoxidil is a peripheral vasodilator that is used to reduce blood pressure. Applied under skin patches, this chemical causes the growth of fine, baby-like hair anywhere hair

Figure 17-6
Bleaching of the hair by hydrogen peroxide. There are several chemical intermediates between the amino acid tyrosine and the hair pigment melanin, which is partly protein. Hydrogen peroxide oxidizes melanin back to colorless compounds, which are stable in the absence of the enzyme tyrosinase (found only in the hair roots). Melanin is a high-molecular-weight polymeric material of unknown structure. The structure shown here is only a segment of the total structure.

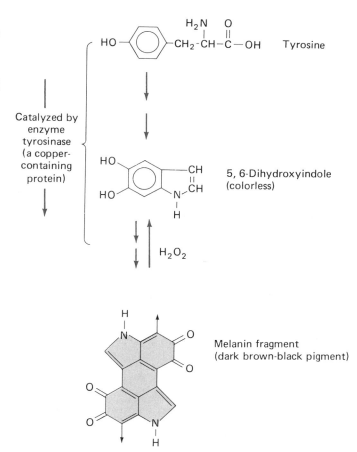

Figure 17-7
Film of hair spray. The hair spray was allowed to dry on a white surface and was then pulled up to reveal the film.

follicles exist on the skin. Oral doses produce the effect bodywide. The effect, though temporary—hair growth continues only as long as the drug is administered—offers some possibilities for the controlled growth of human hair.

Depilatories

The purpose of a depilatory is to remove hair chemically. Since skin is sensitive to the same kind of chemical attack as hair, such preparations should be used with caution, and even then some attack on the skin is almost unavoidable. Because of this, intervals between applications of a depilatory should be in the order of a week or so. Depilatories should never be used on skin that is infected or that has a rash, and their use should not be followed by application of a deodorant with an *astringent* (contracting) action. If the sweat pores are closed by the deodorant, the caustic chemicals are retained and can do considerable harm. If the sweat pores are open, body fluids will dilute and wash the caustic chemicals to the outside of the body.

The chemicals used as depilatories include sodium sulfide, calcium sulfide, and strontium sulfide (all water-soluble sulfides), as well as calcium thioglycolate ($Ca[HSCH_2COO]_2$), the calcium salt of the compound used to break S—S bonds between protein chains in permanent waving. A typical cream depilatory contains calcium thioglycolate (7.5%), calcium carbonate (filler, 20%), calcium hydroxide (provides basic solution, 1.5%), cetyl alcohol ($[CH_3(CH_2)_{15}OH]$, skin conditioner, 6%), sodium lauryl sulfate (detergent, 0.5%), and water (64.5%).

The water-soluble sulfides are all strong bases in water, as indicated by the hydrolysis of the sulfide ion:

$$S^{2-} + H_2O \longrightarrow HS^- + OH^-$$
$$\text{Sulfide} \qquad\qquad\qquad \text{Hydroxide}$$

0.1 M means 0.1 mole of Na_2S (7.81 g) dissolved in sufficient water to make 1 liter of solution.

For example, a 0.1 M solution of Na_2S has a pH of about 13, making it a strongly basic solution. The compounds act chemically on the hair to break bonds in the protein chains and cause the protein to break down into soluble amino acids and small peptides, which may be removed.

The area on which a depilatory has been used should be washed with soap and water, dried, and then treated with small amounts of talcum powder.

SELF-TEST 17-A

1. Wet hair stretches because of the weakening of some of the _____ and a swelling of the _____.
2. What acid is used to reduce hair chemically (break the cross links between protein molecules)? _____
3. The last step in the production of a permanent wave in hair is ( ) oxidation or ( ) reduction.
4. What color is imparted to hair by phenylenediamine dyes? _____
5. What common oxidizing agent is used to bleach hair or to destroy the pigmentation in the hair structure? _____
6. Hair sprays leave a film of _____ on the hair.
7. What is the medical use of the drug minoxidil?

What is the relationship between this drug and the control of body hair?
_____

8. Name three chemicals that are used as depilatories. _____, _____, and _____
9. Keratin is a protein found in _____, _____, and _____.
10. Bridges between protein chains may be _____ linkages or _____ bonds.
11. The surface of the skin epidermis is known as the _____ layer.
12. The oily secretion of skin is _____.

Skin Preparations for Health, Beauty, and Fragrance

For skin to remain healthy, its moisture content must stay near 10%. If it is higher, microorganisms grow too easily; if it is lower, the corneal layer flakes off. Washing skin removes fats that help retain the right amount of moisture. If dry skin is treated with a fat after washing, it will be protected until enough natural fats have been regenerated.

Lanolin, an excellent skin softener **(emollient),** is a component of many cosmetics. It is a complex mixture of esters from hydrated wool fat. The esters are derived from 33 different alcohols of high molecular weight and 37 fatty acids. Cholesterol, a common alcohol in lanolin, is found both free and in esters. Cholesterol appears to give fat mixtures the property of absorbing water. This is one factor that makes lanolin an excellent emollient. With its high proportion of free alcohols, particularly cholesterol, and hydroxyacid esters, lanolin has the structural groups (—OH) to hydrogen-bond water (to keep the skin moist) and to anchor within the skin (the fat and ester hydrocarbon structures).

Skin with a low fat content tends to be dry.

Lanolin is grease from wool.

Creams

Creams are generally emulsions of either an oil-in-water type or a water-in-oil type. An **emulsion** is simply a colloidal suspension of one liquid in another. The oil-in-water emulsion has tiny droplets of an oily or waxy nature dispersed throughout a water solution (homogenized milk is an example). The water-in-oil emulsion has tiny droplets of a water solution dispersed throughout an oil (natural petroleum and melted butter are examples). An oil-in-water emulsion can be washed off the hands with tap water, a water-in-oil emulsion gives the hands a greasy, water-repellent surface.

Cold cream originally was an emulsion of rosewater in a mixture of almond oil and beeswax. Subsequently, other ingredients were added to produce a more stable emulsion. An example of a modified cold cream composition is almond oil, 35%; beeswax, 12%; lanolin, 15%; spermaceti (from whale oil), 8%; and strong rosewater, 30%. Other oils can be substituted for some or all of the almond oil. Lanolin stabilizes the emulsion. Any oil preparation that holds moisture in the skin is a *moisturizer.*

Vanishing cream is a suspension of stearic acid in water to which a stabilizer (emulsifier) has been added to prevent the ingredients from separating. The stabilizer may be a soap, such as potassium stearate. These creams do not actually vanish; they merely spread as a smooth, thin covering over the skin.

Creams of various sorts may be used as the base for other cosmetic preparations; other ingredients are added to produce the desired properties. As an example, hydrated aluminum chloride can be added to a cream to produce a deodorant.

Colloids are particles intermediate in size between small molecules and clumps of molecules sufficiently large to precipitate.

Creams add oil or fat content to surface skin.

Lipstick

The skin on our lips is covered by a thin corneal layer that is free of fat and consequently dries out easily. Lip moisture is normally maintained from the mouth. In addition to being a beauty aid, lipstick can be helpful under harsh conditions that tend to dry lip tissue.

Lipstick consists of a solution or suspension of coloring agents in a mixture of high-molecular-weight hydrocarbons or their derivatives, or both. The material must be soft enough to produce an even application when pressed on the lips, yet the film must not be too easily removed, nor may the coloring matter run. Lipstick is perfumed to give it a pleasant odor. The color usually comes from a dye, or "lake," from the eosin group of dyes. A **lake** is a precipitate of a metal ion (Fe^{3+}, Ni^{2+}, Co^{3+}) with an organic dye. The metal ion enhances the color or changes the color of the dye and keeps the dye from dissolving.

Two suitable dyes, used in admixture and with their lakes, are dibromofluorescein (yellow-red) and tetrabromofluorescein (purple):

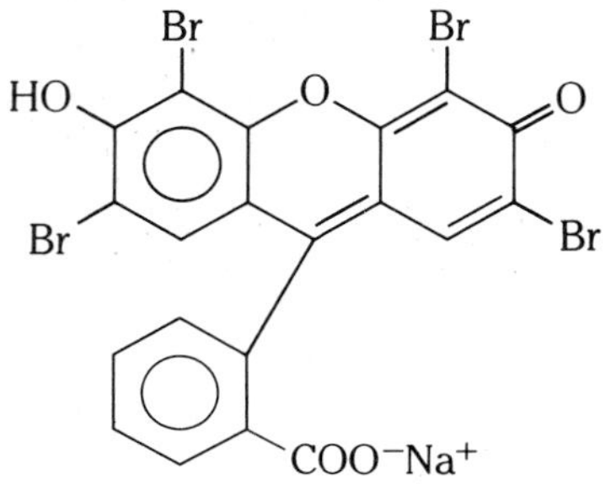

A lake is a coloring agent made up of an organic dye adhering to an inorganic substance called a mordant. Some lakes are also approved as food colors.

Tetrabromofluorescein (Eosin)
(Sodium Salt)

The ingredients in a typical formulation of lipstick include the following:

Ingredient	Function	Amount (%)
Dye	Furnishes color	4–7
Castor oil, paraffins, or fats	Dissolve dye	50
Lanolin	Emollient	25
Carnauba wax or beeswax*	Makes stick stiff by raising its melting point	18
Perfume	Imparts pleasant odor	Small amount

Face Powder

Face powder is used to give the skin a smooth appearance by covering up oil secretions, which would produce a shiny look. The powder must have some hiding ability without being so opaque as to be obvious. A powder that has the proper appearance, sticking properties, absorbence for oily skin secretions, and spreading ability usually requires several ingredients. A typical formulation is shown below:

Ingredient	Function	Amount (%)
Talc	As absorbent	65
Precipitated chalk	As absorbent	10
Zinc oxide	As astringent	20
Zinc stearate	As binder	5
Perfume	For odor	Trace
Dye	For color	Trace

Astringents shrink tissue, restricting fluid flow.

* Carnauba wax and beeswax are high-molecular-weight esters.

Perfume

A typical perfume has at least three components of somewhat different volatility and molecular weight. (Recall that low-molecular-weight compounds are generally more volatile than high-molecular-weight compounds.) The first, called the **top note**, is the most volatile and is the most obvious odor when the perfume is first applied. The second, called the **middle note**, is less volatile and is generally a flower extract (violet, lilac, etc.). The last, or **end note**, is least volatile and is usually a resin or waxy polymer.

Most perfumes contain many components, and chemically they are often complex mixtures. As the analysis of natural perfume materials has progressed, the use of pure synthetic organic compounds to duplicate specific odors has become very common. An example is civetone (see structure below), a cyclic ketone from civet, a secretion of the civet cat of Ethiopia. It is highly valued for perfumes. Civetone is now available in a synthetic form. It is prepared by formation of 8-hexadecene-1,16-dicarboxylic acid into a ring. The thorium ion (Th^{4+}) catalyzes closure of the ring.

Perfumes are complex mixtures of odorous compounds.

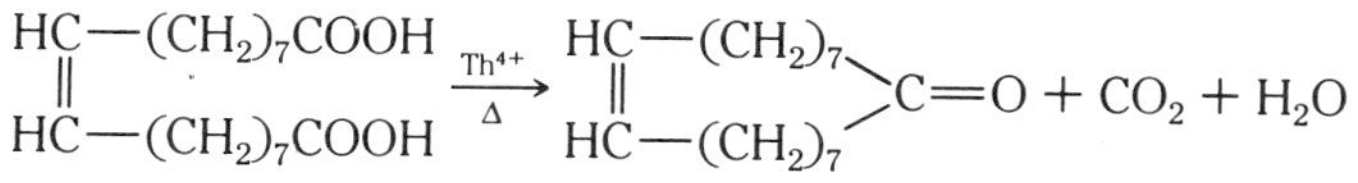

$$\begin{array}{c} HC-(CH_2)_7COOH \\ \parallel \\ HC-(CH_2)_7COOH \end{array} \xrightarrow{Th^{4+} \atop \Delta} \begin{array}{c} HC-(CH_2)_7 \\ \parallel \\ HC-(CH_2)_7 \end{array} \!\!\! C=O + CO_2 + H_2O$$

Civetone

Civet is a collection of sex attractants, as in musk. In perfumes that "work," these sex attractants are cleverly masked by herbaceous and floral odors. The initial attraction comes from the pleasant odor, but the "basic effect" is from the civetone or musk.

Musk is obtained from the musk deer.

Other compounds used in perfumes include high-molecular-weight alcohols and esters. An example is geraniol (boiling point, 230°C), a principal component of Turkish geranium oil:

Aftershave lotions and colognes are diluted perfumes, about one-tenth (or less) as strong.

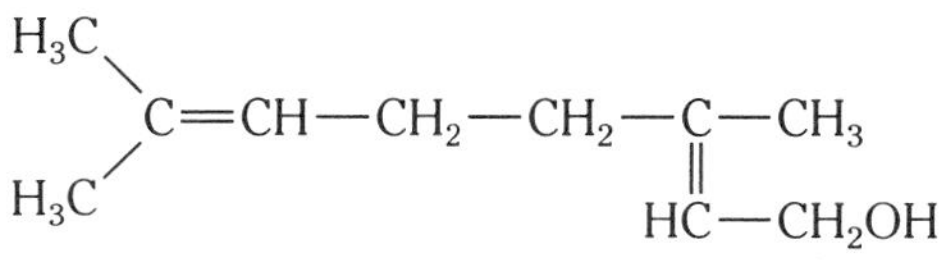

$$\begin{array}{c} H_3C \\ \diagdown \\ C=CH-CH_2-CH_2-C-CH_3 \\ \diagup \parallel \\ H_3C HC-CH_2OH \end{array}$$

Geraniol

Esters of this alcohol are used to make synthetic rose aromas for perfumes. For example, the ester formed by the reaction between geraniol and formic acid has a rose odor.

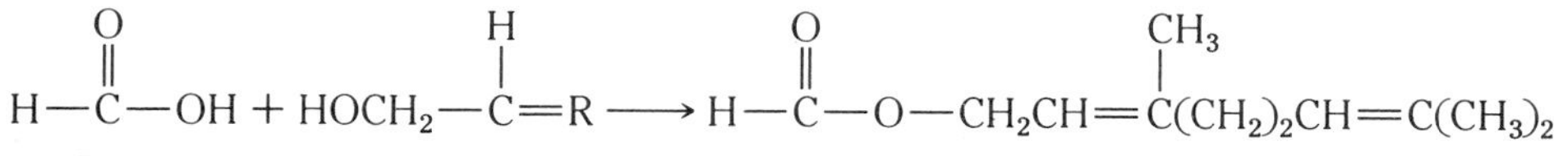

$$\underset{\text{Formic Acid}}{H-\overset{O}{\overset{\parallel}{C}}-OH} + \underset{\text{Geraniol}}{HOCH_2-\overset{H}{\underset{}{\overset{\mid}{C}}}=R} \longrightarrow \underset{\text{Geranyl Formate}}{H-\overset{O}{\overset{\parallel}{C}}-O-CH_2CH=\overset{CH_3}{\underset{}{\overset{\mid}{C}}}(CH_2)_2CH=C(CH_3)_2}$$

Typical perfumes are 10% to 25% perfume essence and 75% to 80% alcohol and a fixative to retain the essential oils. Perfumes are added to most cosmetics to give the products desirable odors; they also mask the natural odors of other constituents such as sex attractants. They are mildly bactericidal and antiseptic because of their alcohol content.

Ethyl alcohol is a major constituent of most perfumes.

Suntan Lotions

Short-wavelength ultraviolet light from the Sun is very harmful to the skin. Persons who want to get a tan should therefore exclude these short wavelengths, while exposing themselves to enough less energetic, longer wavelength ultraviolet light to permit gradual tanning.

Ultraviolet radiation tans skin.

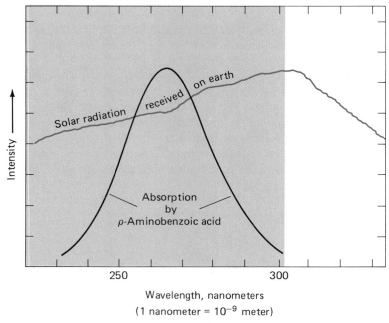

Figure 17-8
Absorption spectrum of p-aminobenzoic acid and its relationship to solar ultraviolet radiation received on Earth. Maximum absorption occurs at 265 nanometers, although absorption also occurs at other wavelengths as shown. The maximum of the deep-burning ultraviolet radiation received on Earth is about 308 nanometers.

The variety of suntan products ranges from lotions, which selectively filter out the high-energy ultraviolet rays of the sun, to preparations that essentially dye light-colored skin tan.

The lotions that filter out the ultraviolet rays are most accurately described as sunscreens, but their ingredients are often mixed with other materials to produce a lotion that both screens and tans. A common ingredient in preparations used to prevent sunburn is p-aminobenzoic acid.

Sunbathers refer to p-aminobenzoic acid as PABA.

$$H_2N-\underset{\text{p-Aminobenzoic Acid (PABA)}}{\bigcirc}-COOH$$

Like most aromatic compounds, PABA absorbs strongly in the ultraviolet region of the spectrum (Fig. 17-8).

In tanning the skin is stimulated to increase its production of the pigment **melanin.** At the same time the skin thickens and becomes increasingly resistant to deep burning. Preparations for the relief of sunburn pain are solutions of local anesthetics such as benzocaine.

Increased amounts of melanin protect sensitive lower layers of skin.

Deodorants

The 2 million sweat glands on the surface of the body are primarily used to regulate body temperature. They do this by secreting water, the evaporation of which produces a cooling effect. This evaporation also leaves solid constituents, mostly sodium chloride, as well as smaller amounts of proteins and other organic compounds. Body odor results largely from amines and hydrolysis products of fatty oils (fatty acids, acrolein, etc.) emitted from the body and from bacterial growth within the residue from sweat glands. Sweating is both normal and necessary for the proper functioning of the human body; sweat itself is quite odorless, but the bacterial decomposition products are not.

Body odor is promoted by bacterial action.

There are three kinds of deodorants: those that directly "dry up" perspiration or act as astringents, those that have an odor to mask the odor of sweat, and those that remove odorous compounds by chemical reaction. Among those that act as astringents are hydrated aluminum sulfate, hydrated aluminum chloride ($AlCl_3 \cdot 6\ H_2O$), aluminum chlorohydrate (actually aluminum hydroxychloride, $Al_2[OH]_5Cl \cdot 2\ H_2O$ or $Al[OH]_2Cl$ or $Al_6[OH]_{15}Cl_3$), and alcohols. Those compounds that act as deodorizing agents include zinc peroxide, essential oils and perfumes, and a variety of mild antiseptics. Zinc peroxide removes odorous compounds by oxidizing the amines and fatty acid compounds. The essential oils and perfumes absorb or otherwise mask the odors, and the antiseptics are generally oxidizing or reducing agents that kill the odor-causing bacteria.

An astringent closes the pores, thus stopping the flow of perspiration.

SELF-TEST 17-B

1. The _____ note is the most volatile part of a perfume.
2. Lanolin is a skin softener, or a(n) _____.
3. A skin cream is either an oil-in-water or a water-in-oil _____.
4. An example of a cosmetic that is colored with a "lake" is _____.
5. The major mineral ingredient in face or body powder is _____.
6. Civetone is likely to be the attracting odor in a perfume. True ( ) or False ( )
7. An example of an active chemical that absorbs ultraviolet light in a sunscreen lotion is _____.
8. Skin darkens because of an increased concentration of the skin pigment _____.
9. Name three kinds of deodorants. _____, _____, and _____.
10. Explain the action of aluminum chlorohydrate in a deodorant.
_____

Cleansing Agents

Dirt has been defined as matter in the wrong place. Tomato catsup is esteemed as a palatable food, but on your shirt it is dirt. There are a large number of cleansing, or **surface active,** agents that are capable of removing dirt without harming the shirt (Fig. 17-9). Indeed, radio and television advertising might lead us to believe that the soaps and detergents we have today are unique and vastly superior to the products of a year or a century ago. This is not always so. Soap, for example, continues to be made by a time-tested recipe that dates back at least to the 2nd century A.D. Galen, the great Greek physician, mentions that soap was made from fat, ash lye, and lime and that it served not only as a medicament but also to remove dirt from the body and clothes.

Surface active agents stabilize suspensions of nonpolar materials in polar solvents or vice versa. Examples include soaps, detergents, wetting agents, and foaming agents.

 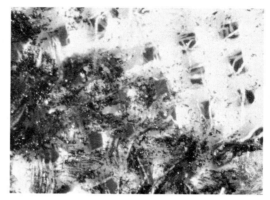

Figure 17-9
Photomicrograph of clean cotton cloth (left) and soiled cotton cloth (right). The proper application of surface active agents should return the soiled cloth to its original state.

What *is* new is the greater purity of soap, improvement in its cleaning action by numerous additives, and the advent of the relatively new synthetic detergents.

Soaps

Fats and oils can be hydrolyzed in strongly basic solutions to form glycerol and salts of the fatty acids. Such hydrolysis reactions are called **saponification** reactions; the sodium or potassium salts of the fatty acids formed are **soaps.** Pioneers prepared their soap by boiling animal fat with an alkaline solution obtained from the ashes of hard wood. The resulting soap could be "salted out" by the addition of sodium chloride, a method based on the fact that soap is less soluble in a salt solution than in water.

Review the molecular structures of fats and oils in Chapter 10.

Soaps can be made by treating fats or oils with sodium hydroxide.

$$\begin{array}{c} CH_3(CH_2)_{16}COO-CH_2 \\ | \\ CH_3(CH_2)_{16}COO-CH \\ | \\ CH_3(CH_2)_{16}COO-CH_2 \end{array} + 3\ NaOH \longrightarrow 3\ CH_3(CH_2)_{16}COO^-Na^+ + \begin{array}{c} HO-CH_2 \\ | \\ HO-CH \\ | \\ HO-CH_2 \end{array}$$

Tristearin (Glyceryl Tristearate) — Sodium Stearate (A Soap) — Glycerol

The cleansing action of soap can be explained in terms of its molecular structure. Material that is water-soluble can be readily removed from the skin or a surface by simple washing with an excess of water. To remove a sticky sugar syrup from one's hands, one dissolves the sugar in water and rinses it away. If the material to be removed is oily, water will merely run over the surface of the oil. Since the skin has natural oils, even substances such as ordinary dirt that are not oily themselves can cover the skin with a greasy layer. The cohesive forces (forces between like molecules that tend to hold them together) within the water layer are too large to allow the oil and water to intermingle (Fig. 17-10). When present in an oil-water system, soap molecules such as sodium stearate move to the interface between the two liquids.

Soap, water, and oil together form an emulsion, with the soap acting as the emulsifying agent.

$$CH_3CH_2CH_2CH_2CH_2CH_2CH_2CH_2CH_2CH_2CH_2CH_2CH_2CH_2CH_2CH_2C\begin{smallmatrix}\diagup O \\ \diagdown O^-Na^+\end{smallmatrix}$$

Principal Fats and Oils Used for Making Soap

- Tallow or animal fat from beef or mutton is primarily an ester of stearic acid — $CH_3(CH_2)_{16}COOH$. It is usually mixed with coconut oil in making soap to prevent the product from being too hard.
- Coconut oil is a low-melting solid. It is primarily an ester of lauric acid — $CH_3(CH_2)_{10}COOH$. A soap made from coconut oil alone is very soluble in water and will lather even in sea water.
- Palm oil contains a very high concentration of free fatty acids, about 45% to 50% of which is oleic acid, $CH_3(CH_2)_7CH=CH(CH_2)_7COOH$. It is an important constituent of toilet soaps.
- Olive oil is used in making Castile soap. It has a larger percentage (70–85%) of esters of oleic acid than does palm oil.
- Bone grease is an animal fat of somewhat lower melting point than tallow, and it comes from a variety of sources. It is a relatively inexpensive source of fat. The esters of oleic acid (41–51%) are most prominent.
- Cottonseed oil is also an inexpensive source of glycerides for making soap. Its esters are mostly of linoleic acid, $CH_3(CH_2)_4(CH=CHCH_2)_2(CH_2)_6COOH$.

The carbon chain, which is a nonpolar organic structure, mixes readily with the nonpolar grease molecules, whereas the highly polar $—COO^-Na^+$ group enters the water layer because it becomes hydrated (Fig. 17–10b). The soap molecules will then tend to lie across the oil-water interface. The grease is then broken up into small droplets by agitation, each droplet being surrounded by hydrated soap molecules (Fig. 17–10c). The surrounded oil droplets cannot come together again, since the exterior of each droplet is covered with $—COO^-Na^+$ groups that interact strongly with the surrounding water. If enough soap and water are available, the oil will be swept away, forming a clean, water-wet surface.

Toilet Soaps. Toilet soaps generally have little or no filler and a minimal amount of free base, if any. Often much of the glycerol released in the saponification process is left in the soap. Perfumes, dyes, and medicinal agents may be added before the soap is cast into a solid form. Floating soaps have air beaten into them as they solidify. A hard soap is produced by the addition of a high percentage of a sodium salt of a relatively long-chain fatty acid, such as stearic acid. A soft or liquid soap is obtainable by saponification with potassium hydroxide, with the liquidity increasing as the chain length of the fatty acid decreases. Fatty acids with chains as short as C_{12} or shorter are not used because the resultant soaps irritate the skin. They are also volatile and have an unpleasant odor. Fatty acids with chains longer than that of stearic acid (18 carbon atoms) tend to make very insoluble soaps.

Sodium — hard soap.
Potassium — soft soap.
Ammonium — liquid soap.

Synthetic Detergents

Synthetic detergents (**"syndets"**) are derived from organic molecules that have been designed to have the same cleansing action as soaps, but less reaction than soaps with the cations found in hard water, such as Ca^{2+}, Mg^{2+} and Fe^{3+}. As a consequence, synthetic detergents are more effective in hard water than is soap, which gives a precipitate in the presence of Ca^{2+}, Mg^{2+}, or Fe^{3+} ions. Since such precipitates have no cleansing action and tend to stick to laundry, their presence is undesirable.

There is a synthetic detergent for almost every type of cleaning problem.

Hard water contains metal ions that react with soaps and give precipitates.

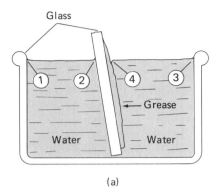

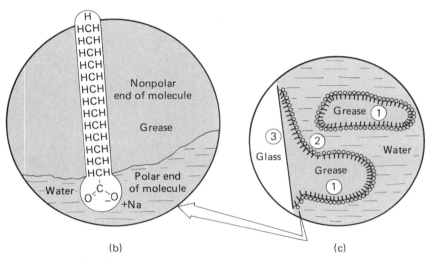

Figure 17-10
The cleaning action of soap. (a) A piece of glass partially coated with grease inserted in water gives evidence for the strong adhesion between water and glass at 1, 2, and 3. The water curves up against the pull of gravity to wet the glass. The relatively weak adhesion between oil and water is indicated at 4 by the curvature of the water away from the grease against the force tending to level the water. (b) A soap molecule, having oil-soluble and water-soluble ends, will become oriented at an oil-water interface such that the hydrocarbon chain is in the oil (with molecules that are electrically similar, or nonpolar) and the COO^-Na^+ group is in the water (highly charged polar groups interact electrically). (c) In an idealized molecular view, a grease particle, 1, is surrounded by soap molecules, which in turn are strongly attracted to the water. At 2 another droplet is about to break away. At 3 the grease and clean glass interact before the water moves between them.

$$R-O-\overset{O}{\underset{O}{S}}-O^-$$

Sulfate Group

$$R-\overset{O}{\underset{O}{S}}-O^-$$

Sulfonate Group

The preparation of all synthetic detergents begins with naturally occurring raw materials.

There are many different synthetic detergents on the market. Their molecular structure consists of a long oil-soluble (hydrophobic) group and a water-soluble (hydrophilic) group. The hydrophilic groups include the sulfate ($-OSO_3-$), sulfonate ($-SO_3-$), hydroxyl ($-OH$), ammonium ($-NH_3^+$), and phosphate ($-OPO[OH]_2$) groups.

The early synthetic detergents were mostly sodium alkyl sulfate. The preparation of sodium lauryl sulfate is given here to illustrate the chemical processes involved. The principal starting material is a suitable vegetable oil, such as cottonseed oil or coconut oil. The first step is hydrogenation:

$$\begin{array}{c}\text{RCOOCH}_2\\|\\\text{RCOOCH}\\|\\\text{RCOOCH}_2\end{array} + 6\,\text{H}_2 \xrightarrow{\text{Catalyst}} 3\,\text{RCH}_2\text{OH} + \begin{array}{c}\text{CH}_2\text{OH}\\|\\\text{CHOH}\\|\\\text{CH}_2\text{OH}\end{array}$$

Coconut Oil   Hydrogen   (Mainly Lauryl Alcohol, $CH_3[CH_2]_{11}OH$)   Glycerol

The second step involves esterification of the —OH group on the end of the lauryl alcohol hydrocarbon chain. This is accomplished by treatment of the lauryl alcohol with sulfuric acid:

$$CH_3(CH_2)_{11}OH + H_2SO_4 \longrightarrow CH_3(CH_2)_{11}OSO_3H + H_2O$$

Lauryl Alcohol   Sulfuric Acid   Lauryl Hydrogen Sulfate

The final step involves neutralization of the acidic lauryl hydrogen sulfate with sodium hydroxide:

$$CH_3(CH_2)_{11}OSO_3H + NaOH \longrightarrow CH_3(CH_2)_{11}OSO_3^-Na^+ + H_2O$$

Sodium Lauryl Sulfate

Shampoos. Shampoos are often mixtures of several ingredients designed to satisfy a number of requirements. In addition to soaps, condensation products from diethanolamine and lauric acid are often used. These are essentially a type of detergent obtained by the following reaction:

$$HN(CH_2CH_2OH)_2 + CH_3(CH_2)_{10}COOH \longrightarrow CH_3(CH_2)_{10}\overset{\overset{O}{\|}}{C}-\overset{\overset{H}{|}}{N}(CH_2CH_2OH)_2$$

Diethanolamine   Lauric Acid   An Amide Detergent

Some shampoos contain anionic detergents, which are less damaging to the eyes than cationic detergents. Sodium lauryl sulfate is an example of an anionic detergent. Hair is most manageable and has the best sheen if all the shampoo is removed. An anionic detergent can be removed by rinsing with a cationic detergent (about a 1% solution), which neutralizes the anions and facilitates their removal. Caution should be exercised with rinses, since cationic detergents are damaging to the eyes. Types of cationic detergents are described in the next section.

Anions are negative ions.

Shampoos also contain compounds to prevent the calcium or magnesium ions in hard water from forming a precipitate; EDTA is often used for this purpose. Lanolin and mineral oil are often added to keep the scalp from drying out and scaling. The presence of these oils is indicated by a cloudy appearance.

A metal ion chelated by EDTA is shown on p. 315.

Fillers or Builders. A number of materials are added to soap powders for laundry purposes. These materials are often quite basic, and their addition gives the soap a greater detergent action. Commonly added materials include sodium carbonate, sodium phosphates, sodium polyphosphates, and sodium silicate. Rosin neutralized with sodium hydroxide is also commonly added to laundry soaps in large amounts. The rosin is mostly abietic acid. The neutralized acid has the nonpolar (hydrocarbon) part and polar end required for a soap. Such soaps are not to be recommended for use on the human skin. Phosphates, carbonates, and silicates hydrolyze to give OH^- ions, which react with grease to make soaps.

$$PO_4^{3-} + H_2O \longrightarrow HPO_4^{2-} + OH^-$$

$$CO_3^{2-} + H_2O \longrightarrow HCO_3^- + OH^-$$

$$SiO_3^{2-} + H_2O \longrightarrow HSiC_3^- + OH^-$$

Builders also assist in negating the effect of the ions (Mg^{2+}, Ca^{2+}, Fe^{3+}) that cause water to be hard. Since the phosphate, carbonate, and hydroxide compounds of these ions are insoluble, the ions are precipitated. It is fortunate that these precipitates are powdery and easily rinsed away; in contrast, soap precipitates form scum that sticks to the material being washed.

$$HCO_3^- + OH^- \longrightarrow CO_3^{2-} + H_2O$$

$$Ca^{2+} + CO_3^{2-} \longrightarrow CaCO_3\downarrow$$

$$Mg^{2+} + 2\,OH^- \longrightarrow Mg(OH)_2\downarrow$$

> Soft water—less than 65 mg of metal ion per gallon. Slightly hard water—65–228 mg. Moderately hard water—228–455 mg. Hard water—455–682 mg. Very hard water—above 682 mg.

> Soaps containing pumice (finely powdered volcanic ash) will wash out ground-in dirt.

Other Cleansers. A very large number of special cleaners or cleansing agents are available. Simple abrasive cleansers contain a large percentage of an abrasive such as silica (SiO_2) or pumice (65–75% SiO_2, 10–20% Al_2O_3), a variable amount of soap, and generally some polyphosphates. They may also contain some synthetic detergent and a bleaching agent. All-purpose solid cleansers may contain one or more of a variety of salts, which react with water to produce a basic solution: trisodium phosphate, sodium carbonate, sodium bicarbonate, sodium pyrophosphate, or sodium tripolyphosphate, plus a detergent and perhaps pine oil to give an attractive odor. Metal cleansers may contain a strong acid or strong base to dissolve impurities. Many cleaning liquids contain organic solvents such as tetrachloroethylene, 1,1,1-trichloroethane, and the like. Because the vapors of these solvents are quite toxic, cleansers containing them should be used only in ventilated areas.

Bleaching Agents. Bleaching agents are compounds that are used to remove color from textiles. Most commercial bleaches are oxidizing agents such as sodium hypochlorite. Optical brighteners are quite different; they act by converting a portion of the invisible ultraviolet light, which impinges on them, into visible blue or blue-green light, which is emitted. Together or separately, these two classes of compounds are present in a large number of commercial laundry and cleaning preparations.

Textiles used to be bleached by exposure to sunlight and air. In 1786 the French chemist Berthollet introduced bleaching with chlorine, and subsequently this process was carried out with sodium hypochlorite, an oxidizing agent prepared by the passage of chlorine into aqueous sodium hydroxide:

$$2\,Na^+ + 2\,OH^- + Cl_2 \longrightarrow \underset{\text{Sodium Hypochlorite}}{Na^+ + OCl^-} + Na^+ + Cl^- + H_2O$$

Shortly after this, hydrogen peroxide was introduced as a textile bleach. Later, a number of other oxidizing agents based on chlorine were developed and introduced.

One way to decolorize materials is to remove or immobilize those electrons in the material that are activated by visible light. When such electrons are activated, a portion of the visible light is absorbed and the reflected light is consequently colored. Household bleaches are oxidizing agents that make whites look whiter by removing loosely bonded electrons from the fabric. A typical household bleach contains 4.25% sodium hypochlorite. The hypochlorite ion is capable of removing electrons from many colored materials. In this process, it is reduced to chloride and hydroxide ions:

> Chlorine produces hypochlorite when it reacts with water:
> $$H_2O + Cl_2 \longrightarrow HOCl + HCl$$

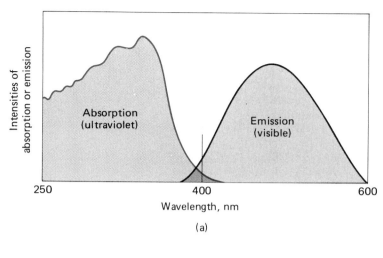

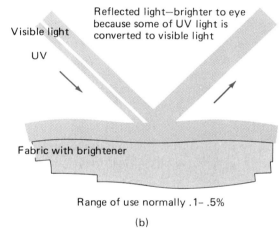

Figure 17–11
Optical brighteners absorb unseen ultraviolet light and emit the energy as visible light.

$$ClO^- + H_2O + 2\ e^- \longrightarrow Cl^- + 2\ OH^-$$

As stated previously, optical brighteners are compounds that transform incident ultraviolet light into emitted visible light; this is a type of fluorescence. When optical brighteners are incorporated into textiles or paper, they make the material appear brighter and whiter. An example of such a brightener has the structure shown below, and its absorption and emission spectra are presented in outline form in Figure 17-11.

A fluorescent material absorbs shorter wavelength light and emits light of a longer wavelength.

Spot and Stain Removers

Many stains can be removed by an appropriate solvent or chemical reagent.

To a large extent, stain removal procedures are based on solubility patterns or chemical reactions. Many stains, such as those due to chocolate or other fatty foods, can be removed by treatment with typical dry-cleaning solvents such as 1,1,1-trichloroethane, $H_3C—CCl_3$. Stain removers for more resistant stains are almost always based on a chemical reaction between the stain and the essential ingredients of the stain remover. A typical example is an iodine stain remover, which is simply a concentrated solution of sodium thiosulfate. The reaction here is shown below:

$$I_2 + 2\,Na_2S_2O_3 \longrightarrow \underline{2\,NaI + Na_2S_4O_6}$$

Iodine   Soluble in Water (Colorless)

Citric acid and tartaric acid also remove iron stains and are less toxic than oxalic acid.

Iron stains are removed by treatment with oxalic acid, which forms a soluble complex ion with the iron:

$$Fe_2O_3 + 6\,H_2C_2O_4 \longrightarrow \underline{3\,H_2O + 2\,Fe(C_2O_4)_3^{3-} + 6\,H^+}$$

Oxalic Acid   Soluble in Water

Mildew stains can be removed by hydrogen peroxide or laundry bleach (sodium hypochlorite), which oxidizes the fungus responsible for the mildew. Blood stains on cotton can be removed by hypochlorite solution. Bleach should not be used on wool because it reacts chemically with the nitrogen atoms present in the peptide chains. The chemicals used to remove a few common stains are listed in Table 17–1.

Table 17–1
Some Common Stains and Stain Removers*

Stain	Stain Remover
Coffee	Sodium hypochlorite
Lipstick	Isopropyl alcohol, isoamyl acetate, Cellosolve ($HOCH_2CH_2OCH_2CH_3$), chloroform
Rust and ink	Oxalic acid, methyl alcohol, water
Airplane cement	50/50 amyl acetate and toluene or acetone
Asphalt	Benzene or carbon disulfide
Blood	Cold water, hydrogen peroxide
Berry, fruit	Hydrogen peroxide
Grass	50/50 amyl acetate and benzene or sodium hypochlorite or alcohol
Nail polish	Acetone
Mustard	Sodium hypochlorite or alcohol
Antiperspirants	Ammonium hydroxide
Perspiration	Ammonium hydroxide, hydrogen peroxide
Scorch	Hydrogen peroxide
Soft drinks	Sodium hypochlorite
Tobacco	Sodium hypochlorite

* Before any of these stain removers are used on clothing, the possibility of damage should be checked on a portion of the cloth that ordinarily is hidden.
    Some stain removers, such as benzene and chloroform, are suspected carcinogens, and some, such as methanol and carbon disulfide, are acutely toxic.

Toothpaste

The mineral content, or the hard part, of bones and teeth consists of two compounds of calcium. Calcium carbonate ($CaCO_3$) is present in bones and teeth in the crystalline form, known to mineralogists as aragonite. The second calcium compound found in teeth is calcium hydroxyphosphate — $Ca_5(OH)(PO_4)_3$ — or apatite (Fig. 17-12). Such structures are readily attacked by acid. Since the decay of some food particles produces acids and since bacteria convert plaque, a deposit of dextrins, to acids, it is important to keep teeth clean and free of prolonged contact with acids if the hard, stonelike enamel is to be preserved.

The two essential ingredients of toothpaste are a detergent and an abrasive. The abrasive serves to cut into the surface deposits, and the detergent assists in suspending the particles in a water medium to be carried away in the rinse. Abrasives commonly used in toothpaste formulations include hydrated silica (a form of sand, $SiO_2 \cdot nH_2O$); hydrated alumina ($Al_2O_3 \cdot nH_2O$); and calcium carbonate ($CaCO_3$). It is difficult to select an abrasive that is hard enough to cut the surface contamination and yet not so hard as to cut the tooth enamel. The choice of detergent is easier; any good detergent such as sodium lauryl sulfate will do quite well.

Since the necessary ingredients in toothpaste are not very palatable, it is not surprising to see the inclusion of flavors, sweeteners, thickeners, and colors to appeal to our senses.

One addition to the toothpaste mixture has made a significant difference in the amount of tooth decay in our population; it is stannous fluoride (SnF_2), which provides a low level of fluoride ion concentration. The fluoride ion replaces the hydroxide ions in the hydroxyapatite structure — $Ca_{10}(PO_4)_6(OH)_2$ — to form fluoroapatite — $Ca_{10}(PO_4)_6F_2$ (Fig. 17-12). Because the fluoride ion forms a stronger ionic bond in the crystalline structure, fluoroapatite is harder and less subject to acid attack than is hydroxyapatite; hence, there is less tooth decay. The fluoride ion has also been introduced into drinking water on a widescale basis.

Most teeth are lost as a result of gum disease, which results from the lack of proper massage, deposits of plaque below the gum line, and bacterial infection in these deposits. Increasing attention is being given to toothpastes containing disinfectants such as peroxides in addition to a soap and an abrasive.

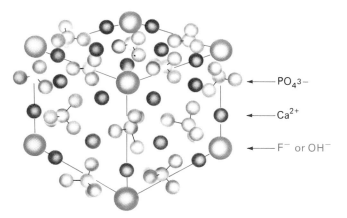

Figure 17-12
Structure of apatite and fluoroapatite. The dark circles denote Ca^{2+} ions, the groups of four circles tied together by lines represent the PO_4^{3-} groups, and the largest circles represent OH^- groups in apatite or F^- groups in fluoroapatite.

← PO_4^{3-}

← Ca^{2+}

← F^- or OH^-

Paints

All paints involve polymers in one form or another. Popular latex and acrylic paints contain addition polymers, which serve as **binders.** A paint binder forms a molecular network to hold the **pigment** (coloring agent) in place and to hold the paint to the painted surface. In oil-based paints a drying oil (such as linseed oil) or a resin is the binder. All paints also have a volatile solvent or thinner; this is water in water-based paints and turpentine (or mineral spirits, or both) in oil-based paints. A varnish is a mixture of a drying oil (such as linseed oil), rosin, and a thinner. Lacquers contain cellulose esters and ethers.

Early latex paints were emulsions of partly polymerized styrene and butadiene in water (Fig. 17–13). Some type of emulsifying agent (such as soap) was present to keep the small drops of nonpolar styrene and butadiene dispersed in the polar water.

Immediately after the application of a latex paint, the water begins to evaporate. When some of the water is gone, the emulsion breaks down, and the remaining water evaporates quickly, leaving the paint film. Further polymerization of the styrene and butadiene follows slowly, but the paint appears to be dry in a few minutes. The pigment is trapped in the network of the polymer. If the paint is white, the pigment is probably titanium dioxide (TiO_2); this has replaced the poisonous compound "white lead," ($Pb(OH)_2 \cdot 2\ PbCO_3$), which was used in paints prior to World War II.

The styrene-butadiene resin is the least expensive binder material used, but it has a relatively long curing period, relatively poor adhesion, and a tendency to yellow with age. Polyvinylacetate is only a little more expensive and is an improvement over the styrene-butadiene resin. When it was introduced, it quickly captured 50% of the latex market for interior paints. Other increasingly popular types, though about one-third more expensive, include the acrylic resins and the "acrylic latex" paints. These are more washable and much more resistant to light damage than the first latex paints. They are especially useful as exterior paints.

The fluoropolymers, similar to Teflon, are especially promising as surface coatings because of their great stability. Fluorine atoms are substituted for hydrogen atoms in the organic structure. Metals covered with polyvinylidene fluoride carry up to a 20-year guarantee against failure from exposure.

In the past few years paint manufacturers have begun to blend linseed oil emulsions with latex emulsions in order to take advantage of the penetrating ability of the triglyceride molecules in the linseed oil. Some "latex" paints now contain as much as 75% linseed oil emulsion but still have the desirable characteristics of latex. Table 17–2 lists various additives to emulsion paints and gives the rationales for their use.

Side notes:

The first commercial water-based latex paint was Glidden's Spred Satin, introduced in 1948.

Water-based latex paints reduce fire hazards and air pollution associated with the handling and application of oil-based paints.

Monomer of polyvinylacetate:

$$\begin{array}{c} H \\ \diagdown \\ C=C \\ \diagup\diagdown \\ HO-\overset{\overset{\displaystyle O}{\|}}{C}-CH_3 \end{array}$$

Acrylic polymers have a sheen that allows latex paint to compete in the exterior gloss market, which has traditionally been monopolized by oil-based coatings. Acrylics, which adhere well and control corrosion, are polymers of acrylonitrile:

$$\begin{array}{c} HCN \\ \diagdown\diagup \\ C=C \\ \diagup\diagdown \\ HH \end{array}$$

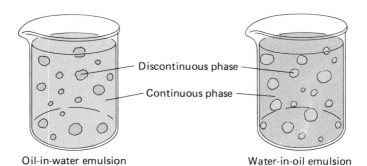

Oil-in-water emulsion — Discontinuous phase / Continuous phase — Water-in-oil emulsion

Figure 17–13
Two kinds of emulsions. An emulsion is composed of two immiscible liquids, one dispersed as tiny droplets in the other. An emulsifying agent is required to stabilize an emulsion.

The drying of modern oil-based paints involves much more than the evaporation of the mineral spirits or turpentine solvent. The chemical reaction between a drying oil and oxygen from the air completes the drying process. Common drying oils are soybean, castor, coconut, and linseed; the most widely used is linseed, which comes from the seed of the flax plant. All of these oils are glyceryl esters of fatty acids, as discussed in Chapter 10. Hydrolysis yields the following percentages of fatty acids in a typical linseed oil:

Mineral spirits are petroleum fractions of moderate volatility.

Palmitic acid (16 C atoms, saturated)	4%–7%
Stearic acid (18 C atoms, saturated)	2%–5%
Oleic acid (18 C atoms, unsaturated)	9%–38%
Linoleic acid (18 C atoms, unsaturated)	3%–43%
Linolenic acid (18 C atoms, unsaturated)	25%–58%

The chemical action of oxygen on a drying oil is to replace a hydrogen atom on a carbon atom next to a C=C double bond in an unsaturated fatty acid chain. When oxygen reacts with two fatty acids on two oil molecules, the result is cross linking between the two molecules.

Part of One Molecule

$$-CH_2-CH_2-CH=CH-CH_2- \qquad\qquad -CH_2-CH-CH=CH-CH_2-$$
$$+ O_2 \longrightarrow \qquad\qquad | \qquad\qquad\text{Ether Linkage} \qquad + H_2O$$
$$-CH_2-CH_2-CH=CH-CH_2- \qquad\qquad -CH_2-CH-CH=CH-CH_2-$$

Part of Another Molecule

Table 17–2
Additives Used in Water-Based Emulsion Paints

Additive	Reason for Use
Dispersing agents for pigments	Example: tetrasodium pyrophosphate ($Na_4P_2O_7$) causes like-charged pigment particles to repel; this keeps the pigment from settling to the bottom of the container.
Protective colloids and thickeners	A thicker paint is slower to settle and drips and runs less than a thinner paint. A protective colloid tends to stabilize the organic-water interface in the emulsion (same mechanism as soap dispersing oil in water). Examples: sodium polyacrylates, carboxymethylcellulose, clays, gums
Defoamers	Foaming presents a serious problem if not corrected. Chemicals used: tri-n-butylphosphate, n-octyl alcohol and other higher alcohols, silicone oil
Coalescing agents	As the water evaporates and the paint dries, an agent is needed to stick the pigment particles together. As the resin film forms, the agent evaporates. Coalescing agents must volatilize very slowly. Examples: hexylene glycol and ethylene glycol
Freeze-thaw additives	Freezing will destroy the emulsion. Antifreezes such as ethylene glycol are used.
pH controllers	The effectiveness of the ionic or molecular form of the emulsifier depends on the acid or alkaline conditions (pH). The wrong pH will break down the emulsion. Most paints tend to be too acidic. Ammonia (NH_3) is added to neutralize the acid.

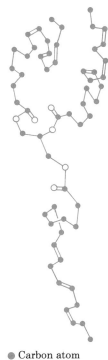

Typical molecule in linseed oil.

● Carbon atom
○ Oxygen atom

The polymeric network produced by the cross linking hardens the paint, traps the pigment, and secures the paint in the crevices of the painted surface.

Metal ions such as Zn^{2+}, Co^{2+}, Fe^{3+}, Mn^{2+}, and Ca^{2+} are added to oil-based paints to catalyze the drying process. These ions decompose peroxides (compounds containing the —O—O group) formed during the cross-linking process and precipitate free acids as salts of the ions.

Nail Polish and Polish Remover

Nail polish is essentially a lacquer or varnish. It can be made of nitrocellulose, a plasticizer, a resin, a solvent, and perhaps a dye. The nitrocellulose can be replaced by another polymer molecule that possesses similar qualities. Evaporation of the solvent leaves a film of nitrocellulose, plasticizer, resin, and dye. The nitrocellulose furnishes the shiny film; the plasticizer is added to make the film less brittle; and the resin is added to make the film adhere to the nail better and to prevent flaking. Perfumes are added to cover the odor of the other constituents. A typical formulation is shown below:

Nail polish and hair sprays are formulated very much alike.

Plasticizers fit between the molecular layers of plastic and make the plastic less brittle.

Nitrocellulose	15%
Acetone (solvent)	45%
Amyl acetate (solvent)	30%
Butyl stearate (plasticizer)	5%
Ester gum (resin)	5%

Perfumes and colors are added as needed.

Ester gum is a combination of esters, mainly glyceryl, methyl, and ethyl esters of rosin. Rosin is the sticky resin remaining after distillation of turpentine from pine exudate. It is 80% to 90% abietic acid. Rosin is slightly toxic to mucous membranes and slightly irritating to the skin. The ester gum is prepared by the heating of rosin and alcohol under pressure until esterification occurs. The gums are soluble in nonpolar solvents.

Nail polish removers are simply solvents that dissolve the film left by the nail polish. They consist largely of acetone or ethyl acetate or both, to which small amounts of butyl stearate and diethylene glycol monomethyl ether have been added to reduce the drying effect of the solvent. However, some formulations contain combinations of amyl acetate, butyl acetate, ethyl acetate, benzene, olive oil, lanolin, and alcohol. Both nail polish and nail polish removers are very flammable, and care should be taken never to use them in the presence of open flames or lighted cigarettes.

Cuticle softeners are primarily wetting agents and alkalies used to soften the skin around the fingernails so it can be shaped as desired. The use of alkali to soften and swell protein is well known. A typical cuticle softener contains potassium hydroxide (3%), glycerol (12%), and water (85%). A cuticle softener may also contain sodium carbonate (an alkali), triethanol amine (a detergent, used as wetting agent), and trisodium phosphate (an alkali).

Abietic Acid (Main Component of Rosin)

Diethylene Glycol Monomethyl Ether

SELF-TEST 17-C

1. What is the oldest surface active agent used for cleansing purposes? _____
2. To make soap, a fat is treated with a(n) _____.
3. Which group in the soap structure is more soluble in water, the hydrocarbon chain or the ionic salt end? _____

4. What oil is used in making Castile soap? _____
5. What causes floating soap to float? _____
6. Name three common metal ions that cause water to be hard. _____, _____, and _____
7. Which are more likely to precipitate in hard water, traditional soaps or synthetic detergents? _____
8. Give two reasons for adding a detergent builder to a laundry product.
   _____
   _____
9. Optical brighteners transform ultraviolet light into _____ light.
10. What are the two fundamental ingredients of a toothpaste? _____ and _____
11. A compound of what element is added to toothpaste to replace some of the hydroxide ions in apatite? _____
12. What acid is effective in removing iron stains in clothing? _____
13. Dried paints have pigments locked into place in the binders. What is a suitable chemical binder? _____
14. What solvent is most often used in nail polish? _____

MATCHING SET

_____	1. Keratin	a. Salt of fatty acid
_____	2. Melanin	b. Reducing agent in wave lotion
_____	3. Sodium lauryl sulfate	c. Hair spray resin
_____	4. Polyvinylpyrrolidone	d. Abrasive in toothpaste
_____	5. Alcohol	e. Holds moisture in skin
_____	6. Hydrated aluminum chloride	f. Common alcohol in lanolin
_____	7. Soap	g. Ultraviolet absorber in suntan lotion
_____	8. Fat	h. Alcohol produced in saponification of fat or oil
_____	9. Cholesterol	i. Detergent builder
_____	10. Thioglycolic acid	j. Radiate a different wavelength of light than that absorbed
_____	11. p-Aminobenzoic acid	k. Synthetic detergent
_____	12. Glycerol	l. Laundry bleach
_____	13. Sodium tripolyphosphate	m. Dark pigment
_____	14. Whiteners	n. Deodorant component
_____	15. Sodium hypochlorite	o. Furnishes fluoride for stronger teeth
_____	16. Calcium carbonate	p. Dehydrates skin microbes
_____	17. Tin (II) fluoride (stannous fluoride)	q. Skin and hair protein

QUESTIONS

1. You read in the newspaper about a new compound that will break disulfide bonds in proteins. What potential use might it have?
2. a. What is the purpose of an emulsifier?
   b. In which of the following cosmetics is an emulsifier important: suntan lotion, hair spray, cold cream?
3. Which one of each of the following pairs of properties would be appropriate for a hair spray propellant? Why?
   a. High or low boiling point
   b. Soluble or insoluble in the active ingredients
   c. Capable or incapable of chemical reaction with the active ingredients
   d. Odorous or odorless
   e. Toxic or nontoxic
4. Describe an ionic bond that holds protein chains together.
5. What is the purpose of each of the following?
   a. Detergent in toothpastes
   b. Polyvinylpyrrolidone in hair sprays
   c. Aluminum chloride in deodorants
   d. *p*-Aminobenzoic acid in suntan lotion
6. What specific substance is broken down during the bleaching of hair?
7. Use the structures of the constituents of lanolin to justify its ability to emulsify face creams.
8. Hydrogen bonding is a very handy theoretical tool. Name three applications of hydrogen bonding in cosmetics and cleansing agents.
9. If you were going to formulate a suntan lotion, what particular spectral property would you look for in choosing the active compound?
10. Why are detergents better cleansing agents than soaps in regions where the water supply contains calcium or magnesium salts?
11. Why is a soap made from coconut oil more soluble in water than a soap made from palm oil?
12. Suggest ways of removing each of the following from clothing:
    a. Motor oil
    b. Iodine
    c. Lard
    d. Copper sulfate
13. Explain why vinegar is able to remove some stains that are soluble in weak acids.
14. Name and give the functions of four chemical ingredients of perfume.
15. Explain why Grandma's lye soap produced rough, red hands.
16. Explain how an optical brightener in a detergent works.
17. What do skin, hair, and nails have in common?
18. What is the major difference between keratin in the hair and other proteins?
19. Why do the lips dry so easily?
20. What is the structure of the monomer unit in polyvinylpyrrolidone?
21. Describe in chemical terms what happens when a person gets a permanent.
22. What three types of chemical bonds hold hair proteins together?
23. Is hair curl the result of altered protein structure? Explain.
24. What happens to the solar energy absorbed by *p*-aminobenzoic acid in a suntan lotion or oil?
25. If a substance is astringent, what is its action?
26. What is the purpose of talc in face powder?
27. Name an astringent widely used in deodorants.
28. What is the major ingredient of lipstick?
29. What is the chemical action of hydrogen peroxide on hair?
30. Commercial lanolin comes from what animal?
31. Is a fat an acid, an alcohol, an ester, or an alkane?
32. Can vegetable oils be used as effectively as animal oils to make soap?
33. Is the hydrocarbon end of the soap molecule polar or nonpolar? Explain.
34. Which is more soluble in water, calcium stearate or sodium stearate?
35. The oxide stains of what metal can be removed with a solution of oxalic acid?
36. How important is advertising in the sale of cosmetics? What part does advertising play in consumer satisfaction?

EPILOGUE

What does the future hold, chemically speaking? No one knows. However, this will not keep thoughtful persons from extrapolating from past and current events to the future for a variety of reasons. Some enjoy the intellectual challenge, others are dedicated to the solution of current and developing human problems, and many simply want to put themselves in a strong financial position in the ebb and flow of the appetite expressed by our society.

Barring catastrophic disasters such as a thermonuclear war or a contagious, uncontrolled disruption of "normal" cellular chemistry, molecular manipulations and chemical understandings will cause or allow us to:

1. Produce vastly greater amounts of food through both the modification of the food-producing organism and the chemical environment in which it lives, as well as to chemically transform previously unacceptable food raw materials into usable foodstuffs.
2. Use to near extinction the petroleum reserves of the world for its burnable energy content, necessitating renewed and increasing interest in the clean use of coal. (Engineers are presently working on new designs for coal-fired trains.)
3. Generate a larger percentage of electricity with nuclear power, and thereby also producing ever larger stockpiles of radioactive wastes, which require centuries of protected storage.
4. Proceed slowly with alternative sources of energy, such as solar, geothermic, and wind, as long as fossil fuels can be burned to produce cheaper energy.
5. Move toward environmental controls that will include the cost of waste disposal and cleanup in the cost of the item or material produced.
6. Produce an almost endless array of new materials that will revolutionize structural and facade materials in construction and in transportation and that will offer many new choices in the materials we use in clothing, personal tools, and surroundings for pleasure and comfort.
7. Continue the explosion in our ability to store and process information by the controlled molecular changes on the surface of semiconductor materials. The transmission of knowledge will be increasingly cheap, allowing more human energy to be devoted to understanding and value choices; this will expand human creativity.
8. Expand our knowledge and control of genetic engineering and genetic coding of life-controlling information, and thereby produce microbes designed to control specific chemical changes in and out of other life forms, and modify the chemistry of complex organisms to affect gross functions. Considerable environ-

mental risk will be taken in this area as we try to sort out "cause and effect" at the same time as we are modifying organisms that have the ability to reproduce at an exponential rate.
9. Continue to improve health and conquer disease by using recently acquired chemical knowledge to design and synthesize drugs to alleviate cancer, atherosclerosis, hypertension, and disorders of the central nervous and immune systems.

Human control of chemical change is neither good nor bad in and of itself; rather, it is the uses made of the controlled changes that can be classified as good or bad.

Perhaps the most important question of all is the choice of the chooser. Who should be entrusted with these fateful choices? Should it be the person or company that stands to make a financial profit from the change? Should it be a government agent or agency that is properly schooled to make such selections for those who are represented or controlled? Should it be a group of scientists who gain their position in history by advancing new ideas (correct ones, we hope) and by having them accepted? The answers to these questions are not altogether obvious because of limitations in our knowledge and understanding. However, we believe that the most important scientific attribute — after intelligence, of course — is skepticism. Scientific skepticism calls for relatively little regard for human authority in explaining nature and total acceptance of natural displays as the final authority in understanding what is and what is not. Theory from any source, although it often predicts fact, must always be subservient to observable phenomena. If the majority of society can understand this most fundamental working in chemistry and the other sciences, the public will realize that it cannot depend on any vested interest group to make societal chemical choices and will, through government, seek to force the common consensus. Our future depends on this high level of societal chemical responsibility.

We have presented to you the story of chemistry, how it works, and what its potentialities are. We hope you are convinced that you do not have to be a chemist to participate, through good citizenship and personal choices, in the control of the unfolding chemical story.

APPENDIX A

The International System of Units (SI)

A coherent system of units known as the Système International (SI system), bearing the authority of the International Bureau of Weights and Measures, has been in effect since 1960 and is increasingly gaining acceptance among scientists. It is an extension of the metric system that began in 1790, with each physical quantity assigned a unique SI unit. An essential feature of both the older metric system and the newer SI is a series of prefixes that indicate a power of 10 multiple or submultiple of the unit.

Units of Length

The standard unit of length, the **meter,** was originally meant to be 1 ten-millionth of the distance along a meridian from the North Pole to the equator. However, the lack of precise geographical information necessitated a better definition. For a number of years the meter was defined as the distance between two etched lines on a platinum-iridium bar kept at 0°C (32°F) in the International Bureau of Weights and Measures at Sevres, France. However, an inability to measure this distance as accurately as desired

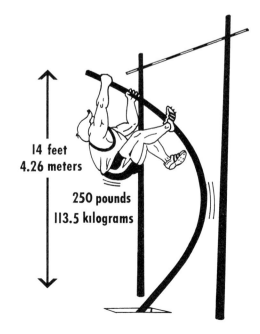

Figure A–1
The pole-vaulter is easily recognized as hefty when described as weighing 250 pounds and making a jump something less than 14 feet. To persons in the habit of using the system of international measurements, 113.5 kilograms and 4.26 meters produce similar conceptualizations.

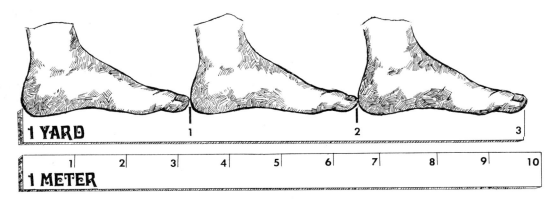

Figure A-2
A meter equals 1.094 yards.

prompted a recent redefinition of the meter as a length equal to 1,650,763.73 times the wavelength of the orange-red spectrographic line of $^{86}_{36}$Kr.

The meter (39.37 inches) is a convenient unit with which to measure the height of a basketball goal (3.05 meters), but it is unwieldy for measuring the parts of a watch or the distance between continents. For this reason, prefixes are defined in such a way that, when placed before the meter, they define distances convenient for the purpose at hand. Some of the prefixes with their meanings are shown below:

nano	1/1,000,000,000 or 0.000000001
micro	1/1,000,000 or 0.000001
milli	1/1,000 or 0.001
centi	1/100 or 0.01
deci	1/10 or 0.1
deka	10
hecto	100
kilo	1000
mega	1,000,000

The corresponding units of length with their abbreviations are the following:

nanometer (nm)	0.000000001 meter
micrometer (μm)	0.000001 meter
millimeter (mm)	0.001 meter
centimeter (cm)	0.01 meter
decimeter (dm)	0.1 meter
meter (m)	1 meter
dekameter (dam)	10 meters
hectometer (hm)	100 meters
kilometer (km)	1000 meters
megameter (Mm)	1,000,000 meters

Since the prefixes are defined in terms of the decimal system, the conversion from one metric length to another involves only shifting the decimal point. Mental calculations are quickly accomplished.

How many centimeters are in a meter? Think: Since a centimeter is the one-hundredth part of a meter, there would be 100 centimeters in 1 meter.

Table A-1
Conversion Factors*

Length	1 inch (in.)	= 2.54 centimeters (cm)
	1 yard (yd.)	= 0.914 meter (m)
	1 mile (mi.)	= 1.609 kilometers (km)
Volume	1 ounce (oz.)	= 29.57 milliliters (mL)
	1 quart (qt.)	= 0.946 liter (L)
	1.06 quart (qt.)	= 1 liter (L)
	1 gallon (gal.)	= 3.78 liters (L)
Mass (weight)†	1 ounce (oz.)	= 28.35 grams (g)
	1 pound (lb.)	= 453.6 grams (g)
	1 ton (tn.)	= 907.2 kilograms (kg)

* Common English units are used.
† Mass is a measure of the amount of matter, whereas weight is a measure of the attraction of the Earth for an object at the Earth's surface. The mass of a sample of matter is constant, but its weight varies with position and velocity. For example, the space traveler, having lost no mass, becomes weightless in earth orbit. Although mass and weight are basically different in meaning, they are often used interchangeably in the environment of the Earth's surface.

Conversion of measurements from one system to the other is a common problem. Some commonly used English-SI equivalents (conversion factors) are given in Table A-1.

Units of Mass

The primary unit of mass is the **kilogram** (1000 grams). This unit is the mass of a platinum-iridium alloy sample deposited at the International Bureau of Weights and Measures. One pound contains a mass of 453.6 grams (a five-cent nickel coin contains about 5 grams).

Conveniently enough, the same prefixes defined in the discussion of length are used in units of mass, as well as in other units of measure.

Units of Volume

The SI unit of volume is the **cubic meter** (m^3). However, the volume capacity used most frequently in chemistry is the liter, which is defined as 1 cubic decimeter (1 dm^3). Since a decimeter is equal to 10 centimeters (cm), the cubic decimeter is equal to $(10\ cm)^3$, or 1000 cubic centimeters (cc). One cubic centimeter, then, is equal to one milliliter (the thousandth part of a liter). The mL (or cc) is a common unit that is often used in the measurement of medicinal and laboratory quantities. There are 1000 liters in 1 kiloliter or cubic meter.

Units of Energy

The SI unit for energy is the **joule** (J), which is defined as the work performed by a force of 1 newton acting through a distance of 1 meter. A newton is defined as that force which produces an acceleration of 1 meter per second per second when applied to a mass of 1 kilogram. Conversion units for energy are:

1 calorie = 4.184 joules
1 kilowatt-hour = 3.5×10^6 joules

Other SI Units

Other SI units are listed below:

Time	second (s)
Temperature	Kelvin (K)
Electric current	ampere (A) = 1 coulomb per second
Amount of molecular substance	mole (mol) = 6.022×10^{23} molecules
Pressure	pascal (Pa) = 1 newton per square meter
Power	watt (W) = 1 joule per second
Electric charge	coulomb (C) = 6.24196×10^{18} electronic charges
	= 1.036086×10^{5} faradays

Further information on SI units can be obtained from H.F.R. Adams: *SI Metric Units — An Introduction.* Toronto, McGraw-Hill Ryerson Ltd., 1974.

Topics for Themes

APPENDIX B

These are suggested topics for short themes in which you can express your views. The depth of research is to be determined by you or your instructor. If a topic is to be researched, the *Readers' Guide* is a good place to start, and *Scientific American* (well-indexed!), *Chem Tech, Environmental Science and Technology, Discover, High Technology,* and *Science Digest* are good places to continue. Rather than discuss bona fide solutions to a particular problem, it might be more interesting to state the questions that must be answered before you could vote informatively on the issue. A primary requirement for an informed citizenry is that we ask enough good questions to ensure that all sides of a given issue have been covered. Asking investigative questions is our privilege and responsibility.

1. Contrast knowledge gained by scientific methods with knowledge in philosophy, religion, art, and other disciplines.
2. What role should scientists play in deciding the moral uses of their discoveries?
3. Would it be a good idea to limit the number of people on Earth in order to have more material and energy per person?
4. Explore some ways in which technology has changed our occupational patterns.
5. Find examples in which a technological development at first was considered more good than bad, but with passage of time was looked upon as more bad than good.
6. Is technological "progress" always obtained at some cost? Explain.
7. Discuss ways in which side effects of chemicals can be determined before a product is put on the market.
8. Should every conceivable side effect of a product be tested for and eliminated before the product is put on the market? Justify your answer.
9. Techniques have been developed to analyze for some elements even when only 1 nanogram (10^{-9} g) is present in the sample. What effect may this have on our knowledge of the side effects of chemicals?
10. Modern society demands more energy with less pollution and less expenditure of money. From your point of view, is this a "pipe dream"? Discuss reasons for your answer.
11. To what degree should food chains be considered when eradication of insects via insecticides seems to be the only practical way to protect a food supply and health?
12. What should be done with radioactive wastes? What problems are involved?
13. Discuss the need to recycle chemicals in view of limited resources.
14. What can one do about waste materials to make a better, safer, more livable world for everybody?

15. To what extent should political decisions influence scientific investigation?
16. Should scientists be morally responsible in their investigations of nature?
17. Defend or deny the following: Education is the solution to problems concerning energy, pollution, and the proper use and reuse of materials.
18. Discuss one or more ways in which food production creates a pollution problem and wastes energy.
19. Cite a recent technological advance; state its benefits, and briefly discuss problems arising from it.
20. How should wastes be dealt with? In your discussion, consider minimum use of energy, minimum pollution, and maximum recycling.
21. Discuss significant differences and similarities between science and technology.
22. Discuss an important practical problem facing us with respect to how science and technology have helped solve the problem and in turn have created other problems.
23. Describe evidence of technological advances that have caused a decline of civilization. (Consider the article in the *Journal of Occupational Medicine*, Vol. 7, pp. 53–60, 1965.)
24. Defend or deny the following: Technology is more beneficial than detrimental to humanity.
25. Discuss areas or problems in which technology is most needed today.
26. Discuss the limitations of science.
27. Discuss the limitations of technology.
28. What factors should be considered before a chemical is banned for a particular use? Defend each factor.
29. Chemical dumps have developed into a national shame. Do you think that the users of chemical products should pay for the long-term disposition of associated waste materials when they purchase the product?
30. Do you think it possible or likely that chemicals will ultimately cause our destruction in a sea of chemical poison?
31. Debate the following statement: Scientific and technological knowledge is dangerous in the absence of human values.
32. How dangerous do you consider the possibility that scientists and others will "play God" in the control of new human life? Do you consider genetic control a threat to you?
33. Should we refrain from seeking certain types of new chemicals for fighting insects?
34. The term *chemophobia* has been coined to describe a certain public phenomenon. Can you explain?
35. Should we refrain from seeking certain types of new knowledge for fear of the consequences? Can people be trusted to understand and show proper restraint? Does nature have a trap for us waiting to be sprung?
36. Should a proliferation of fruit flies be eradicated by spraying of malathion? What facts should be known and weighed before we make this kind of decision?
37. From a chemical standpoint, what do we owe future generations?
38. Among the following, which are the three most pressing problems? Justify your choices. Conservation of metals, pollution, energy, production of food and feeding the world, controlling pests, storing nuclear wastes, transportation, biogenetic engineering
39. What technological advances would you most like to see?
40. What is the proper balance or equilibrium point between the use of petroleum

as a fuel and the use of petroleum as a source of medicine, plastics, paints, and other consumer products?

41. Is there really a "miracle fuel" that is not currently popular but will solve our energy problems? Speculate on the prospects of such a fuel and on its existence or nonexistence.

42. Before you could vote informatively on the following items, what questions would you need to have answered?
    a. Clean Air Act
    b. Nuclear energy for electrical power plants
    c. What to do with solid refuse

43. In other parts of the world it is acceptable to buy fruit that might contain a few worms. In the United States insecticides have been used to produce commercial quantities of insect-free fruit. Do you think it wise to spray large sections of California or Florida to have absolute control over the medfly?

Answers to Self-Test Questions and Matching Sets

Chapter 1

Self-Test 1–A
1. observed experimental facts
2. the same
3. the integrated circuit
4. *E. coli*
5. (a) theories, (b) laws, (c) facts

Matching Set
1. i
2. d
3. b
4. a
5. h
6. c
7. f
8. g

Chapter 2

Self-Test 2–A
1. wood, soil, air, and concrete
2. operational definition
3. rain water, table salt, table sugar, and baking soda
4. false
5. solution
6. (a) 3, (b) 1, (c) 4, (d) 2
7. true
8. false
9. pure substances

Self-Test 2–B
1. (a) metals: iron, copper, gold, silver, chromium, magnesium
   (b) nonmetals: oxygen, silicon, carbon, nitrogen, chlorine, fluorine
2. 109
3. false
4. burning of natural gas, burning of sulfur, production of steel from iron ore, corrosion of iron
5. cutting diamonds, blowing glass, producing electricity from falling water, plowing ground, slicing bread
6. nuclear change
7. pure substance
8. (a) oxygen, (b) iron, (c) hydrogen
9. (a) a sodium atom
   (b) two sodium atoms
   (c) a molecule of hydrogen chloride containing one atom of hydrogen and one atom of chlorine
   (d) yields or produces
   (e) a molecule of hydrogen containing two atoms of hydrogen
   (f) two units of sodium chloride, each containing one sodium atom and one chlorine atom
   (The word mole can be substituted for atom, molecule, or unit in the above answers. In other words, the symbols and formulas also represent a mole of the substance)
10. (a) the element
    (b) an atom of the element
    (c) a mole of the atoms
11. (a) the elements in a substance
    (b) the relative number of atoms of each element (if two or more) in the substance
12. a coefficient
13. 7, 45, 21
14. True

Self-Test 2–C
1. false
2. macroscopic, microscopic, molecular
3. up and down a column
4. silicon, or any element in Group IVA
5. (a) lithium, or any element in Group IA
   (b) Mg, Ca, Sr, or Ba
   (c) F, Cl, Br, I, or At
6. Mendeleev, Meyer
7. No. The present working theory allows only one element for each atomic number.
8. Yes. Present experiments and theory predict nuclear changes that will extend the list of elements.
9. seven
10. They increase.

Matching Set

1. b
2. e or o
3. d
4. a
5. c
6. f
7. i
8. k
9. g
10. h
11. j
12. n
13. l
14. m
15. e or o

Chapter 3

Self-Test 3–A

1. Leucippus, Democritus
2. (b) philosophy
3. gained, chemical
4. CO, CO_2
5. (a) new compound, (b) 2:4:1, (c) law of multiple proportions
6. (d) Atoms are recombined into different arrangements.
7. (a) the same, (b) atoms
8. repel, attract
9. alpha (α), beta (β), gamma (γ), gamma (γ)

Self-Test 3–B

1. protons, neutrons
2. small
3. nucleus
4. electrons, protons, neutrons
5. electrons, protons
6. atomic
7. about 1836
8. 33, 33, 42
9. different, identical
10. electrons
11. electrons, protons
12. false

Self-Test 3–C

1. particles, waves
2. spectrum
3. farther from, closer to
4. sun, planets
5. ground state
6. 18

Matching Set

1. e
2. a
3. j
4. i
5. d
6. l
7. b or m
8. f
9. k
10. g
11. h

Chapter 4

Self-Test 4–A

1. Becquerel, Curie
2. alpha (α), beta (β), gamma (γ)
3. gamma (γ), alpha (α)
4. transmutation
5. half-life
6. carbon-14
7. uranium-238

Self-Test 4–B

1. Rutherford
2. kinetic energy
3. no
4. transuranium
5. 109
6. roentgen
7. curie
8. rad
9. radium

Self-Test 4–C

1. sterilization, tracers, medical diagnosis
2. (a) ^{60}Co
3. (d) all of these
4. (b) imaging
5. (c) metastable
6. (a) one eighth of the original dose
7. (b) 6-hour half-life isotope

Matching Set

1. b
2. i
3. a
4. k
5. e
6. c
7. m
8. d
9. p
10. o
11. h
12. n
13. f
14. l
15. g
16. j

Chapter 5

Self-Test 5–A

1. ions
2. ionic
3. one
4. CaI_2
5. Cl^-
6. valence
7. losing
8. gaining

9. Rb: one electron lost
   Ca: two electrons lost
   K: one electron lost
   S: two electrons gained
   Mg: two electrons lost
   Br: one electron gained

Self-Test 5–B

1. (a) H—H, (b) H—Cl
2. 6
3. 3
4. fluorine
5. (a) 8
   (b) octet
   (c) most of the time
6. fluorine
7. nitrogen, oxygen, fluorine
8. HCl
9. $MgCl_2$
10. H_2O
11. angular
12. 2, 2
13. 3, 1

Matching Set

1. k
2. d
3. l
4. f
5. i
6. g
7. b
8. e
9. j
10. c
11. h

Chapter 6

Self-Test 6–A

1. disorder, or entropy
2. bonds
3. 1 kg
4. 85 kcal
5. independent
6. temperature, concentration, and catalysis
7. The oxides of nitrogen catalyze the oxidation of sulfur dioxide to sulfur trioxide.
8. (a) true, (b) false
9. (c) The rate of the forward reaction is equal to the rate of the reverse reaction.
10. false

Self-Test 6–B

1. solution
2. solvent
3. nonelectrolyte
4. 2
5. dissociates

6. 0.5
7. hydrogen or hydronium, hydroxide
8. (a) nitric, sulfuric, hydrochloric
   (b) acetic acid
   (c) sodium or potassium hydroxide
   (d) ammonia
9. 7
10. 0.01 times as large
11. hydroxide, constant
12. hydrogen phosphate, dihydrogen phosphate

Self-Test 6–C

1. electrolysis
2. cathode
3. hydrogen, oxygen
4. chemical cell or battery, chemical
5. lead storage or nickel-cadmium battery, dry cell
6. protective coatings, galvanization, cathodic protection
7. battery. A fuel cell constantly draws fresh chemicals for electrochemical change.
8. hydrogen, oxygen
9. cathode: $Cu^{2+} + 2\ e^- \longrightarrow Cu$
   anode: $Zn \longrightarrow Zn^{2+} + 2\ e^-$

Matching Set

1. j
2. m
3. k
4. h
5. f
6. c
7. g
8. l
9. e
10. b
11. i
12. n
13. d
14. a

Chapter 7

Self-Test 7–A

1. oxygen
2. aluminum
3. Minnesota
4. limestone
5. copper
6. copper, aluminum, and magnesium
7. slag
8. reduced
9. cathode
10. magnesium
11. positive ions

Self-Test 7–B

1. nitrogen, oxygen
2. liquid oxygen
3. icy cold substances
4. oxidation in steel making—oxygen
   freezing of tissue in cryosurgery—nitrogen

creation of inert atmospheres — nitrogen
preservation of frozen foods — nitrogen
5. silicon, oxygen
6. lead
7. silicon
8. clay, sand, feldspar
9. Portland cement

Self-Test 7–C

1. sulfur
2. the manufacture of fertilizers, in the petroleum industry, and in the production of steel
3. chlorine, hydrogen, sodium hydroxide
4. air
5. sodium carbonate decahydrate, sodium hydrogen carbonate (sodium bicarbonate)
6. New compounds will be made, and chemical change and purifications in gravity-free space will produce new materials.

Matching Set

1. i
2. j
3. b
4. g
5. h
6. e
7. f
8. a
9. c
10. d

Chapter 8

Self-Test 8–A

1. petroleum
2. natural gas
3. 33%
4. coal, petroleum, natural gas
5. oxygen, water
6. true
7. work
8. quantitatively
9. disorder (entropy)
10. quality
11. calorie, joule, Btu, kilowatt-hour
12. watt, kilowatt, joule/sec
13. aluminum, aluminum

Self-Test 8–B

1. fission
2. fusion
3. containment of reactants at high temperature
4. hydrogen
5. 42
6. plutonium ($^{239}_{94}$Pu), uranium ($^{233}_{92}$U)
7. water, air, greenhouse
8. by a primary source of energy
9. electricity

10. coal
11. natural gas
12. arsenic, boron (or gallium; see Chapter 7)

Matching Set

1. f
2. l
3. b
4. j
5. c, e
6. e
7. g
8. a
9. p
10. h
11. k
12. m
13. o

Chapter 9

Self-Test 9–A

1. tetrahedral
2. Urea
3. ethylene
4. benzo(α)pyrene
5. structural
6. Acetylene
7. cyclobutane
8. benzene
9. hydrogen, carbon
10. —C_2H_5
11. double bond
12. geometric
13. 2,4-dimethylhexane

Self-Test 9–B

1. fractional distillation
2. methane
3. methyl tertiary-butyl ether (MTBE)
4. catalytic reforming
5. c

Matching Set

1. d
2. e
3. a
4. c
5. f
6. g
7. b
8. h

Chapter 10

Self-Test 10–A

1. (a) diethyl ether
   (b) ethanol
   (c) acetic acid
   (d) acetone
2. (a) ester, carboxylic acid groups
   (b) carboxylic acid, alcohol groups
   (c) amine, carboxylic acid groups

3. 42%
4. acetaldehyde
5. denatured
6. isopropyl
7. fatty
8. acetic acid

Self-Test 10-B

1. four
2. rotation of polarized light
3. (a) A fat is a solid at room temperature, whereas an oil is a liquid.
   (b) by hydrogenation (the addition of hydrogen to C=C double bonds)
4. atherosclerosis
5. alkaloids

Self-Test 10-C

1. pyrolysis
2. Aromatics
3. natural gas, petroleum
4. monomers
5. (a) $CH_3CH_2=CH_2$ or (H)(CH_3)C=C(H)(H)
   (b) $C_6H_5CH=CH_2$ or 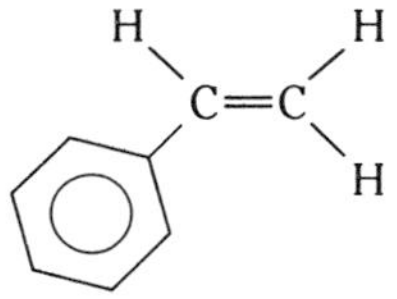 (C_6H_5)(H)C=C(H)(H)
   (c) $CF_2=CF_2$ or (F)(F)C=C(F)(F)
6. cis-isoprene
7. condensation
8. water
9. condensation

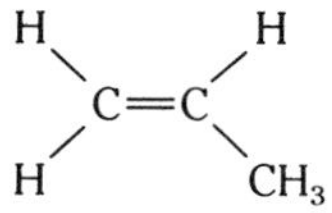

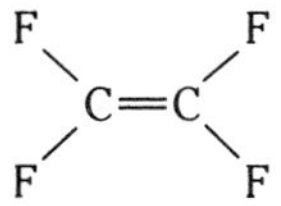

Matching Set

1. g   7. b
2. e   8. c
3. k   9. f
4. i   10. m
5. a   11. l
6. d   12. j

Chapter 11

Self-Test 11-A

1. carbon, hydrogen, oxygen
2. monosaccharides
3. D-glucose, D-fructose
4. D-glucose
5. alpha-D-glucose
6. hydrogen bonding

Self-Test 11-B

1. amino acids
2. essential amino acids
3. 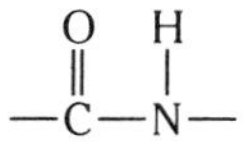 $-\overset{O}{\underset{}{C}}-\overset{H}{\underset{}{N}}-$
4. 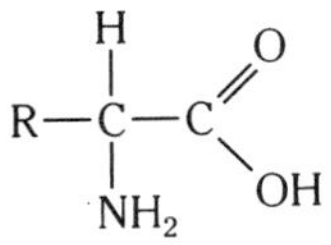 $R-\overset{H}{\underset{NH_2}{C}}-C\overset{O}{\underset{OH}{}}$
5. (a) sequence of amino acids
   (b) hydrogen bonded structures to form helices or sheets
   (c) arrangements of helices into super helices or balls (globs)
   (d) positioning of the tertiary structures (globs)
6. (a) 27, (b) 6
7. Hydrogen bonds form the helices (H-bonds between each third amino acid in the chain) and the sheets (H-bonds between chains).
8. *catalyst*
9. key, lock
10. enzyme
11. enzyme
12. active site

Self-Test 11-C

1. Sun
2. ATP
3. ADP, H_3PO_4, energy
4. carbon dioxide, water, energy
5. chlorophyll
6. hydrolysis
7. emulsifying agents
8. ATP, lactic acid
9. carbon dioxide, water, ATP

Self-Test 11-D

1. (a) DNA, (b) t-RNA
2. hydrogen
3. ATP
4. (a) thymine or uracil, (b) guanine, (c) cytosine, (d) adenine, (e) adenine
5. false
6. true
7. ribose, deoxyribose
8. phosphoric acid, a sugar (ribose or deoxyribose), a nitrogeneous base.
9. double helix

Matching Set

1. g
2. f
3. a
4. e
5. c
6. b
7. d
8. k
9. l
10. o
11. q
12. i
13. j
14. n
15. r
16. s
17. p
18. m

Chapter 12

Self-Test 12–A

1. clays, silts, sandy soils, loams
2. sour
3. acidic, basic
4. size of soil particles, chemical composition of soils
5. a trivalent ion like Fe^{3+} (hydrolyzes more than Na^+)
6. humus
7. subsoil, or inorganic
8. calcium, magnesium, sulfur
9. oxidation (usually) to nitrogen-oxygen species
10. true

Self-Test 12–B

1. false
2. nitrogen, phosphate, potash
3. potassium nitrate for potassium
4. yes
5. gas
6. nitric acid (HNO_3), nitrous acid (HNO_2)
7. denitrification
8. K_2CO_3
9. magnesium
10. 33%
11. DDT
12. chlordan
13. selective herbicide
14. 2,4-D
15. plant growth regulator, or plant hormone

Matching Set

1. g
2. j
3. b
4. l
5. p
6. c
7. n
8. e
9. h
10. m
11. k
12. o
13. q
14. a
15. i
16. r
17. f
18. d

Chapter 13

Self-Test 13–A

1. proteins, fats, carbohydrates, vitamins, minerals, water
2. oxygen, calcium
3. USRDA
4. partially true
5. fat
6. 1, 1
7. basal metabolic rate, weight in pounds

Self-Test 13–B

1. apoenzyme
2. protein
3. false
4. urea
5. lecithin, cholesterol
6. triglycerides
7. linoleic acid
8. pork lard, chicken fat
9. fat, carbohydrate
10. fat
11. peroxides, polymers, free radicals
12. false
13. sugar, flour

Self-Test 13–C

1. variety, whole, different places
2. iron, hemoglobin
3. manganese, skin pigmentation
4. zinc
5. zinc
6. sodium, potassium
7. greater than 1
8. magnesium
9. calcium, bones, teeth
10. kidney
11. iodine
12. false
13. fat, water
14. provitamin
15. carrots, liver
16. A
17. antioxidant
18. vitamin E
19. D
20. B_6
21. vitamin D
22. scurvy
23. coenzymes, protein production
24. B_1, beriberi
25. niacin (B_3)

Self-Test 13-D

1. synergistic, potentiation
2. drying, salting
3. false
4. volatile compounds
5. more easily oxidized
6. complexing agent (sequestrant)
7. false
8. flavor enhancer
9. dehydrating
10. Generally Recognized As Safe
11. saccharin
12. conjugated chains with aromatic rings
13. citric acid, lactic acid
14. carbon dioxide (CO_2)
15. breaks
16. active

Matching Set I

1. e
2. c
3. l, a
4. f
5. d
6. m
7. n
8. i
9. b
10. k
11. j
12. g
13. h

Matching Set II

1. b
2. c
3. d
4. e
5. f
6. a
7. h

Chapter 14

Self-Test 14-A

1. dehydration, hydrolysis
2. oxidizing
3. hemoglobin, oxygen
4. false
5. CN^-, cytochrome oxidase, oxygen
6. fluoroacetate ion
7. heavy metal poisons, chelate
8. true
9. liver

Self-Test 14-B

1. neurotoxins
2. synapses
3. acetylcholine
4. genes, chromosomes
5. false
6. chimney sweeping
7. liver
8. bioactivation
9. cancer, heart disease
10. teratogen

Matching Set

1. b
2. f
3. h
4. i
5. j
6. n
7. d
8. e
9. c
10. k
11. a
12. m
13. g

Chapter 15

Self-Test 15-A

1. true
2. water
3. 60 gallons
4. radioactivity, heat, toxic chemicals
5. Superfund
6. groundwater

Self-Test 15-B

1. troposphere
2. stratosphere
3. decreasing
4. adsorbed
5. London
6. photochemical
7. nitric oxide, sulfur dioxide
8. true

Matching Set

1. g
2. k
3. m
4. h
5. b
6. c
7. o
8. a
9. n
10. d
11. e
12. i or n
13. p
14. i
15. f
16. j
17. l

Chapter 16

Self-Test 16-A

1. generic
2. high blood pressure
3. (a) magnesium hydroxide, $Mg(OH)_2$
4. analgesic

5. poppy
6. true
7. antipyretic
8. allergy
9. germicide
10. antibiotics
11. *p*-aminobenzoic acid
12. mold
13. fungus

Self-Test 16–B

1. cholesterol, triglycerides
2. faster
3. fatigue, lethargy, depression, confusion
4. relaxing
5. low-density lipoproteins
6. true
7. true
8. DNA
9. antimetabolites, childhood leukemia
10. testosterone
11. progesterone
12. muscle building
13. liver damage, liver cancer, acne, baldness, changes in sexual desire, enlargement of breasts

Self-Test 16–C

1. heroin
2. cocaine
3. AIDS
4. phosphate
5. T cells
6. designer drug

Matching Set

1. g
2. s
3. p
4. c
5. i
6. r
7. o
8. d
9. b
10. q
11. a
12. j
13. n
14. l
15. k
16. h
17. f

Chapter 17

Self-Test 17–A

1. ionic bonds, keratin
2. thioglycolic
3. oxidation
4. black
5. hydrogen peroxide
6. resin
7. to reduce blood pressure; Minoxidil can be used to produce hair growth on skin already containing hair follicles.
8. the sulfides of sodium, calcium, and strontium, or calcium thioglycolate
9. skin, hair, nails
10. disulfide, ionic
11. corneal
12. sebum

Self-Test 17–B

1. top
2. emollient
3. emulsion
4. lipstick
5. talc
6. false
7. *p*-aminobenzoic acid
8. melanin
9. astringent, odor mask, chemical that removes the odor-causing compound
10. Aluminum chlorohydrate is an astringent that closes the pores of the skin.

Self-Test 17–C

1. soap
2. alkali or base
3. ionic salt end
4. olive
5. air in the soap bar
6. calcium, magnesium, iron
7. traditional soaps
8. greater detergent action in converting grease to soap, precipitation of the ions that cause water to be hard
9. visible
10. detergent, abrasive
11. fluorine
12. oxalic
13. linseed oil
14. acetone

Matching Set

1. q
2. m
3. k
4. c
5. p
6. n
7. a
8. e
9. f
10. b
11. g
12. h
13. i
14. j
15. l
16. d
17. o

INDEX

Note: d following a page number indicates a definition; i indicates an illustration or figure; s indicates a structure; and t indicates a table.

Abelson, P.H., 73
Abietic acid, 442
Abrasive cleaners, 436
Abrasives in tooth paste, 439
Absorption, 270
    of pollutants by particulates, 367
Acaricide, 284dt
Accelerators, 72
Acetamide, 206t
Acetaminophen (Tylenol), 391, 392s
    biotransformation of, 349
Acetate ion, 100
Acetic acid, 118, 119, 206t, 213, 216t, 227t
    in vinegar, 214
Acetic anhydride
    synthesis of, 227, 228i
    uses of, 227
Acetone, 206t, 213, 227t
    in nail polish remover, 442
Acetylcholine, 336, 337s, 390, 391i
    cycle, 338i
Acetylene, 193
    bonding in, 99i
    plant growth enhancer, 283
    uses of, 225t
Acetylsalicylic acid. See Aspirin.
Acid rain, 374–376, 375i
Acid skin, 119
Acid soil, 270
Acid(s), 119d, 120d
    amino, 244–248, 245t
        isomers of, 222
    aspartic, 245st
    DNA, 258
    as food additives, 313t
    gibberellic, 283s
    glutamic, 245st
    lactic, 255
    nitric, 273
    nitrous, 273
    nucleic, 238, 256–265
    orthophosphoric, 274
    phosphoric, 258
    RNA, 257
Acid-base buffers, 123d
Acid-base reactions, 109
    peptide bond formation as, 244
Acidity
    of soils, cause of, 271
Acidulant, 317d
Acquired immunodeficiency syndrome. See AIDS.
Acrilan, 233t
Acrolein
    in body odor, 431

Acrylics
    in paints, 440
Acrylonitrile, 227t, 233t
    as carcinogen, 346
    in paints, 440
Actinide series, 35, 68
Actinium series, 35, 68
Activation energy, 248, 249i
Active sites
    enzymes and substrates, 250
Addition polymers. See Polymers, addition.
Adenine, 257s
    nitrous acid effect on, 340s
Adenosine diphosphate. See ADP.
Adenosine monophosphate. See AMP.
Adenosine triphosphate. See ATP.
Adipic acid, 227t, 232s, 233t
Adipose tissue, 300
ADP, 253s
    hydrolysis of, 253
Adrenal gland
    aldosterone from, 308
Adrenalin, 222
Adsorption, 270
    of pollutants by particulates, 367
Advil, 391, 392s
Aerobic, 254d
Aerosol(s), 367d
Aflatoxin(s)
    as carcinogen, 346
    as mutagen, 341
Agent orange, 287–288
Aging
    vitamin E and, 310
Agrichemicals, 268d
Agriculture, 268–288. See also Food, production of.
    basic philosophy of, 268
    laws of, 268
Ahlquist, Raymond, 400
AIDS, 414–416
    retrovirus cause of, 415–416
Air
    composition in soil, 270
    fractionation of, 144, 145i
    liquid form, 145
    as mixture, 19
    in soaps, 433
Air pollution, 366–379
    costs of controls, 383
Alanine, 245st
    optical isomers of, 222i
Alanylglycine, 246s
Alchemists, 70
    and ceramics, 151

Alcohols
    classification of, 208–209
    denatured, 212
    in deodorants, 431
    dietary recommendations for, 296
    distillation of, 21i
    effect on human body, 293
    functional group for, 206t
    in perfumes, 429
    polyhydric, food additives, 313t, 319
Alcohol, Tobacco and Firearm Branch
    regulation of dynamite, 325t
Alcoholic beverages, 210t
    blood levels of, 211t
    breathalyzer test for, 211
Alcoholics
    red nose of, 306t
Alcoholism, 211
Aldehyde(s), 212
    functional group for, 206t
Aldicarb oxime, 286s
    Bhopal, India and, 286
    chemical plant in West Virginia and, 286
Aldose, 239d
Aldosterone
    sodium in blood and, 308
Aldrin, 285s
    use of, 285t
Algae
    hydrogen from, 181, 183i
Algicide, 284dt
Alice in Wonderland, 333
Alkali metal hydroxides, 119
Alkali metal(s), 31d
    electronic arrangement of, 93
Alkaline battery, 130t
Alkaline earth metals, 31d
    electronic arrangement of, 93
Alkaline fuel cells, 134
Alkaloid(s), 220, 389
Alkanes, 189–192, 191t
    cyclic, 195–196
    source of, 225t
    structural isomers of, 190, 192
Alkenes, 192–193
    cyclic, 196
Alkyl halides, 206t, 207–208
Alkylating agent(s)
    and cancer treatment, 404
Alkynes, 193–194
    cyclic, 196
Allergen(s), 392d
Allergies, 392–394
    treatment of, 393
Allergy, 392d

463

Alloy, 19, 140
Allura Red AC
  cost of introducing, 312
Almond oil
  in cold creams, 427
Alpha particle(s), 46t
  and the gold foil experiment, 53
  penetrating power of, 66i
alpha-2-deoxy-D-ribose, 256s
alpha-D-ribose, 256s
Alumina, 151
Aluminum, 33t
  in crust of Earth, 137, 138t
  electronic arrangement, 61t, 94s
  reduction of, 140
Aluminum chloride, hydrated, 431
  in deodorants, 427
Aluminum chlorohydrate, 431
Aluminum oxide, 140
  in glass, 150t
Aluminum reduction, 141i
Aluminum sulfate, 227t, 431
Aluminum-28, 71
Alzheimer's disease
  treatment of, 416
Amalgam, 333d
  sodium-mercury, 157
Americium
  nuclear reaction for, 74t
Amide(s), 219–221
  functional group for, 206t
Amine groups
  oxidized in hair, 423
Amine(s), 219–221
  in body odor, 431
  functional group for, 206t
Amino acid(s), 244–248, 245t
  base code for transfer RNA, 263t
  conversion to glucose and glycogen, 299
  essential, 244d, 298d
  kinds in human protein, 298
  optical isomers of, 222
  in protein of skin, 419
  synthesized from urea, 299
p-Aminobenzoic acid (PABA), 396
  in suntan lotions, 430s
Aminoguanidine, 416s
  treatment of atherosclerosis, 416
Ammonia, 21, 120, 227
  anhydrous, danger of, 279
  bonding in, 99i
  first U. S. plant for, 278
  in hair curling solution, 422
  injection into soil, 279i
  molecular structure of, 103i
  production by Haber process, 278i
  production of nitrates from, 279
Ammonia solution(s), 118
  pH of, 122i
Ammonium carbamate, 281s

Ammonium ion, 100, 279
Ammonium nitrate, 227t
  in atmosphere, 373
  danger of, 280
  synthesis of, 280
Ammonium sulfate, 227t
Amoxil and ten most prescribed drugs, 387t
AMP, hydrolysis of, 253
Amphiprotic species, 120
Ampicillin, 387t
Amylopectin, 240, 242i
  hydrolysis of, 242i
Amylose, 241si
  condensation polymer, 240
  test for, 240
Anabolic steroids, 409
  fluoxymesterone, 410s
  methyltestosterone, 410s
  oxomethofone, 410s
  side effects of, 409
Anabolic, 409d
Anaerobic, 254d
Analgesics, 388–392
Anemia
  deficiency of vitamins B6, B9, and B12, 309t
  iron and, 307
  pernicious, 306t
  and pica, 335
  sickle cell, 248
  vitamin B6 and, 307
  vitamin B9 and, 307
  vitamin B12 and, 307
Angular molecular structure, 103i
Aniline, 345s
Anode, 126d
Anode slime, 143
Antacids, 119, 387–388, 388t
Anthracene, 197s, 226t
Antibiotics, 394
  invention incubation period, 3t
Antibody, 394d
Anticaking agents, 3l3t
  food additives, 318
Anticancer drugs, 403–406
  5-fluorouracil, 405s
  cisplatin, 5–6, 405s
  cyclophosphamide, 404s
  mechanism of, 404–405, 405i
Anticholinesterase poisons, 337d, 339t
Anticodon, 263d
Antigen(s), 392–393
Antihistamines, 392–394
Antimetabolite(s), and cancer treatment, 405–406
Antioxidants
  food additives, 313t, 314
  in food, 314
Antiseptics, 398–399
  in deodorants, 431
Anuria, 335d

Apatite, 305, 439s
Apatite minerals, 158
Apoenzyme, 249d, 298
Apollo spacecraft fuel cells, 133
Apples, 119
Aqueous solution, 115d
Aquifer, 358d
Arachidonic acid, 218, 219s, 300
Aramite, for mite control, 285t
Arginine, 245st
  infant requirement of, 298
Argon
  electronic arrangement, 61t
  percentage in air, 144t, 145
Argonne National Laboratory, 177
Aristotle, 40
Aromatic(s). See also Hydrocarbons, aromatic.
  from coal tar, 226t
  source of, 225t
Arrhenius, Sevante, 120
Arsenates
  ion, 100
  weed killers, 287
Arsenic
  as carcinogen, 346
  as insecticide, 285t, 286
  limits in foods, 331
  removal of in phosphoric acid production, 159
  in solar battery, 182
  as teratogen, 344
  toxicity of, 331–332
Arsenic oxide
  in glass, 150t
Arsenicals. See Arsenic.
Arsenites
  weed killers, 287
Artificial nuclear changes, 70–72
Asbestos, 22i
  as carcinogen, 346
  regulation of, 325t
Ascorbic acid, 308s
Aspartame, 317s
Aspartic acid, 245st
Asphalt, 199t, 200i
Aspirin, 390–391, 392s
  lethal dose, 325
Astatine, 31
Asthma, bronchial, 392–393
Asthtabula, Ohio, 87
Astringents
  in deodorants, 431
  in face powder, 428
Asymmetric, 221d, 223, 224
Asymmetry
  in amino acids, 244
Atherosclerosis, 218, 400
  carbohydrates and, 303
  cholesterol and, 400
Atherosclerotic plaque, 302

Athletes
    high protein diets and muscle, 299
    muscle from protein or exercise, 294
Atmosphere(s), 22
Atmosphere, the, 364–366
Atmospheric oxidation
    of food, 314
Atom, 27d
    Bohr model, 57–62
    Dalton's definition, 43
    the Greeks' definition, 40
    Rutherford model, 53
Atomic bomb(s)
    Niels Bohr and, 60
    plutonium and, 74
Atomic dating, 68–70
Atomic mass, 54d
Atomic number, 35d, 54d
Atomic pile, 175
    University of Chicago and, 175
Atomic theory, 39
    early Greeks, and, 39
Atomic weights
    early development, 43–45
    fractional, 56
    idea, 31
    modern, 45
Atoms for Peace Prize, 60
ATP, 252s
    hydrolysis of, 253si
    Krebs cycle and, 255
    phosphorus and, 306t
    photosynthesis and, 251–253
    protein synthesis and, 263, 264i
Atrazine, 288s
    no-till farming and, 288
Atropine, 338
Augustus, King of Saxony, 151
Aureomycin, 398s
Aurora borealis, 22
Australia, iron ore in, 139
Automobile(s)
    catalytic converters on, 202, 229
    engine knock and, 199
Auxins
    hormone-type weed killers, 287
    plant growth enhancers, 283
    plant hormones, 283
Aviation safety
    and radioisotopes, 83
Avicide, 284dt
Avogadro, Amedeo, 44
Azothymidine, 415s
AZT, 415s
    treatment of AIDS patients, 415–416

B-complex vitamins, 310d
Baby food
    MSG and, 316

Bacon, Francis, 40
Bacon, Roger, 40
Bacteria
    intestinal, source of vitamin B12, 309t
    oil spills and, 264
    in tooth decay, 439
Bactericide, 284dt
Baking powder, 119, 158, 319
Baking soda, 119, 319
BAL. See British Anti-Lewisite.
BaP. See Benzo(alpha)pyrene.
Barium hydroxide, 119
Barium, 31, 138t
Barrels of oil
    equivalences of, 164t
    gallons per, 171d
Basal metabolic rate (BMR), 297–298
    influences on, 297d, 298
    how to estimate, 298
Base code on messenger RNA, 263t
Base pairs
    in DNA and RNA, 262
Base(s), 119d, 120d
    nitrogenous, 257si
Basic oxygen process, 141i
Batavia, Illinois, 72
Battery, 129d
    acid, 119
    car, 119
    characteristics, 130t
    primary, 130d
    secondary, 130d
Battery acid pH, 122i
Bauxite, 138t, 140
Bayer process, 140
Becquerel, Henri, 46, 65
Beeswax
    in cold creams, 427
    in lipstick, 428
Benzaldehyde, 212s
Benzene, 196i-197, 226t, 227t
    as carcinogen, 346
    indoor air pollution, 379
    octane rating of, 201t
    uses of, 225t
Benzo(alpha)pyrene, 196, 197s
    as air pollutant, 376–377
    as carcinogen, 344s, 346
    as mutagen, 341
Benzoyl butanoate, 216t
Benzoyl peroxide
    hazards, 324t
Benzyl alcohol, 216t
Beriberi
    thiamine deficiency, 309t
Berkelium
    nuclear reaction for, 74t
Berthollet, Compte Claude Louis, 41
Beryllium, 31, 33t
    as carcinogen, 346
    electronic arrangement of, 61t

Beta blockers
    heart disease and, 400–401
Beta particle, 46t
    penetrating power, 66i
Beta receptors
    beta blockers and, 400–401, 400t
    heart disease and, 400–401
Beta-amylase, 248
Beta-carotene, 309, 317
Beverages
    alcoholic, 210t
BHA (butylated hydroxyanisole), 314s
    antioxidant in food, 314
Bhopal, India tragedy, 11, 14, 286, 381
BHT (butylated hydroxytoluene), 314s
    antioxidant in food, 314
Bile salts, 254
Binders
    in paints, 440
Binding energy, 81d
Bioactivation, 349d
Biochemical systems
    energy and, 251–256
Biochemical(s)
    as polymers, 238
Biochemistry, 238d, 238–267
Biogenetic engineering, 264d
    diseases and, 265
Biomass
    future energy from, 185
Biosynthesis
    of human growth hormone, 10
    of human insulin, 10
    interferon and, 10
Biotechnology, 9–11
    benefits of, 10
    ethics and, 11
Biotransformation, 348d
Birth control pills, 408–409
Birth defects, 343
Bismuth-210
    half-life, 68t
Bismuth-214
    half-life, 68t
Blast furnace, 139i
Bleach solution
    pH of, 122i
Bleaching agents, 436
Bleaching hair, 425i
Blindness
    night, 309
Blister copper, 142
Blood
    glucose content of, 254
    pH of, 122i
Blood cells
    replacement rate of, 292
Blood pressure
    prostaglandins and, 301
    salt and, 308
    sodium and, 308

Blood sugar. *See* D-glucose.
Blue vitriol. *See also* Copper (II) sulfate.
   algae killer, 99
BMR. *See also* Basal metabolic rate.
   thyroxine and, 307
Body odor, 431
Bohr Model of atom, 57–62
Bohr, Niels, 57–59, 59i
   personal side, 60
Boiling temperature
   hydrogen bonds and, 104
   molecular mass and, 104
Boise, Idaho
   hot springs in, 174
Boltwood, Bertram, 68
Bond angle, 92i
Bond energy, 166d, 167t
   of methane, 166
Bond(s), 91–198
   bridge, 102
   charge separation in, 101
   coordinate covalent, 100
   covalent, 98–101. *See also* Covalent bonds.
   disulfide, in proteins, 248
   examples of, 91
   hydrogen, examples of, 91. *See also* Hydrogen bonds.
   intermolecular, 91, 102
   ionic, 92–98
   ionic, properties of, 92
   London, 105i
   metallic, 105
   nonpolar, 100
   peptide, 244d, 245s
   polar, 100
Bonding principles, 91
Bonding. *See* Bond(s).
Bone grease, 433
Borates, weed killers, 287
Borax, a plant nutrient, 277t
Borax solution, pH of, 122i
Boron, 33t
   electronic arrangement, 61t
   in glass, 150t
   plants' need for, 275
   in solar battery, 182
   toxicity to plants, 282
Boron steel, a neutron absorber, 175
Boron trichloride, bonding in, 99
Borosilicate glass, 150t
Bottger, Johann Friedrich, 151
Bottled gas, 191t
Brackish water, 355d
Bran, dietary fiber and, 303
Branched-chain hydrocarbons, 194–195
Brass, 19d
Bread, leavened, 319
Breath, and acetone, 302
Breathalyzer test, 211

Breeder nuclear reactors, 177, 179i
Brine, electrolysis of, 158i
Bristlecone pine, age of, 69
British Anti-Lewisite, 332s
   chelating effect of, 333s
British Association for the Advancement of Science, 43
British Thermal Unit (Btu), 163d
   equivalences of, 164t
Brittleness, ionic compounds and, 96
Bromine, 31
Bromobenzene, biotransformation of, 349
Brönsted, J.N., 120
Bronze age, 124
Brooklyn Union Gas Company, 174
Brown sugar, minerals in, 304
Browning of fruit, decrease of, 283
Buffer solution(s), 123d
Buffers, food additives, 313t
Builders, in soap products, 435
Burning, 124d
Butadiene, 227t
Butanoic acid, 216t
Butter, 217
   as emulsion, 427
Butyl butanoate, 216t
Butyl stearate, in nail polish remover, 442
Butylene, uses of, 225t
1-Butyne, 194s
   properties of, 194
2-Butyne, 194s
   properties of, 194
Butyric acid, 213t
Byron, Illinois, 362

Cadaverine, 220s
Cadmium
   as carcinogen, 346
   neutron absorber, 175
   poisoning by, 13
   as teratogen, 344
Cadmium oxide
   in glass, 150t
Caffeine, 220s
   BMR and, 298
   as mutagen, 341
   as teratogen, 344
Calan, 401, 403s
Calciferol, 305
Calcium, 31, 33t, 124
   absorption by the human body, 305
   bone and teeth and, 305
   calciferol and, 305
   cost of, 26t
   in crust of Earth, 137, 138t
   effect on heartbeat, 305
   electronic arrangement, 61t
   fertilizers and, 281

Calcium *(Continued)*
   metabolism in birds, 285
   secondary nutrient in plants, 274
   soil acidity and, 271
Calcium carbonate, 114, 140, 151
   in glass making, 147
   in tooth structure, 439
Calcium channel blocker(s)
   heart muscle and, 401
   mechanism of action of, 401, 402i
Calcium chloride, 227t
Calcium cyanamide
   weed growth retarder, 287
Calcium fluoride
   color in glass, 148t
Calcium hydroxide, 112, 119
   as depilatory, 426
Calcium hydroxyphosphate, 439
Calcium ion(s)
   EDTA complex, 335
   in hard water, 433
   muscle action and, 401
Calcium oxide, 114, 124, 141i
   in glass, 147
Calcium silicate, 140
Calcium sulfate, 152
Calcium thioglycolate
   as depilatory, 426
Calcium-binding protein, 305
California, University of, 73
Californium
   nuclear reaction for, 74t
Calmodulin, 305
Caloric needs
   human body and, 296
Calorie(s), 35, 163d
   equivalences of, 164t
   food, 218d, 296d
   scientific, 218d, 296d
Canada, acid rain and, 374, 375
Canada, Montreal, 46
Canal rays, 51, 52i
Cancer(s), 344d
   childhood, 406
   geographic distribution, 345
   initiation period, 347i
   lack of dietary roughage and, 303
   lung, and radon in homes, 77
   molybdenum and, 306t
   survival rates, 404t
   treatment of, 403–406
Candle, 35d
Caproic acid, 213t
Captan, as mutagen, 341
Car battery acid, 119
Carbazole, 226t
Carbohydrates, 238–244, 302–304
   cellulose, 242
   combustion of, 167t
   consumed indigestible, 302

Carbohydrates *(Continued)*
  cotton, 243i
  daily needs of, 302
  dextrins, 242
  digestion of, 253
  food sources of, 300t
  functions of in human body, 302
  monosaccharides, 239
  oligosaccharides, 239
  polysaccharides, 240
  problems with, 302–303
  starches, 240
Carbon, 24, 33t, 125, 154
  in crust of Earth, 137, 138t
  electronic arrangement, 61t
  in iron ore reduction, 139
  in steel, 140
Carbon black, 227t
Carbon dioxide, 21, 28, 140, 227t
  as air pollutant, 382
  in basic oxygen process for steel, 141
  bonding in, 99i
  chemical energy deficiency of, 168
  formula for, 109
  heat of formation of, 110
  percentage in air, 144t, 145
  photosynthesis and, 251
  produced by heating calcium carbonate, 114
  produced in lungs, 114i
  released by leavening agents in bread, 319
  in soil, 270
Carbon monoxide
  bonding in, 99i
  in fuel cells, 134
  hazards, 324t
  heat of formation of, 110
  mode of toxicity, 328, 329i
  in steel making, 140
  toxic action, 327–330
Carbon tetrachloride, 207s
  as carcinogen, 346
  toxicity of, 335
Carbon-12, 45, 46
Carbon-14, 85t
  dating, 69
  half-life, 69
Carbonate ion, 100
Carbonated beverages, 158
Carbonates, in soils, 270
Carbonic acid, as sink for carbon dioxide, 382
Carboxylic acid(s), 213–214
  functional group of, 206t
Carboxylic group, in ionic protein bond, 420, 421s
Carbyl, 339s
Carcinogen(s), 5d, 344–348, 346t
Carcinogenic, 196d
Carcinoma, 344d
Carnauba wax, in lipstick, 428

Carol, Lewis, 333
Carrageenin, 302
  emulsifier in salad dressing, 318i, 319
CAS ONLINE, 7
Cassiterite, 138t
Castile soap, 433
Castor oil, in lipstick, 428
Catalyst(s)
  catalytic converter, 229
  effect on reaction rate, 111
  industrial, 229
  platinum, 229
  surface, 230i
  zeolite, 230
  ZSM-5, 230–231
Catalytic converter, automobiles and, 202, 229
Catalytic cracking unit, 200i
Catalytic reforming, 229
  gasoline and, 201–202
Cathode rays, 47
Cathode, 126
Cathode-ray tube, 48, 49i, 50i, 52i
Cathodic protection, 133i
Cato the Elder, seed selection and, 268
Cavendish Laboratory, 259
Cell, electrochemical, 129
Cell, living, 257i
Cells, red blood, renewal rate of, 250
Cellulose, 242i
  glucose from, 243
Cellulose acetate, uses of, 227
Celsius
  degree, 35d
  scale of temperature, 36d
Cement kiln, 153i
Cement, Portland, 151
Centrioles, 257i
Cephalexin, 387t, 398s
Cephalosporins, 398
Ceramics, 149
Cesium, 31
  cost of, 26t
Cetyl alcohol, in depilatory, 426
CFC. *See* Chlorofluorohydrocarbon.
CFC-11. *See also* Chlorofluorohydrocarbons.
CFC-11, 205s
CFC-12, 205s
CFC-12. *See also* Chlorofluorohydrocarbons.
CFC-22, 205s
Chadwick, James, 52
  neutron discovery, 71
Chain reaction, 79d, 80i
Chalcopyrite, 142
Chalcosite, 138t
Chalk, 151
  in face powder, 428
Challenger, 11
Charge(s)
  on electron, 51

Charge(s) *(Continued)*
  on proton, 51
  unlike attraction, 91
Charging batteries, 131
Chelate, 282di
Chelating agent, 332d
Cheilosis, 309dt
  riboflavin deficiency, 309t
Chemical Abstracts, 6–7
  CAS ONLINE, 7
  example of, 7i
  growth rate, 6i
  indexes of, 6
  Registry numbers and, 7
Chemical bond(s), 91–108. *See also* Bond(s).
  peptide, 232s
Chemical reaction(s), reversibility of, 112
Chemical energy, 109
Chemical formula(s), 27d
  predicted by ion charges, 95
Chemical formulations, 419
Chemical literature, 6–8
Chemical radicals, and smog production, 370i, 371
Chemical spills
  kepone in James River, 286
  in Rhine River, 353
Chemical symbol, 27d
Chemicals
  top 50 in 1986, 227t
Chemistry, 1d, 18d
  attitudes about, 15
  fields, 2i
Chemophobia, 15
Chernobyl, 11–12, 14, 176, 178i
Chicago, University of, 69
Chinaware, 151
Chinese restaurant syndrome, 315
Chip, 154, 155i
  computer, 8
  invention of, 9
  microprocessor, 8
  original, 9i
  photomicrograph of, 9i
  silicon, 154, 155i
  size of, 8i
  uses of, 8
Chloramine, as corrosive poison, 328t
Chlorate ion, 100
Chlorates, as weed killers, 287
Chlordane, 285t, 286s
Chlordimeform, as insecticide, 285t
Chloride ion(s), 118
Chlorine, 31, 138t, 157, 227t
  as corrosive poison, 328t
  electronic arrangement, 61t
  as a laundry bleach, 436
Chlorine atom, role in ozone layer depletion, 378

Chlorofluorohydrocarbon(s), 205–207
  effect on ozone layer of, 206–207
  role in ozone layer depletion, 378–379
  uses of, 205
Chloroform, 207s
  carcinogen suspect, 207
  how regulated, 325t
  toxicity of, 335
Chlorophyll, 252s
  magnesium and nitrogen requirements of, 273
  why green, 252
Chloroplasts, 252
Chloroprene, as mutagen, 341
Chlorosis, 274d
Cholecalciferol, IU requirement for, 296
Cholesterol, 306t, 407s
  atherosclerotic plaque and, 400, 402
  brains and, 301
  dietary recommendations for, 296
  egg yolk and, 301
  in food, 300
  high-density lipoproteins and, 402
  in lanolin, 427
  low-density lipoproteins and, 402
  lowering of, 402, 403t
Cholestyramine resin, 403s
Cholic acid, 319s
  surface active agent in food, 319
Cholinergic nerves, 337i
Cholinesterase, 337, 338i
Chromate ion, 100
Chromatography, 20i
  paper, 20i
Chromite, 138t
Chromium, as carcinogen, 346
Chromium complexes, in hair dyes, 422
Chromosomes, DNA and, 264
Chronic toxicity, 333d
Churchill, Winston, 60
Cigarette smoke, and carbon monoxide levels in blood, 343
Cimetidine, 392s
Cinnamaldehyde (cinnamon), 212s
Circulatory system, 348i
Cirrhosis of the liver, 211
Cisplatin, 405s
  anticancer drug, 6
  discovery of, 5–6, 6s
  invention incubation period, 3t
Citric acid, 214t
Citrus fruit, 119
Civet, sex attractant in perfume, 429
Civetone, 429s
Clay(s), 149, 150d, 151, 270d
  in crust of Earth, 137, 138t
Clean Air Act, 14, 366
  changes in air pollution since, 366i
Cleaning action of soap, 434i
Cleansing agents, 419, 431

Cloning recombinant DNA, 9
Clover, source of humus in soil, 272
Coal
  air pollution and, 173
  chemicals from, 226–229
  combustion of, 167t
  composition of, 172
  Eastman Kodak plant and, 227–229
  electricity and, 173
  future demand and use, 185
  gasification of, 226–229
  hydrogenation of, 173
  liquefaction of, 173
  mining problems and, 173
  projected depletion of, 226
  pyrolysis of, 226
  requirements in U. S., 173
  reserves in U. S., 173
  sulfur in, 172
  sulfur recovery from, 229
  synthesis gas from, 226–227
  types of, 172t
  uses of, 225t
  world production of, 173i
Coal tar, distillation fractions of, 226t
Coalescing agents
  in paints, 441
Coatings, surface, 419
Cobalt, as teratogen, 344
Cobalt (II) oxide, color in glass, 148t
Cobalt complexes, in hair dyes, 422
Cobalt ions, in lipstick color, 428
Cobalt-60, 83, 84i
Coca Cola, sugar in, 303t
Cocaine, 390st
  crack and, 410–411
Coconut oil, 218t, 433
  in detergent making, 434
Codeine, 389s
Codon, 263d
Coefficient, in chemical equations, 28
Coenzymes, 249d
  minerals as, 249
  vitamins as, 249
Coffee
  flavor of, 315
  pH of, 122i
Coffeemate, sugar in, 303t
Coke, 139, 154, 158
Cold cream, 427
Cold waving of hair, 421
Collagen, 247, 248t, 306t
Collision rates, to explain reaction rates, 113i
Colloidal suspension, in skin creams, 427
Colloids, 19d
  in paints, 441
Color of hair, 422
Colored glass, 148
Coloring hair, 421
Combustion, 124d, 166d, 167t

Combustion (Continued)
  of organic substances, 167t
  of fossil fuels, 166
Committee on Nutritional Misinformation, 301
Common cold, vitamins A and C and, 310
Compost pile, source of humus in soil, 272
Composting, 268
Compound(s)
  covalent, examples of, 91
  ionic, examples of, 91
Computer, information retrieval and, 7
Concentration
  effect on reaction rate, 111, 112i, 113i
  of solutions, 118
Concrete, 151
Condensation polymers. See Polymers, condensation.
Conductance, in electrical solution, 117
Conductivity of water, 120
Conjugated structures, 302d
  food colors and, 317
Connine, 220s
Conservation of matter, Law of, 40d
Constant Composition, Law of, 41d
Contact herbicide, paraquat as, 288s
Cooking, digestion and, 320
Copenhagen, University of, 60
Coplay, Pennsylvania, and first cement kiln, 153i
Copper, 21, 24, 125, 137, 142
  in brass, 19
  cost of, 26t
  used in battery, 129
Copper (I) oxide, color in glass, 148t
Copper (II) sulfate, 127
  algae killer, 99
Copper Basin, Tennessee, 364i
Copper oxide, 125, 142
Copper sulfide, 142
Copper-zinc battery, 129i
Copying, Xerox, 3t
Corgard, 400s, 401
Cori, Carl Ferdinand, 255
Cori, Gerty Theresa, 255
Corn, nutrients required to produce, 277t
Corn oil, 218t
Corneal layer, of skin, 419
Corning Glass Company, 152i
Corrosion, 131
Corrosive poisons, 326–327
Cortisone, 407s
Cosmetics, 419
Cosmic rays, and carbon-14 dating, 69
Cotton, 243i
  hydrogen bonding in, 243i
Cotton fabrics, hydrogen bonds delay drying of, 102
Cottonseed oil, 218t, 433
Cottrell electrostatic precipitator, 367, 368i

Coulomb, 35d, 51
County Extension Office, 272
Covalent bonds, 98–101
   in covalent compounds, 98
   depiction of, 98
   double, 98
   electron dot structures of, 98–102
   examples of, 91
   formation of, 98
   line representation of, 98
   strength of, 98
   triple, 98
Covalent compounds, examples of, 91, 98
Covellite, 142
Crack, 410–411
Creams, skin, 427
Creosote oil, 226t
Cresols, 226t
Cretinism, iodine and, 306t
Crick, Francis H. C., 258, 259i
Critical mass, 79d
Crop yield explosions, 276
Crops
   annual loss of, 283
   exports of, in U. S., 268
   water requirements of, 271
Cross links, disulfide
   in protein, 420, 421s
Crust of Earth, 137
Cryogen, 146d
Cryolite, 141
Cryosurgery, 146
Crystal formation, in batteries, 131
Crystal glass, 150t
Cultivation, acreage used for, 276
Cumene, 227t
Curare, 338
Circuits, integrated, 155
Curie, unit of measure, 73d
Curie, Marie, 46, 255
   personal side, 65
Curium, nuclear reaction for, 74t
Curling hair, 421
Cuticle softeners, 442
Cutin, 320
Cyanate(s)
   ammonium, 188
   silver, 188
Cyanide ion, 330–331
   lethal dose, 325
   mode of toxicity, 330i
Cyclic hydrocarbons, 195–197
Cycloalkanes, 195–196
Cyclobutane, 196s
Cyclohexane, 196s, 227t
Cyclohexene, 196s
Cyclooctyne, 196s
Cyclopentane, 196s
Cyclophosphamide, 404s
Cyclopropane, 195

Cyclosporin A, 414
Cyhexatin
   mite control, 285t
Cysteine, 245st
   disulfide bonds and, 248
Cystine, 245st
   disulfide bonds and, 248
   in hair, 419s
Cytochrome C, 330i
Cytoplasm, 257di
Cytosine, 257s
   nitrous acid effect on, 340s

2,4-D (2,4-Dichlorophenoxyacetic acid), 208t, 287s
   agent orange and, 287
Dacron, 232
Dalton, John, 42i
   atomic theory, 42–45
   Law of Multiple Proportions, 42
Damage
   by neutrons, 74
   caused by radiation, 73–76
Darvon, 389s
Databases
   CAS ONLINE, 7
   DIALOG, 7
   equipment for searching of, 7
Davies, Humphrey, humus theory and, 273
DDT, 207–208, 208s, 284s, 285t
   banning of in U. S., 285
   bird eggs and, 269
   eagles and, 285
   first production of, 284
   half-life of, 284
   ospreys and, 285
de Saussure, 273
de Voeux, Dr. Harold, 367
Death, leading causes of, 395i
Deductive reasoning, 3d
Deer, source of musk, 429
Definitions, operational and theoretical, 22d
Defoamers, in paints, 441
Degree Kelvin, 35d
Dehydration, of food, 312
Delaney Clause, 14
Delphene, insect repellent, 285t
Demerol, 389s
Democritus, 39i
   and atomic theory, 39
Denatured alcohol, 212
Denitrification, 273i, 280d
Denmark, 60
Density, of ice, 354
Dentistry, 333
Denver, Colorado, 359
Deodorants, 426, 430
   types of, 431
Deoxyribonucleic acids. See DNA.

Department of Energy, 87–88
Department of Health and Human Services, 296
   dietary recommendations of, 296
Department of Transportation, regulation of chemicals, 325t
Depilatories, 426
Dermatitis, vitamin B7 (biotin) deficiency, 309t
DES, as teratogen, 344
Designer drugs, 410–413
   China White, 411
   comparison with heroin, 411
   Ecstasy, 413s
   fentanyl derivatives, 411s
   meperidine derivatives, 412s
   MPPP, 412s
   MPTP, 412s
Detergents, 158
   in tooth paste, 439
Deuterium, 56
   nuclear fusion and, 180
   in sea water, 82
   shielder as deuterated water, 175
Dextrins, 242i
   in tooth decay, 439
   use of, 242
Dextrose, 21. See also D-glucose.
Diabetes mellitus, 254
   carbohydrates and, 303
Diagnostics, radioisotopes used in, 86t
DIALOG, 7
Diamond, 21, 24, 103i
Diazinon, as insecticide, 285t
1,2-Dibromoethane. See Ethylene dibromide.
Dichlorodiethyl sulfide
   as mutagen, 341
Dichlorodiphenyltrichloroethane. See DDT.
Dichloromethane, 206t, 207s
   toxicity of, 335
   use of, 206t
2,4-Dichlorophenoxyacetic acid. See 2,4-D.
Dieldrin, 285s
   use of, 285t
Diet
   alcohol recommendation for, 296
   bran and fruit pulp and, 302
   cholesterol recommendation for, 296
   decline in intake of roughage, 302
   dietary fiber recommendation for, 296
   fat recommendation for, 296
   fats in, 218
   muscle from protein, 294
   roughage, functions of, 303
   sodium recommendation for, 296
   starch recommendation for, 296
   sugar recommendation for, 296
Diet recommendations
   Department of Health and Human Services, 296

Diet recommendations *(Continued)*
  United States Department of Agriculture, 296
Dietary Guidelines for Americans, 302
Diethanolamine, 435
Diethyl ether, 206s
  disadvantages of as anesthetic, 212
Diethylene glycol monomethyl ether, 442
Diethylstilbestrol
  as carcinogen, 346
  as teratogen, 344
Digestion, 253–254, 253d
  of carbohydrates, 253
  enzymes and, 253
  of fats and oils, 253, 254
  hydrolysis and, 253
  of proteins, 253
  role of acid in, 119
Digoxin, 387t
Dihydrogen phosphate ions, 123
Dihydroxyindole, in bleaching hair, 425i
Dimethyl sulfate, as mutagen, 341
Dimethylbenzene(s). *See* Xylene(s).
2,4-dimethylhexane, octane rating of, 201t
Dinoseb, mildew and fungi control, 285t
Dioxin(s), 208
  impurity in agent orange, 288
  in Times Beach, 13
Diphenylaminesulfonic acid, dye, hair, 422s
Dipole, 105d
  instantaneous, 105
Dirt, 431d
Disaccharides, 240, 241si
Disasters
  Bhopal, 11
  Challenger, 11
  Chernobyl, 11
Discoveries, methods of, 5
Disinfectants, 398–399
Disorder, 109
  entropy and, 168
Dispersing agents, in paints, 441
Dissociation, ionic, 117i
Distillation, 20i, 21i
Distilled water, conductivity of, 120
Disulfide bonds
  in hair, 421
  in skin and hair, 419–420s
DNA
  base pairing in, 262
  code alteration by nitrous acid, 342i
  code for protein synthesis, 261
  effects of radiation on, 76
  function of, 258
  genes and, 264
  heredity and, 258–264
  hydrogen bonding in, 91
  monomers of, 258i
  primary structure of, 258, 259i
  recombinant, 3t, 264–265

DNA *(Continued)*
  replication of, 259–262, 261i
  secondary structure of, 258, 259i
  transcription of, 261i, 262i
Döbereiner, Johann Wolfgang, 31
Dolomite, 143, 281d
Domagk, Gerhard, 396
Doonesbury cartoon, 412i
Dopamine, 412s
p-Doped semiconductor, 154
Doping silicon, 154
Dose, 325d
DOT. *See* Department of Transportation.
Double bonds, 98
Double helix, 259i, 260i, 261i
Drain cleaner, 119
Drug abuse, 410–413
Drugs
  classification of, 388t
  in combination, 413
  list of widely prescribed, 387t
  top ten in 1986, 387t
Dry cell, 130t
Dry ice, 124
  London forces and, 91
Dubos, Rene, 397
Ducktown, Tennessee, 364i
Duggan, N., 398
Dwarfism, zinc and, 306t
Dyazide, and ten most prescribed drugs, 387t
Dyeing hair, 422, 424i
Dye(s)
  in face powder, 428
  in nail polish, 442
Dynamite, how regulated, 325t
Dyspepsia, carbohydrates and, 303

E. coli
  platinum effect on, 5
  recombinant DNA and, 9
Earth, population of, 268, 269i
Earth's gravity, effect on purification, 160
Eastman Kodak, 227
  chemicals from coal plant, 227–229, 228i
Ecdysone, insect growth regulator, 285t
Ecstasy, 413
Eczema, 302d
EDB. *See* Ethylene dibromide.
Edema, 299d, 307d
Edema, pulmonary, 327d
Edison, Thomas, 5
Edison storage battery, 130t
EDTA
  chelation of iron, 282i
  lead complex, 335i
  metals chelated by, 315i
  in shampoos, 435
Efficiency, 169d
  automobiles and, 255

Efficiency *(Continued)*
  electricity production and, 180
  energy and, 168
  heat engines and, 255
  human body and, 255
Egyptians, and static electricity, 45
Ehrlich, Paul, 395
Einstein, Albert, 168
  equation, 80
Einsteinium, nuclear reaction for, 74t
Elastomers, 234
Electric arc furnace, purifying silicon, 154
Electrical bill, what we pay for, 180
Electrical conductivity, ionic compounds and, 96
Electricity, 179–180
  coal and, 173
  efficiency of production of, 181i
  in magnesium production, 143
  primary energy source of, 180i
  production by coal, 179
  production in U. S., 179
  as secondary source of energy, 179
  static, 45
Electricity transmission, with superconductors, 160
Electrochemical cell, 129
Electrode reactions in electrolysis, 127
Electrolysis, 126d
  of brine, 158i
  in purification of copper, 143
  of water, 128i
Electrolyte(s), 116d
Electromagnets, and superconductors, 160
Electron dot formulas, how to draw, 95
Electron loss, 125
Electron micrograph, scanning, 22i
Electron pair, G. N. Lewis and, 99
Electron volt, 71d
Electron(s), 48d
  charge-to-mass ratio of, 49, 50i
  discovery of, 48–50
  loss of, by metals, 93
  mass of, 51
  metal loss of, 94
  nonbonding, 98d
  nonmetal gain of, 95
  pairing of, 91
  properties, 54t
  stability of pair of, 93
  valence, in bonding, 91
Electronegativity, 100d, 101t
  sorption of water in soils and, 270
Electronic arrangements, 61t
Electronic structures
  aluminum, 94
  fluorine, 94
  magnesium, 93s
  oxygen, 95
  sodium, 93s
Electroscope, 45, 46i, 66i

Electrostatic precipitator, 367, 368i
Element(s), 23d
  free in nature, 24
  in human body, 293, 294t
  new predicted, 160
Electroplating of copper, 127i
Embden, Gustav, 255
Embden-Meyerhof pathway, 255
Emergency planning quantity, 381
Emission spectrum, 57d
Emollient, skin softener, 427
Empire State Building, 91
Emulsification, of oil and water, 254
Emulsifying agents, food additives, 313t, 319
Emulsion(s)
  in skin creams, 427
  in paints, 440i
End note, in perfume, 429
Endothermic, 166d, 253d
Endrin, use of, 285t
Energy, 163d
  analogy of types of, 168
  biochemical systems and, 251-256
  body actions requiring, 255-256
  bond, 166d
  chemical, 109
  coal and, 172-173
  consumption of, in U. S., 165i
  efficiency and, 168
  electricity, 179
  endothermic, 166d
  exothermic, 166d
  expended by activity of human body, 297t
  fate of Sun's, 256
  fertilizer costs and, 279
  fertilizer production and, 276
  food production and, 276
  fundamental principles of, 166
  fusion, controlled, 180
  garbage and, 174
  geysers and, 174
  hot springs and, 174
  kinetic, 163d
  losses in transformations, 168
  natural gas and, 174
  nonmainstream sources of, 174
  nuclear, 78-82
  nuclear, controlled, 175-181
  petroleum and, 171-172
  potential, 163d
  primary sources of, 163, 169d
  prognosis of, 185
  requirements to produce some common products, 170t
  secondary sources of, 163, 169d
  societal use of, 164i
  solar, 181
  solar battery, 182
  sources of, 163
  from Sun, 165
  transformation of, 168

Energy (Continued)
  from trash, 174
Energy changes, 110
Energy of activation, 248, 249i
England, iron ore in, 139
Enkephalins, 390
Enrichment of uranium, 80
Entropy, 168d
Environmental problems, 13
Environmental Protection Agency, 13, 14, 358
  classes of industrial pollutants, 362t
  DDT and, 285
  gasoline blends and, 202-203
  kepone in James River, 286
  lead-free gasoline and, 202
  regulation of chemicals, 325t
Enzymes, 243d, 248d, 248-250. See also specific enzymes.
  action of, 250i
  apoenzyme, 249d
  coenzyme, 249d
  composition of, 249
  function of, 248, 249i
  lock and key analogy, 248, 249i
  meat tenderizers, 254, 320
  number of, 299
  stain removers, 254
  structure of, 248i, 249
P-450 enzymes, 349
Epidermis of skin, 419s
Epinephrine. See Adrenalin.
Epsom salts, plant nutrient, 277t
Equal, 317
Equation, chemical, 27, 28d
Equilibrium
  chemical, 113
  glucose structures and, 239
  in hemoglobin-carboxyhemoglobin reaction, 330
Erosin dyes, 428s
Erythromycin, 387t
Escherichia coli. See also E. coli.
  protein synthesis in, 263, 262i
Essential amino acids, 244d
Essential fatty acid(s), 218-219, 300
  linoleic, 218s
Essential oils, in deodorants, 431
Estradiol, 407s
Ester(s), 215, 216t
  functional group for, 206t
Esterification, protein synthesis and, 263
Estrogen(s), 407-409
  bone dissolution and, 306
  estradiol, 407s
  estrone, 407s
Ethane, 190, 191t, 192i
  bonding in, 99i
1,2-Ethanediol. See Ethylene glycol.
Ethanoic acid. See Acetic acid.
Ethanol, 209s

Ethanol (Continued)
  in alcoholic beverages, 210t
  blood levels of, 211
  in bread making, 319
  breathalyzer test for, 211
  combustion of, 167t
  denaturation of, 212
  detoxification of, 211
  industrial uses of, 212
  gasohol and, 202
  from glucose fermentation, 210
  lethal dose, 325
  metabolism of, 211
  octane rating of, 201t
  oxidation of, 296
  proof of, 210
  synthesis of, 212
  toxicity of, 211
Ethene, 192s, 193. See also Ethylene.
  1,1-dichloro, 193s
  cis-1,2-dichloro, 193s
  trans-1,2-dichloro, 193s
Ether(s)
  functional group for, 206t
  synthesis of, 212
Ethics
  biogenetic engineering and, 265
  genetic engineering and, 11
Ethyl acetate, 206t
  in nail polish remover, 442
Ethyl alcohol. See Ethanol.
Ethyl chloride, 205s
Ethyl ethanoate. See Ethyl acetate.
Ethyl ether. See Diethyl ether.
Ethyl group, 205s
Ethylbenzene, 227t
Ethylene, 227t, 233t
  bonding in, 99i
  in Dalton's atomic theory, 42
  plant growth enhancer, 283
  uses of, 225t
Ethylene dibromide, 207s
  carcinogen suspect, 207
Ethylene dichloride, 227t
Ethylene glycol, 209s, 227t, 233t
  uses of, 212
Ethylene oxide, 227t
  as carcinogen, 346
Ethylenediaminetetraacetic acid. See EDTA.
Ethyne. See Acetylene.
Exercise, muscle growth and, 299
Exothermic, 166d, 253d
Experimental methods, 5
Exports, crops from U. S., 268

Face powder, 428
Fahrenheit, temperature scale, 36d
Families of elements, 31
Famines, 268

Faraday, Michael, 126
Farm Chemical Handbook, 283
Farming
 acreage used for, 276
 no-till, 288d
 organic, 284d
 primitive, 276
Fat(s)
 combustion of, 167d
 daily needs of, 301
 dietary recommendations for, 296
 digestion of, 254
 edible from petroleum, 171
 energy from, 218
 food sources of, 300t
 functions of in human body, 300
 kinds of, 300
 monounsaturated, 301d
 oxidation of, 296, 297t
 oxidation of by atmosphere, 314
 peroxides from heated, 302
 polyunsaturated, 301d
 problems with, 301
 rancidity of, 302
 range in human body, 293
 replacement rate of, 292
 saturated, 217–218, 301d
 in soap making, 431
 unsaturated, 217–219
 use in lipstick, 428
 used to control moisture of skin, 427
Fat cells, number, 301
Fatty acid(s), 213, 217s
 in body odor, 431
 essential, 218, 300
 in paints, 441
 oxidation of, 296, 297t
 precursors of prostaglandins, 301
 saturated, 217–218
 unsaturated, 217–219
Fatty oils, in body odor, 431
Feldspar, 150
Fentanyl, 411s
Fermentation, of garbage to produce methane, 174
Fermi National Laboratory, 72i
Fermi, Enrico, 175
Fermium, nuclear reaction for, 74t
Fernald, Ohio, 87
Fertilizer(s), 158
 chemical, basis for, 269
 complete, 276d
 grade of, 276d
 manure, 276
 meaning of 6–12–12, 276i
 mixed, first production of, 276d
 nitrogen, synthesis of, 280i
 petroleum costs and, 278
 phosphates, synthesis of, 280i
 quick-release, 276d

Fertilizer(s) *(Continued)*
 slow-release, 276d
 straight, 276d
Fiber, dietary recommendations for, 296
Fibrinogen, blood clotting and, 299
Field-effect transistor, 154
Fillers, in soap products, 435
Filtration, 20i
Fingernails, composition, 421
First law of thermodynamics, 167
Fission, 78d, 175d
 of plutonium, 177
 uranium and, 175
Fission reactions, 79–80
Fixed nitrogen, 278d
Flavin adenine dinucleotide (FAD), 249
Flavor enhancers, food additives, 313t, 315
Flavorings
 artificial, 215
 food additives, 313t
Fleming, Sir Alexander, 396, 397i
Flint glass, 150t
Floating soaps, 433
Flotation process, 142i
Flour
 bleaching of, 304
 refined, 304
Flower extract
 in perfume, 429
Fluorescence, in optical brighteners, 437
Fluoride ion(s), in drinking water, 439
Fluorine, 31, 33t, 138t
 electronic arrangement, 61t, 94s
Fluorine-18, 70
Fluoroacetic acid, 331s
Fluoroapatite, 306t, 439s
Folic acid, 406s
 and sulfa drugs, 396
Food
 additives for, 311–320
 canning of, 312
 carbohydrates in, 300t
 dehydration of, 312
 energy from, 300t
 fat in, 300t
 irradiation of, 82, 83t, 84i
 metals in, 314
 plant requirements of, 269
 preservation of, 312
 production of, 268–288. *See also* Agriculture.
 protein in, 300t
 salt in, 307
 source of minerals, 306t
Food additives, 311–320
 anticaking agents, 318
 coloring agents, 317
 flavor enhancers, 315
 flavoring agents, 315
 GRAS list, 311, 313t

Food additives *(Continued)*
 meat tenderizers, 320
 pH control by, 317
 polyhydric alcohols, 319
 preservatives, 312
 purposes of, 312i
 sequestrants, 314
 stabilizers, 318
 suface active agents, 319
 sweeteners, 316
 thickeners, 318
Food and Drug Administration, 294, 311, 414
 limits on toxics in foods, 331
 and sodium nitrite ban, 342
 thalidomide ban, 343
Food and Nutrition Board, 294, 301
Food colors, conjugated systems and, 317
Formaldehyde, 206t, 212, 227t
 uses of, 210
Formic acid, 213, 216t
 in perfumes, 429
Formulas
 chemical, 27d
 chemical, predicted by ion charges, 95
 electron dot, how to draw, 95
Formulations, chemical, 419
Fossil fuels, 171–174, 198–199
 limits on, 168
 organic chemicals from, 224, 225t
Fractional distillation, of petroleum, 199, 200i
Fractionation of air, 144, 145i
France, iron ore in, 139
Francium, 31
Franklin, Benjamin, 45
Frasch process, 156i
Free radical, 310d
Freeze-thaw additives in paints, 441
French Revolution, 40, 42
Freon(s), 205
Fresh Kills (Staten Island, NY), 174
Fire, brush, 124i
Frisch, Otto, 79
D-Fructose, 240si. *See also* Fructose.
 sweetness of, 239
Fruit pulp, dietary fiber and, 303
Fruit(s), odor and flavor of, 215, 216t
Fuel, synthetic, future use of, 185
Fuel cells, 133, 134i, 182
Functional groups, 205d
 listing of, 206t
Fungal protease, meat tenderizer, 320
Fungicide, 284dt
Fungus organisms, in moist skin, 419
Fusion, 78d, 81–82
 nuclear, controlled, 180, 182i
 Sun and, 163

G. D. Searle Co., 317
Gabon Republic, 80

D-Galactose, 224s
Galactose, 240, 241si
Galen, Greek physician, 431
Galena, 138t
Galileo, 40
Gallium, 33
   cost of, 26t
   as teratogen, 344
Gallium-67, 86t
Gallons, barrel equivalences, 171d
Galvanizing, 132d
Gamma radiation, mutations and, 265
Gamma rays, 46t
   penetrating power, 66i
Garbage, energy from, 174
Gas, bottled, 191t
Gases, 22d
Gasohol, 202
Gasoline, 199–203
   additives for, 201–203
   blends of, 202
   combustion of, 167t
   EPA regulations of, 202
   how regulated, 325t
   lead content of, 202
   from methanol, 229–231
   methanol blends with, 202–203
   octane enhancers in, 201
   octane rating of, 201
   straight-run, 199
Gastric juice pH, 122i
Geiger, Hans, 52
Gemini spacecraft fuel cells, 133
Gene(s), 264d
   cloning of, 9, 10i
   recombinant DNA and, 9–11
   splicing of, 9, 10i
Genentech, 10
General Electric Company, patent for production of life, 264–265
Generic name, drug and, 387d
Genetic code, 264
   DNA and, 264
Genetic effects, of radiation, 75
Genetic engineering. See Biotechnology.
Geometric isomerism, 193
Geraniol, 429s
Geranium oil, 429
Geranyl formate, 429
Germanium
   prediction by Mendeleev, 34
   some properties of, 34t
Germinal cell(s)
   effect of mutagens on, 340
Geysers, in Boise, Idaho, 174
Ghana, 299
Gibberellic acid, 283s
Gibberlins
   plant growth enhancers, 283
   plant hormones, 283

Gilfbaar plant, 331
Glass, 147d
   colors in, 148
Glass ceramics, 151
Glazed porcelain, 151
D-Glucose, 239si, 244s. See also Dextrose.
   from amino acids, 299
   from cellulose, 243
   conversion to lactic acid, 255
   metabolism of, aerobic, 255–256
   metabolism of, anaerobic, 255–256
   metabolism, efficiency of, 254
   optical isomers of, 224
   oxidation of, 296
   from photosynthesis, 251
   in polysaccharides, 240
   sweetness of, 239
   source of energy, 239
Glucose phosphate, 254
Glucuronic acid, 349s
Glue solvents, as mutagen, 341
Glutamic acid, 245st
   in ionic protein bond, 420, 421s
Glycerin. See also Glycerol.
   from soap making, 432s
Glycerol, 209s, 216, 217s
   from detergent manufacture, 435
   esters of, 217s
   from fat, 302
   from petroleum, 171
   from soap making, 432s
   uses of, 212
Glycine, 222, 244, 245st
   saccharin and, 316
Glycocholic acid, sodium salt of, 254s
Glycogen, 240, 254
   use of, 242
Glycylalanine, 246s
Glycylglycine dipeptidase, 250i
Glycylglycine, 244s
Gneiss rock, 150
Gofman, John, 402
Goiter, iodine and, 306t
Gold, 21, 24, 125, 127
   14-carat, 42
   catalyst, 229
   color in glass, 148t
   from copper purification, 143
   cost of, 26t
   in crust of Earth, 137
Gold foil experiment, 53i
   alpha particles and, 71i
   neutrons and, 71
Gold oxide, 125
Goodyear, Charles, 234
Gore, Albert, Jr., 15
Goulian, Mehran, 264
Grain alcohol. See Ethanol.
Granite, 150
Grape sugar. See D-glucose.

Graphite, 24
   moderator in atomic reactor, 175
GRAS list of food additives, 311, 313t
Gravity, effect on purification, 160
Greenhouse effect, 382d
Germanium, 33
Growing hair, 421
Ground state, 60d
Groundwater, 270
   leaching of phosphates and, 274
   in water cycle, 361i
Groups of elements, 31
Gruel, 299
Guanine, 257s
   nitrous acid effect on, 340s
Guano, 276
Gutta-percha, 234
Gylceryl tristearate, fat for making soap, 432
Gypsum, 152
   plant nutrient, 277t

Haber process, 159
Haber, Fritz, 278
Hahn, Otto, 79
Hair, 421
   brittle and lifeless, 422
   chemical view of, 419
   curling, 423i
Hair sprays, 424, 425i
Half-life, 67d, 284d
Halide(s), organic, 206t, 207–208
Hall process, 141
Hall, Charles, 140
Halogenated hydrocarbons, 378t
Halogenated solvents, toxicity of, 335
Halogens, 31
   ions of, 94
Hamlet, 353
Hard candy, hydrogen bonds as cause of stickiness, 102
Hard water, 433, 436
Hardness
   ionic compounds and, 96
Hazardous waste(s), 13
   disposal methods, 363t
Hazards, regulation of, 324, 325t
Heart disease, 399–402
   atherosclerotic plaque and, 400, 402
   beta blockers and, 400t
   calcium channel blockers and, 401–402
   deaths from, 399
   discovery of beta blockers, 400
Heat, food generated, 297t
Heat capacity, of water, 354
Heat of chemical reaction, 110
Heat of vaporization, of water, 355
Heat requirements, human body and, 296, 297t

Heat stroke, 308
Heavy metals, toxicity of, 331–335
Helium, 24, 33t
   from alpha particles, 68
   blimps and, 92
   composition of atom, 54
   electronic arrangement, 61t
   liquid, 160
   molecule, 27
   percentage in air, 144t, 145
Helix, double, 259i, 260i, 261i
Hellriegel, nitrogen fixation and, 274
Hematite, 138t
Hemicellulose, 302
Hemlock, poison from, 220i
Hemoglobin
   action on by carbon monoxide, 327–329i
   function of, 248
   illustration of reversible reaction, 112, 114i
Hemolysis, vitamin E deficiency, 309t
Heptachlor, 285, 286s
Herbicides, 284dt, 287–288
   contact, 288
   nonselective, 287
   selective, 287
Heredity, DNA and, 259–264
Heroin, 389s, 390
Hess, G.H., 110
Hevea brasiliensis tree, source of rubber, 233
Hexamethylenediamine, 232s, 233t
Hill, Archibald Vivian, 255
Histamine, 393s
   role in allergy, 393
Histidine, 245st
   wound healing and, 298
Homologous series, 190d
Horizons, soil, 270
Hormone(s), 299d
   female, 407
   male, 407
Horowitz, Jerome, 416
Hot springs, in Boise, Idaho, 174
Hughes, John, 390
Human body
   elements in, 293, 294t
   energy expended by physical activity, 297t
   fat tissue range in, 293
   muscle tissue range in, 293
Human growth hormone, 10
   synthesized by bacteria, 265
Humus, 271d
   friable soil and, 271
Hydrated ions, 117i
Hydrocarbons, 189d
   alkanes, 189–192
   alkenes, 192–193
   alkynes, 193–194
   aromatic, 196–197
   branched chain, 194–195
   cyclic, 195–198

Hydrocarbons *(Continued)*
   nomenclature, 194–195, 197
   saturated, 189d
Hydrochloric acid, 119, 227
   as corrosive poison, 328t
Hydrogen, 24, 33t, 109, 125, 157
   from algae, 181, 183i
   from aqueous electrolysis, 18
   bright-line emission spectrum, 58i
   from catalyzed water, 182
   from coal and steam, 182
   composition of atom, 54
   in crust of Earth, 137, 138t
   electronic arrangement, 61t
   in fuel cells, 133
   hydrogenating coconut oil, 434
   isotopes of, 56
   nuclear fusion and, 180
   production from propane, 278
Hydrogen bond(s), 102–105
   boiling points and, 104
   in cotton, 243i
   cotton fabric drying and, 102
   in DNA, 258, 260i, 261i
   enzymes and, 102
   examples of, 91
   hard candy and, 102
   ice and, 102, 104i
   lanolin and, 102
   between lanolin and water, 427
   in proteins, 102, 246i, 247
   in skin proteins, 419
   sugar with water, 239
   surface active agents in food, 319
   water and, 102–105
Hydrogen chloride, 120, 157
   polar nature of, 102i
Hydrogen cyanide, toxicity, 330
Hydrogen economy, 184
Hydrogen fluoride, polar nature of, 101i, 102i
Hydrogen molecule, 27
Hydrogen peroxide
   in bleaching hair, 425i
   in coloring hair, 423
Hydrogen phosphate ions, 123
Hydrogen sulfide, H_2S, 159
Hydrogen-3, 85t
Hydrogenation
   catalytic, 217
   of fats and oils, 302
   of fats, 217
Hydrolysis, 238d
   of ADP, 253
   of AMP, 253
   of amylopectin, 242i
   of ATP, 253
   in cooking, 320
   digestion and, 253
   of disaccharides, 241i
   of parathion, 286

Hydrolysis *(Continued)*
   of starch, 240, 242i
   of urea, 281
Hydronium ion(s), 119d
   concentration in water, 121
Hydrophilic group
   in detergent, 434s
Hydrophobic group
   in detergent, 434s
Hydroxide ion, 100
   concentration in water, 121
Hydroxides, 119
Hydroxyapatite, 305
Hydroxyl group, 208s
Hydroxyproline, 245st
Hyperglycemia, 254d
Hypertonic solution, 312d
Hypochlorite ion, 100
   as corrosive poison, 328t
Hypoglycemia, 254d, 303
Hypothesis, 3d, 5

Ibuprofen, 391, 392s
Ice
   hydrogen bonding in, 104i
Ice cream
   sugar in, 303t
Imaging
   in nuclear medicine, 85i
Inderal, 400s, 401
   and ten most prescribed drugs, 387t
Indium phosphide
   catalyst for decomposition of water, 182
Indoor pollution, 379–380
Inductive reasoning, 3d
Industrial pollutants
   classes of, 362t
Inert atmosphere, 146
Information retrieval, 7
   CAS ONLINE and, 7
   computer use, 7
   from databases, 7
   DIALOG and, 7
Inosinic acid, 316s
   flavor enhancer, 315
Insect repellent, 285t
Insecticide, 284–287, 284dt
   use or not use, 286
Insects
   number of, 283
Insulators
   R values of, 169
   thermal, 151
Insulin, 302, 316d
   diabetes mellitus and, 303
   function of, 254, 303
   human, 10

Insulin *(Continued)*
   quaternary structure of, 248
   synthesized by bacteria, 265
   zinc and, 306t
Integrated circuit, 115. *See also* Chip.
Interferon, 10
Intermolecular bonds, types of, 91
International System of Units, 35
International Union of Pure and Applied Chemistry, element names, 74t
Intestinal bacteria, source of vitamin B12, 309t
Intestines, toxicity of wastes in, 304
Inventions, incubation period of, 3t
Iodine, 31
   complex with polyvinylpyrrolidone, 399s
   as corrosive poison, 328t
   cost of, 26t
   stains, 438
   thyroid and, 306
Iodine-123, 86t
Iodine-131, 85t
Iodized salt, 307d
Iodophor(s), 398, 399s
Ion exchange, nitrogen cycle and, 273i
Ion formation, periodic table and, 96i
Ion(s), 22, 92d
   formation of, 92–95
   electronic configurations of, 95t
   hydrated, 117i
   polyatomic, 99, 100t
   properties of, 96
   stabilization of, 94
Ionic bond(s)
   examples of, 91
   in hair, 421
   in proteins of skin and hair, 420
   properties of, 92
Ionic compounds
   brittleness of, 96
   electrical conductivity of, 96
   examples of, 91, 92
   hardness of, 96
   properties of, 96
Ionic dissociation, 117i
Ionic solution(s), 116i
Ionization
   of acetic acid solutions, 118
   of ammonia solutions, 118
Iridium
   cost of, 26t
Iron
   anemia and, 307
   chelation of by EDTA, 282
   corrosion, 131
   cost of, 26t
   in crust of Earth, 137, 138t
   plant yellowing and, 275
   role in cytochrome C, 330
   soil acidity and, 271

Iron (II) compounds, color in glass, 148t
Iron (III) compounds, color in glass, 148t
Iron age, 125
Iron ion(s)
   in corrosion, 132
   in hard water, 433
   in lipstick color, 428
Iron oxide, 125, 139, 140, 151
Iron pigments, in hair, 422
Iron stains, 438
Iron sulfide, 142
Irradiation, of food, 82, 83t, 84i, 313
Islets of Langerhans, 303
Isobutane, 192. *See also* Methylpropane.
   properties of, 192
Isobutyl alcohol, 216t
Isobutyl formate, 216t
Isobutyl propionate, 216t
Isoleucine, 245st
Isomer(s), 189d
   geometric, 193
   optical, 221–224
   structural, 190, 192
Isooctane, 201
Isopentyl acetate, 216t
Isopentyl alcohol, 216t
Isopentyl pentanoate, 216t
Isoprene, 233s
Isopropyl alcohol, 227t
Isoptin, 401, 403s
Isotopes, 55d, 56i
Itai-Itai disease, 13
IUPAC (International Union of Pure and Applied Chemistry), nomenclature and, 194
IUs (International Units), 295d

James River, Virginia, chemical spill and, 286
Jello, sugar in, 303t
Jelly making, 318
Joliet, Irene, 255
Joule, 163d
   equivalences of, 164t
   fundamental units of, 166
Joule-Thompson effect, 145
Jupiter, ammonia on, 120
Justice Department. *See* Alcohol, Tobacco and Firearm Branch.

Kane, Pennsylvania, 374
Kaolin, 151
Keflex, *See* Cephalexin.
Kelsey, Frances O., 414
Kelvin degree, 35d
Kepone, 286

Keratin
   hardness and pH, 421
   protein of corneal layer, 419
Kerosene, 174, 199t, 200i
   combustion of, 167t
Ketchup
   sugar in, 303t
Ketoacidosis, 302
Ketone(s), 212–213
   functional group for, 206t
   ketosis, ketonemia, ketonuria and, 301
Ketonemia, 301
Ketonuria, 301
Ketose, 239d
Ketosis, 301
Khorana, Gobind, 264
Kidney stones, 305
Kilby, Jack, 9
Kilocalories
   equivalences of, 164t
Kilogram, 35d
Kilowatt hour(s), 166d
   equivalences of, 164t
Kimax, 150t
Kinetic energy, 163d
Knocking in gasoline engines, 199
Knowledge
   pursuit of, 1
   scientific, 1, 3
Kornberg, Arthur, 264
Kosterlitz, Hans, 390
Krebs cycle, 255
   disruption by fluoroacetate ion, 331
   thyroxine and, 307
Krebs, Hans Adolf, 255
Krypton, 24
Kwashiorkor, 299

L-Dopa, 412s
   Parkinson's disease and, 412
Lacquer, 442
Lactation, 294d
Lactic acid, 214t
   muscle soreness and, 255
   optical isomers of, 223s
Lactose, 240, 241i
Laetrile, FDA and, 414
Lake, coloring material, 428d
Lands, Alonzo, 400
Language of chemistry, 27
Lanolin
   hydrogen bonds with, 102
   skin softener, 427
   use in lipstick, 428
Lanoxin, and ten most prescribed drugs, 387t
Lanthanide series, 35
Lard, 218t

Larvicide, 284dt
Lasix, and ten most prescribed drugs, 387t
Latex paints, 440
Latex. See Rubber.
Latin symbol, 27
Lauric acid, 435
Lauryl alcohol, 435
Lauryl hydrogen sulfate, 435
Lavoisier, Antoine Laurent, 40i, 110
Law of Conservation of Matter, 40d
Law of Constant Composition, 40d
Law of the Minimum
  in agriculture, 268d
  plant nutrients and, 268
Law(s)
  chemical, 4
  natural, 4
  risk-based, 14
  scientific, 3d, 4
  technology-based, 14
Lawes, John, superphosphate and, 281
Lawrencium, nuclear reaction for, 74t
LD50, 325, 326d
  Carbyl, 339
  comparison to human lethal doses, 326t
  Malathion, 339
  paraoxon, 339
  Parathion, 339
  Saran, 339
  Sevin, 339
  Tabun, 339
Leaching, of soils, good and bad effects of, 271
Lead, 138
  in air, 334i
  in batteries, 130
  in beverages, 334i
  cost of, 26t
  EDTA complex, 335i
  equilibrium, 334i
  in foods, 334i
  gasoline and, 202
  in paints, 334–335
  as teratogen, 344
  toxic dangers of, 333–335
  in water, 334i
Lead arsenate, 332t
Lead dioxide, in batteries, 130
Lead glass, 150t
Lead oxide, in glass, 150t
Lead paint, 441
Lead sulfate, in batteries, 130
Lead-206, 67
  half-life, 68t
Lead-210, half-life, 68t
Lead-214
  half-life, 68t
Leavened bread, 319
Lecithin, 300
Lemon juice pH, 122i

Lethane
  insecticide, 285t
Leucine, 245st
Leucippus, 39
Leukemia
  caused by radiation, 75
  treatment of, 406
Lewis, Gilbert Newton
  electron dot formulas and, 95
  electron pair and, 99
Lewisite, 332
Libby, Willard F., 69
Lidocaine, 390st
Liebig, Justus von, agriculture and, 268, 269
Life, patent for production of, 264–265
Life expectancy, 386t
Light
  formation of according to atomic theory, 59i
  polarized, 222
  wave properties, 58–60
Lightning, nitrogen fixation and, 273
Lignin, 302
Like begets like, 259d
Lime, 119, 227t
  absorbent for carbon dioxide, 145
  from calcium hydroxide, 112
  iron depletion in soil by, 275
  from limestone, 114
  in magnesium production, 143
  to neutralize acid rain, 376
  slaked, plant nutrient, 277t
  in soap making, 431
Limestone, 114, 138t, 151
  in iron ore reduction, 139
  sweet soil and, 270
Limeys, derivation of name, 310
Linde process, 145
Linoleic acid, 218s, 219s
  from cottonseed oil, 433
  essential fatty acid, 300
  sources of, 301
Linseed oil, in paints, 440, 441s
Lipid(s), 217, 300d
  combustion of, 167t
Lipman, Fritz Albert, 255
Lipoprotein(s)
  high-density, 402
  low-density, 402
Lipstick, 428
Liquid air, 145
Liquid nitrogen, 146
  composition of, 281
Liquid oxygen, 146
Liquid soap, 433
Liquids, 22d
Liter, 35d
Lithium, 31, 33t
  electronic arrangement, 61t
  as teratogen, 344

Litmus, 119
Liver, 348i-349
  ammonia converted to urea in, 299
  cirrhosis of, 211
  detoxification of pollutants in, 254
  glycogen storage and, 302
  as nutrient bank of the body, 245
  protein replacement in, 261
  synthesis of fat in, 301
Loam, 270
Lock and key analogy
  for enzymes, 248, 249i
London forces, 91, 105i, 145
London, Fritz, 105
Lopressor, 400s, 401
Los Alamos, New Mexico, 60
Los Angeles, California, air pollution in, 366
Lotions, sunlight protection, 429
Love Canal, 13, 359
Lowry, T.M., 120
LOX, 146
LSD, as mutagen, 341
Lubricants, 199t, 200i
Lucite, 233t
Lucretius, 39
Lunar rock, 18i
Lung cancer, and radon in homes, 77
Lye, 119
  in soap making, 431
Lysergic acid diethylamide, as mutagen, 341
Lysine, 245st
  in ionic protein bond, 420, 421s
  removed in milling flour, 304

Macroscopic samples, 30
Mad Hatter, 333
Magnesium, 31, 33t
  in crust of Earth, 137, 138t
  electronic arrangement, 61t
  electronic structure of, 93s
  fertilizers and, 281
  and sea water, 139, 143, 144i
  secondary nutrient in plants, 274
  soil acidity and, 217
Magnesium ion, in hard water, 433
Magnesium silicate, anticaking agent, 318
Magnetic bottles, 180
Magnetic ceramics, 151
Magnetic resonance imaging, 160
Magnetic separation, of a mixture, 19
Magnetite, 138t
Magnets, ceramic, 149
Malathion, 286s, 339s
  use of, 285t
Maleic acid, 214t
Maleic hydrazide, as mutagen, 341
Malnutrition, 293d
Maltase, 248

Maltol, 315s
  flavor enhancer, 315
Maltose, 240, 241si
Manganese, 138t
Manganese (IV) oxide, color in glass, 148t
Mannitol, 319s
  polyhydric alcohol, sweetener, 317, 319
D-mannose, 224s
Manure, 268
Margarine, 217
Marijuana, destruction of fields by paraquat, 288
Marl, 270d
Marsden, Ernest, 52
Mass defect, 80d, 81d
Mass number, 55d
Mass spectrometer, 55i
Mast cells, 393
Matter, 1d, 18
  Law of Conservation of, 40d
  structure of, 29
Mayer, Julius Robert, 168
McCollum, E. V., 293
McGill University, 46
McMillian, E.M., 73
MDA, 413s
MDMA, 413s
Measurement, 35
Meat, salted, 312
Meat tenderizers, 320
  protein enzymes as, 254
Medfly, 339
Medicine, cost of, 386
Meiosis, 259d
Meitner, Lise, 79
Melanin
  manganese, skin pigment and, 306t
  pigment in hair, 422
  in skin tanning, 430
Mendeleev, Dmitri, 31
Mendelevium, nuclear reaction for, 74t
Menopause
  BMR and, 298
  calcium and, 306
Menstranol, 408s
Meperidine, 389s, 412s
Meprobamate, 387t
Mercury, 157
  cost of, 26t
  dangers of, 332
  in food chain, 333
  as teratogen, 344
Mercury battery, 130t
Mercury oxide, and the Law of Conservation of Matter, 40
Mercury poisoning, 13
Meta-, 197d
Metabolic poisons, 327–336
Metabolism of glucose. See Glucose, metabolism.

Metabolite, 330d
Metal ores, 124
Metallic bonding, 91, 105
Metalloids, 35
Metallurgy, 139d
Metals
  catalysts for fat oxidation, 314
  chelated and sequestered by EDTA, 315i
  in periodic chart, 30
  properties of, 105
Metastases, 345d
Meteors, 160
Meter, 35d
Methanal. See Formaldehyde.
Methane, 124, 189, 190i, 191t
  combustion of, 167t
  in Dalton's atomic theory, 42
  from garbage, 174
  hazards, 324t
  in magnesium production, 143
  molecular structure of, 103i
  oxidation of, 166
  percentage in air, 144t, 145
Methanol, 206t, 209s, 227t
  as fuel, 230
  lethal dose, 325
  gasoline from, 229–231
  octane rating of, 201t
  synthesis of, 227
  from synthesis gas, 210, 227
  uses of, 210
Methionine, 245st
  amino acid required in greatest amount, 299,
Methotrexate, 406s
  treatment of leukemia and, 406
Methyl alcohol. See Methanol.
Methyl bromide, insecticide, 285t
Methyl ethyl ketone, 213s
Methyl group, 194d
Methyl isocyanate, 286s
  Bhopal, India and, 286
  properties, 381s
  toxicity of, 11, 381
Methyl mercury, 13
Methyl propyl ether, 212
Methyl t-butyl ether, 227t
2-Methyl-2-propanol, 209s. See also Tertiary butyl alcohol.
Methyl-tertiary-butyl ether, 202
Methylamine, 206t, 220s
Methylene chloride. See Dichloromethane.
Methylmethacrylate, 233t
Methylpropane, 192s
  properties of, 192
Metoprolol, 400s, 401
Metric system, 35
  prefixes, 35
Meyer, Lothar, 31
Meyerhof, Otto Fritz, 255

Microballoons, nuclear fusion and, 180–181
Microcurie, unit of measure, 73d
Microfibril, in cotton, 243
Micronutrients
  applications of, 282
  difficulties in using, 282
Microprocessors, 155
Microscopic samples, 29, 30i
Microwave oven, chip and, 8
Middle note, in perfume, 429
Milk
  as emulsion, 427
  lactose in, 240
  pH of, 122i
Milk of magnesia solution, pH of, 122i
Millikan, Robert Andrews, 50–51
  oil drop experiment, 50i
Minamata Bay, 13
Mine drainage, 119
Mineral oil
  as carcinogen, 346
  in shampoos, 435
Mineral spirits, in paints, 440
Minerals
  in brown sugar, 304
  as coenzymes, 249
  deficiency effects of, 306t
  in food, 305d, 305–308
  functions of, 306t
  in molasses, 304
  RDA for, 295t
  removal during purification, 293
  removed by refining sugar and flour, 304
  sources of, 306t
  USRDA for, 295t
Minnesota, iron ore in, 139
Minoxidil, hair growth agent, 424
Mitochondria, 257i
  oxidation of sugars in, 254
Mitosis, 259d
Mixture(s), 19
  heterogenous, 19
  homogeneous, 19
Moisture, in healthy skin, 427
Moisturizer, for skin, 427
Molar solutions, 118
Molarity, 118d
Molasses, minerals in, 304
Mole, 27d, 35d, 110
Molecular solution(s), 116i
Molecular structures
  angular, 103i
  tetrahedral, 102, 103i
  triangular pyramidal, 103i
Molina, M.J., 379
Molluscicide, 294dt
Mono-, 238d
Monatomic molecules, 27
Monomer(s), 231

Monosodium glutamate (MSG), 315, 316s
   flavor enhancer, 315
Monosodium methanearsenate, 332t
Monosaccharides, 238, 239
Montreal, Canada, 46
Moon, 160
Morphine, 220s, 389s
   lethal dose, 325
Mother of vinegar, 214
Motion, perpetual, 168
MPPP, 412s
MPTP, 412s
MTBE. See Methyl-tertiary-butyl ether.
Mulder, G. T., 244
Multiple Proportions, Law of, 42d
Muriate of potash, 281d
Muriatic acid, 326
Murray, James, 281
Muscle
   exercise and, 299
   protein or exercise, 294
   range in human body, 293
   soreness, 255
Musk, in perfume, 429
Mustard gas, as mutagen, 341
Mutagens, 340–342
Mutation, 265d
   effects of radiation on, 76
   gamma radiation and, 265

n-Butane, 190, 191t, 192s
   properties of, 192
n-Butyl alcohol, 216t
n-Decane, 191t
n-Doped semiconductor, 154
n-Heptane, 191t
n-Hexane, 191t
n-Nonane, 191t
n-Octane, 191t
n-Pentane, 191t
   octane rating of, 201t
Nadolol, 400s, 401
Nail polish remover, 442
Nail polish, 442
Nails, animal, chemical view of, 419
Naphthalene(s), 197s, 226t
2-Naphthylamine, as carcinogen, 345s
1-Naphthylamine, 345s
Nashville Thermal Transfer Corporation, 174, 175i
National Academy of Sciences, 301
National Aeronautics and Space Administration, 380
National Research Council, 301
Natural gas, 174, 198–199
   bond energy of, 166
   cubic feet equivalences of, 164t

Natural gas (Continued)
   in magnesium production, 143
   projected depletion of, 226
   world reserves of, 174
Natural gas wells, 145
Nebulae, Orion, 22i
Nematicide, 284dt
Neon, 24, 33t
   electronic arrangement, 61t
   isotopes of, 55, 56i
   percentage in air, 144t, 145
Neosporin, 399
Neothyl. See Methyl propyl ether.
Nephrotoxins, 335d
Neptunium
   discovery of, 73, 74t
   nuclear reaction for, 74t
Neurotoxins, 336–340
Neutralization, between acids and bases, 119, 120
Neutron, 51
   discovery of, 52
   in fission reactions, 79
   mass of, 52
   properties, 54t
New Organon, 40
New Zealand Synthetic Fuels Co., 230
Niacin, 387t, 403st. See also Vitamin B3.
   cholesterol lowering, 402
   coenzyme, 249
   prevents pellagra, 249
Niagara Falls, New York, 359
Nickel ions, in lipstick color, 428
Nickel-cadmium (NiCad) battery, 130t
Nicolar, 403s
Nicotinamide, 221s
Nicotinamide adenine dinucleotide (NAD+), 249
Nicotine, 220s, 340
Nicotinic acid, 387t. See also Niacin.
Nifedipine, 401, 403s
Night blindness, 309
Niobium alloys, as superconductors, 160
Nitrate ion, 100
Nitrates, ammonia production of, 279
Nitric acid, 119, 227t
   nitrogen fixation and, 273
   how regulated, 325t
   synthesis of, 280
Nitric oxide
   formation by combustion, 372
   formation by lightning, 372
   nitrogen fixation and, 273
Nitrification, 273i, 280d
Nitrite ion, 100
Nitro compound(s), produced in hair coloring, 423
Nitrocellulose, in nail polish, 442
Nitrogen, 21, 24, 33t, 109, 138t, 227t
   in air, 19

Nitrogen (Continued)
   electronic arrangement, 61t
   fertilizers and, 278–281
   fixation by Haber process, 278i
   fixed, 278d
   liquid, 146, 160, 281
   percentage in air, 144t, 145
   primary nutrient for plants, 273
   protein and, 299
   quantity above each acre, 273
Nitrogen cycle, 273i, 374i
Nitrogen dioxide, 372
   as corrosive poison, 328t
   effect on vegetation, 374
   nitrogen fixation and, 273
Nitrogen fixation, 273
Nitrogen mustard(s), 404
Nitrogen-14, and carbon-14 dating, 69
Nitrogenous bases, 257si
Nitroglycerine, 387t
Nitrophenols, 287s
   organic herbicides, first selective, 287
Nitrosyl chloride, as corrosive poison, **328t**
Nitrous acid
   as mutagen, 340–341
   nitrogen fixation and, 273
Nobel prize
   Bohr, Niels, 59
   Curie, Marie, 65
   Roentgen, Wilhelm, 65
   Rutherford, Ernest, 54
Nobelium, nuclear reaction for, 74t
Noble gases, 24
   as basis for bonding, 92
   compounds of, 92
Nomenclature, organic, 194–195, 197
Nonbonding electrons, 98d
Nonelectrolyte(s), 116d
Nonmetals
   gain of electrons by, 94
   in periodic chart, 30
Nonpolar bonds, 100d
Nonpolar molecules, emulsifiers and, **254**
Nonspontaneous reaction(s), 109
Norethindrone, 408s
North Pole, 379i
Northwestern States Company, 153i
Novocain, 390st
Noyce, Robert, 9
No-till farming, 288d
Nuclear age, beginnings, 65
Nuclear energy, 78–82
   controlled, 175–181
   invention incubation period, 3t
Nuclear medicine, 85d, 87d
Nuclear meltdown, 178i
Nuclear particle accelerators, 72, 160
Nuclear power plant, 176i
Nuclear reactors
   breeder, 177, 179i

Nuclear reactors *(Continued)*
   location in U. S., 177i
Nuclear wastes, 87–88, 137
Nucleic acid(s), 256–265
   condensation polymers, 238, 256
   as esters, 258
   synthetic, 264–265
Nucleotide, 258si
Nucleus
   discovery of, 52–54
   size of, 54
NutraSweet, 317
Nutrients. *See also* Soil, nutrients in.
   corn requirements of, 277t
   plant, chemical sources for, 277t
   proteins, 298–299
   RDA requirements, 294, 295t
   USRDA requirements, 294, 295t
Nutrition, 292–320, 292d
   anatomical and physiological differences in requirements for, 293
   balance and, 307
   dietary recommendations for, 296
   evolvement of, 292
   modern setting of, 292
Nutritional status, 293d
Nylon, 232, 233t
   invention incubation period, 3t

Oak Ridge, Tennessee, 87
Obesity
   carbohydrates and, 303
   fat intake and, 301
   skinfold test for, 301
Occupational Safety and Health Administration, permissible exposure limits, 380t
Octane enhancers, 201–202
Octane rating, 201
Octet rule, 98d
Oil
   from coal tar, 226t
   used, how regulated, 325t
Oil drop experiment, 50i
Oil shale rock, 171
Oil spills, bacteria that eat, 264
Oil(s) (glyceryl esters), 217s, 218t
   cooking, 217
   digestion of, 254
   hydrogenated, 217
   problems with hydrogenation of, 302
   vegetable, 217–218
Oil, crude. *See also* Petroleum.
   California crude, 199
   Pennsylvania crude, 199
Oleic acid, 213, 214t, 433
   from bone grease, 433
Oligo-, 238d

Oligonucleotide, 258si
Oligosaccharides, 238, 239, 253
Olive oil, 218t, 433
Operational definitions, 22d
Opium, 389
Optical brighteners, 437i
Optical isomerism, 221d
   measurement of, 223i
Optical isomers. 221–224. *See also* Isomer(s), optical.
   amino acids and, 244
   biological properties and, 222
   galactose and glucose, 240
Oral contraceptive(s), 408–409
   health risks associated with, 408
Orbit
   of electron in atom, 57
   maximum number of electrons in, 59
Orbitals, 62d
Order, 109
Ores, 137d
Organic acid(s), 213. *See also* Carboxylic acid(s).
   salts of, 214
Organic chemicals
   as air pollutants, 376–377
   uses of, 224–234
Organic chemistry, 188d
Organic farming, 284d
Organic peroxides, and smog production, 368, 370i
Orion nebula, 22i
Orlon, 233t
Ortho-, 197d
Ortho-novum
   oral contraceptive, 408
   and ten most prescribed drugs, 387t
Orthophosphoric acid, 274
OSHA. *See* Occupational Safety and Health Administration.
Osmosis, 312d
Ovaries, effects of radiation on, 75
Ovicide, 284dt
Oxalic acid, 214t
   stain remover, 438
Oxidation, 124d, 125d, 251d
   alanine, 296
   atmospheric, of food, 314
   fats, 296, 297t
   fatty acids, 296, 297t
   palmitic acid, 296
   photosynthesis and, 251
   proteins, 296, 297t
Oxidation dyes, in hair coloring, 422
Oxidation of iron, 131, 132i
Oxidation theory, 125
Oxidation-reduction reactions, 109
   of nitrogen, 273
Oxidizing agent, 126d
   in curling hair, 422

Oxygen, 24, 33t, 109, 227t
   in air, 19
   in atmosphere, 109
   from aqueous electrolysis, 18
   in crust of Earth, 137, 138t
   electronic structure of, 61t, 95s
   in fuel cells, 133
   liquid, 146
   percentage in air, 144t, 145
   in soil, 270
   in steel making, 140
Oxygen gun, 141i
Oxyhemoglobin, 112, 114i
Oyster shells, 143
Ozone
   as corrosive poison, 328t
   health effects, 377
   as mutagen, 341
   reaction with silver, 377
   as secondary air pollutant, 377
   in stratosphere, 378–379
Ozone layer, 378
   chlorofluorohydrocarbons and, 206–207
   over the poles, 379i

PABA, 430
Pacemaker, 87
Paducah, Kentucky, 87
Paints, 440
Palm oil, 218t, 433
Palmitic acid, oxidation of, 296
PAN, 370i
Pancreas, 303
   enzyme source, 253
Papain, meat tenderizer, 320
Paper chromatography, 20i
Para-, 197d
Paraffin(s), 199t, 200i
   and the discovery of the neutron, 52
   solid sinks in liquid, 102i
   use in lipstick, 428
Pataoxon, 339s
   by bioactivation, 350i
Paraquat, 288s
   contact herbicide, 288
   marijuana destruction by, 288
Parathion, 286s, 339s
   bioactivation of, 350i
Paris green, 332t
Parkinson's disease, MPTP and, 412
Particulates, 367d
Parts per million, 328d
Pasteurization, 313
Patent, for production of life, 264–265
Pauling, Linus, 247i
Paycheck, analogy to energy, 168
PCB(s), 208
   as teratogen, 344

Peaches, sugar in, 303t
Peanut butter, sugar in, 303t
Peanut oil, 218t
Peat, from humus, 271
Pediculicide, 284dt
PEL. *See* Permissible Exposure Limit.
Pellagra
   niacin and, 249
   niacin deficiency, 309t
Penicillin, 396, 397i
Penicillin G potassium, 387t
Penicillin G, 396s
Penicillin VK, 387t
Pentanoic acid, 216t
Pentazocine, 389s
1-Pentene, octane rating of, 201t
Pepsin, 254
Peptide bond, 232s, 244d, 245s
Peptide link
   action on by corrosives, 326
   action on by sodium hydroxide, 327
Perborates, oxidizing agents in curling hair, 422
Perchlorate ion, 100
Percolation, in soils, 271d
Perfume(s), 429
   in deodorants, 431
   in face powder, 428
   use in lipstick, 428
Periodic table (chart), 30d, 32i
   by Mendeleev, 34i
   electronegativity and, 101i
   ion formation and, 96i
Periods, in periodic table, 34
Permeability, of cell wall, 336d
Permanent dyes, for hair coloring, 422
Permissible Exposure Limit, 380dt
Pernicious anemia, 306t
Peroxide ion, as corrosive poison, 328t
Peroxides
   heated fats and, 302
   oxidizing agents in curling hair, 422
Peroxyacyl nitrate, 370i
Perpetual motion, 168
Peruvian guano. *See* Guano.
Pesticides, 284d
   classes of, 284t
   number of, 284
   persistent, 285d
   types of, 284t
Petrochemicals, 224–226
Petroleum
   California crude, 199
   combustion of, 167t
   composition of, 171, 172
   conservation of, 171
   consumption of by the U. S., 171
   as emulsion, 427
   fats from, 171
   fertilizer costs and, 278
   fractions of, 199t

Petroleum *(Continued)*
   future demand and use, 185
   glycerol from, 171
   Pennsylvania crude, 199
   projected depletion of, 226
   recoverable, 171
   refining of, 171, 199–201
   uses of, 225t
   world use of, 172i
pH, 121d
   of acid rain, 374
   of common materials, 122i
   control of, in food, 317
   control with buffers, 123
   of depilatory, 426
   scale, 122t
   of soil, 270
   phosphates and, 274i
pH controllers, in paints, 441
Phenobarbital, 387t
Phenol, 208, 227t, 266t
   uses of, 225t
Phenylalanine, 245st
Phenylenediamine, dye for hair, 422s
Phosgene, 327s
   as corrosive poison, 328t
Phosphate ion, 100
Phosphate ore, 158
   conversion to superphosphate, 281
Phosphate runoff, pollution and, 269
Phosphates
   availability in soil, function of pH, 274i
   plant yellowing and, 275
   in soils, 270
   uses of in food, 319
Phospholipids, 300
Phosphoric acid, 158, 22t, 258
Phosphorus, 33t, 138t, 158
   cost of, 26t
   electronic arrangement, 61t
   fertilizers and, 281
   insecticide, 286
   plant tissue and, 274
   primary nutrient for plants, 273
Phosphorus-32, 85t
Photon, chlorophyll and, 252
Photosynthesis, 109, 251–253
   dark reactions of, 252
   energy producer, 166
   light reactions of, 252
   oxidation-reduction and, 251
   reactions of, 251
   stentor coeruleus and, 251
   Sun and, 163
Photovoltaic device. *See* Solar battery.
Physical activity, energy expended by, 297t
Pica, 335d
Pigment(s)
   in hair, 422
   in paints, 440
Piketon, Ohio, 87

Pimples, 119
Pine oil, in cleaning products, 436
Piscicide, 284dt
Placenta, and teratogens, 343
Planck, Max, 58
Planets, source of new materials, 160
Plant growth enhancers
   acetylene, 283
   auxins, 283
   ethylene, 283
   gibberlins, 283
Plant nutrients
   chemical sources for, 277t
   law of the minimum and, 268
Plants
   annual loss of, 283
   fundamental food requirements of, 269
   growth enhancers, 283
   legumes, 273i
   protection of, 283–288
   protection from insects, 283
   protection from weeds, 283
   yellowing of, 275
Plaque
   atherosclerotic, 302
   sugar and, 304
   on teeth, 304, 439
Plasma, 22d, 82d, 180d
Plasmids, recombinant DNA and, 10i
Plasticizer(s), 208d
   in hair sprays, 424
Plastics, 231. *See* Polymers.
Platinum, 24
   as catalyst, 229, 230i
   catalyst in nitric acid production, 280
   catalytic converter in cars, 202
   cost of, 26t
   effect on E. coli, 5
Plato, 40
Plexiglas, 233t
Plutonium
   discovery of, 73, 74t
   fission of, 177
   nuclear reaction for, 74t
   toxicity of, 177
Plutonium-238, 87
Plutonium-239, half-life, 73
Poison detoxifier, vitamin E, 310
Poisons, anticholinesterase, 286
Polar bear liver, retinol content, 309
Polar bonds, 100d
   in hydrogen fluoride, 101i, 102i
Polar groups, on B vitamins, 308
Polar molecules
   bile salts and, 254
   emulsifiers and, 254
   sorption of water in soils and, 270
   surface active agents as, 319
Polarized light, 222
Pollen, distribution of, in U.S., 394i
Pollutants, detoxified in the liver, 254

Polonium-210, half-life, 68t
Polonium-218
  half-life, 68t
Poly-, 238d
Polyacrylonitrile, 233t
Polyatomic ions, 99, 100t
Polychlorinated biphenyls. See also PCB(s).
  as teratogen, 344
Polyester, 232, 233t
Polyethylene(s), 91, 231–232, 233t. See also Polymers, addition.
Polyhydric alcohols, food additives, 313t, 319
Polyisoprene
  cis, 234s
  trans, 234s
Polymer(s), 231–235
  addition, 231–232, 233t
  cellulose, 238
  condensation, 232, 238
  glycogen
  natural, 231
  nucleic acids as, 256
  protein, 238
  starch, 238
Polynucleotides
  function of, 258
  primary structure of, 258, 259i
  secondary structure of, 258
Polypropylene, 233t
Polysaccharides, 238, 240, 242–243
  glycogen, 240
  starch, 240
Polystyrene, 233t
Polyunsaturated fats, 218t. See also Unsaturated fats.
Polyuria, 335d
Polyvinyl chloride, 233t
Polyvinylacetate
  in hair sprays, 424
  in paints, 440
Polyvinylpyrrolidone, 398
  complex with iodine, 399s
  in hair sprays, 424
Population, of Earth, 268, 269i
Porcelain, 151
Portland cement, 151
Potable water, 356d
Potash, 277t
  fertilizers and, 281
Potassium, 31
  cost of, 26t
  in crust of Earth, 137, 138t
  electronic arrangement, 61t
  fertilizers and, 281
  function in living matter, 274
  primary nutrient for plants, 273
Potassium chloride
  fertilizers and, 276
Potassium hydroxide
  in cuticle softeners, 442
  in fuel cells, 134

Potassium ions in solution, 130
Potassium stearate, in vanishing cream, 427
Potassium sulfate, 130
Potassium/sodium ratio, 307
  in human body, 307
Potential energy, 163d
Potentiation, food and, 315
Pottery, 150
Powder, face, 428
Power plant, nuclear, 176i
Power transmission, with superconductors, 160
Power, 166d
Precipitator, electrostatic, 367, 368i
Predicide, 284dt
Prednisone, 387t
Pressor amines, 413
Primary battery, 130d
Primary pollutants, 370d
Primary structure, of proteins, 247
Procaine, 390st
Procardia, 401, 403s
Progesterone, 407s
Proline, 245st
Prontosil, 396
Propane, 190, 191t, 192i
  production of hydrogen from, 278
1,2,3-Propanetriol. See Glycerol.
1-Propanol, 209s
2-Propanol, 209s. See also Isopropyl alcohol.
  uses of, 212
Propanolol, 400s, 401
Propanone. See Acetone.
Properties
  examples of, 91
  of selected elements, 31t
Propionic acid, 213t, 216t
Propoxyphene, 389s
Propylene, 227t, 233t
  uses of, 225t
Propylene oxide, 227t
Prostaglandin(s), 219
  E1, 219s
  E2, 219s
  fatty acids, precursors of, 301
  functions of, 301
Protactinium-234, 67
  half-life, 68t
Protective colloids, in paints, 441
Protein(s), 244–248
  amount per human body, 299
  calcium binding, 305
  combustion of, 167t
  condensation polymers, 238
  daily needs of, 299
  digestion of, 253
  fibril structure of, 247, 248i
  food sources of, 299
  function of, 244, 298
  globular structure of, 247, 248i
  helical structure of, 246i

Protein(s) (Continued)
  high quality, 299d
  lifetime of, 260
  loss of, 299
  low quality, 299d
  muscle tissue and, 294, 299
  natural synthesis of, 260–264
  origin of name, 244
  oxidation of, 296, 297t
  percentage in food, 300t
  primary structure of, 247
  problems with, 299
  quaternary structure of, 248
  RDA for, 295t
  replacement of, 260
  secondary structures of, 247
  sheet structure of, 247i
  sulfur in, 244
  synthesis of, 262i
  synthesized by bacteria, 265
  tertiary structures of, 247
  USRDA for, 295t
  why so many, 246
Protein bridges, in skin, 419
Protium, 56
Proton
  discovery of, 51
  mass of, 51
  properties of, 54t
Proton, (hydrogen ion), in acid-base neutralization, 120
Proust, Joseph Louis, 41
Public Health Service, 376
Pulmonary edema, 327d
Pumice, in abrasive cleaners, 436
Pure Food and Drug law, 292
Pure substance(s), 19, 20d
  operational definitions, 22
  theoretical definition, 23
Pure water, conductivity of, 120
Purification methods, 20i
Putrescine, 220s
PVC. See Polyvinyl chloride.
Pyrethrum, insecticide, 285t
Pyribenzamine, 394s
Pyridine, 220s, 226t
Pyrex, 150t
Pyroceram, 151, 152i
Pyrolusite, 138t
Pyrolysis, of coal, 226
Pyrosulfuric acid, 157
Pyruvate, from glycerol, 302

Quad, 163d
Quaker faith, 42
Quantative energy changes, 110
Quantative weight changes, 110
Quantum, 57d

Quartz, 147i
Quaternary structure, in proteins, 248
Questran, 403s
Quinidine sulfate, 387t

R value, 169d
Rad, 75d
Radiation
  average dose in the United States, 76t
  damage by, 74-76
Radiation damage, 73-76
  factors governing, 73
Radioactive carbon, as tracer in equilibrium study, 115
Radioactive decay series, 67, 68t
Radioactive nuclear wastes, 137
Radioactivity
  natural, 46
  useful applications, 82-87
Radiocarbon dating, 70t
  assumptions made, 69
Radioisotopes, in medical diagnostics, 86t
Radium, 31
Radium-226, half-life, 68t
Radon
  in homes in the United States, 77i
  production, 76
Radon-222, half-life, 68t
Ragweed pollen, 392
Rain, acid, 119
Raleigh, Sir Walter, 338
Rancidity, from heated fats and oils, 302
Rate-determining step, 69d
Raw materials, 137
RCRA. See Resource Conservation and Recovery Act.
RDA (Recommended Dietary Allowances), 294, 295t
Reaction rates, 110, 111d
Reaction(s)
  acid-base, 109
  nonspontaneous, 109
  nuclear, 66-68
  oxidation-reduction, 109
  spontaneous, 109
Reasoning, deductive, 3d
Recharging batteries, 131
Recombinant DNA, 264-265
  cloning of, 9
  E. coli and, 9-11
  invention incubation period, 3t
  process for, 9-10
  uses of, 10-11
Recrystallization, 20i
  KCl and NaCl, 28l
Red blood cells, renewal rate of, 250
Redox reactions, 125

Reducing agent, 126d
  in curling hair, 421
Reduction, 124d, 126d, 251d
  photosynthesis and, 251
Refined flour, 304
Refined sugar, 302-304
Registry number, 7
Regulation, of chemicals hazards, 324, 325t
Removing hair, 421
Replication, DNA and, 259-262
Research
  applied, 2
  basic, 1
Reserpine, 387t
Resin(s)
  in hair sprays, 424
  in perfume, 429
Resource Conservation and Recovery Act, 358
Retinal equivalent, 296d
  vitamin A requirement for, 296
Retinol, 308s
Retrovirus, AIDS and, 415-416
Reversibility of chemical reactions, 112
Reye's syndrome, 391
Rhine River, 353
Rhodanase, 331
Rhodium
  catalyst, 229
  cost of, 26t
Rhodopsin
  night blindness and, 309
Riboflavin. See also Vitamin B2.
  as coenzyme, 249
  function of, 250
Ribonucleic acids. See RNA.
Ribosome
  protein synthesis and, 262
Rickets
  vitamin D deficiency, 309t
Risk(s)
  assessment of, 14
  biogenetic engineering and, 265
  laws and, 14
  management of, 14
  of technology, 11-15
  technology-based laws and, 14
RNA
  base pairing in, 262
  messenger, 257, 261, 262i, 264
  monomers of, 258i
  ribosomal, 257
  transfer, 257, 262i, 263, 264
  types of, 257
Rock
  age of, 68-69
  lunar, 18i
Rodenticide, 284dt
Roentgen, unit of measure, 74d
Roentgen, Wilhelm, 65

Roland, F.S., 379
Rosenberg, Barnett, 5
Rosewater, in cold creams, 427
Rosin, in laundry products, 435
Roughage
  decline of dietary intake of, 302
  dietary, functions of, 303
Rubber
  natural, 233-234, 234s
  synthetic, 234
  typical formulation in tires, 235t
  vulcanized, 234
Rubidium, 31
Ruby, 91
Running, energy expended by, 297t
Russia, iron ore in, 139
Rust, 132
Rutherford, Ernest, 52, 54i
  atomic dating and, 68
  first artificial nuclear change, 70
  and naming natural radioactivity, 46
  and the nuclear atom, 53, 54
  and radiation, 65
Rutile, 138

Saccharin, 316s
Safe Drinking Water Act, 14
Safflower oil, 218t
Saffrole, 315s
  banned by FDA, 315
Salad dressing
  emulsified by carrageenin, 318i
Salt, 21
  blood pressure and, 308
  derivation of name, 308
  food content of, 307
  iodized, 307d
Salt bridge, 129i
Salt solution, 115, 116i
Salt tablets, 308
Salted meat, 312
Samples
  macroscopic, 30i
  microscopic, 30i
  submicroscopic, 30i
Sand, 150, 151
  in crust of Earth, 137, 138t
Saponification, 431
Sapphire, 91
Sarcoma, 344d
Sarin, 339s
Saturated fats, 217-218
Saylor, David, 153i
Scandium, 33
  electronic arrangement, 62
Scanning electron micrograph, 22i
Scattering, of alpha particles, 53i

Index   483

Science(s), 1d
　applied, 2d
　basic, 1d
　classification, 2i
　fields, 1
　image, 2
　interdisciplinary, 1
　Latin derivation, 1
　natural, 1
　policy, 14–15
Scientific discoveries, 2
Scientific fact, 4
Scientific literature, 6
Scientific method, 3
Scurvy, vitamin C deficiency, 309t
Sea water
　energy in, 82
　pH of, 122i
　source of magnesium, 143
Searles Lake, California, first potash production, 281
Sebaceous glands, 419
Sebum, 419
Second law of thermodynamics, 168d
Second, 35d
Secondary battery, 130d
Secondary pollutants, 370d
Secondary structure
　of proteins, 247
Selenium
　cost of, 26t
　in glass, 150t
　leaching and, 271
　poison, 271
　in soils, 271
Semiconductor(s), 154d
　chip and, 9
　n-type, 183i
　p-type, 183i
　solar battery and, 182
　transistors as, 8
Semimetals, 35
Separates
　in soil, 270
Sequester, 314d
Sequestrants, food additives, 313t, 314
Sequoia trees, age of, 69
Serendipity, 5
Serine, 245st
Sesamex, synergist for insecticides, 285t
Set point weight, 301d
Sevin, 339s
　use of, 285t
Sexual impotency, molybdenum and, 306t
Shake'N Bake, sugar in, 303t
Shakespeare, 353
Shale, 151
Shale oil, future use of, 185
Shampoos, 435
Shark skin, riboflavin deficiency, 309t

Shaw, Sir Napier, 365
Sick-building syndrome, 380
Sickle cell anemia, 248
Silica, 147, 151, 158
　in abrasive cleaners, 436
Silia gel, 145
Silicate ion, 100
Silicate rocks, 137
Silicates, 148s
　soil and, 270, 272
Silicon, 33t
　central element in soil, 272
　cost of, 26t
　in crust of Earth, 137, 138t
　electronic arrangement, 61t
　materials, 147
　in solar battery, 182
　use of, in chip, 9
Silicon chip, 154, 155i
Silicon dioxide, 140, 147, 154, 272
Silicone oils, in hair sprays, 424
Silk, composition of, 244
Silt, 271
Silver, 24, 127
　from copper purification, 143
　cost of, 26t
　in crust of Earth, 137
Silver battery, 130t
Silver cyanate, 188
Silvicide, 284dt
Skatole, 220s
Skin
　acid, 119
　chemical view of, 419
　cross section, 421i
　on lips, 428
　physical view of, 420i
　replacement rate of, 292
Slag, 140d
Slaked lime, plant nutrient, 277t
Slash-burn-cultivate cycle, soils and, 275
Slimicide, 284dt
Smithsonite, 138t
Smog, 367d, 368, 369i
　London type, 368d
　photochemical type, 368d
　production of, 370
Smoking, and cancer, 347
Snyder, Solomon, 390
Soap, 431, 432d
　in vanishing cream, 427
Soda, baking, 119
Soda-lime glass, 147, 148
Sodium, 31, 33t
　in crust of Earth, 137, 138t
　dietary recommendations for, 296
　electronic structure of, 61t, 93s
　heat transfer medium in atomic reactors, 175
　isolation of, 126

Sodium *(Continued)*
　soil acidity and, 271
　urinary excretion of, 307
Sodium alkyl sulfate, 434
Sodium benzoate, 214s
Sodium bicarbonate solution, pH of, 122i
Sodium bicarbonate, 21
Sodium bromate, oxidizing agent in curling hair, 422
Sodium carbonate, 159, 227t
　in glass, 147–148
　in laundry products, 435
Sodium chloride, 21, 126
　crystal structure of, 97i
　in perspiration, 431
　solution, 117i
Sodium hydrogen carbonate, 158
Sodium hydroxide, 119, 157, 227t
　as corrosive poison, 327, 328t
　to neutralize rosin, 435
Sodium hypochlorite, in laundry bleach, 436
Sodium ion(s), 118
Sodium lauryl sulfate, 435s
　in depilatory, 426
Sodium methanearsenate, 332t
Sodium nitrate, 280
　as mutagen, 342
Sodium perborate, as corrosive poison, 328t
Sodium phosphate, in laundry products, 435
Sodium pyrophosphate, in laundry products, 435
Sodium silicate, 227t
　in laundry products, 435
Sodium stearate, in soap, 432s
Sodium sulfate, 227t
Sodium sulfide, as depilatory, 426
Sodium thiosulfate, stain remover, 438
Sodium tripolyphosphate, 227t
Sodium/potassium ratio, in human body, 307
Soft soaps, 433
Soft water, 436
Soil
　acidic, 270–271
　air in, 270
　alkaline, 270
　black, red, white, 272
　chemical composition of, 272
　denitrification and, 280
　flooded, 280
　friable, 270
　horizons of, 270
　leaching of, 271
　micronutrients, 272
　nonmineral nutrients, 272, 273
　nutrients in, 272–275, 272t
　primary nutrients in, 272, 273
　secondary nutrients, 272, 274
　separates, 270
　silts, 271
　slash-burn-cultivate cycle, 275

Soil *(Continued)*
  sour, 270
  sweet, 270, 271
  types of, 270
  water in, 270
  water-logged, 271
Soil samples, analysis of, 272
Solar battery, 182
Solar energy, 181
  algae and, 181–183i
  collectors, 184i
Solids, 22d
Solute, 115d
Solution, 115d
Solvay process, 159
Solvents, 115d
  as mutagen, 341
  in nail polish, 442
Somatic effects, 75d
Soot, as carcinogen, 346
Sorbitol, 319s
  polyhydric alcohol, sweetener, 317
Sorption, cause of, in soils, 270
Sources of new materials, 159
South pole, 379i
Soybean oil, 218t
Space shuttle fuel cells, 133
Spectrum, 57d, 58i, 59i
Spermaceti wax, in cold creams, 427
Sphalerite, 138t
Spiegelman, Sol, 264
Spontaneous reaction(s), 109
Spot removers, 438t
Spray, hair, 424
Stabilizers, food additives, 313t, 318
Stain removers, 438
  protein enzymes as, 254
Stannous fluoride, in tooth paste, 439
Staphylococcus aureus, 396
Starch, 240
  combustion of, 167t
  condensation polymers, 238
  dietary recommendations for, 296
Starvation, 268
Static electricity, 45
Stearic acid, 213, 214t
  in vanishing cream, 427
Steel, 139, 140d
  corrosion, 131
Steel storage tanks, 133
Stellarator, 180
Stentor coeruleus, photosynthesis and, 251
Steroid(s), 300, 406–409
  anabolic, 409
  sports and, 409
Stomach fluids pH, 122i
Strassman, Fritz, 79
Stratosphere, 365i
Stratum, of skin, 419
Streptomyces aureofaciens, 398

Streptomyces griseus, 397
Streptomycin, 397
Stress, BMR and, 298
Stretch marks on skin, zinc and, 306t
Strong acid, 119d
Strontium, 31, 138t
Strontium sulfide, as depilatory, 426
Structural aluminum, 140
Structural isomers, 190, 192
Structure of matter, 29
Styrene, 225t, 227t, 233t
Sublimaze. *See* Fentanyl.
Sublime, 92d
Submicroscopic samples, 30
Subscripts in formula, 27, 28
Subsoil, 270
Substrate, 250d
Sucrase, 248
Sucrose, 109, 240, 241si
  production of per year, 240
  saccharin and, 316
  sweetener in food, 316
  sweetness of, 239, 316, 317
Sudbury, Ontario, Canada, 373
Sugar, 24, 240
  composition of, 42, 151, 303t
  in consumer products, 303t
  dietary recommendations for, 296
  as pure substance, 21
  refined, 302–304
Sugar solution, 115, 116i
Sugarless gum, polyhydric alcohols and, 317, 319
Sulfa drugs, 395–396
Sulfanilamide, 396s
Sulfate group, in detergent, 434s
Sulfate ion, 100
  in solution, 130
Sulfates, weed killers, 287
Sulfhydryl group, reaction with heavy metals, 331
Sulfite ion, as corrosive poison, 328t
Sulfonate group, in detergent, 434s
Sulfur, 19, 21, 24, 125, 138t, 155
  in coal, 172
  cost of, 26t
  electronic arrangement, 61t
  fertilizers and, 281
  in fuels, 371
  oldest insecticide, 285t
  in proteins, 244
  secondary nutrient in plants, 274
Sulfur dioxide, 142
  in acid rain, 374–375
  action on marble and limestone, 371–372
  as air pollutant, 371
  effect on vegetation, 364i
  from elements, 111
  removal methods, 372
Sulfur molecule, 27

Sulfur trioxide, 111, 157
Sulfur-35, 85t
Sulfur-coated urea, 281
Sulfuric acid, 119, 156, 157, 227t
  in batteries, 130
  as corrosive poison, 328t
  in detergent manufacture, 435
  formula, 27
  hazards, 324t
  superphosphate synthesis and, 281
  weed killer, 287
Sulfurous acid
  in acid rain, 374
  from sulfur dioxide, 371
Sumner, James B., 248
Sun, 109
  energy from, 165
  fusion on, 163d
  photosynthesis and, 163, 251
  supplier of all energy, 251
Sunflower oil, 218t
Sunlight, danger to skin, 429
Suntan lotions, 429
Superchips for computers, 160
Superconductivity, 160d
Superconductors, electricity and, 180
Superfund, 358
  priority sites, 359i
Superphosphate, 281
Surface active agents, 431
  food additives, 313t, 319
Surface coatings, 419
Surface tension, of water, 355
Surfactant, 288d
Sweat glands, 431
Sweden, iron ore in, 139
Sweeteners
  food additives, 313t, 316, 319
Sweetness
  of sugars, 239
Swimming
  energy expended by, 297t
Symbiosis, 2d
  plants and bacteria, 273
  science and technology and, 2
Synapse, 336d
Syndets, 433
Synergism, 315d, 406d
  in food, 315, 318
  insecticides and Sesamex, 285t
Synthesis gas, 210, 226–227
Synthetic detergents, 433
Synthetic fuel(s), future expectations from, 185

T cells, 415d
2,4,5-T, 287
  agent orange and, 287

Table salt, as pure substance, 21
Tabun, 339s
Tagamet, 392s
   and ten most prescribed drugs, 387t
Talcum powder
   in face powder, 428
   use with depilatory, 426
Tall stacks, 372, 373i
Tallow, 433
   combustion of, 167t
Talwin, 389s
Tanning rays, 429
Tar, 226t
   as carcinogen, 346
Tartaric acid, 214t
Tea solution, 115
Teague, Tennessee, 359
Technetium-99m, 86t
Technology, 1, 2d
   risks and, 14
   society and, 15
Teflon, 233t
Teisserenc de Bort, Leon Phillipe, 365
TEM, as mutagen, 341
Temperature, effect on reaction rate, 111, 113i
Template(s), DNA and, 261i
Tenormin, and ten most prescribed drugs, 387t
Teratogens, 343–344
Teratology, 343d
Terephthalic acid, 227t, 232s
Termites, digestion of cellulose, 243
Termiticide, 284dt
Terramycin, 398s
Tertiary structure
   of proteins, 247
Tertiary-butyl alcohol, octane rating of, 201t
Testes, effects of radiation on, 75
Testosterone, 407s
Tetrabromofluorescein, sodium salt, 428s
Tetrachloroethylene, in cleaning solutions, 436
Tetracycline HCl, 387t
Tetracyclines, 397–398
Tetraethyllead, 202, 334
Tetrafluoroethylene, 233t
Tetrahedral angle, 102d
Tetrahedral molecular structure, 103i
Tetrahydroaminoacridine, 416s
Textiles, fire resistant, 158
THA, 416s, Alzheimer's disease and, 416
Thalidomide, 414
   as teratogen, 343s
Thallium, as teratogen, 344
Thallium-201, 86t
Thallium-210, half-life, 68t
Theoretical definitions, 22d
Theory, 3d, 4, 5
   model, 4

Theory of oxidation, 125
Thermal insulators, 151
Thermal inversion, 368d, 369i
Thermodynamics, 167d
   first law of, 167
   second law of, 168
Thickeners
   food additives, 313t, 318
   in paints, 441
Thiocyanate ion, 331
Thioglycolic acid, in curling hair, 422s
Thiosulfate ion, 331
Thompson, Sir Joseph John, 49
   experiment with electrons, 50i
   personal side, 50
Thorium series, 67
Thorium-230, half-life, 68t
Thorium-234, 67
   half-life, 68t
Three Mile Island, 87
Threonine, 245st
Thymidine, 415s
Thymidine-5'-phosphate, 415s
Thymine, 257s, 415s
Thyroid, 86i, 387t
Thyroid glands, iodine and, 306–307
Thyroxine, 307s
Tillage, weed control and, 287
Times Beach, Missouri, 13
   dioxins in, 208
Tin, 138
Tin (IV) oxide, color in glass, 148t
Tire(s), typical formulation of, 235t
Tiredness, lactic acid and, 255
Tissue, adipose, 300
Titanium, 138t
Titanium dioxide, 227t
   paint pigment, 440
Tobacco, moistness of, 319
Tobacco smoke, indoor air pollution, 379
Toenails, composition of, 421
Toilet bowl cleaner, 119
Toilet soaps, 433d
Tokamak, 180
Toluene, 197s. 226t, 227t
   uses of, 225t
Tomatoes, pH of, 122i
Ton, short and long, 173d
Toone, Tennessee, 359
Tooth plaque, sugar and, 304
Toothpaste, 439
Top note, in perfume, 429
Topsoil, 270
Toxaphene, insecticide, 285t
Toxic substances, 324d
Toxic Substances Control Act, 14
Toxicity
   of plutonium, 177
   of wastes in intestines, 304
Toxins, in food, 313

Trace air pollutants, 383t
Tracers, radioisotopes, 84–85
Transcription
   DNA, RNA and, 262
   protein synthesis and, 262i
Transistor(s), 8–9, 154d
Transmutation, 70–72, 76d
Transportation accidents
   involving nuclear wastes, 88
Transuranium elements, 72–73
Trash, energy from, 174
Tri-Sweet, 317
Triacylglycerol. See Triglyceride.
Triangular pyramidal molecular structure, 103i
1,3,5-Triazine, 288s
Triazines, herbicides, 288
Trichloroethane
   in cleaning solutions, 436
   drycleaning solvent, 438
1,1,1-Trichloroethane, 207s
Triethylenemelamine, as mutagen, 341
Triglycerides, 300
2,2,4-Trimethylpentane. See Isooctane.
Triple bonds, 98
Trisodium phosphate
   as corrosive poison, 328t
   floor cleaner, 99
Tristearin, fat for making soap, 432
Tritium, 56, 85t
   nuclear fusion and, 180
Trona ore, 159
Troposphere, 365i
Trypsin, 248
Tryptophan, 245st
Turpentine, in paints, 440
TV Till PM HA, 298
Tylenol, 391, 392s
Tylenol with codeine, and ten most prescribed drugs, 387t
Tyrosine, 245st
   in bleaching hair, 425i

Ulcers, 219
   anti-ulcer drugs, 392
   prostaglandins and, 301
Ultraviolet light, effect on skin, 429
Ultraviolet radiation, and smog production, 370i
Underground storage tanks, 133, 354i
United States Department of Agriculture, dietary recommendations of, 296
Units, of SI system, 35
Unnilennium, nuclear reaction for, 74t
Unnilhexium, nuclear reaction for, 74t
Unnilpentium, nuclear reaction for, 74t
Unnilquadium, nuclear reaction for, 74t
Unnilseptium, nuclear reaction for, 74t

Unsaturated fats, 217–219
  hydrogenation of, 217
Uracil, 257s, 405s
Uranium, 21
  atomic pile and, 175
  color in glass, 148t
  enrichment, 80
  fissionable isotopes, 79–80
  reserves of, 176
Uranium potassium sulfate, 65
Uranium series, 67, 68t
Uranium-234, half-life, 68t
Uranium-238, 67
  half-life, 68t
  use to determine age of rocks, 68–69
Urea, 188, 221s, 227t, 281s
  fertilizer and, 276
  hydrolysis of, 281
  plant nutrient, 277t
  product of protein oxidation, 296
  synthesis of, 280–281
Urease, 248
Uremia, 299
Used motor oil, how regulated, 325t
USRDA (United States Recommended Dietary Allowances), 294, 295t
  for minerals, 306t
  values for vitamins, 309t

Valence electrons, 93d
  bonding and, 91
  stability of eight, 93
Valence shell electron pair repulsion theory, 102
Valeric acid, 213t
Valine, 245st
Valium, and ten most prescribed drugs, 387t
Valley of the Drums, 361i, 362
Vanilla bean. See Vanillin.
Vanillin, 212s
Vanishing cream, 427
Varnish, in nail polish, 442
Vegetable oils, 217–218
Venezuela, iron ore in, 139
Verapamil, 401, 403s
Vertigo, 299
Vinegar, 119, 214. See also Acetic acid.
  pH, 122i
Vinyl acetate, 227t
Vinyl chloride, 208t, 227t, 233t
  as carcinogen, 346
Virus, 264
  disease in cherries, 283
  heliothis, insecticide, 285t
Viscosity modifiers, in food, 318

Vitamin A, 308s
  bacterial infection and, 309
  common cold and, 310
  food color, 317
  RE for, 296
Vitamin B6
  anemia and, 307
  master vitamin, 310
Vitamin B9
  anemia and, 307
Vitamin B12, 306t
  anemia and, 307
  supplied by intestinal bacteria, 309t
Vitamin C, 308s
  common cold and, 310
  not required in diet of dogs and rats, 293
  required by human beings, 293
Vitamin D, 305
  IU for, 296
  toxicity of, 308
Vitamin E
  aging and, 310
  antioxidant, 309
  destroyed by freezing, 309
  IU for, 296
  poison detoxifier, 310
Vitamins, 308–310, 308d
  B-complex, 310
  classes of, 308
  as coenzymes, 249
  deficiency effects of, 309t
  discovery of, 293
  IUs, 295
  oil soluble, 308, 309t
  RDA for, 295t
  removed by refining sugar and flour, 304
  removal during purification 293
  sources of, 308, 309t
  USRDA values for, 295t, 309t
  water soluble, 308, 309t
Vitrified porcelain, 151
Volatility, 91d
Volcanic ash, in abrasive cleaners, 436
Volcanoes, 367
Vulcanization, 234

Waksman, Selman, 397
Walking, energy expended by, 297t
Washing soda, 159
Washington monument, 91, 142
Washington, George, use of marl, 270
Waste(s)
  hazardous, 13, 358t
  nuclear, 87–88, 137
Water, 354–357
  abundance, 355

Water (Continued)
  average daily usage, 356t
  basis for temperature scale, 36
  bonding in, 99i
  catalysis decomposed, 182
  chemical energy deficiency of, 168
  composition of, 42
  conductivity of pure, 120
  content of, for different organisms, 355t
  crop requirements of, 271
  electrical conductivity in, 116i
  electrolysis of, 128i, 278
  food and, 300t
  how held in soil, 270
  hydrogen bonding in, 103i
  moderator in atomic reactor, 175
  molecular structure of, 103i
  produced by fuel cells, 134
  public use of, 356
  removal of, from soils, 271
  standards for, 357t
  unique properties, 354–355
Water cycle, 360i
Water molecule, geometry, 117i
Water pollutants, classes of, 357t
Water quality, industrial wastes, effect on, 357–363
Water supplies, future of, 363
Water-soluble dye, for hair coloring, 422
Watson, James D., 258, 259i
Watt, 166d
Waxes, 300
  fertilizer dispersion and, 282
Weak acid, 119d
Weak base, 120d
Weed killers. See Herbicides.
Weeds, number of, 283
Weight, set point for human body, 301
Weight changes, 110
Weightlifting, muscle tissue and, 294
Wet hair, 421
Wheat germ, rancidity of, 304
Wheeling, West Virginia, 374
White lead, paint pigment, 440
Wilkins, Maurice H. F., 259
Wine pH, 122i
Wishbone Dressing, sugar in, 303t
Wöhler, Frederick, 188
Wood alcohol. See Methanol.
Wood burning, 113
Wound healing, histidine and, 298

X ray(s)
  discovery of, 65
  used in oil drop experiment, 51
Xenon, 24

Xerophthalmia, vitamin A deficiency, 309t
Xerox copying, invention incubation period, 3t
Xylene(s), 226t
   meta, 197s
   octane rating of, 201t
   ortho, 197s
   para, 197s
   uses of, 225t
p-Xylene, 227t
Xylocaine, 390st

Y-12 plant, 88
Young, Arthur, 269

Zamenhof, Dr. Stephen, 341
Zeolites, 230
Zinc
   in brass, 19
   cost of, 26t

Zinc *(Continued)*
   electronic arrangement, 62
   as teratogen, 344
   used in battery, 129
Zinc oxide, in face powder, 428
Zinc peroxide, 431
Zinc-copper battery, 129i
Zinn, Walter, 177
Zirconia, 151
Zone refining, 154di

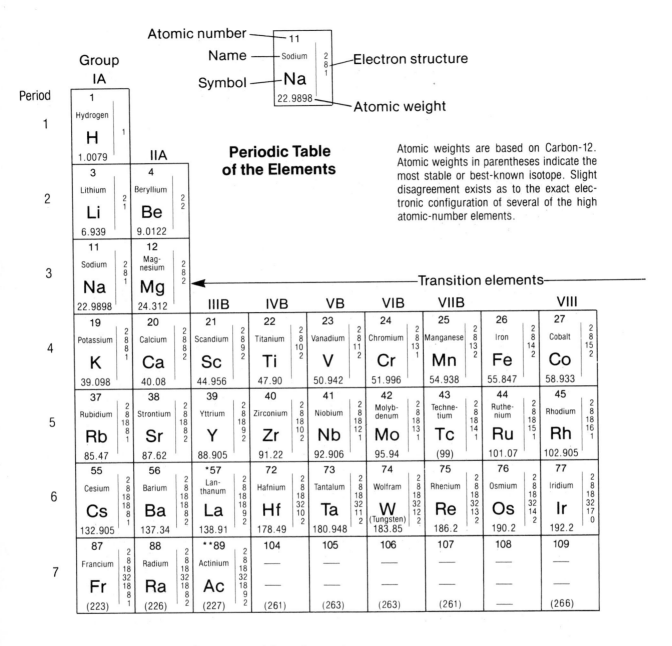